Statistics Explained

Statistics Explained

John Parry Lewis
Alasdair Traill

Addison-Wesley

Harlow, England ● Reading, Massachusetts
Menlo Park, California ● New York
Don Mills, Ontario ● Amsterdam ● Bonn
Sydney ● Singapore ● Tokyo ● Madrid
San Juan ● Milan ● Mexico City ● Seoul ● Taipei

© Addison Wesley Longman 1999

Addison Wesley Longman Limited
Edinburgh Gate
Harlow
Essex CM20 2JE
England

and Associated Companies throughout the World.

The rights of John Parry Lewis and Alasdair Traill to be identified as authors of this Work have been asserted by them in accordance with the Copyright, Designs and Patents Act 1988.

Cover designed by OdB Design & Communications, Reading
Text design by Claire Brodmann
Typeset in Times and Optima by 32
Printed by Longman Singapore Publishers (Pte) Ltd.
Printed in Singapore

First printed 1998

ISBN 0-201-17802-8

British Library Cataloguing-in-Publication Data
A catalogue record for this book is available from the British Library

Trademark Notice
Sharp EL556G is a trademark of Sharp
Casio fx-6300G is a trademark of Casio Instruments

The publishers wish to thank the following for permission to reproduce the following copyright material.

American Cyanamid Company
Table D7; Adapted from: Wilcoxon, F. (1949) Some Rapid Approximate Statistical Procedures, American Cyanamid Company, New York, p.13.

Routledge Ltd
Table D11; Elementary Statistical Tables by H.R. Neave 5% significance levels of sums of squares of rank difference.

Oxford University Press
Table D13; Adapted from: Shapiro, S.S. and Wilk, M.B. (1965) An analysis of variance test for normality (complete samples) Biometrika 52, 591–611.

To
G W M
and
L J T

Contents

List of How to ... boxes

List of noteboxes

Preface

College courses in many subjects have statistics as a component. This textbook is based on an examination of courses provided by all British universities and many other colleges, and of the requirements of major professional bodies. We have kept in mind that very few of its readers will know or like mathematics. Almost all of them simply want to know what is meant by various statistical terms and how to use techniques. Providing that knowledge has been our main purpose, but we know from long and varied experience of teaching and examining that all too often the terms and techniques are misunderstood and misused.

We have tried to ensure that users of this book have a proper understanding of what they are doing, and know both when and when not to use various techniques. We have stressed the need to have respect for both data and techniques, understanding their possibilities and limitations. Our emphasis is on explanation so that students know not only what they are doing, but why they are doing it. More detail of our approach is given in Chapter 1, which should be read carefully. Explanation, examples and exercises will be updated and kept topical on http://www.statisticsexplained.aston.ac.uk.

Many people have helped us. In particular, we have to thank the anonymous reviewers for their comments and suggestions, Dr John Elgy for providing many of the diagrams and tables, Linda Traill for three patient years and hours of help in the restructuring and presentation of certain sections, Joy Atkins and Joan Domone for unflagging aid in moments of crisis, Joyce Davies and Marion Yates for valued comment and encouragement, Ian Walton for patient practical assistance and our publishers and others involved in design and production, without whose genuine partnership this would have been a lesser book.

J. Parry Lewis
Alasdair Traill
January 1998

How to use this book

 General guidance

The main aim of this book is to explain statistical ideas and techniques to students who need them in support of their main study. If you are such a student you will probably not be happy with a lot of mathematics or explanations that are too concentrated to be digestible. But you will want to be sure that you know both how to apply a technique, and when and when not to use it. Misapplied techniques can give a false status to quite invalid conclusions. If you read this book from cover to cover and do the exercises, you should end up with a useful and reliable knowledge of statistical methods and be able to tackle a wide range of problems. If you go further and study the subject at a higher level, you will have a good grounding and will not need to 'unlearn' anything.

The other aim is to provide a reliable quick reference to statistical techniques, which is done by the chapter summaries and over one hundred How to ... boxes.

The main chapters begin with a few examination-type questions indicating the main concern of the chapter. This is stated more fully in a few paragraphs before the main treatment begins. At the end of the chapter a detailed summary of the main argument highlights the more important techniques and results.

Most chapters contain How to ... boxes. There are also Noteboxes with relevant material that can be read separately from the main text. This makes a first reading easier and facilitates reference by highlighting their contents.

Some readers will find that reading the chapter summaries, the How to ... boxes and the Noteboxes will provide a short cut to an overview of the subject and an ability to perform a wide range of tests. But please recognise that, until you have read the fuller account, you will be in danger of misapplying a technique or failing to spot some important point. The wisest thing to do is to read the whole chapter and then, if necessary, use the chapter summaries and the boxes to refresh your memory or to emphasise its main points, but do read the whole chapter first.

Read the worked examples carefully. Then cover the working and attempt to solve the problem for yourself, checking your solution against the worked example as necessary.

Further understanding and practice can come through doing the exercises. These exercises are important, especially the relatively simple consolidation exercises that come at key points in most chapters; they are designed to test that you have understood and absorbed the main ideas of the past few pages. Most of the answers are in the preceding text. Answers to most calculations are given at the end of the book. The main problem exercises are collected at the end of the book, in convenient chapter groups.

1.2 The order of reading

The chapters need not be read in sequential order, but this is recommended for most readers. All readers should begin with Chapter 2, which introduces the subject and deals with basic concepts, including a few words that have special meanings for statisticians. If these words are not understood at the outset, there will be difficulties later on.

The remaining chapters can be grouped into eleven sections. The main logical links between them are indicated approximately in Figure 1.1. Almost all of any section can be understood by first reading the sections that precede it. If you wish to depart from the suggested order of reading, or want to know more about the coverage of the various chapters, you will find further information in the following summary of section contents, which also gives a useful overview of the plan of the book.

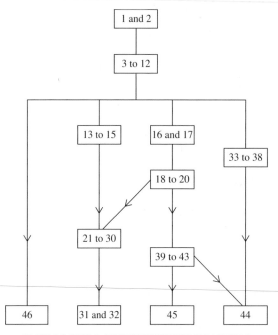

MAIN LOGICAL LINKS BETWEEN CHAPTERS

Figure 1.1

Summarising and describing statistical information (Chapters 3 to 12).

Chapter 3 discusses how to examine (or analyse) data with the help of tables and diagrams. Much of the book depends on ideas explained in this chapter, especially frequency distributions. Chapter 4 introduces methods for describing them; this usually involves calculating some kind of average. Chapters 4 and 5 discuss the most common average – the arithmetic mean. We emphasise that this should not be calculated for data measured in certain ways, such as ranks. This important point ties up with the distinction between parametric and non-parametric statistics, already indicated in Chapter 2, and it will become clearer as you proceed. Anybody who forgets may sometimes apply techniques under circumstances that make them invalid. The major part of this book is concerned with parametric statistics.

Chapters 5 to 9 describe other averages and discuss their use. Detailed knowledge of these averages is not essential for understanding the rest of the book, but some familiarity with them will be useful. Chapter 9 ends with a brief account of index numbers.

Chapters 10 to 12 consider another aspect of statistical description – measures of spread. The standard deviation is a particularly important measure and it should be thoroughly understood.

Regression and correlation (Chapters 13 to 15)

Frequently we have two related sets of data, such as marks gained by pupils in each of two examinations, and we wish to explore and measure the strength of the relationship. Chapters 13 to 15 discuss how to do this for parametric statistics. A common error is to use these techniques when only non-parametric techniques should be used. Chapter 15 ends by asking questions that introduce the idea of statistical significance.

Chapters 13 to 15 can be postponed until immediately after Chapter 30. The techniques they describe should not be used for prediction or to make generalisations until Chapters 31 and 32 have been read, and for that you will need to understand most of the intervening chapters.

Sample, population and probability (Chapters 16 and 17)

These are key chapters. Chapter 16 introduces the idea of sampling, discussed more fully in Appendix C. It shows how the mean, standard deviation and other parameters of populations are estimated by using the sample values. Instead of a point estimate, we may want an interval estimate that has perhaps a 95% chance of containing the true value. This introduces the idea of confidence intervals in estimation. Another use of sample results is in testing the chance of a hypothesis being right. Before we can make further headway with estimation, confidence intervals and hypothesis testing, we have to study chance.

Chapter 17 provides the basic tools. It defines probability and develops some important results that are used a great deal in most of the remaining chapters. More advanced ideas of probability, which we do not use in the main part of the book, are considered in Chapter 45, but some of this requires other reading, especially Chapters 18, 39 and 40.

Random variables: binomial and normal distributions (Chapters 18 to 20)

Chapter 18 introduces the idea of a discrete random variable and considers how to find the probability that it has a specified value. This depends on how the probabilities are distributed. An important discrete probability distribution is the binomial distribution, which enables us to tackle a wide range of statistical problems and contributes towards estimation and hypothesis testing.

Chapter 19 shows how this discrete probability distribution ties up with the very important continuous normal distribution, discussed further in Chapter 20.

Estimation, confidence intervals and hypothesis testing (Chapters 21 to 30)

The ideas covered in Chapters 18 to 20 lie at the heart of confidence limits, significance and hypothesis testing. Chapters 21 to 23 use them to deal with confidence intervals. Chapter 24 begins to tackle the problem of hypothesis testing, using one example to introduce and illustrate the main points. Chapters 25 to 27 extend these ideas and show how to test different kinds of hypotheses for data from a 'large' sample. (The meaning of 'large' depends to some extent on the test, and is specified from place to place, but as a rough guide we may here take it to be at least 40.)

Chapter 28 is concerned with the precise significance of a sample result, not necessarily as part of hypothesis testing. It shows how to use the continuous normal distribution to provide quick, good, approximate answers for problems about the discrete binomial distribution, as long as the sample is large enough.

Small-sample tests are considered in the next two chapters. They can be used for samples of any size, but the tests in the earlier chapters are more appropriate for large samples and they are easier to use.

Significance of regression and correlation coefficients (Chapters 31 and 32)

The regression and correlation coefficients described in Chapters 13 and 14 may be based on sample data. Chapters 31 and 32 take the ideas of the previous section and extend them to testing the significance of these coefficients and the differences between them. They show how to establish confidence intervals for the coefficients and estimates (or predictions) based on them.

Non-parametric tests (Chapters 33 to 38)

Chapter 33 introduces non-parametric tests and indicates the plan of the next few chapters. Noteboxes explain levels of measurement. Chapter 34 uses two non-parametric tests to determine whether a sample was taken randomly. Several non-parametric tests that use the median, some of them with data from matched pairs, are considered in Chapter 35, then Chapter 36 discusses contingency tables and the very important χ^2 test. Two 'before and after' tests on the effect of some experience or treatment are considered in Chapter 37, and Chapter 38 explains how to measure the correlation between two non-parametric variables, such as ranks.

Other distributions, including Poisson and exponential (Chapters 39 to 43)

We then return, temporarily, to parametric statistics. The next few chapters can be read immediately after Chapter 20, since they do not involve estimation, hypothesis testing or any of the non-parametric tests. In Chapter 39 we consider two more discrete probability distributions – the discrete uniform distribution and the geometric distribution – leading to a statement of some important results about means and variances. Chapter 40 gives a similar treatment of two continuous distributions – the rectangular distribution and the triangular distribution. These chapters pave the way for an account of two frequently cited and related distributions.

The first is considered in Chapters 41 and 42. An important discrete probability distribution is the Poisson distribution; it is used for analysing events that occur at random intervals in time or space (Chapter 41), and to provide a good approximation to the binomial distribution in the study of rare events (Chapter 42). A closely related continuous probability distribution, the (negative) exponential distribution, is described in Chapter 43. The intervals between Poisson-distributed events have an exponential distribution, which also arises in a wide range of decay studies.

Goodness of fit (Chapter 44)

We now come to the point where non-parametric tests become an essential tool of the parametric statistician. If we wish to check whether some observed data conform sufficiently well to a theoretical distribution, such as the binomial distribution, the normal distribution or one of the distributions described in Chapters 39 to 43, we need to use non-parametric tests for goodness of fit. The most widely used test is the χ^2 test of Chapter 36, and Chapter 44 explains how it is used in this context. Another test, due to Shapiro and Wilk, is described at the end of Chapter 44.

Further probability (Chapter 45)

Chapter 45 contains ideas in probability that are used by statisticians but are not essential to an understanding of the earlier chapters. It includes probability trees, Venn diagrams, conditional probability, a little on covariance, and various results about probabilities and expectations.

Time series (Chapter 46)

Chapter 46 provides an elementary treatment of time series, with a hint at more difficult ideas beyond the scope of this book.

Appendices and problem exercises

Appendices deal with the use of pocket calculators, tricks to speed up calculations, and sampling. Problem exercises are collected immediately after the appendices.

 1.3   Mathematical symbols

We have kept the use of mathematics to a minimum. But if we tried to avoid all use of mathematical symbols, we would be making things harder rather than easier. The mathematical symbols that may be new to you, or the cause of horror, are explained at the time of their first use or in Noteboxes. Here are the main ones:

$\sum$	explained in Noteboxes 5.1 and 6.3
X, x	explained in Chapter 2
$\bar{x}$ and μ	explained in Chapter 5
x_i	explained in Notebox 6.3
s and σ	explained in Chapter 11
$n!$	explained in Chapter 17
^nC_r and ^nP_r	explained in Chapter 17
$0!$	explained in Notebox 17.1
e	explained in Notebox 31.2
$\log$ and $\ln$	explained in Notebox 31.2
v	explained in Chapter 29
$e^{-\lambda}$	explained in Notebox 41.1
y^0	explained in Notebox 41.1

We refer to equations by using square brackets. Thus [37.2] is equation 2 of Chapter 37. To save space, divsion is sometimes indicated by /. Thus

$$\frac{(a-b)}{(c+d)}$$

may appear as $(a-b)/(c+d)$ and

$$\sqrt{\frac{s}{n-1}}$$

may appear as $\sqrt{s/(n-1)}$ or $\sqrt{[s/(n-1)]}$.

Introduction

 The emergence of statistics

In *The Elements of Universal Erudition*, published in 1770, *statistics* was defined as 'the science that teaches us what is the political arrangement of all the modern states of the known world.' It concerned knowledge about the political state.

Slowly the description of matters of state began to contain numerical information about population, trade, and so on. As time went on, this tendency increased and the numerical information was often presented in tables.

In 1838 the first volume of the *Journal of the Statistical Society of London* proclaimed that *statistics* was 'the ascertaining and bringing together of those facts which are calculated to illustrate the condition and prospects of society' and that 'the statist commonly prefers to employ figures and tabular exhibitions.'

The emphasis rapidly changed from 'the state' to 'figures and tabular'. There were articles on statistics of mental characteristics in man, and on statistics of children under the headings of bright–average–dull. Before the end of the nineteenth century one examined the characteristics of Latin verse 'with statistics'. The first stage in the change of meaning of the word had been accomplished; instead of 'matters of state' it now meant 'numerical information'.

Ways of presenting this information are discussed in Chapter 3. How it is obtained is a question whose answer we postpone because of another change in the meaning of *statistics*.

As more and more numerical information appeared, mathematicians became interested in the methods for handling it:

- collecting it in ways that reduce the chances of bias or error,
- summarising it in meaningful ways,
- examining relationships between one set of numerical data and another,
- exploring the reliability of measurements and estimates,
- testing hypotheses,
- other related purposes.

Statistical theory and techniques dealing with general cases by using algebra rather than numbers were developed, and the word *statistics* had its meaning extended to embrace this study. It now means

- numerical information,
- the theory behind the analysis of numerical information,

- techniques for collecting and analysing numerical information,
- the subject that is concerned with these matters.

Our treatment of these matters is helped by examples drawn from many fields, and frequent references to the university and city of Statingham, whose location we mention below.

A person with a good command of words can describe many things to other people who understand those words. Some things can be described accurately and in detail, but things like beauty and pain may be difficult to describe with any precision.

Words can often describe quantities with great clarity. We convey a clear picture of quantity when we speak of a sackful of potatoes. But if we want more detail, we have to weigh or count them, and then we need numbers. Be they written in letters or figures, numbers are the words used to describe quantities. Just as language is a means of manipulating words to convey meanings and describe ideas, so statistics is a means of manipulating numbers for the same purposes. A person unable to cope correctly with statistics cannot cope correctly with the description and analysis of quantities. A large part of the world of ideas and information is shut off, or so misunderstood that wrong conclusions are reached or accepted.

2.2 Respecting numbers

The word *respecting* means 'about' and 'having respect for'. The need to know something about numbers is obvious, but respect for numbers is something less tangible. Careful readers of this book will gradually acquire this respect, and so become better statisticians.

We draw an analogy with a good woodworker who has respect for his or her tools. The good woodworker does not misuse them, perhaps by treating a chisel as a screwdriver. But he or she also has respect for the material – the wood. Every piece of wood has its qualities and its limitations. The good woodworker considers them and chooses a piece that is suitable for its purpose. Then he or she works on it in a way that takes account of its grain and other properties, and uses tools appropriate for the purpose and for the wood. All the way until it is finished, he or she examines the work carefully for imperfections. And the good woodworker does not expect too much of a finished item; he or she does not expect a long slender column to carry an enormous weight, or to lever a heavy load.

Statisticians use statistical techniques to work on the material of numbers. They must have respect for their techniques, never misusing them. They must note the limitations and possibilities of their data. Technique and data have to be suited to each other. And technique must also be related to purpose – to the question that was asked. When statisticians have an answer, they must consider its plausibility; if they accept an answer as plausible, they must not use it wrongly.

We now consider a few things that need to be part of the knowledge and thinking of the statistician who wishes to develop respect for his or her tools and material – techniques and numbers.

2.3 A few words

Like many other subjects, statistics sometimes uses words in a different way than in everyday language. Here are a few examples. We shall come across some more later on.

Accurate and precise

The university city of Statingham is located somewhere in England. That statement is **accurate** (in that it is correct), but not **precise** (since it is vague, allowing for many possible locations). We can add to precision by giving the National Grid reference NY450173. This locates Statingham to within 1 km east–west and 1 km north–south, which is fairly precise (or exact). But although it is precise, this reference is not accurate, since NY450173 is in Ullswater, a large expanse of water in the Lake District, and we have no underwater universities. If we were rash enough to state the location of Statingham both precisely and accurately, some characters mentioned in our examples would possibly generate libel actions!

The front gate of the university of Statingham is exactly 473 metres from the main doorway. Unaware of this, the professor of geography states that the distance is 500 metres, and the head porter says it is 500.032 metres. Neither of these distances is accurate. The professor's statement is slightly *more accurate* than the head porter's (in that it is nearer the truth). On the other hand the head porter's is *more precise* (in that it goes to more figures).

Variables and constants

Another word that frequently arises in statistics is **variable**. Any measurement that can *in principle* have more than one value is a variable. The phrase *in principle* is important. For example, Vine Close has houses numbered 1 to 54. Statistics lecturer Anna Liszt lives in that street but we do not know her street number. It could be anything between 1 and 54. We therefore say that the street number is a variable (often denoted by the capital letter X, which stands for the name of the variable). Anna's number is a selected value of this variable (and is denoted by the lower case x). If it is 32 then we say that $x = 32$. People frequently abbreviate this kind of statement by saying that Anna's number is a variable, which is not quite right since it is in fact one value of a variable.

A number (or algebraic symbol representing a quantity) that is not a variable is a **constant**. Something may be a variable in one context and a constant in another. For example, if we are considering the spending patterns of many households, the number of children in a household is a variable, since we will want to know how household spending depends on the number of children. But if we are considering the spending patterns of households with two children, the number of children is a constant.

There may also be non-quantitative variables and constants, such as the colour of a car or the nationality of an airline passenger. They are often called **attributes**.

Discrete and continuous

Quantitative variables may be discrete or continuous. A variable is **discrete** if it can take only a countable number of values, so that all possible values are separate from each other. An example is the number of people in a stated room, which must be an integer. The average number of people per room in an eight-room hostel is also discrete. It must be a whole number divided by 8, so if we know that the average is greater than $13/8$, which is 1.575, then we also know that it must be at least $14/8$, which is 1.75. It cannot be 1.576.

An example of a **continuous** variable is the mass of a specified chair. If we know it lies between 10 kg and 14 kg, we cannot say it must be 11, 12 or 13 kg; nor can we say it must be 11.1, 11.2 or 11.3 kg, etc. It could, for example, be somewhere between 12.3789 and 12.3790 kg. Any mass between the limits of 10 kg and 14 kg is possible. (The fact that no scale could be calibrated finely enough to record all the possible masses is a different matter.)

It could be argued that mass presented to three decimal places (such as 12.379 kg) is discrete, since the next highest possible value has to be 12.380; it could not be 12.3796 because this has four decimal places, which is one place too many. However, here the different values are separated by the degree of accuracy in the measurement or the computation, not by logical necessity.

Sometimes the distinction between discrete and continuous is less clear. For example, a person's *actual age* is a continuous variable. If we say that at some moment a person's age is 32.4768 years, we are measuring age more accurately than most people know their ages (it is given to 0.0001 of a year, which is less than one hour) but that age does exist, and all people older than it have been it. But a person's *stated age* is a discrete variable, usually expressed in whole years and rounded down.

Another example is the world population. This can increase only by integers, so in a strict sense it is discrete. But the differences between consecutive integers are so small, compared with the values themselves, that we can treat the world population as continuous.

Ordinal and cardinal

Another distinction is between **ordinal numbers** that simply describe position – first, second, third, etc. – and **cardinal numbers** – such as 1, 2, 3, etc. – which can meaningfully be added and multiplied. This distinction is close to our discussion of levels of measurement in Chapter 33. Here we simply note that most techniques devised for the analysis of quantities (and described in the earlier chapters of this book) are not applicable to ordinal data. For example, as we emphasise in Chapter 5, it is meaningless and misleading to average ranks. Ordinal data should be analysed by methods such as those of Chapter 33.

Population and sample

In statistics the word **population** has a much wider meaning than in everyday language. Any collection or assembly of people, items, things or numbers that *is of interest in its own right*, rather than because it may be representative of something larger, is a population. The number of items in a collection is its size.

For example, if I want to learn something about all the cars first registered in 1996 then those cars form my population. As another example, suppose I want to learn something about the 40 passengers on my bus. I am interested in those passengers in their own right, not as representatives of other bus passengers, or other people more generally. In that case those 40 passengers are the population that I am studying.

On the other hand, if I do a statistical analysis of the 40 people on the bus in order to reach a (possibly tentative) conclusion about (a) all bus passengers on that route, or (b) all bus passengers on any route that day, then the 40 passengers form a **sample** that is being used to indicate something about population (a) or (b). Whether it is a good sample is a matter to be considered later.

Always be clear about whether you have sample or population data. Much of this book concerns the use of sample data to make statements about the population from which the sample comes.

Consolidation exercise 2A

1. Which of the following statements are (a) accurate, (b) precise?
   (a) Sweden is in the northern hemisphere.
   (b) The tallest building in Europe is 20 metres high.
   (c) The tallest building in Europe is 20.37 metres high.
   (d) The population of Germany is more than 2 million.

2. What is the essential difference between (a) a discrete variable and a continuous variable, (b) a cardinal number and an ordinal number?

3. Say in each of the following cases whether the italicised group is a sample or a population. If you have doubts, say why.
   (a) *Ten customers* taken as representative of shoppers.
   (b) *Eleven members* of a football team whose hotel requirements need to be known by their manager.
   (c) The ability of a teacher to teach form IV mathematics is assessed by examining the work of *all of his or her form IV pupils.*
   (d) The ability of a teacher is assessed by examining the work of *all of his or her form IV pupils.*
   (e) The ability of a teacher is assessed by examining the work of *some of his or her form IV pupils.*
   (f) In order to investigate the existence of any link between eating carrots and seeing well at night, a doctor looks at *his or her patient records.*
   (g) In planning a Christmas menu, the owner of a restaurant asked *all customers on seven consecutive nights* whether they liked simmered tripe.

2.4 **Respecting techniques**

Many people have an imperfect idea of when addition cannot be used. Few people would add two telephone numbers and give any meaning to the answer. Yet other kinds of numbers (especially ranks) are often added when this is meaningless and misleading.

Like numbers, techniques have to be respected, and this usually means understanding not only the procedures but also the assumptions on which they are based. Statistics is both a science and an art. A large part of the art is choosing the right technique. At times you may come across a problem for which no technique known to you seems to be appropriate. Resist the temptation to make do with an inappropriate technique. It is like cooking chips in cart grease – they get cooked, but using them is disastrous.

If no suitable technique for the description or analysis of your data is known to you, perhaps consult a professional statistician or look at other books in the library. Alternatively, provided you are careful not to violate the data, you may be able to invent a technique (especially a graphical technique). But if you do this, seek comment on it from a better statistician than yourself.

Remember that our task is to reveal, to the best of our abilities, the story *truly* told by the data. If technique and data do not match (such as when using them together violates an essential assumption), the only honest thing to do is to say that you do not know how to extract the story.

 2.5 **Units**

Be careful to look at the units in which data are given. Often a question may give some measurements in centimetres and others in metres; time in seconds and in minutes; prices per item and per pack of ten items; and so on. Convert the data into consistent units, so that all lengths are measured in the same units; all times in the same units; and so on. When you reach your answer, remember that it should usually be in the same units. If it is not, consider whether it should be, then state the units. A mean mass is not 37, but it may be 37 kg.

 2.6 **Plausibility**

People often produce answers that are obviously wrong. A mean (or ordinary average) mass of 80 kg for the members of an infant school class is unlikely to be right. Most people notice that kind of error, but a statement that 69, 63, 59, 68, 64 have 60 as their mean is typical of the kind of error that is passed by. With four numbers well above 60 and one just below it, how can the average be 60? There is no need to repeat all your calculations; even if you did, you might be repeating the mistake. But always look at your answer and the data together and ask whether the answer looks sensible, credible or plausible.

2.7 Stating an answer

A calculation may be right, but a wrong impression can be created if the answer it not properly stated. Measurements of height, weight, age or any other continuous variable are almost always stated with some degree of approximation. We may say that a line measures 324 mm, but we are unlikely to mean 324 mm exactly. Conventionally a measurement recorded as 324 mm means anything between 323.500 mm and not quite 324.5 mm. The measurement has been **rounded** to the nearest whole number. Some measurements, such as people's ages, are usually rounded downwards, so that an age of 32.9 years is recorded as 32 years. Some banks round up their charges so that a charge of £24.30 is recorded as £25.

Rounding does not apply only to decimals. The annual sales of a large company over the last few years may be tabulated in millions of pounds, with rounding up or down to the nearest million.

An implication of this for statisticians is that the answers to calculations have to be stated with a precision that is compatible with the data. Most people realise that if the data are inaccurate they cannot normally expect an accurate result, but there is less appreciation of the danger of an answer being too precise. If seven people are weighed correct to the nearest kilogram, and have a total mass of 569 kg, their mean mass, obtained from dividing 569 by 7, would show up on a calculator as 81.28571 . . . kg. If we want to be no more precise than the data, this would have to be recorded as 81 kg. As we discuss in Chapter 5, we might sometimes be justified in quoting the mean as 81.3 kg, which is more precise than the original data, but we may be skating on thin ice when we do so. If we state it as 81.29 kg, we are implying that if measured a little more precisely the number would lie between 81.285 and 81.294. Since this is an average of 7 items, it means that we are also implying that the total lies between $7 \times 81.285 = 568.995$ kg and 7×569.058 kg. This means that we have measured masses correct to 0.001 kg, which is 1 gram. Possibly we can do this, but we have not – we have already said that the data are correct to the nearest kilogram.

Consolidation exercise 2B

1. When you have a problem involving masses, distances, times or other measured quantities, what should be your first action?

2. When you finish your calculations, what should you check, apart from the arithmetic?

3. State, without doing anything besides rough mental calculations, which of the following statements is implausible, and say why.
   (a) The square root of 37 is 6.9?
   (b) The mean of 17, 18, 23, 32, 27, 16, 19, 18, 23, 26, 19 is 26.
   (c) The usual average speed of the 12.15 mainline train between London and Birmingham is 32.37 mph. ➲

(d) Travelling at a an average speed of 7 mph through central London, a car took 9 hours 37 minutes to get from Buckingham Palace to the Houses of Parliament.

(e) The Sun is 148.8×10^8 cm from the Earth.

4. Say how precisely you can measure (a) distance across a living-room with a metre rule; (b) the time now on a kitchen clock; (c) the time between two loud bangs, using a kitchen clock; (d) weight on a set of bathroom scales; (e) speed on a car speedometer; (f) distance travelled in a car by using its trip meter.

Summary

1. The meaning of *statistics* has changed throughout history.

2. Both numbers and techniques have to be respected. This means knowing the limitations of the data, and the assumptions underlying the techniques. Number and technique have to match. If they do not, either in the sense that using them together violates an essential assumption or in some other way, the only honest and acceptable thing to do is to say that you do not know how to extract the story. Telling a story you know to be based on an invalid application of a technique is to lose respect for yourself.

3. Make sure you understand the definitions of *accurate* and *precise*, *variable* and *constant*, *discrete* and *continuous*, *ordinal* and *cardinal*, *population* and *sample*. These words are used frequently throughout the book.

4. Always keep in mind the units in which your data are given, and state them in your answers.

5. Always consider the plausibility of your answers.

6. Do not present your answer in a way that gives a false precision.

3

Tables and diagrams

3.1 Introduction

Example 3.1 Harry, the new landlord of the Calculators' Arms (which is Statingham's best non-alcoholic pub), has found a record of the numbers of customers between 8.00 and 11.00 on the 31 evenings of last March (Table 3.1). He finds the figures a bit bewildering and would like some kind of diagram that will tell him at a glance what things were like, and perhaps how things are going.

Table 3.1

Numbers of customers

32	28	31	29	20	27	30	29	29	30	33	26	29	31	27	28
30	27	28	28	23	28	27	32	26	29	31	28	30	25	32	

Example 3.2 Winnie, president of the Statingham Association of Working Women, wants female office assistants working for the council to have comfortable typing seats. She arranges with 60 clerical assistants to sit on an adjustable typing seat while she alters the height until she is satisfied that no adjustment will increase the sitter's comfort. She records the heights for these 60 workers correct to the nearest 0.1 cm, heedless of the criticism that if the alteration is so small nobody will know. The meter on the adjustable seat will measure to this degree of precision, so she is going to take advantage of it. The recorded heights are given in Table 3.2. Although put off by the

Table 3.2

Seat heights (cm) for 60 workers

71.0	72.8	75.7	75.9	71.4	71.6	71.7	73.2	72.1	64.3
72.6	72.5	74.4	67.8	73.6	73.8	75.3	75.5	70.8	71.8
71.8	74.2	73.8	77.4	74.3	68.9	69.3	69.3	69.8	76.8
77.5	79.3	81.7	70.1	73.1	73.1	73.4	68.7	72.1	72.4
60.3	63.2	72.2	73.4	73.3	72.4	74.2	71.1	71.4	74.5
72.8	70.5	74.6	74.6	74.7	70.3	70.3	72.9	76.3	76.4

amount of data, she knows that she can now set about persuading the council to buy new typing seats that will suit her fellow members of the Association of Working Women, but she is not certain about how many different heights she can order. Her first task is to try to get a better picture of what her fellow members need.

Example 3.3 Table 3.3 shows data about the level of GDP and its major components for India during 1988–89 (in billions of rupees) and Pakistan during 1990 (in millions of Pakistan rupees). How can this information be presented graphically in a way that shows clearly both the difference in the totals and the different importances of the various components?

Table 3.3

Gross domestic product

	1988–89 India (Rs billion)	1990 Pakistan (Pak Rs million)
Agriculture	1 137	230
Manufacturing	630	154
Trade	437	152
Public administration and defence	205	75
Total	3 489	1 017

Few things are more disheartening than a mass of unorganised statistical data. But if the data are put into some kind of order, patterns and meaning often begin to emerge. A clearer impression of the information buried in the statistics will frequently be obtained by tabulating them, or by putting them into a diagram. In this chapter we describe techniques for doing this.

We look mainly at techniques that help us to analyse – or to squeeze meaning out of – the data. But we also comment briefly on techniques used in the presentation of data – in telling others a story that the analysis has already squeezed out. And we describe one of these techniques in detail – the pie chart.

It is always important to decide whether the data are discrete or continuous, as the choice of technique may depend on this. We have to be especially careful if we have continuous data recorded as discrete data, such as when weights are rounded to the nearest kilogram. We describe a few techniques that strictly are valid only for continuous data but are often used with discrete data, noting appropriate cautions about their use and interpretation.

3.2 The ordered bar chart for discrete data

Look at Table 3.1. The number of customers has to be an integer, so the data are discrete. The figures rise and fall haphazardly. As things stand, the data are unorganised. The landlord decides to get a better idea by drawing a **bar chart** of 31 horizontal lines (or bars) with lengths proportional to the numbers, as in Figure 3.1(a). A glance at this bar chart gives a better idea than a glance at the table, but the effect is still somewhat jumbled. To get a clearer picture, the landlord rewrites the table with the numbers arranged in order of magnitude (Table 3.4) then produces an **ordered bar chart** (Figure 3.1(b)) showing the numbers of evening customers in order of magnitude. The numbers are discrete, and there is a bar for every number. The chart prompts the landlord to ask questions like these:

Table 3.4

Numbers of customers

20	23	25	26	26	27	27	27	27	28	28	28	28	28	28
29	29	29	29	29	30	30	30	30	31	31	31	32	32	32	33

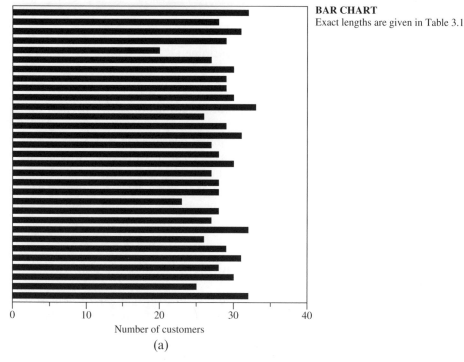

BAR CHART
Exact lengths are given in Table 3.1

Number of customers

(a)

Figure 3.1 (a)

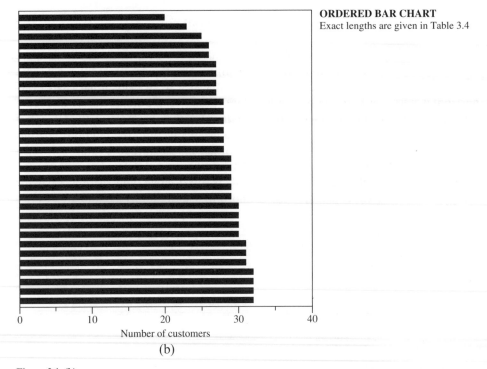

ORDERED BAR CHART
Exact lengths are given in Table 3.4

Number of customers

(b)

Figure 3.1 (b)

- What was the most common number of customers?
- How much more custom was there on the five busiest nights taken together than on the five slackest nights?

It also helps him to answer them. The landlord keeps his ordered bar chart on the wall. After a few days he tries a new way of arranging the bar, putting all the Sunday bars at the top, with the second Sunday beneath the first Sunday, and a large space beneath them for further Sundays. Then he draws the Monday bars, and so on, as in Figure 3.1(c). Every morning he draws a new bar for the night before. Soon he sees that his Wednesday bars are getting shorter, but his Tuesday bars are getting longer. His diagram is becoming useful, enabling him to see the latest information against the pattern of the past.

 3.3　Frequency distributions and polygons for discrete data

The landlord takes a closer look at Table 3.4 and the bar charts. On several evenings the same numbers of customers were present. Economically minded, he decides to compress the information in a **frequency table** as in the first two columns of Table 3.5. This table is an example of a **frequency distribution**. Column 2 shows the frequency f with which each value of the variable x occurs.

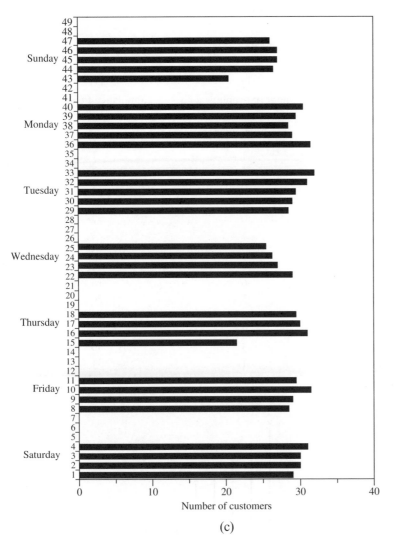

(c)

Figure 3.1 (c)

We will discuss column 3 in the next section. Notice that the data are **ungrouped**. There is a frequency for every single value of x. (We discuss grouping below.) Sometimes the data are presented in rows rather than columns, saving space.

The data in columns 1 and 2 can be represented graphically by plotting points for each pair of values (x, f), with x measured horizontally and f vertically. Vertical lines drawn from each point to the corresponding point on the x-axis finish the presentation (Figure 3.2). We have a vertical bar chart, with bar length proportional to frequency. There are two main variations of this technique.

Table 3.5

Evening customers: frequency distribution

(1) Number of customers, x	(2) Number of evenings with this number of customers (or frequency), f	(3) Cumulative frequency
20	1	1
21	0	1
22	0	1
23	1	2
24	0	2
25	1	3
26	2	5
27	4	9
28	6	15
29	5	20
30	4	24
31	3	27
32	3	30
33	1	31

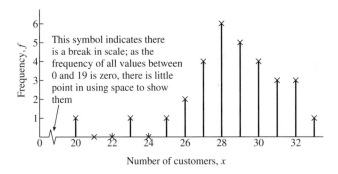

Figure 3.2

Replacing the bars by columns

Essentially this is a presentational technique, known as a **column graph**. If we have discrete data the only points on the x-axis that have any meaning in terms of measurement are the points marked 0, 1, 2, 3.... The spaces between have no measurement associated with them. If columns are being used to represent ungrouped discrete data, each column should have its base centred on the value whose frequency is shown by its height (Figure 3.3). They can be of any width

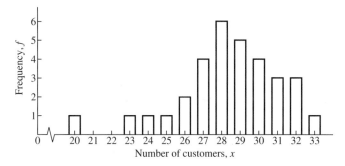

Figure 3.3

(as long as all columns are of the same width). It is better for them not to touch, in case they are misinterpreted as continuous data.

Joining consecutive points by straight lines

This is also a presentational device that has the two merits of guiding the eye from one point to the next and being easier to draw than a column graph. The graph (Figure 3.4) is sometimes called a **frequency polygon**. A major objection to drawing a frequency polygon for discrete data is that points along the line have no meaning. There is no value of x equal to 25.3; and even if there were, no good reason would exist for asserting that the corresponding value of f lies on the straight line vertically above 25.3.

The vertices of the frequency polygon coincide with the tops of the vertical bars, or the centres of the tops of the columns.

If we construct a diagram such as Figure 3.2 or 3.3, then at the end of each day one of the frequency lines (or columns) will have to be made one unit higher (or a new line will have to be started if that day has a number of customers that has not previously occurred). After another 31 days, some of the lines will have grown much more than others, and the landlord can see whether the last month differs importantly from the preceding month.

Figure 3.4

How to ... construct a frequency table and diagram for ungrouped discrete data

1. In column 1 (or row 1) write all the different values of the variable x in ascending order, with the lowest number at the top of the column (or the left of the row).

2. Form column 2 by writing opposite each entry in column 1 the frequency f with which that value occurs.

3. Draw two axes at right angles. Mark a horizontal scale extending to the highest value of x and a vertical scale extending to the highest frequency. Break the horizontal scale near the origin if the smaller values of x all have zero f, and the vertical scale if the smallest value of f is so high that, without breaking the scale, the whole diagram would be at the top of the page.

4. Lightly mark the first value of x along the horizontal axis and the associated value of f (in column 2) along the vertical axis. Make a small cross at a point where a vertical line through the x-mark would cut a horizontal line through the f-mark.

5. Do the same for all other values of x.

6. Then do one or more of the following:
   (i) Draw vertical lines from the centre of every cross down to the x-axis.
   (ii) Draw the vertical lines of (i) lightly and then for each line construct a vertical column with sides parallel to the line and the same distance from it, so that the line runs up the centre of the column. Draw a horizontal line across the top of the column. Make certain there is a column for every line, all the columns are of the same width, and they do not overlap. Warn users not to misinterpret the graph.
   (iii) Moving from left to right, join every cross to the next by a straight line. Warn users not to misinterpret the graph (see text).

Column 3 of Table 3.5 contains the **cumulative frequencies**. Every entry is the sum of all entries in column 2 from the top down to and including the entry opposite it. Thus, the fifth entry in column 3 is 9. This is the sum of the first five entries in column 2. The corresponding entry in the first column is 27. The entry in column 3 therefore means there are 9 items with a value of x less than or equal to 27.

   Cumulative frequency graphs for ungrouped discrete data are usually based on the column graph approach (Figure 3.5) but sometimes appear as polygons. In the column graph approach there is a series of columns, each higher than its predecessors. The height of each column is given in the cumulative frequency column of the table.

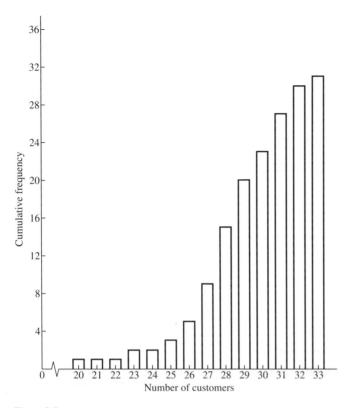

Figure 3.5

The cumulative frequency polygon rises to the right. The vertices are in the same positions as the centres of the column tops. Only the heights of these vertices have meaning. The straight lines joining the vertices simply guide the eye.

Box 3.2

How to ... **construct a cumulative frequency table and diagram for ungrouped discrete data**

1. In column 1 (or row 1) write all the different values of the variable x in ascending order, with the lowest number at the top of the column (or the left of the row).

2. Form column 2 by writing opposite each entry in column 1 the frequency f with which that value occurs.

3. Column 3 contains the cumulative frequency (CF). The top entry is the same as the top entry of column 2. The next entry is the sum of the first two entries in column 2. ⊙

4. Obtain the third entry in column 3 by adding the *third* entry in column 2 (to the side of the space you are about to fill) to the *second* entry in column 3 (immediately above the space). Continue by forming the fourth and subsequent entries in the same way.

5. Check that the final entry in column 3 is the total of column 2, which is the total frequency.

6–9. Proceed as in steps 3–6 of Box 3.1, reading 'column 3' for 'column 2' and 'cumulative frequency' for 'frequency'.

Consolidation exercise 3A

1. What is a bar chart, an ordered bar chart, a column graph?

2. How do you construct a frequency table for ungrouped discrete data?

3. In Table 3.4, what is the meaning of 15 in the last column?

4. In a column graph for ungrouped discrete data, what does the width of a column show?

5. What is a frequency polygon? What is shown by the midpoint of a straight line between two vertices of a frequency polygon?

6. Table 3.6 shows the numbers of puppies in 50 litters. This information is to be shown in a column graph with columns 1 cm wide. One column will be 12 units high. Where should its base be?

Table 3.6

Number of puppies	0–2	3	4	5	6	7	8	9	10+
Number of litters	0	4	8	12	8	10	3	5	0

7. What are (a) a cumulative frequency table and (b) a cumulative frequency polygon?

3.4 Grouped discrete data and histograms

The information given in Table 3.5 could be condensed for convenience into something like Table 3.7. In this form the data are quite as *accurate* as in Table 3.5 but they are less *precise*. They are accurate in that they are correct, but it would be more useful to us if things could be narrowed down, and presented more precisely, which means in narrower class intervals. As we shall see, the lack of precision will lead to possible inaccuracy if we attempt to calculate an average to a level of precision greater than the precision of the data.

Table 3.7

Evening customers: grouped frequency distribution

Number of customers, x	Number of evenings with this number of customers (or frequency), f	Cumulative frequency
20–24	2	2
25–27	7	9
28–30	15	24
31–33	7	31

In Table 3.7 the discrete data have been **grouped**. In this example the widths of the groups, or the **class intervals**, are not all the same. The first group covers five integers (20 to 24 inclusive) and the others cover three. It would have been quite in order to have chosen different class intervals. They should be chosen with an eye on condensing the data in a way that highlights the main features. An example is discussed a little later. The choice of class intervals can also facilitate comparison within the distribution. But this becomes difficult if there are many different class widths.

Box 3.3

How to ... **condense a frequency distribution into a grouped frequency distribution**

1. Decide on roughly how many groups are needed if your grouped distribution is to show the main features of the ungrouped distribution.

2. Try combining a few consecutive rows of the frequency distribution table into a group. Each row of column 1 in the grouped table will contain a range of values, from the lowest to the highest in the group. Column 2 will contain the total of the frequencies from those rows.

3. Repeat the procedure until you have roughly your selected number of groups, preferably of equal width.

4. Consider whether a better representation of the data would be achieved if you made slight alterations to the groupings.

5. Consider whether consecutive groups with small frequencies can be combined into a larger group without important loss of detail.

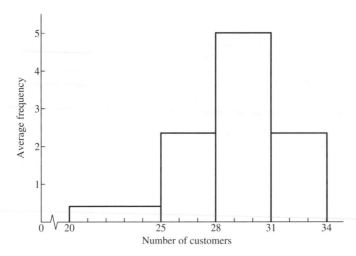

Figure 3.6

The information in Table 3.7 is shown graphically in Figure 3.6. This is an example of a **histogram**. The columns have been drawn so that each has a width proportional to the class interval. This means drawing columns that touch each other. The column representing the interval 20–24 customers has a width of 5 units, going from the 20 mark right up to (but not quite reaching) the 25 mark. Note that the *areas* (not the lengths) of the columns are proportional to the frequencies. Only if the widths are equal should the lengths be proportional to the frequencies (Notebox 3.1). It follows that the height of the column represents the average of the numbers of evenings on which 20, 21, 22, 23 or 24 customers were present. It is an average frequency. The calculations needed for the construction of Figure 3.6 are shown in Table 3.8. They follow the procedure outlined in Box 3.4.

Table 3.8

Evening customers: construction of histogram

(1) Number of customers, x	(2) Number of evenings with this number of customers, f	(3) Class width	(4) = (2)/(3) Column height
20–24	2	5	2/5 = 0.40
25–27	7	3	7/3 = 2.33
28–30	15	3	15/3 = 5.00
31–33	7	3	7/3 = 2.33

In Figure 3.6 the second column shows that on seven evenings there were 25, 26 or 27 customers. If we take 26 as an average this means that the total number of customers on those seven evenings can be estimated at $7 \times 26 = 182$. This is only an approximation, since we have assumed that the average is 26.

In this explanation we have derived the table from Table 3.5 and we can go back to it as a check. It tells us that on one evening there were 25 customers, on two evenings there were 26 and on four evenings 27, so in all there were 185 customers. This is three more than our estimate. To save work, we have approximated by collecting our data into groups. But in doing this we have lost some of the detail and become slightly inaccurate.

Subject to this inaccuracy due to grouping, each column area gives the total number of customers on the evenings represented. Thus the sum of the column areas (or the total area under the graph) tells us the total number of customers on all 31 evenings.

Note that when we measure the area of a column, we are multiplying height by breadth, and consideration of the units involved quickly indicates what the area is measuring. Suppose we had been drawing a histogram showing how many customers had drunk 5 pints, how many 6 pints, and so on:

Area of column = height × breadth

= number of customers × pints drunk by a customer

= total pints drunk by all customers

Box 3.4

How to ... **draw a histogram**

1. Construct a grouped frequency distribution table.

2. Complete a table column showing the class widths. This will be the difference between the group's lower bound (or its lowest possible value) and the lower bound of the next highest group.

3. Divide each group frequency by the class width to obtain the height of the column to be drawn in the histogram. If it is an open-ended group, divide the group frequency by rather more than the width of the group and follow the conventions indicated in Notebox 3.3.

4. Consider whether the horizontal scale can properly be broken by a zigzag.

5. Choose horizontal and vertical scales that will enable your diagram to fill the space you have available for it. The vertical scale should be enough to accommodate the tallest column.

6. On the horizontal scale, mark off the class boundaries.

7. Draw columns of appropriate heights (as found in step 3) between these boundaries. The columns will touch each other.

Consolidation exercise 3B

1. What are (a) grouped data, (b) class intervals and (c) histograms?

2. How are frequencies represented in a histogram? Should consecutive columns of a histogram touch each other?

3. Draw a histogram to show the data of Table 3.9 for the numbers of passengers carried by 830 buses. How many units high are your columns?

Table 3.9

Number	13–15	16–19	20–24	25–29	30–39	40+
Frequency	90	120	180	200	180	60

3.5 Stem and leaf diagrams

We turn now to Example 3.2. Winnie wants to squeeze the pips out of her data about the desirable heights of seats. Mr Venn, an amateur statistician, advises her that the figures are values of a continuous variable, so the stem and leaf approach is particularly useful, although it can also be used for discrete data.

The figures in Table 3.2 show heights correct to the nearest tenth of a centimetre. They are distributed between 60.3 cm and 81.7 cm – a span of 21.4 cm. We consider aggregating (or collecting together) the data into groups of width 4 cm, then drawing a histogram of six columns. We will start the first column at 60.0, the second at 64.0, and so on. To draw the histogram, we have to know how many seat heights there are in the various groups. A useful way of doing this automatically produces a table that can also be viewed as a diagram. It is to construct a **stem and leaf diagram**.

We begin by writing the starting points of the six groups in a column called the 'stem', as below. Then we take the first entry in the table, 70.0. It is in the interval beginning 68 and is $70.0 - 68.0 = 2.0$ units 'along' it. We therefore denote the entry 70.0 by writing **2.0** in the 'leaf' row alongside the 'stem' entry of 68. This is shown in bold in Table 3.10.

Similarly the second item in Table 3.2, which is 72.8, is represented by an entry of **0.8** in the leaf, corresponding to the stem entry of 72. The third item, 75.7, is represented by an entry of **3.7** in the same leaf. Proceeding in this way we get the stem and leaf diagram (Table 3.10). Notice how there is no need to write the entries in the leaf in ascending (or descending) order, although this may be useful later.

Box 3.5

How to ... construct a stem and leaf table and then a grouped frequency distribution table

1. Find the lowest value of the variable (60.03 in the example) and write on the left side of the page the largest integer less than it (60). ▶

2. Find the highest value and roughly calculate the **range** (the difference between the highest and lowest values), which is about 22 in the example.

3. Decide on a convenient number of intervals spanning this range (or a little more) and write the starting points of each interval in a column, called the **stem**, as in the example.

4. Pick out the entries that go in the first interval (60.3 and 63.2). Subtract the starting point of the interval (60) from each entry and write the remainders (0.3, 3.2) in a row opposite the starting point; this row is called a **leaf**.

5. Pick out the entries (64.3 and 67.8) that go in the next interval (between 64.0 and 67.9). Subtract the start of the interval (64) from them and write the remainders (0.3 and 3.8) in the row.

6. Repeat this for all intervals, with entries in an orderly manner, equally spaced out.

7. To construct the grouped frequency table, write out the stem entries as column (1) of the table, and obtain the frequencies by counting the number of entries in the leaves.

If the entries in the leaves are written evenly spaced across the page, a quick impression of the relative frequencies is easily obtained. More accurately, counting the items in each leaf leads to the grouped frequency distribution shown in panel A of Table 3.11. The third column will be considered later.

We can now use this table in the construction of a histogram, as in Figure 3.7(a), which has all columns of equal width. So much of the total frequency is squeezed into two groups (or columns of an associated histogram), there may be some sense in decomposing them so that we can examine the distribution a little more closely.

Table 3.10

Seat heights for 60 workers: stem and leaf

Stem	Leaf										
60	0.3	3.2									
64	0.3	3.8									
68	**2.0**	3.4	3.6	3.7	2.8	3.8	3.8	0.9	1.3	1.3	1.8	2.1
	0.7	3.1	3.4	2.5	2.3	2.3					
72	**0.8**	**3.7**	3.9	1.2	0.1	0.6	0.5	2.4	1.6	1.8	3.3	3.5
	2.2	1.8	2.3	1.1	1.1	1.4	0.1	0.4	0.2	1.4	1.3	0.4
	2.2	2.5	0.8	2.6	2.6	2.7	0.9				
76	1.4	0.8	1.5	3.3	0.3	0.4					
80	1.7										

Table 3.11

Grouped frequency distribution for 60 seat heights

Height	Frequency	Cumulative frequency	Frequency/group width
A. Original distribution			
60.0–63.9	2	2	2/4 = 0.5
64.0–67.9	2	4	2/2 = 0.5
68.0–71.9	18	22	18/4 = 4.5
72.0–75.9	31	53	31/4 = 7.75
76.0–79.9	6	59	6/4 = 1.5
80.0–83.9	1	60	1/4 = 0.25
B. New distribution			
60.0–63.9	2	2	2/4 = 0.5
64.0–67.9	2	4	2/4 = 0.5
68.0–69.9	5	9	5/2 = 2.5
70.0–71.9	13	22	13/2 = 6.5
72.0–73.9	19	41	19/2 = 9.5
74.0–75.9	12	53	12/2 = 6
76.0–79.9	6	59	6/4 = 1.5
80.0–83.9	1	60	1/4 = 0.25

Let us make new groups beginning with 68.0, 70.0, 72.0 and 74.0. The group 68.0–69.9 will contain all the entries in the 68 leaf that are not more than 1.9; there are 5 of them. The other 13 entries fall into the new group 70.0–71.9.

Similarly the 31 entries in the group 72.0–75.9 are reallocated as in panel B of Table 3.11. Once again, the third column will be considered later.

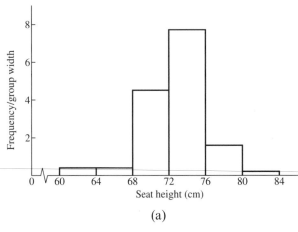

(a)

Figure 3.7 (a)

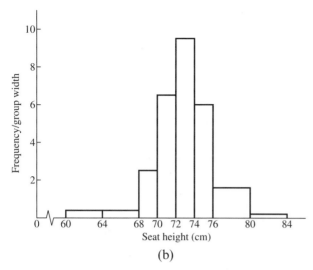

Figure 3.7 (b)

Histograms of the data in Table 3.11 are shown in Figure 3.7. Each is accurate (in the sense that it is correct) but Figure 3.7(b) is more precise than Figure 3.7(a).

In later chapters we shall see how Winnie uses these grouped frequency distributions. Before doing that, we must say a bit more about frequency distributions and their graphs.

3.6   Frequency curves

When we have a histogram of grouped continuous data, we may wish to estimate the frequency with which a precisely stated value of the variable occurs. A first approximation may be obtained by joining the midpoints of the tops of the columns to produce a **frequency polygon** as described for discrete data in Box 3.4. Because we have continuous data, points along the lines have meaning, but they provide only a first approximation to the frequency of an imprecisely specified value of the variable (such as 69.3, which means anything between 69.25 and 69.35). It is better to treat the polygon as an approximation to a **frequency curve**.

We have considered histograms and frequency polygons derived from grouped frequency distributions. The histograms have consisted of columns with flat tops, and the polygons have consisted of straight lines. We now extend these ideas, replacing the flat tops and straight lines by curves. Strictly speaking, they are valid only if our data are for a continuous variable (which is often the case), or we have an infinitely large (or, in practice, a very large) number of observations.

If we have a very large number of groups, the histogram will have a very large number of tightly packed narrow columns, as we try to indicate in Figure 3.8. Figure 3.8(a) shows a histogram with eight fairly wide columns. Figure 3.8(b) splits each column into two narrower columns, with appropriate heights, just as some of

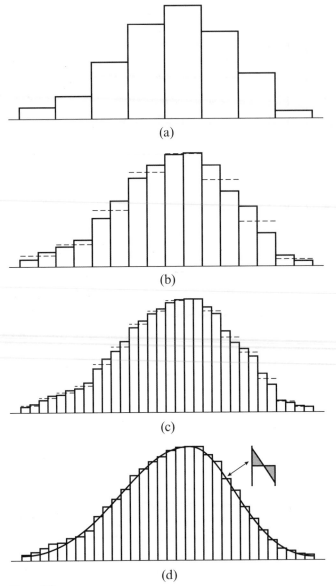

(a)

(b)

(c)

(d)

Figure 3.8

the columns in Figure 3.7(a) were split for Figure 3.7(b). The dashed lines in Figure 3.8(b) show the positions of the tops of the columns in Figure 3.8(a). There has been further subdivision in Figure 3.8(c), with the broken lines showing the column tops of Figure 3.8(b). As the number of groups is increased, these columns get narrower and more numerous. The flat lines at their tops become shorter. The frequency polygon formed by joining their centres becomes almost indistinguishable from a continuous curve, as in Figure 3.8(d).

Before we can properly call this a **frequency curve**, note how this term should be applied only when we have continuous data, in the sense that between any two values of the variable x there exist in principle an infinite number of other possible values. However, if the possible values are closely packed – such as the numbers of words on the many pages of a book – it is usually sensible to look upon it as continuous. We discuss this more fully in later chapters.

An important point about the interpretation of the height of an ordinate of a frequency curve is discussed in Notebox 3.4. Because the variable is continuous, if we seek the frequency of a specified value (such as 22), we have to replace it by a narrow range (such as 21.50–22.49). The frequency for this range is the area beneath the curve between the end-of-range values.

If, as in Figure 3.8(d), the column heights increase more or less regularly before reaching a maximum, then descend more or less regularly, you should try to draw a smooth curve cutting the column tops in such a way that the areas balance. This means that, for each column, the two shaded areas are equal (Figure 3.8(d), inset). A similar procedure is used for the downward-sloping portion. The highest column may have its top cut twice, and the same principle applies. Making these areas equal is not easy, but even approximate equality may produce a more accurate idea of the distribution than the frequency polygon.

If the column heights do not increase more or less regularly, consider whether they might do so if there were more observations – such as customers on many more nights. If this seems to be so, try to produce a smooth frequency curve by a technique such as combining some adjacent columns into a wider one with their average height.

3.7 Cumulative frequency curves (ogives)

We have already shown how to construct cumulative frequency graphs for discrete data. Now we consider the equivalent for continuous data. In Tables 3.7 and 3.11 for grouped data we have columns headed 'Cumulative frequency'. Notice that each entry in the cumulative column represents the number of items with a value of x up to the highest bound of the appropriate group.

Every histogram can be turned into a cumulative histogram simply by adding to every column the total of the heights of all columns to its left. This will produce a diagram similar to Figure 3.5 but all the columns will be touching. If we narrow the columns and increase their number, the points at the top right-hand corners of the columns will eventually form points on the cumulative frequency curve, subject to the same comment about continuous variables as we made when discussing frequency curves.

A cumulative frequency curve is sometimes called an **ogive** (pronounced o-jive). The typical elongated S-shape of an ogive has led to the use of 'S-curve' as an alternative name.

Notebox 3.5 extends these ideas to percentage frequency and cumulative frequency curves.

Box 3.6

How to ... construct a cumulative frequency distribution for grouped data and how to draw an ogive

1. Construct column 1, listing all the groups in ascending order. The upper bound of every group is the lower bound of the next group.

2. Form column 2 by writing opposite each entry in column 1 the frequency f with which a value in that group occurs.

3. Column 3 contains the cumulative frequency (CF). The top entry is the same as the top entry of column 1. The next entry is the sum of the first two entries in column 1.

4. Obtain the third entry in column 3 by adding the *third* entry in column 2 (to the side of the space you are about to fill) to the *second* entry in column 3 (immediately above the space). Continue by forming the fourth and subsequent entries in the same way.

5. Check that the final entry in column 3 is the total of column 2, which is the total frequency.

6. Draw two axes at right angles. Mark a horizontal scale extending to the highest value of x (or the upper bound of the last group), and a vertical scale extending to the total frequency.

7. Plot a point with an x-value equal to the lower bound of the first group, and a CF of zero.

8. Plot a point with an x-value equal to the upper bound of the first group and a CF equal to the first value in column 3.

9. Repeat for the upper bound of the second group and the second value in column 3. Continue in this way for every group.

10. Join the points in order, either with straight lines or with a smooth curve. The lines (or curve) will always point upwards to the right.

3.8 Dot plots

Often we do not need an accurate representation of a frequency curve or a histogram, but we want to get a rough indication of how the data are distributed. Do the values tend to cluster anywhere? Are they symmetrically distributed?

One way of getting quick answers to such questions is to draw a horizontal line with the minimum and maximum values of x marked on it, and a convenient scale drawn between them. For every observed value put a dot at (approximately) the right place on this line. If the same value occurs twice, put one dot above the other.

Figure 3.9 shows a dot plot for the data in Table 3.1. There is clearly a clustering around 27–29 with tailing off in both directions. Values less than 28 are more spread out than values greater than 28.

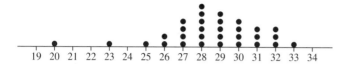

Figure 3.9

Essentially the same information can be obtained quite as quickly by following the early steps in the production of a frequency table, as described in the account of stem and leaf diagrams.

In later chapters we see that certain tests should be applied only if the distribution is of approximately some stated shape. Either technique can quickly indicate whether this is so.

Consolidation exercise 3C

1. What is a stem and leaf diagram? What is shown in the stem? What is shown in the leaves?

2. What is shown by the entry 41 in column 3 of Table 3.11 (panel B)?

3. What is a frequency curve? How does it show the frequency with which $x = 17$?

4. What is an ogive? How can a cumulative frequency curve be used in forming a cumulative percentage frequency curve?

5. When and how should a dot plot be drawn?

 3.9 Pie charts

It is not difficult to interpret Table 3.3. The data immediately give a quick impression of how the two countries compare. However, it is sometimes useful to have a graphical or pictorial presentation.

A frequently used device is the 'pie chart'. This is a circle (or pie) whose total area is taken to represent the total under discussion. It is divided into several sectors, one for each component, with the angles between the bounding radii proportional to the percentages of the total that each sector is to represent. Noting that there are 360° in a circle, this means that a sector representing 10% of the total would have an angle of 36°. More generally, a sector representing $p\%$ would have an angle of $3.6p°$. Since the area of a sector of a circle with a given radius is proportional to the angle, we can compare the sizes of the sectors by comparing either the angles or the areas. Before looking at Example 3.3, we take a simpler example.

⬭ **Example 3.4** A plane has 90 passengers: 30 of them are adult men, 40 are adult women and 20 are children. Show this using a pie chart.

To answer this we draw a circle of any convenient radius. Its total area represents 90 passengers. Since 30 out of the 90 passengers are adult women, one-third (30/90) of the area can be used to represent the adult women. We therefore make two 'cuts' into the pie at an angle that allows them to contain one-third of the pie. Since there are 360° in a circle, one-third of the area will be marked out by an angle of $360°/3 = 120°$. The 40 men can be represented by the slice that has an angle of $(40/90) \times 360° = 160°$, and the 20 children by a slice with an angle of $(20/90) \times 360° = 80°$. Note that the angles add up to 360°, representing the total number of passengers. The resulting pie chart is shown in Figure 3.10.

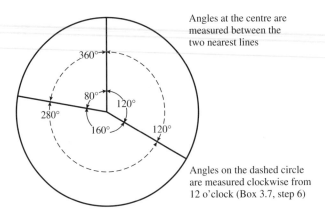

Figure 3.10

The data for India and Pakistan shown in Table 3.3 can be represented in much the same way in two separate pie charts. However, each of them will require a fifth sector, 'other industries', so that the totals add up. For Pakistan the angle cutting out the agriculture slice will be $(230/1017) \times 360° = 81.4°$.

One thing not shown by the pie charts is that India's gross domestic product (GDP) is much bigger than Pakistan's. To adjust for different currencies and various other problems of international trade, we use the official conversions of the two totals into US dollars. These value India's total output at $287 billion and Pakistan's at $40 billion (1 billion = 1000 million). Thus India's output is approximately 7.2 times as big as Pakistan's.

We therefore want the area of the India circle to be 7.2 times as big as the Pakistan circle. Since the area of a circle depends on the square of the radius, this means the India radius should be $\sqrt{7.2} = 2.68$ times as big as the Pakistan radius.

Now we have a choice. One way of proceeding is to draw two circles, the Pakistan circle with a radius of say 1 inch, and the India circle with a radius of 2.68 inches. Then we subdivide each circle. The arithmetic of the subdivision is shown in Table 3.12. The column of degrees is easily calculated from the percentages. In practice, if the sectors are constructed by measuring and drawing these angles in turn, there is likely to be a cumulative error and the last sector goes wrong. It helps to avoid this if a cumulative column is drawn and every angle is measured from the vertical, as illustrated in Figure 3.11 and described in Box 3.7. A little rounding of the figures may be necessary if the last entry in the cumulative column is not exactly 360. This is best done by adjusting a large angle, as indicated by the + signs for 'Pakistan : Other' in Table 3.12.

Table 3.12

	India			
	Value	Percentage of total value, p	Degrees $(p \times 360°)$	Cumulative degrees
Agriculture	1 137	32.6	117.3	117.3
Manufacturing	630	18.0	65.0	182.3
Trade	437	12.5	45.1	227.4
Public admin. and defence	205	5.9	21.2	248.6
Other	1 080	31.0	111.4	360.0
Total	3 489	100.0	360	

	Pakistan			
	Value	Percentage of total value, p	Degrees $(p \times 360°)$	Cumulative degrees
Agriculture	230	22.6	81.4	81.4
Manufacturing	154	15.1	54.5	135.9
Trade	152	14.9	53.8	189.7
Public admin. and defence	75	7.4	26.5	216.2
Other	406	39.9	143.7(+)	359.9(+)
Total	1 017	100.0	360	

The result, consisting of two pie charts drawn to the same scale, is shown in Figure 3.11. It is possible to compare both the magnitudes of the totals and the contributions to these totals of the various sectors.

The other way of proceeding is simply to impose the smaller pie chart on the larger one (Figure 3.12, see page 40). Now we have to imagine the radial lines of the larger circle as continuing into the centre. Some people prefer one way, some the other.

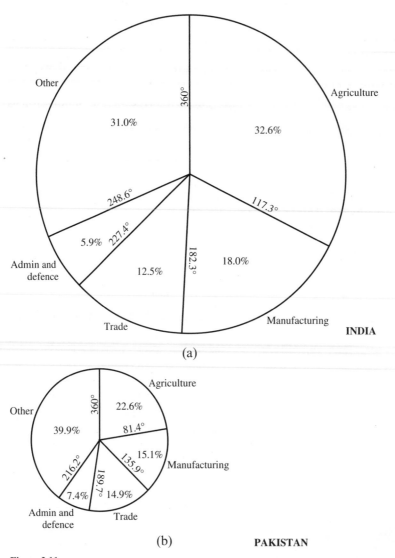

(a)

INDIA

(b)   PAKISTAN

Figure 3.11

Box 3.7

How to ... draw a pie chart

1. If the pie chart is not to be compared with others, draw a circle of any convenient radius. Then go to step 3.
2. If more than one pie chart is to be drawn showing different apportionments of different but comparable totals, work out the radii of the different pie charts as follows:

(i)   Divide all totals by the smallest total.

(ii) Take the square roots of all the answers. The radii of the pies have to be proportional to these square roots, with the smallest radius being 1. Let the largest radius (which is the largest square root) be L.

(iii) Consider whether a circle of radius L cm will be too big or too small for your sheet of paper. If it is too big, choose some fraction of L that will make it a suitable size. If it is too small, choose some multiple.

(iv) Adjust all the radii derived in (ii) by the fraction or multiple chosen in (iii).

(v)   Draw several circles, one for each pie chart, either separately with different centres or superimposed, all with the same centre. Treat every circle as in steps 3–6, being careful to show the different sectors in the same order for every pie chart.

3. Decide in which order you want the sectors to appear, going clockwise from the top of the circle. List them in this order in column 1 of a new table.

4. For each pie chart, write in column 2 the degrees for that sector. Do this by dividing every sector entry by the total entry and then multiplying by 360°.

5. Construct column 3 by cumulating these degrees so that the first entry shows the degrees for the first sector, the second entry the degrees for the first two sectors combined, and so on. The last entry will be 360°.

6. Divide the circle into sectors. Do this by drawing a vertical line from the centre up to the circle. Using a protractor, mark off an angle clockwise from this vertical equal to the first angle in column 3 and draw a line outwards from the centre. Then, again measuring clockwise from the vertical, mark off an angle equal to the second entry in column 3; and so on. When the angle exceeds 180° subtract 180° from the angle and measure the remainder clockwise from the bottom of the circle.

3.10 Pictograms

Quantitative data are sometimes represented by pictorial symbols that represent the subject of the data. A great deal of ingenuity is often used, some of it misguided. Pictograms are used in two main ways.

Modified bar or column charts

Instead of drawing bars of various lengths, we use rows of symbols. For example, if we wish to show the amounts of house building in several different towns, we may use the idea of a bar chart but show that 97 houses were built by drawing 97 houses side by side.

To reduce the work involved in drawing and interpreting this row of 97 identical symbols, we may use one house symbol to represent say 10 houses. We

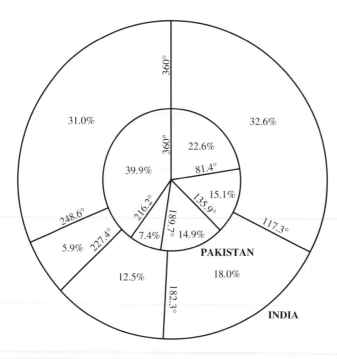

Figure 3.12

then depict 97 houses by having nine symbols side by side and a tenth unfinished house to represent 7 houses. In that case it is usual to provide a key such as a symbolic house broken into 10 parts so that each represents one built house.

This procedure is followed (with different numbers) in Figure 3.13(a). Pictograms of this kind usually present the data properly.

Areal representations

Mistakes are frequently made here. If a symbol is to represent area, its width and height should be proportional to the square root of the area. This is frequently forgotten, so an incorrect picture is presented. Even when correctly drawn, areal representations may be difficult to interpret. For example, it is clear from Figure 3.13(a) that slightly more than twice as many houses were built in B as in A. It is much more difficult to see this by looking at Figure 3.13(b).

Both for accuracy and for ease of interpretation, the modified bar chart presentation is usually to be preferred. Certainly it is to be preferred over a three-dimensional picture, whose volume should be proportional to the quantity, so a doubling of height corresponds to multiplying the quantity by eight if the proportions of the symbol are to be preserved.

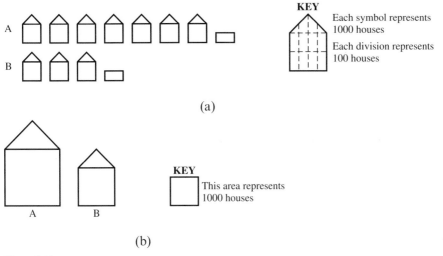

(a)

(b)

Figure 3.13

Consolidation exercise 3D

1. Three tables of figures are to be represented by pie charts. They show the detail making up totals of 100, 900 and 2500. If the radius of the smallest pie is 1 cm, what will be the other radii?

2. A pie chart represents a total of 300 and one of the sectors represents 45. What angle is formed by the edges of the sector?

3. A simple drawing of a square surmounted by a triangle is used to represent housing. The height of the drawing that represents 100 houses is 2 cm, how high should be the drawing for 225 houses?

4. If a drawing for housing shows the front and one side of a house, and a drawing representing 1000 houses is 1 m high, what is represented by a drawing 2 cm high?

Notebox 3.1

Graphing grouped data

Grouped data are usually portrayed graphically by drawing columns. Many people go wrong here. For example, suppose that we had 10 people aged 20–24, 10 people aged 25–29 and 10 people aged 30–34. If we want to represent these three pieces of information by a graph, we can draw three columns, all 5 units wide, so that the widths indicate the span of ages covered by the group. We also make them all 10 units high, arguing that frequency has to be represented by height. We get Figure 3.14(a).   ⊃

Now combine the first and second groups, so we have 20 people aged 20–29 and 10 people aged 30–34. We get two columns, one 10 units wide and the other 5 units wide, to represent the age spans. If we still argue that frequency has to be represented by height, the first column will be 20 high and the second will remain 10 high, as in Figure 3.14(b). Somehow, by merging two groups, we have completely altered the diagram.

On the other hand, if we let the frequency be represented not by the height of the column but by its area, we get Figure 3.14(c), which looks just like Figure 3.14(a) except for the vertical line and the legend.

Here we have chosen the vertical scale so that the column 10 units wide has an area that represents 20, the number of people in the age range 20–29. Its height must be area/width = 20/10 = 2. In the same way the second column, of width 5, has an area representing the 10 people aged 30–34, which means the height is 10/5 = 2.

These three diagrams illustrate how it is the area of the column that must represent the frequency – an important point. If the columns are all the same width, this means that the heights also represent the frequencies; but if the widths are different, the heights don't represent the frequencies. Stick to areas.

It also shows why the vertical scale should not be measured in units of frequency, but in frequency per unit width. Usually it is denoted by numbers without named units.

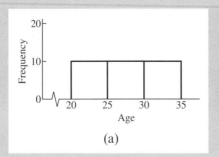

(a)

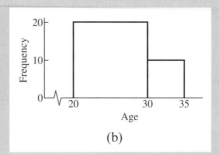

(b)

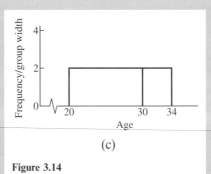

(c)

Figure 3.14

Notebox 3.2

Special case: graphing rounded data

Sometimes the values of the variable have been presented as integers but in fact are rounded values of a continuous variable. In that case the graph has to take account of this. To emphasise the points involved, we use the same numerical data as in Table 3.5 but we alter the labels.

(1) suppose that the values of x in the first column are age in complete years. Now a frequency of $f = 6$ for $x = 28$ means there are six persons with ages between 28 years exactly and almost 29 years. This means that a column of width equal to one unit of x (i.e. 1 year) can be drawn, with its base stretching from $x = 28$ right up to $x = 29$. This could produce a graph such as Figure 3.15(a). The vertices of a frequency polygon for this data should be halfway along the tops of the rectangular columns, as shown, since there are six persons whose average age is $28\frac{1}{2}$.

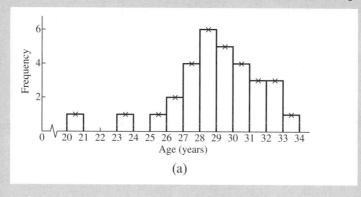

(a)

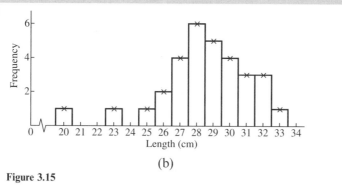

(b)

Figure 3.15

(2) Now suppose x may be length measured to two decimal places and then rounded to the nearest centimetre; $x = 28$ means any length between 27.50 and 28.49. Now a vertical bar (or column) centred on 28, and of unit width, from $x = 27.50$ to $x = 28.49$ is appropriate, as in Figure 3.15(b). Once again the vertices of the frequency column should coincide with the midpoints of the tops of the columns.

Graphing open-ended data

Sometimes a grouped frequency distribution is **open-ended**, as when the first group is described as 'under 25' or the last group as '34 or over'. In these cases we know the areas that should be represented by the first and last columns but we cannot be certain of their widths, so we cannot be certain of their heights. One way of representing this graphically is shown in Figure 3.16, based on data

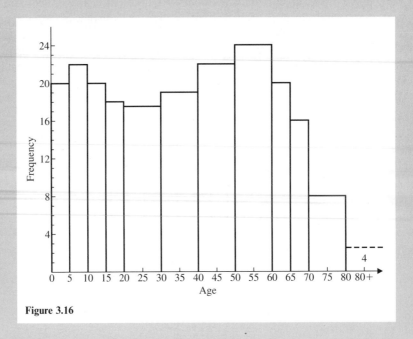

Figure 3.16

in Table 6.5. Here, as described in Chapter 6, we have had to make a reasoned guess at the heights. We indicate this using a broken top to the column, and by omitting the outer edge. We may also indicate the actual frequencies numerically.

The height of an ordinate on a frequency curve

A frequency curve may be such that if we take any value of x to be represented by a point on the horizontal axis, then the frequency with which that value of a continuous variable occurs is denoted by the height of the ordinate (or vertical line from the axis to the curve) at that point. ⊃

However, because a continuous variable has an infinite number of values, this gives rise to problems. It is better to say that the total area under the curve represents the total frequency, and that the area to the left of say $x = 25.3$ in Figure 3.17 represents the frequency with which values of 25.3 or less arise. Even this causes a problem. If we are measuring correct to the second decimal place, we need to think of 25.3 in Figure 3.17 as being between 25.25 and 25.35. Thus the area to the left of 25.25 will represent the frequency of values up to 25.25, and the area to the left of 25.35 the frequency of values up to 25.35. The difference between the areas will represent the frequency of values between 25.25 and 25.35 – the frequency with which 25.3 (to one decimal place) arises.

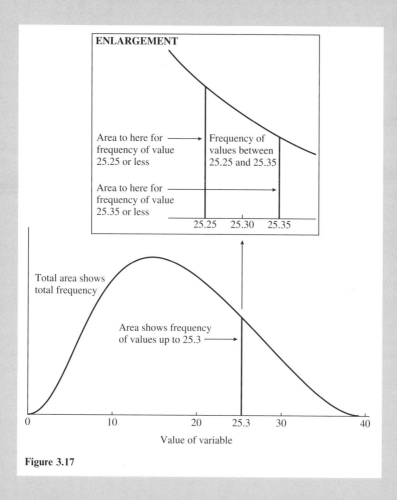

Figure 3.17

Similarly the frequency with which $x = 13$ (correct to the nearest integer) arises is shown by the area between $x = 12.5$ and $x = 13.5$.

Notebox 3.5

Percentage frequency and cumulative percentage frequency distributions

All frequency distributions, for grouped data or otherwise, can be converted into **percentage frequency distributions.** Every frequency is divided by the total frequency and multiplied by 100. The associated diagrams look just the same but the vertical scales have to be renumbered so they record percentages. This is particularly useful when we have continuous distributions.

Sometimes a graph of the **cumulative percentage distribution** is drawn. This looks like a cumulative frequency curve but every cumulative frequency is expressed as a percentage of the total frequency, and the vertical axis is renumbered so that it goes from 0 to 100.

Summary

1. The use of tables and diagrams in analysing data is our main concern, but other diagrams are useful for presenting data.

2. When we have discrete data, a common technique is the bar chart, which is often more useful if it is ordered.

3. Discrete data may lead to the tabular and graphical presentation of a frequency distribution and a frequency polygon (Box 3.1). A disadvantage of the polygon is that intermediate points along the lines have no meaning. Vertical lines, or vertical bar charts, are to be preferred. There are also cumulative versions (Box 3.2).

4. Sometimes it is useful to condense (or group) the data. In that case the frequency distributions (Box 3.3) are usually displayed as histograms (Box 3.4), where the grouped frequencies are shown by columns. It is important to ensure the columns have areas (rather than heights) proportional to the amounts they represent.

5. When the data are continuous the construction of a grouped frequency distribution and histogram is helped by first drawing a stem and leaf diagram (Box 3.5).

6. The frequency polygons for continuous data may be smoothed in frequency curves. Their interpretation is described in Notebox 3.4.

7. Frequency distributions can lead to cumulative frequency polygons, and to cumulative frequency distributions, sometimes called ogives (Box 3.6). Often we use percentage frequency distributions.

8. Dot plots are simple devices for getting a quick indication of the shape of a distribution.

9. A useful presentational device is the pie chart which represents the component parts of a total by sectors of a circle, each with an angle proportional to the size of the component. Two totals may be compared with two pie charts, each having an area proportional to the total (Box 3.7).

10. Modified bar charts are simpler to draw than areal representations and they can be interpreted more easily.

11. Noteboxes deal with (1) graphing grouped data, (2) graphing rounded data, (3) open-ended data, (4) the height of an ordinate of a frequency curve, and (5) percentage frequency and cumulative percentage frequency distributions.

Describing frequency distributions

Example 4.1 The teaching hospital of the University of Statingham is in Upper Sternum. Many of its staff belong to the Society of Ladies of Upper Sternum, whose chair is Lady Agatha. On hearing of the antics of Winnie and her Working Women, Lady Agatha decided to get some reliable data. She arranged for one of her hospital friends to repeat the measurements of the 60 Statingham workers, then graphed the two sets of measurements as in Figure 4.1(a) and (b).

She also arranged for two sets of measurements to be taken for the 50 members of her Society of Ladies who ever sat on typing seats. These were graphed as in Figures 4.1(c) and (d).

Lady Agatha needs to describe and compare these graphs on the telephone. She wonders how best to do it. Having been told that all new office seats for public authorities must be of the same height, she wonders what height to recommend.

In this chapter we introduce the main ideas that are used in describing and comparing frequency distributions, which is usually easier to do if they are graphed, but can sometimes be done just by studying the table.

An obvious part of any description is **shape,** and the more common shapes have been given names. However, shape is not enough. Figure 4.1(a) and Figure 4.1(b) have roughly similar shapes, but Figure 4.1(b) seems to be like Figure 4.1(a) shifted to the left (as if everybody had been shorter). In describing this, Lady Agatha has to refer to the **location** of the graph.

She notes how Figure 4.1(a) and (b) suggest that Winnie's measurements are too high. Also for Upper Sternum Figure 4.1(c) is less **spread out** than Figure 4.1(d). It is as if the lady who has done the measuring has undermeasured the heights for the taller ladies and overmeasured them for the shorter ladies.

Lady Agatha also notes that, although Figure 4.1(c) and (d) show fewer heights than Figure 4.1(a) and (b), the highest point on Figure 4.1(d) is higher than the

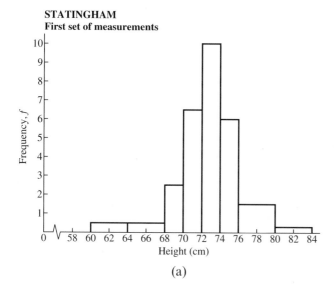

(a)

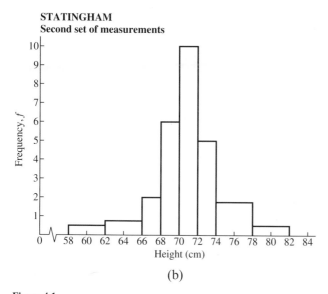

(b)

Figure 4.1

highest points on Figure 4.1(a) and (b), but the comparison is complicated by the different numbers. However, as far as heights are concerned, it does look as if the women of Statingham go to more extremes than the ladies of Upper Sternum. We discuss these comparisons later. In this chapter we devote most of our time to shape, location and spread.

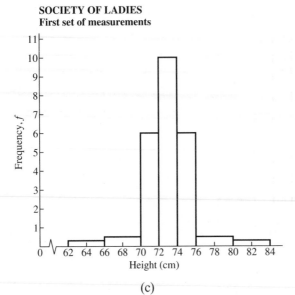

(c)

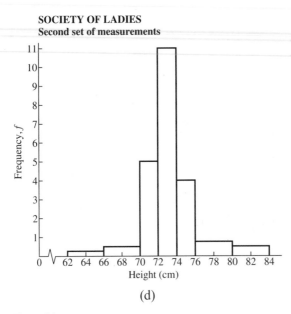

(d)

Figure 4.1

4.2 Shape

Since a single line is easier to draw than a lot of columns, it is easier to show a continuous distribution than a discrete distribution. We therefore illustrate some of the more common shapes mainly by drawing continuous frequency distributions. It is always possible to derive the corresponding discrete distribution simply by putting a few crosses on the curve and using them to denote the tops of columns.

Some of the more important shapes are shown in Figure 4.2. The **bell-shaped** distribution in Figure 4.2(a) is a **symmetrical** distribution. It is possible to draw a vertical line that divides the distribution into two parts, one an exact mirror image of the other.

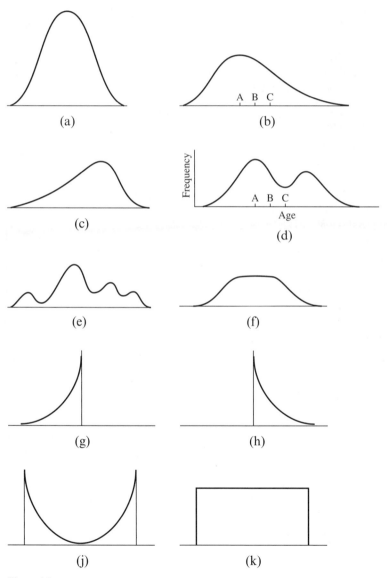

Figure 4.2

Figures 4.2(b) and (c) show distributions that are skewed to the right (the skewer is pointing to the right) and skewed to the left. We say more about skewness later.

Figures 4.2(a), (b) and (c) are **unimodal** distributions; they have one peak. Figure 4.2(d) shows a **bimodal** distribution; the two peaks (or modes) are not of equal height, but they could be. Figure 4.2(e) shows a **multimodal** distribution, and Figure 4.2(f) has a **distributed mode**.

Figures 4.2(g), (h) and (j) respectively show a **J-shaped**, a **reverse J-shaped** and a **U-shaped** distribution. They have their highest points at their left or right extremes (or both). Some writers call them **end-moded**. And some use the term *unimodal* to include the J-shaped and reverse J-shaped curves, but others restrict it to curves where the mode is not at an end of the distribution.

Figure 4.2(k) shows the **rectangular**, or **uniform**, distribution.

Most of these distributions are discussed in later chapters, which also describe special and more precisely defined examples.

Consolidation exercise 4A

1. Which of the following terms can be used to describe the four distributions of Example 4.1?
   (i)   symmetrical
   (ii)   asymmetrical
   (iii)   skewed to the right
   (iv)   skewed to the left
   (v)   unimodal
   (vi)   bimodal
   (vii)   multimodal
   (viii) having a distributed mode
   (ix)   J-shaped
   (x)   reverse J-shaped
   (xi)   U-shaped
   (xii)   rectangular

  Measures of location

We indicate the location of a distribution by locating some point (or value) around which the observations are to some extent clustered, such as the points A, B and C in Figures 4.2(b) and (d).

Suppose Figure 4.2(b) represents the ages of the 1200 members of the Statingham Association of Working Women. We seek some numerical idea of these ages but we do not want a list of all 1200 ages. We therefore ask the secretary to select one age that (i) is less than the ages of some members but more than the ages of others, and (ii) can be taken, so far as any single age allows, as a fair representation of the members' ages. Without yet attempting a more rigorous definition we will call this single age an **average** age. Note the word 'an'.

There are several different ways of defining, measuring and calculating an average, and usually they lead to different numerical results. Unfortunately, an average defined in one way may be helpful for one purpose but misleading for another. Whenever we use an average, we need to consider carefully which definition is appropriate.

The most common average is the one that is normally used to calculate an average mark or an average cricket score. We add the individual marks (or scores) and divide the total by the number of marks (or scores); it is called the **arithmetic mean**. Chapters 6 and 7 define the arithmetic mean more precisely and show how to calculate it.

Sometimes, the arithmetic mean is inappropriate. For example, if a car travels 10 miles at 20 mph (taking 30 minutes to do so) and then 10 miles at 40 mph (taking 15 minutes) then the arithmetic mean of these two speeds is

$$(20 + 40)/2 = 30 \, \text{mph}$$

However, the car travels the first 10 miles in 30 minutes and the second 10 miles in 15 minutes. This means that it travels a total of 20 miles in 45 minutes, equivalent to

$$(20/45) \times 60 = 26.67 \, \text{miles}$$

in an hour, which cannot be reconciled with an average speed of 30 mph. We consider such matters in the next few chapters.

One feature of the arithmetic mean is that if any single value is altered in the set whose mean is being calculated, then the mean is also altered. As we shall see shortly, this precise dependence of the result on the exact value of every single item is a feature of some averages but not of others. Those for which this dependence exists are called **means**. We have already mentioned the arithmetic mean. Two other means that will be considered in Chapter 9 are the geometric mean and the harmonic mean, each being the most appropriate to use under certain circumstances. 'The mean' almost always implies the arithmetic mean.

Other averages do not depend on the precise value of every single item. One of them is the **median**, which may be defined as the middle item when all the items are listed in numerical order. For example, 18 is the median of 12, 13, 18, 19 and 20, since there are two values lower than 18 and two values higher. If any of the values were altered but remained on the same side of 18, the median would be unchanged, as in 11, 16, 18, 27 and 43.

Sometimes the median is more appropriate than the arithmetic mean. The median is considered in Chapter 8, where we also look at how to find the 'middle item' when there is an even number of items.

The **mode**, or most fashionable value, is another average that is not a mean, as it does not depend on every value. It is the subject of Chapter 7. The mode is the value of the variable corresponding to the peak of the frequency distribution. Other averages are described in Chapter 9.

One of our tasks will be to consider which average is most appropriate to Lady Agatha's decison about what height to recommend for office chairs.

4.4 Measures of spread (or dispersion)

We may also be interested in knowing whether most of the members of Statingham Association of Working Women are roughly the same age, or whether the ages are widely spread.

To answer such questions we use a measure of **dispersion.** There are two main classes of dispersion measure. One contains measures that summarise the extent to which individual values deviate from the arithmetic mean. The principal measures in this class are the **mean deviation**, the **standard deviation** and the **variance**, which we consider in Chapters 10 to 12. All of them depend on the precise values of every item.

The other measures of spread tell us that a specified fraction of the values lies within a certain range, or above or beneath a certain value. These measures do not depend on the exact values of all the items. We consider them in Chapter 10.

Consolidation exercise 4B

1. What are the essential properties of an average?

2. When can an average be called a mean? Name two averages that are not means.

3. What is meant by 'a measure of dispersion'?

Summary

1. To describe a distribution we have to refer to its shape, location and spread. This chapter introduces these concepts, treated more fully later.

2. In discussing shape we refer to symmetry, skewness and modality. We also mention J, reverse J and U distributions.

3. Location is measured by an average, which is a value less than some of the values in the distribution but more than others, and having some claim to being representative. An average may be defined in many ways. An average that is chosen to answer one question about a specific distribution will often be quite misleading if used to answer a different question.

4. Averages whose calculation requires knowledge of all the values are called means. Examples are the arithmetic mean, the geometric mean and the harmonic mean. Averages that are not means include the mode and the median.

5. There are several measures of spread, or dispersion.

5

The arithmetic mean: existence and meaning

5.1 Introduction

Example 5.1 The seven bar staff at the Calculators' Arms have the following masses, measured in kilograms: 56, 75, 60, 58, 69, 72 and 80. The landlord wants to know their average.

Example 5.2 Harry, the landlord, plays cricket for Statingham Publicans. Over a series of eleven matches his scores are 42, 37, 0, 84, 34, 47, 57, 0, 9, 102 and 54 runs. What are his arithmetic mean score, his median score and his modal score, and what interpretations may be placed on them?

Example 5.3 In the last hour before their midday break, 60 bricklayers laid 1727 bricks. No bricklayer laid a fraction of a brick. What interpretation is to be attached to the arithmetic mean number of bricks per bricklayer?

Example 5.4 When Guzzling Greg declared to Erratic Eric, Steady Steve and Tippling Tommy that he did not like his nickname, the landlord declared that he deserved it because he was the pub's heaviest imbiber of NonTox. 'I've got proof!' he said, 'I've listed you in order for every night of last week. Low score means heavy imbibing.' He produced Table 5.1. 'You see,' he said, 'You are second heaviest every night. There are nights when Tommy or Eric beat you – look at their 1 scores – but some nights they imbibed least of all – look at the 4s. If we take it

Table 5.1

	Rank of drinker (heaviest drinker = 1)							Average score	Position for week
	Mon	Tues	Wed	Thurs	Fri	Sat	Sun		
Tommy	1	3	1	4	1	4	1	15/7 = 2.14	2
Greg	2	2	2	2	2	2	2	14/7 = 2.00	1
Eric	3	1	3	1	3	1	3	16/7 = 2.29	3
Steve	4	4	4	3	4	3	4	25/7 = 3.57	4

all in all and average it out, you come out with the lowest score, which makes you the heaviest imbiber.' Greg told him he was talking nonsense. Why?

The most frequently used average (or measure of location) is the arithmetic mean. It is easy to calculate and has important mathematical properties that make it the basis of much advanced statistical analysis. Here we consider some simple examples of its calculation, interpretation and misuse. More detail appears in later chapters. Check that you are familiar with the use of $\sum$, explained in Notebox 5.1.

5.2   Definition and interpretation of the arithmetic mean

The arithmetic mean of a set of values is defined as their total divided by the number of values. More formally, if there are n values of X and we denote the sum of these values by $\sum x$, then the **arithmetic mean** value of X is defined to be

$$\frac{\text{Total of all values}}{\text{Number of values}} = \frac{\sum x}{n} \qquad [5.1]$$

The arithmetic mean of all the values of X in a population is usually denoted by μ (pronounced 'mew'), the Greek m, but $\bar{x}$ (read as 'ex bar') is used for a **sample mean**. Since our present emphasis is on populations, we shall use μ in the first few chapters, but our present remarks about μ also apply to $\bar{x}$.

Applying this definition to Example 5.1, we have that the total mass of the seven bar staff comes to 470 kg, so the arithmetic mean mass is $470/7 = 67\frac{1}{7}$ kg. Note that none of the seven bar staff has this mass. This illustrates one reason why we sometimes need to use a different kind of average, but for the moment we put such considerations aside and concentrate on how to calculate the arithmetic mean. The next chapter considers more complicated cases and refers to some tricks (described in Appendix B) for making the calculations easier.

 In answering **Example 5.2** we use the same method to calculate Harry's arithmetic mean score. The total of the 11 scores is 466 runs. Thus the arithmetic mean score per match is

$$\mu = \sum x/n = 466/11 = 42.3636 \text{ runs}$$

Obviously it is impossible to score 42.3636 runs, and this arithmetic mean cannot be looked upon as representative or typical. But we can use it to estimate how many runs Harry would score in 24 matches if he played on average just as well as in those 11 matches. We would estimate a total score of $24 \times 42.3636 = 1016.726$ or 1016 complete runs.

We can also say that 'on average' Harry performs better than Bert, landlord of the Skew Distribution, whose scores in nine matches are 31, 2, 48, 45, 50, 31, 9, 112 and 34 runs. These give an average of $362/9 = 40.222$ runs, which is lower than Harry's average in these matches. We now consider the example about the bricklayers.

In **Example 5.3** the arithmetic mean number of bricks laid is $1727/60 = 28.783$ bricks. All the bricklayers are recorded as having lain whole numbers of bricks, and this result cannot be taken as representative or typical. We can, however, invent an 'average bricklayer' that lays 28 bricks per hour and has also done 0.783 of the work involved in laying a further brick.

Suppose we want to estimate how many bricks would be lain in 3 hours by a team of 200 bricklayers who are 'on average' just as productive as our 60 bricklayers. On average one bricklayer working for 1 hour lays 28.783 bricks, so 200 working for 3 hours would lay $200 \times 3 \times 28.783 = 17\,269.9$ bricks, which we might round up to 17 270 or down to 17 269, according to our willingness to make certain assumptions about part-laid bricks.

For example, if bricklayers always finish laying a brick once it has been started, then in a period of 3 hours the numbers of bricks lain could be rather more than the average multiplied by 3. But if no bricklayer starts to lay a brick unless he or she is sure of finishing it before clocking-off time, then the number may be smaller. If we do not know which of these practices is followed, the estimate can be rounded down to reduce the risk of overestimating the work done, or up to reduce the risk of not providing enough bricks.

Box 5.1

How to ... **find the arithmetic mean (basic method)**

1. If there are many observations, consider grouping them and using one of the techniques described in Chapter 6. Otherwise proceed as below.

2. Total the values and divide by the number of observations. This gives the arithmetic mean.

3. Check that your answer is not too precise (Noteboxes 5.2 and 5.3) and, if appropriate, that you have the number of significant figures or decimal places specified by the question.

Consolidation exercise 5A

1. Forty-seven men weigh themselves at a drugstore. The total of their masses is 8609 lb. What is the arithmetic mean mass?

2. We want to convert the mean mass of these 47 men into kilograms. The conversion factor is 0.453 592 (to 6 d.p.). Why is it unnecessary to be so precise? How many decimal points do you need for an answer of realistic precision?

3. If x takes the values 6, 7, 8, 10 and 12. What are the values of $\sum x$, $\sum x^2$, $(\sum x)^2$, $\sum (x + 1)$? (Read Notebox 5.1.)

A wrong use of the mean

In Example 5.4 Guzzling Greg, rebelled because he had Table 5.2, his own record of the amounts of NonTox actually imbibed by himself and his friends.

'So which of us buys most from you?' asked Greg. 'That's who you ought to call the heaviest imbiber! That's Eric, who is the second lightest imbiber according to you.'

The landlord's analysis is a common and serious misuse of the statistical method. Plans, policies or people are frequently ranked first, second, third, etc., according to some test or criterion. Then the ranks are averaged and the 'best' declared to be the one with the lowest average rank. This procedure is invalid and can lead to strange results. One reason is illustrated in this example. On Monday, when Eric and Greg were ranked 3 and 2, Eric imbibed half a pint less than Greg. Thus a difference of 1 in the rank corresponded to half a pint. But on Saturday, when Eric and Greg were ranked 1 and 2 – again a difference of 1 in the rank – Eric imbibed 2 pints more than Greg. If we consider just these two days, the sum of Eric's ranks is the same as the sum of Greg's, but Eric imbibed $1\frac{1}{2}$ pints more than Greg.

⬭ **Example 5.5** The Planning and Road Traffic Committee of Statingham City Council asked the chief planner to consider four versions of a town centre traffic plan. The planner applied three tests (cost, popularity and environmental impact) to each plan and gave them marks out of 100 under each heading, with 100 indicating perfection. Then, by looking at the marks, the planner ranked the plans as first, second, third or fourth under each of three headings. Next the planner added the ranks and arranged the plans in order of preference, with the lowest total rank coming first (Table 5.3). Accordingly, the chief planner recommended plan B.

Councillor Barker managed to get the actual marks that had been given in the three tests (Table 5.4), and checked that they were consistent with the ranks used by the chief planner. Then Councillor Barker totalled the marks and used these totals to draw up a new order of preference; it was very different. Councillor Barker demanded the adoption of plan A.

The trouble is that the data in Table 5.3 are all ranks and conceal a great deal of detail, but it is the detail that tells us in which order the four plans should be placed. In Table 5.4, being first in test 1 scores many more marks than being first in any other test, but Table 5.3 does not show this. Moreover, the difference between being first and second in test 3 (25) is much greater than the difference between being third and fourth. Table 5.3 also fails to show this. It is not difficult to find other examples of why Table 5.3 is inadequate and misleading.

Summing up, when we work from Table 5.3 and list the plans according to the totals of their ranks, we are implicitly making two assumptions. One is that being first (or second, etc.) in one test is exactly as important as being first (or second, etc.) in any other. The other assumption is that the gap in performance between first and second in any test should be taken as exactly equivalent to the gap in performance between second and third in that or any other test. If these

Table 5.2

	Pints drunk							Average consumption (pints)	Position for week
	Mon	Tues	Wed	Thurs	Fri	Sat	Sun		
Tommy	$3\frac{1}{2}$	3	$3\frac{1}{2}$	$\frac{1}{2}$	$3\frac{1}{2}$	$\frac{1}{2}$	$3\frac{1}{2}$	$18/7 = 2.57$	3
Greg	3	$3\frac{1}{2}$	3	3	3	3	3	$21\frac{1}{2}/7 = 3.07$	2
Eric	$2\frac{1}{2}$	5	$2\frac{1}{2}$	5	$2\frac{1}{2}$	5	2	$24\frac{1}{2}/7 = 3.50$	1
Steve	2	$2\frac{1}{2}$	2	$2\frac{1}{2}$	2	$2\frac{1}{2}$	$2\frac{1}{2}$	$16/7 = 2.29$	4

Table 5.3

	Rank of plan according to			Total of ranks	Order of preference
Plan	Test 1	Test 2	Test 3		
A	1	3	4	8	3
B	2	1	3	6	1
C	4	4	1	9	4
D	3	2	2	7	2

Table 5.4

	Marks according to			Total of marks	Order of preference
Plan	Test 1	Test 2	Test 3		
A	90	50	39	179	1
B	80	52	40	172	3
C	60	49	65	174	2
D	70	51	45	166	4

assumptions do not hold then calculations based on the marks may not lead to the same listing as calculations based on the ranks.

There may also be another problem. In the NonTox example it is reasonable to assume that a pint on Tuesday is equivalent to a pint on Wednesday. But in the planning example it is less reasonable to assume that a single mark in test 1 is equivalent to a single mark in test 2.

It follows that adding ranks leads to conclusions which may not be valid, so procedures based on these additions should not be used. Decision makers take note!

This illustrates that the definition of the arithmetic mean as the sum of the values of the variables divided by their number presupposes that we are dealing with variables which can be summed meaningfully. Ranks cannot be summed meaningfully. Are there other variables to which this or similar strictures apply?

The answer depends on 'the level of measurement'. There is no need for us to say much about this here, but we should note that it is wrong to add (and to calculate the arithmetic mean) if the numbers are being used simply to classify things or to place them in order.

Almost all the theory and techniques described in Chapters 4 to 32 depend on our being able to make certain assumptions about the arithmetic mean or some other parameter (which we define in Chapter 16). We therefore describe this part of statistical theory as **parametric statistics**. We consider this matter more fully later on. If the data cannot meaningfuly be added, these parameters canot be calculated. But there are other techniques, known as **non-parametric techniques**, that may be applicable. Some of these techniques are described in Chapters 33 to 37.

Consolidation exercise 5B

1. Consider whether it is valid to calculate the arithmetic mean for (a) the telephone numbers of ten friends, (b) the route numbers of the next five buses, (c) the prices in six shops of a kilogram of bananas, (d) the finishing positions of four sisters competing in a race with ten runners, (e) the numbers of times that each of five sisters has been first in a race.

Notebox 5.1

Sigma notation, $\sum$

The Greek capital S, written $\sum$ and pronounced 'sigma', is simply a shorthand for 'the sum of all the values of ...'.

Thus $\sum x$ (read as 'sigma ex') means the sum of all the values of x. It is assumed that we know what these values are, or how to find them. Sometimes it is made clear by writing something such as

$$\sum x \, (x = 3, 4, 5, \ldots, 14, 15)$$

which means $3 + 4 + 5 \ldots + 14 + 15$.

The sign is also used with more complicated expressions. We explain them as we go along, but here are a few of the more common ones. To keep things simple, x takes the values 1, 2, 3 and 5.

$\sum x$	means $1 + 2 + 3 + 5 = 11$
$\sum x - 1$	means $(1 + 2 + 3 + 5) - 1 = 11 - 1 = 10$
$\sum (x - 1)$	means $(1 - 1) + (2 - 1) + (3 - 1) + (5 - 1) = 0 + 1 + 2 + 4 = 7$
$\left(\sum x\right)^2$	means $(1 + 2 + 3 + 5)^2 = 11^2 = 121$
$\sum x^2$	means $1^2 + 2^2 + 3^2 + 5^2 = 39$

Notebox 6.3 discusses the meaning of things such as

$$\sum_{i=24}^{32} x_i$$

Notebox 5.2

The precision of the stated mean

Does it make sense to express an arithmetic mean as $67\frac{1}{7}$? In Example 5.1 all seven masses are given as integers (whole numbers), so the actual masses have probably been rounded. Even if we assume the rounding has been systematic, we do not know whether the masses have been rounded all upwards or all downwards, or possibly each to the nearest integer. In the last two cases our calculated average of $67\frac{1}{7}$ would have to be quoted as 67 unless we wanted to produce an average that was more precise than the data.

One occasion when it may be right to go beyond the precision of the data is when the average is being used as the basis of a 'multiplying up' calculation, as in the estimate of Harry's total score over 24 matches, and the number of bricks lain by 300 bricklayers in 2 hours.

We may also use an arithmetic mean expressed to a few decimal places if we wish to compare two batsmen who have played in different numbers of matches.

In putting the mean score as 42.3636, the last few decimal figures are quite useless, and writing them down is a waste of time. If we had put the mean as 42.36, our estimate for 24 matches would have been 1016.64, which still means 1016 complete runs; but using either 42.3 or 42.4 would have led to different answers. Usually it is best to work with two or perhaps three decimal places more than the answer requires, but to end the calculation by rounding to the same degree of precision as the original data (or to a greater degree if appropriate).

Notebox 5.3

To be able to add

Statistical data arise out of some form of measurement, including simple classification. Whether or not a particular statistical technique is valid will depend on how the data have been measured.

The techniques described in Chapters 4 to 16 and 18 to 32 should be used only if the arithmetic mean can be meaningfully calculated. This will be so if, and only if, they can be meaningfully added and counted. The conditions for this are (i) the different numbers can all be placed in order of magnitude, (ii) the magnitudes of the gaps (or intervals) between every pair of numbers are known, and (iii) it must be possible to specify numerically the ratio of the magnitudes of every selected pair of intervals.

An example is temperature measured on either the Fahrenheit or the Celsius scale. Note that (i) the temperatures can be placed in order of magnitude, (ii) the intervals between any two of them can be measured, and (iii) the ratios of pairs of intervals can be found – the difference between two temperatures of 30 and 16 ($= 14$) is twice as big as the difference between two temperatures of 37 and 30 ($= 7$).

We look more carefully at different levels (or types) of measurement in Chapter 33.

Summary

1. We define the arithmetic mean of a set of values as the sum of the values divided by the number of values. This is stated more formally in [5.1] and summarised in Box 5.1.

2. Whether the calculated mean needs rounding up or down depends on why we want it.

3. The arithmetic mean of ranked data is nonsense. The definition of the arithmetic mean presupposes that the values can be added meaningfully. Ranks cannot be added meaningfully. This is discussed a little further in Notebox 5.3, which draws attention to the treatment of levels of measurement in Chapter 33.

4. When data cannot be meaningfully added the parametric techniques to which most of this book is devoted cannot be applied. Instead we need to use non-parametric techniques, such as those introduced in Chapter 33.

5. Noteboxes refer to (1) the Σ notation, (2) precision in the statement of the mean, and (3) the conditions that must hold for the calculation of the arithmetic mean.

6

The arithmetic mean: limitations and calculation

6.1 Introduction

Example 6.1 Knowing that she can recommend only one height of office seat, Lady Agatha decides to cooperate with Winnie and to begin by finding the average height of the entries in Table 3.2. The only average she understands is the arithmetic mean, and she sets about calculating it.

Example 6.2 Nine bank customers have balances of \$30 (credit), \$204 (overdrawn), \$78 (overdrawn), \$47 (credit), \$35 (overdrawn), \$33 (credit), \$129 (credit), \$14 (credit) and \$52 (overdrawn). What is the arithmetic mean balance for the nine customers?

Example 6.3 Statingham's Steady Steve is thinking of moving to a new housing estate, but has heard rumours that lots of big families live there. Harry, whose wife knows people who know everything, compiles Table 6.1 about the 291 families on the estate. Steve decides that first of all he will work out the average family size, using the arithmetic mean.

Table 6.1

Family size distribution

Family size (persons)	1	2	3	4	5	6	7	8	9+
Number of families	30	65	40	55	50	30	12	5	4

Example 6.4 Steve is also interested in the ages of his neighbours. Harry's wife now compiles Table 6.2 about the ages of the 301 people living in the part of the estate that interests Steve the most. Steve decides that he has another average to work out.

Example 6.5 Harry says that over Christmas his regular customers spent an average of £110 on non-alcoholic drink, whereas the regular customers of his

Table 6.2

Age distribution

Age	0–4	5–9	10–14	15–19	20–29	30–39	40–49	50–59	60–64	65–69	70–79	80+
Number	20	22	20	18	35	38	44	48	20	16	16	

brother's pub, the Skew Distribution, spent an average of £130. He said that the average spending by the customers in the two pubs was therefore £120. Was he right?

In Chapter 5 we defined the arithmetic mean and calculated it in a few simple cases. Our method was to sum all the values of the variable x, and to divide the total by the number of values. Even Lady Agatha found the addition of 60 heights tiring. Tapping 60 numbers, each of three digits and a decimal point, into her calculator was tedious; and when she did it a second time, she got a different answer. In the end she confirmed that the total of the 60 heights was 4353.2 cm, so the arithmetic mean was 72.55 cm, which she decided to call 72.6 cm

In this chapter we look at some limitations of the arithmetic and indicate a few time-saving tricks. Some of them are described more fully in Appendix B.

Limitations of the arithmetic mean

We have already seen that an arithmetic mean may be a value that could not possibly occur in the original data. For example, a mean family size may be calculated as 3.45 persons, or the mean number of bottles of wine bought as 2.71. Although means such as these are appropriate for some purposes, for others they are not.

Another defect of the arithmetic mean is that it is affected by extreme values. Consider, for example, a village in which 99 families have incomes of £100 a week and one family has an income of £5000 a week. The arithmetic mean income per family can be estimated as

$$\frac{\text{Total income}}{\text{Number of families}} = \frac{99 \times £100 + £5000}{99 + 1} = £149$$

If the one rich family doubles its income, the arithmetic mean rises to

$$\frac{99 \times £100 + £10\,000}{99 + 1} = £199$$

In neither case can the arithmetic mean be taken as representative of the incomes in the village. Nor can the increase in the mean from £149 to £199 be taken to indicate anything about the incomes of most of the villagers. It may be better in such circumstances to use the mode or the median, which will be more representative of the population.

Use of the arithmetic mean may be criticised for another reason. We have already seen that it is meaningless and misleading when applied to ranks, for which it is not defined. It may also be misleading when applied to certain kinds of composite data such as speeds, prices, ratios and growth rates. We consider these problems later on.

However, it has important mathematical properties that apply to no other form of average and place it at the centre of a great deal of statistical theory and technique.

6.3 Calculating the mean: ungrouped data

Use of frequency tables

We consider first how to calculate the arithmetic mean when we are able to compile a frequency table such as Table 3.5, which we use in Table 6.3 below. All the values of x are listed individually in the first column. The second gives the frequencies with which they occur – the numbers of evenings on which there were various numbers of customers. The total of this column, represented by $\sum f$, is the total number of evenings.

Table 6.3

Calculation of arithmetic mean using a frequency table

(1) Number of customers, x	(2) Number of evenings with this number of customers (or frequency), f	(3) = (1) × (2) Product, fx
20	1	20
21	0	0
22	0	0
23	1	23
24	0	0
25	1	25
26	2	52
27	4	108
28	6	168
29	5	145
30	4	120
31	3	93
32	3	96
33	1	33
	$n = \sum f = 31$ evenings	$\sum fx = 883$ customers

In column 3 we multiply each number of customers x by the number of evenings f that had this number of customers. Thus the entry 52 in column 3 arises because there were 26 customers on each of 2 evenings. The total of this column, represented by $\sum fx$, is the total number of customers arriving during the 31 evenings.

Since the total number of customers is represented by $\sum fx$ and the total number of evenings by $\sum f$, the arithmetic mean number of customers per evening (denoted by $\bar{x}$) is

$$\frac{\text{Total number of customers}}{\text{Total number of evenings}} = \bar{x} = \frac{\sum fx}{\sum f} = 883/31 = 28.48 \qquad [6.1]$$

Obviously there cannot be a fraction of a customer. Unless the answer is required for multiplying up (Notebox 5.2) or for some kind of hypothesis testing (as described later), it should be rounded to 28 or 29, depending on the purpose for which it is to be used (Chapter 5).

The technique just used is set out in Box 6.1. Appendix A shows how to use calculators to find the mean quickly from the first two columns and equation [6.1].

If there are so many different values of x that the frequency table would be excessively long – say more than 20 – perhaps it would be wise to combine them into groups, such as 20–24 and 25–29, along the lines of a stem and leaf approach, then use the technique described in Section 6.4 below and Box 6.2.

Before the advent of computers, various tricks were devised for reducing the effort of calculation. Some of them are described in Appendix B. Two particularly important ones are using an **assumed mean** (and then finding how wrong you are) and using **new units.** Sometimes the tricks are combined. Data adapted in this way are called **coded data.** Although most useful if calculators are unavailable, these techniques can also save effort if used in conjunction with a calculator.

Box 6.1

How to ... find the arithmetic mean for ungrouped data using a frequency table

1. Examine the data to see whether some values occur several times. If so, compile a frequency table as in the first two columns of Table 6.3. If this is too long consider grouping the data and using Box 6.2. Otherwise obtain a third column by multiplying each value of x by the corresponding f to obtain an entry fx, taking account of the sign.

2. Total the entries in the second column to obtain the total frequency $\sum f$ (which should equal n) and those in the third column to obtain $\sum fx$. (If some entries are positive and some negative, add them separately then take the negative total from the positive.)

3. Divide the value of $\sum fx$ by n to obtain the mean, as in equation [6.1].

4. Consider the precision of your answer as in Noteboxes 5.2 and 5.3.

Example 6.2 introduces the idea of negative values. Every credit balance can be prefaced by a + sign, indicating a positive balance in the bank, and every overdrawn account has a negative balance, so it carries a − sign. Thus we seek the mean of +30, −204, −78, +47, −35, +33, +129, +14 and −52. The easiest way of coping with this is to add up the + terms, getting +239, and then to add up the − terms, getting −369. Thus the people with credit balances have a total of $239 whereas the people with overdrafts owe the bank a total of $369. Putting the two groups together, they are collectively in debt to the bank to an amount of $369 − $239 = $130. Another way of putting this is that their total balance is −$130. This is shared between nine people, so the arithmetic mean balance is −$130/9 = $14.44 to the nearest cent. If the bank rounds up, we need to record it as −$14.45.

The trick of dealing with positive and negative entries separately for the purpose of addition is particularly useful when there is a frequency table, as in one of the exercises below.

Consolidation exercise 6A

1. With what kind of data is it wrong to use the arithmetic mean?

2. What are the main disadvantages of the arithmetic mean?

3. Calculate the arithmetic means for the following distributions:

(a)		(b)		(c)		(d)		(e)		(f)		(g)	
x	f	x	f	x	f	x	f	x	f	x	f	x	f
0	2	3	2	23	4	323	4	17	5	12	3	−7	4
1	3	500	1	24	6	324	6	18	6	−5	4	−8	9
2	5	9	6	25	9	325	9	19	7	18	8	3	15
3	1	30	2	28	2	328	2	20	100	−9	3	4	7

4. What are coded data?

Open-ended Data

A special kind of problem arises when a table that is otherwise precise begins and/or ends with an imprecise entry. Such a table is described as **open-ended**. An example arises in the case of Steady Steve's interest in family size in (Example 6.3). The data are reproduced in Table 6.4.

To get the average number of persons per family, we need to divide the number of persons by the number of families. We know there are 291 families. How many persons are there?

We can begin to answer this by proceeding as in Table 6.4. Every entry in the third column has been obtained by multiplying the corresponding entries in the other two columns, i.e family size is multiplied by frequency. For example, families of two persons occur 65 times, contributing 130 people to the total.

Table 6.4

Family size, x	Frequency, $f(x)$	Number of persons, $xf(x)$
1	30	30
2	65	130
3	40	120
4	55	220
5	50	250
6	30	180
7	12	84
8	5	40
9+	4	39?
	$\sum f(x) = 291$	$\sum xf(x) = 1093?$

A difficulty arises in the last column. The last four families listed here are of nine persons or more. If they were all of nine precisely they would contain a total of 36 persons, but there could be families of 10, 11 or more. How can we obtain an estimate of the total number of persons?

We try to make the best use of what information we have. There are two relevant items. One is that there are 4 families of nine persons or more. The other is that the frequencies listed in column 2 are falling rapidly. The number of families containing seven persons is only 40% of the number of families containing six persons. The number of families containing eight persons is only 5, less than half the number of families containing seven persons. It seems unlikely that the number of families containing precisely nine persons is as high as 4. Let us put it at around 40% of the number in the preceding group, and so get 2 families of nine persons. Then we might also have 1 family of ten persons and 1 family of eleven persons. This would give us a total of 4 families with $(2 \times 9) + (1 \times 10) + (1 \times 11) = 39$ persons.

This is a little better than an uninformed guess, but it may still be wrong. For the moment, if we accept 39 as an estimate of the number of people in really large families, we can complete our calculation of the arithmetic mean of the family sizes by using the values shown in the table. We have

$$\bar{x} = \frac{\sum xf(x)}{\sum f(x)} = \frac{1093}{291} = 3.756 \text{ persons}$$

This answer cannot be guaranteed because we do not know the exact number of people in the families of nine persons or more. We also have to ask what it means. This leads to questions that come more conveniently in an account of how to find the mean if we have grouped data.

6.4 Calculating the mean: grouped data

Steady Steve looks at the table printed in Example 6.2 about the ages of his neighbours. They are all in age groups. He can see how many people are aged 0–4,

5–9, etc., but he knows nothing about the age distributions within these groups. How can he estimate the arithmetic mean age?

Grouped data imply imprecision. Whenever information is summarised (as in this table) some of it is lost, so there is bound to be uncertainty about the correctness of any statistic we try to calculate from it. The standard procedure is to make some assumptions that allow us to convert the imprecise grouped data into precise but not necessarily accurate other data. We can then use the techniques already described in this chapter. However, if our assumptions are wrong, our result may be wrong; we must always keep this in mind. The art is to make assumptions that sound plausible but which will not lead to very wrong results even if they are incorrect.

We now consider the basic procedure. We make two assumptions. One is that the average age of people located within a specified range is the age at its midpoint. (This is most likely to be so if the group contains a large number of people. If great accuracy were essential, we might modify this assumption, but this is rarely done.)

The other assumption concerns the open-ended range of 80+ where, if the frequency is small, a small error will have little effect on the result.

The values printed in column 3 of Table 6.5 reflect these assumptions. They give us values of x which have to be multiplied by the values of f to provide column 4.

The calculation is therefore just as for ungrouped data once the midpoint (or other) assumption has been made. This gives us

$$\bar{x} = \frac{\sum fx}{\sum f} = \frac{11\,467}{301} = 30.096 \text{ years (but see below)}$$

Table 6.5

(1) Age	(2) Frequency, f	(3) Midpoint, x	(4) Product, fx
0–4	20	2.5	50
5–9	22	7.5	165
10–14	20	12.5	250
15–19	18	17.5	315
20–29	35	25.0	875
30–39	38	35.0	1 330
40–49	44	45.0	1 980
50–59	48	55.0	2 640
60–64	20	62.5	1 250
65–69	16	67.5	1 080
70–79	16	75.0	1 200
80+	4	83?	332??
	$\sum f = 301$		$\sum fx = 11\,467$

If the open-ended midpoint is put at 82.5, this gives a mean of 15.236×2.5 years $= 38.090$ years. If we had left the assumed midpoint as 83, the final entry in column 4 would have been 33.2 (in units of 2.5), resulting in an fx of 132.8. This would have led to an estimate of the mean identical to that obtained from Table 6.3.

An experienced statistician may sometimes be able to see whether a convenient assumption would alter the result to any important degree. Unless it is clear that no important error is being introduced, assumptions should be made on the basis of belief in probable correctness rather than on the basis of computational convenience.

Box 6.2

How to ... **find the arithmetic mean for grouped data (uncoded)**

1. Use the midpoint of each range as the value of the variable x

2. Proceed as for ungrouped data (Box 6.1).

3. Be careful not to present a result of spurious accuracy.

Consolidation exercise 6B

1. The data in Table 6.5 are presented in ranges that may be five years or ten years wide. By amalgamating some of the data, represent them in a table with ten-year age ranges 0–9, 10–19, 20–29...60–69, 70–79, 80+. Now recalculate the arithmetic mean age, taking 5, 15, 25, etc., as the midpoints. Compare your answer with that obtained above. Consider why it differs and explain why the sign of the difference is as it is.

2. After reading Appendix B, calculate the mean of the data in Table 6.5 using the groups as listed and taking new units of 2.5 years (a) without and (b) with an assumed mean of 14 new units.

3. What is an open-ended table? How should you deal with one?

Notebox 6.1

The mean of two means

In Example 6.5 we know that the mean spending in one pub was £110, and in the other it was £130. What was the overall average?

To answer this question we assume that the two reported averages were correct statements of two arithmetic means. In that case the arithmetic mean spending per customer would be £120, if and only if the two pubs had the same numbers of regular customers. ⮑

Consider three families each of 2 persons, which is therefore their average size. Now take another six families all of 5 persons, which is therefore their average size. Altogether there will be nine families and $(3 \times 2) + (6 \times 5) = 36$ persons. Thus the mean size of family is $36/9 = 4$ persons, which cannot be obtained by simple averaging of the two averages, 2 and 5.

If two groups of sizes n_1 and n_2 have arithmetic means of $\bar{x}_1$ and $\bar{x}_2$ the combined group of size $n_1 + n_2$ has a mean of

$$\frac{n_1\bar{x}_1 + n_2\bar{x}_2}{n_1 + n_2}$$

Thus, if the Calculators' Arms had 60 regular customers and the Skew Distribution had 40, the arithmetic mean spending in the two pubs was

$$\frac{60 \times £110 + 40 \times £130}{60 + 40} = £118$$

Check the plausibility of this answer. The calculated mean lies between the two separate means of £110 and £130, which is necessary. It is also nearer to the mean (of £110) that is based on the greater number of customers, which makes sense.

Notebox 6.2

Alternative notation

Sometimes the formulae in the main text are rewritten with $f(x)$ instead of f, which simply emphasises that f is the frequency of x. For example, [6.1] becomes

$$\bar{x} = \frac{\sum f(x)x}{\sum f(x)} \tag{6.2a}$$

or more usually

$$\bar{x} = \frac{\sum xf(x)}{\sum f(x)} \tag{6.2b}$$

If we use this notation, we have to be careful not to misinterpret an expression like [B.4] of Appendix B which has to be rewritten

$$\bar{x} = a + \frac{\sum (x - a)f(x)}{\sum f(x)} \tag{6.3}$$

Another notation that looks confusing but is very useful involves subscripts so that, for example, we have $\sum x_i$ instead of $\sum x$. Here, if X is the variable 'age of oldest person in house' defined for all 54 houses in a street, then we can denote the sum of these ages by

$$\sum x_i = (1, 2, \ldots, 54) \quad \text{or by} \quad \sum_{i=1}^{54} x_i \qquad \Rightarrow$$

where the figures beneath and above the sign show the upper and lower values taken by the subscript i, and all values between them (inclusively) are to be summed.

In place of equation [6.3] we then have

$$\bar{x} = a + \frac{\sum (x_i - a)f(x_i)}{\sum f(x_i)} \qquad [6.4]$$

This emphasises that we have to take each value of x_i, subtract the assumed mean from it, and then multiply this difference by the frequency $f(x_i)$ with which that value occurs.

Essentially the subscript notation enables us to state various arguments and formulae more precisely. In very advanced work, beyond the scope of this book, this is essential, but we shall use it only when it seems to assist in the explanation or the use of a technique.

Summary

1. The arithmetic mean may not always be suitable as a measure of location. It often has a value that could not exist in the raw data from which it has been calculated (such as 2.4 children). It is also affected by extreme values, so it may cease to be representative. It is meaningless for ranks and may give incorrect results if applied to composite statistics such as speed or other ratios. On the other hand, it has some important mathematical properties that put it at the heart of statistical theory. The important thing is to know when it is correct to use it.

2. Calculating the arithmetic mean is often simplified by using a frequency table and the modified definition given by equation [6.1] as decribed in Box 6.1. The text mentions the use of coded data, with an assumed mean and new units to save work. There is more detail in Appendix B.

3. The problem of open-ended data is solved by making an assumption which affects the accuracy of the result, and is relevant when we consider how many decimal places should appear in the calculated mean.

4. All of this is for ungrouped data. For grouped data, essentially the same procedures are used but there is an important difference. Grouped data imply imprecision. To overcome it, we have to make assumptions; the correctness of our estimate will depend upon the validity of our assumptions. The assumptions enable us to convert the imprecise grouped data into precise but not necessarily accurate other data, which we then submit to the above procedures. We discusss the accuracy of the results, referring back to the assumptions we have made. Box 6.2 summarises the procedure for grouped data.

5. Noteboxes consider (1) combining the two means and (2) alternative notations.

Most fashionable

7.1 Introduction

Example 7.1 Lady Agatha is still unsure about what height of seat to recommend, and how to calculate it. She knows that her exact calculation of the arithmetic mean has produced a mean that is precisely right for nobody. She hopes that perhaps some other measure of average will be better for her purpose.

Two other averages frequently used to describe the location of a distribution are the mode and the median. When calculated for seat heights, the mode will suit more people than any other height, so it will minimise the number of complaints. The median will be too high for half and too low for half, so the complaints will cancel out. In this chapter we consider the mode a little further, leaving the median to Chapter 8. We show how to calculate it, and then compare it briefly with the mean, but this comparison is best done when we have also studied the median.

Although not as widely used as the arithmetic mean, the mode is sometimes the most appropriate average. However with discrete data, especially grouped data, it is often calculated without enough concern for the limitations of its definition and the method of calculation. We try to remedy this. Much of the chapter is important only to the reader who is going to calculate or use modes. The reader who at this stage simply needs to know what the mode is can skip most of the chapter after Section 7.2.

7.2 Basic ideas

In Chapter 4 we introduced unimodal, bimodal and multimodal distributions. The values of the variables corresponding to the greatest frequencies are the **modal values**, or the **modes.**

In simple cases the mode can be read from the table of data or selected as the value of the variable corresponding to the highest point in a frequency diagram. However, when we define the mode as the value of the variable that occurs most frequently, we may run into problems unless the number of items is large compared with the number of possible values. For example, the heights of seats for 60 office workers (recorded in Table 3.2) range from 60.2 to 81.7 cm, a range of

73

21.5 cm, and are given to the nearest 0.1 cm, which we have already suggested is too precise. Between 60.2 and 81.7 there are 216 possible values, correct to 0.1 cm. Since there are only 60 people, most of these possible values do not occur: 36 values occur once and 12 occur twice; no value occurs more than twice.

We can modify the stem and leaf table in Chapter 3 by listing the numbers in order within each group. If we do this for the two groups that contain most duplicated numbers (68–71.9 and 72–75.9), we obtain Table 7.1. These values and their frequencies could be represented by a histogram with many very short columns, each representing a width of 0.1 cm centred on 68.0. 68.1, 68.2, etc. Drawing it would be tedious and it would tell us no more than an ungrouped frequency table. Part of the ungrouped frequency table would look like Table 7.2.

From one viewpoint every entry that has lower entries (including zero) as immediate neighbours on both sides may be called a **local mode**. The first local mode in Table 7.2 is 68.7. Other local modes are at 68.9, 69.3 and so on. With this definition a completed version of Table 7.2 would show many local modes and 12 **major local modes**. But there is no overall mode in the sense that one of these values (correct to 0.1 cm) occurs more often than any other.

However, if we aggregate the data into the groups used in the stem and leaf diagram and apply the concept of a mode to the groups, we see from Table 7.3 (panel A) that the **modal group** of width 4 cm is the group 72.0–75.9 cm. This contains more heights than any other group of that width.

Table 7.1

Extract from Table 3.10

60–67 rows contain a total of 4 entries

68+	0.7	0.9	1.3	1.3	1.8	2.0	2.1	2.3	2.3	2.5
	2.8	3.1	3.4	3.4	3.6	3.7	3.8	3.8		
72+	0.1	0.1	0.2	0.4	0.4	0.5	0.6	0.8	0.8	0.9
	1.1	1.1	1.2	1.3	1.4	1.4	1.6	1.8	1.8	2.2
	2.2	2.3	2.4	2.5	2.6	2.6	2.7	3.3	3.5	3.7
	3.9									

76–84 rows contain a total of 7 entries

Table 7.2

Extract from ungrouped frequency table

x	68.6	.7	.8	.9	69.0	.1	.2	.3	.4	.5	.6	.7	.8	.9	70.0	.1	.2	.3	.4
f	0	1	0	1	0	0	0	2	0	0	0	0	1	0	1	1	0	2	0

Table 7.3

Grouped frequency distribution for 60 seat heights

Height, x	Frequency, f	Cumulative frequency
A. Original distribution		
60.0–63.9	2	2
64.0–67.9	2	4
68.0–71.9	18	22
72.0–75.9	31	53
76.0–79.9	6	59
80.0–83.9	1	60
B. New distribution		
60.0–63.9	2	2
64.0–67.9	2	4
68.0–69.9	5	9
70.0–71.9	13	22
72.0–73.9	19	41
74.0–75.9	12	53
76.0–79.9	6	59
80.0–83.9	1	60

If we want to locate a narrower modal group, we can use panel B (based on Table 3.2) to note that the group 72.0–73.9 cm contains more than any other of the listed 2 cm groups. It is left to you to confirm that if the 60 heights are all rounded up or down to the nearest centimetre (which means using groups 1 cm wide) there will be a unimodal distribution with a mode of 73. More people need seat heights that round to 73 cm (i.e. recorded between 72.5 and 73.4 cm) than to any other height. But if we seek much narrower groups, we begin to run into the problems already described. Lady Agatha's most useful estimate is 73 cm.

The grouped data just used were introduced because the more precise data were difficult to analyse. But sometimes there are no precise data: only grouped data are available. Later on we consider how to find the mode in these cases.

7.3 Different kinds of mode

Major, minor and local modes

The simple definition assumes that there are no rivals for the title of mode. Table 7.4, about family sizes, illustrates the problem. Here there are two modes at 2 and 4. If you draw a graph showing the frequencies with which families of various sizes occur – a frequency distribution graph or a column graph, as

Table 7.4

Family size (persons), x	1	2	3	4	5	6	7	8	9+
Number of families of this size, f	30	65	40	65	45	25	12	5	4

described in Chapter 3 – you will see that it has two peaks of equal height, each corresponding to a modal value of the variable x. It is **bimodal**. We may also have **trimodal** or even **multimodal** distributions. Thus, the arithmetic mean always provides a unique value, but the mode may not.

The highest mode is called the **major mode** and the other modes are called **minor modes** or, as we have seen, local modes. The term *minor mode* is sometimes reserved for a value that is higher than at least two neighbours on each side, distinguishing it from a local mode that has only one lower neighbour on each side. An example is given by the data in Table 7.5.

Table 7.5

x	f	
0	0	
1	5	
2	12	
3	20	minor mode
4	18	
5	15	
6	21	
7	30	major mode
8	23	
9	14	
10	11	
11	13	
12	17	local mode
13	16	
14	18	minor mode
15	2	

Distributed modes

We can have a **distributed mode** corresponding to a plateau in the frequency distribution curve:

x	1	2	3	4	5	6	7	8	9+
f	30	45	65	65	65	45	12	5	4

End modes

The most frequently occurring value of the variable may be at one end of the distribution:

x	0	1	2	3	4	5	6
f	40	35	28	20	10	3	0

If we define the mode as the most frequently occurring value of x, then we have to say that the mode is 0, even though there is no preceding lower value. We call this an **end mode**. J-curves, negative J-curves and U-curves all have end modes.

 7.4   **The mode for grouped data**

Grouped data can be more informative (subject to two assumptions) when there are large frequencies within the modal group and the groups close to it. For example, suppose a grouped frequency distribution reveals a modal group of 50.0–59.9 with a frequency of 105, as shown in Table 7.6, which is an extract from some larger table, and suppose the frequencies in the immediately neighbouring groups are 66 and 94 (as shown). It is tempting to try to be a little more precise about the mode.

Table 7.6

x	f
⋮	⋮
40.0–49.9	66
50.0–59.9	105
60.0–69.9	94
⋮	⋮

One technique assumes that, since there are more entries in the 'upper' neighbouring group of 60.0–69.9 (with values above the modal value) than in the 'lower' neighbouring group of 40.0–49.9 (with values beneath it) the mode will be towards the upper end of the modal group – nearer to 59.9 than to 50.0.

Two versions of this technique – one algebraic, one graphical – are described in Notebox 7.1 and Figure 7.1. They locate the mode at 57.8 (if we have that much faith in the data and the validity of the assumption) or at 58 (if we have slightly less faith). The point is that if we use imprecise data, we cannot put a great deal of faith in the accuracy of a precise result unless we are confident that the imprecise data 'obey' the assumption embodied in our technique. As we

Table 7.7

x	22	23	24	25	26	27	28	29	30	31	32	33	34	35	36
f_1	1	1	0	1	3	13	12	8	8	6	4	2	1	1	0
f_2	2		1		16		20		14		6		2		0
f_3	2				17			28			12			2	

saw in our discussion of the arithmetic mean, and at the beginning of this chapter, any grouping produces a loss of accuracy. To calculate the mode for grouped data, we usually have to make assumptions, and the 'correctness' of our answers will depend on what we assume. The technique described is not to be recommended unless the total frequency is so high that the variable can be treated as continuous.

One of the dangers associated with grouped data is illustrated by Table 7.7. The frequency distribution shown in row 2 clearly has a mode at $x = 27$. However, the data can be grouped as in row 3, and then the modal group is 28–29, which has a higher frequency than any other group of this size. The conventional argument suggests that a more precisely located mode would be between 28 and almost 30. The interval 26–27, just lower than the modal interval, has a frequency of 16. The interval 30–31, just higher, has a frequency of 14. This pair of neighbouring frequencies locates the mode in the interval 28–29, closer to 27 than to 30.

We can also group the data as in row 4, locating the modal interval of 28–30, and indicating that the mode is closer to 31 than to 27, apparently contradicting what we found from row 3.

The practice of locating a modal group and then comparing neighbouring frequencies to locate a mode more precisely within this group assumes (i) that the coarse modal group (such as in row 4) contains the mode that a finer tabulation (such as in row 2) would reveal, and (ii) that the precise location is given by considering neighbouring frequencies. It is safer simply to locate the modal group and to leave it at that, but the temptation to seek greater precision will always be there.

If the groups are of unequal width, we have to proceed very carefully. We should (as always) aim to make the best use of the data we have, but we must be careful not to make erroneous assumptions.

Consolidation exercise 7A

1. What is (a) a major mode, (b) a minor mode and (c) a local mode?
2. What assumptions are commonly made in calculating the mode for grouped data?
3. Find the mode of the data for family size in Example 6.3.

7.5 The mode for continuous variables

The calculation is just the same when the grouped frequencies are for a continuous variable, but the precision of the estimate may need to be considered further. For example, if we have data about ages and estimate the modal age as 29.28 years, we are implying an accuracy to the nearest $3\frac{1}{2}$ days. Can we honestly do this?

Suppose the mode is in an interval 4 years wide. According to our technique, we divide the range of 4 years into 25 parts then assume that the mode is say 8 of these parts along the range. Each part would be 4/25 of one year, which is about 58 days. Thus the technique limits our possible answers to those that are 58 days apart. Even if every assumption is correct, we cannot be more accurate than that! In other words, even a statement implying correctness to one decimal place (e.g. 29.3) would be unjustified, since one-tenth of a year is 36 days.

But if the frequencies had been much higher and the range of 4 years had been divided into 250 parts, the possible accuracy of the result would have been greater.

7.6 Unequal group widths

When the groups are not all of equal width, the calculation of the mode is essentially the same, but the effect of group width on frequency must first be eliminated. We illustrate the problems by using the grouped age distribution in Table 7.8.

Table 7.8

Age, x	0–	5–	10–	15–	20–	30–	40–	50–	60–	65–	70–	80+
Frequency, f	20	22	20	18	35	38	44	48	20	16	16	4
Cumulative frequency	20	42	62	80	115	153	197	245	265	281	297	301

Somehow we have to allow for the fact that not all groups are of equal width. One way of doing this is to merge adjacent five-year groups into ten-year groups, giving Table 7.9. Now all the groups are of equal width (except perhaps the last, as in this case). The group with the largest frequency has an age range of 50–59. There is no way of telling from this table how frequencies vary within that age range, which is the modal range.

Table 7.9

Age, x	0–	10–	20–	30–	40–	50–	60–	70–	80+
Frequency, f	42	38	35	38	44	48	36	16	4
Cumulative frequency	42	80	115	153	197	245	281	297	301

To illustrate both the problem of unequal group widths and the usual algebraic technique for finding the mode more precisely we now examine this distribution further. Note that the modal frequency of 48 in the interval 50–59 years is 4 higher than the frequency in the immediately preceding group, and 12 higher than the frequency in the immediately following group. We assume that we can use this to tip the balance towards the lower end of the modal group 50–59.

The precise assumption is illustrated in the the same figure; it is that the mode is at a value between 50 and 59 and lying

$$\frac{4}{4+12} = \frac{1}{4}$$

of the group width along it.

Thus we estimate the mode as

$$50 + \left(\tfrac{1}{4} \times 10\right) = 52.5$$

This is more precise than the provided data. The correctness or otherwise of the result depends on our assumptions.

Notice that if we had experimented with subdividing the wider groups, we would have had considerable problems. For example, one of the two five-year age groups 50–54 and 55–59 would necessarily have to contain at least 24 people (half of 48), and perhaps both would. But no other five-year group would *necessarily* contain so many, even though many five-year groups could.

With such possibilities, how do we choose the modal five-year group? All we can say is that the modal ten-year group is 50–59, but there are many possible candidates for mode among five-year groups.

Box 7.1

How to ... find the mode

1. *For ungrouped discrete data* use a frequency table or a graph to pick out the value of the variable that occurs most frequently.

2. *For discrete data grouped into bands of equal width* locate the group with the highest frequency. Denote the lower end of this group by L and the upper end by U. Let the value of the frequency m in the modal group be a above the frequency in the immediately preceding group and b above the frequency in the immediately following group. Then the mode is estimated to be

$$L + \frac{a}{a+b}(U-L)$$

rounded to the nearest whole number. This can also be found geometrically (Notebox 7.1). ◯

3. *For discrete data in groups of unequal width* it may be necessary to split some of the wider groups into narrower groups, making assumptions about the allocation of the frequencies. Alternatively we can merge adjacent groups into larger ones, all of the same size, even though we may lose information in doing so. If either procedure is followed we can then apply the method of paragraph 2. In some cases it may not be practicable to locate a mode.

4. *For continuous variables* proceed in the same way but consider more carefully the precision of the estimate. Alternatively draw a column graph, with the *areas* of the different width columns proportional to the numbers they represent. Then try to fit a smooth curve drawn such that the area under the curve and within the age band is unchanged. The mode can then be taken as the value of the variable corresponding to the highest point of this curve. This is not always a justifiable procedure, and although it may give a more precise figure for the mode, it is at the expense of reliability.

5. The mode is also given by the point of inflexion of the ogive, which may be found by careful drawing and use of a straightedge as (Notebox 7.2).

Consolidation exercise 7B

1. Why is there particular need to consider the precision of the calculated mode for continuous data?

2. The mode may be estimated by fitting a smooth curve to a histogram. How do you choose the widths and heights of the columns? What should you try to do when drawing the curve?

3. What point on the ogive indicates the mode? How do you find it?

4. Find the mode for the age distribution given in Example 6.4.

Notebox 7.1

Techniques for estimating a mode more precisely

A diagrammatic technique using a histogram is shown in Figure 7.1, based on Table 7.6. Here the diagonal lines from corners of the neighbouring columns intersect at a frequency that gives the estimated mode. Use of the formula in Box 7.1 with $L = 50$, $U = 60$, $a = 105 - 66 = 39$ and $b = 105 - 94 = 11$ gives the same answer, 57.8. We should still consider whether it needs rounding.

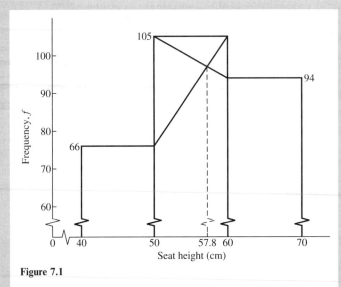

Figure 7.1

Graphical techniques for continuous data

If we believe that more precise information about the variable would fit a smooth curve, we can sometimes get some help by drawing a column graph, with the *areas* of the different width columns proportional to the numbers they represent. Then we can try to draw a smooth curve through the tops of the columns. This should be drawn in such a way that for each column the area under the curve and within the age band is the same as the area of the column. (The shaded portions in Figure 7.2 balance). The mode may then be taken as the value of the variable corresponding to the highest point of this curve, and the modal group is the group that contains it. This is not always a justifiable procedure, and although it may give a more precise figure for the mode, this is at the expense of reliability. Drawing the curve involves making assumptions about the precise position of every point on it. It is neither very easy nor very reliable.

Note that the ogive (or cumulative frequency curve) described in Chapter 3 has a point of inflexion at a value of the variable corresponding to a mode, as illustrated in Figure 7.3.

Remember that the tangent to a curve at a point of inflexion cuts across the curve at the point of touching. One way of finding the mode is therefore to draw the ogive and then slide a (transparent) straightedge along it, just touching it, so that its edge runs along the tangent at the point of contact. The point where the straightedge crosses the curve, just where it touches it, is the point of inflexion. The value of the variable corresponding to that point is the mode. ⊃

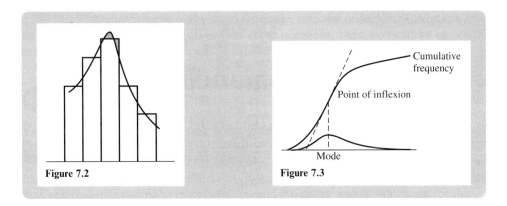

Figure 7.2

Figure 7.3

Summary

1. One average which is not a mean is the mode. This can be loosely defined as the most fashionable value of the variable.

2. A distribution may be multimodal, having several local modes. We may speak of major modes and minor modes. J-shaped and U-shaped distributions have end modes. There are also plateau-shaped distributions with distributed modes.

3. The standard technique for calculating the mode of grouped data has both algebraic and geometric versions, described in Notebox 7.1. It depends on assumptions that need to be spelt out if errors are to be avoided (especially errors of overprecision). The calculated result is also sensitive to the location of the group boundaries.

4. Generally greater reliance can be put on the result if the data are continuous, or the number of items is very large. Sometimes graphical techniques may provide greater precision, but there is a need to keep the assumptions in mind. One graphical technique locates the mode as the point of inflexion of the ogive.

5. The procedures for both ungrouped and grouped data are summarised in Box 7.1

6. Noteboxes concern graphical techniques for (1) estimating the mode more precisely and (2) continuous data.

8

Doing a balancing act

8.1 Introduction

Example 8.1 Lady Agatha knows that having to hear complaints is one of the prices titled ladies have to pay. Having learned that one way of dealing with complaints is to make them balance, she wants to find the height of seat that will be criticised by some for being too low, and by just as many others for being too high.

Example 8.2 In the rarely heard-of Lower Sternum there are 101 households, all with different levels of savings, more or less evenly spread between £200 and £4000. One of the wealthier households wins the national lottery. How can we find an 'average' income that will be unaffected by the win?

Another average that may not be altered if some of the data change (and so is not a mean) is the **median**. It may be defined as the middle value when all the values are arranged in ascending (or descending) order. This means that any or all of the values can change, yet the median will retain its value, provided there is no change in the number of values smaller than it and the number of values larger than it. Thus it is unaffected by extreme values. It also enables Lady Agatha to do her balancing act.

Like the mode, the median is sometimes calculated without due regard for the limitations of the data and the method. Here we describe how to calculate it (i) for ungrouped data and (ii) for grouped data. Some of the computational detail can be skimmed over at a first reading. We end with an important relationship between the arithmetic mean, the mode and the median.

The median is used both in descriptive statistics and in various non-parametric tests (Chapter 33 onwards).

8.2 The median for ungrouped data

If we have an *odd* number of values, it is easy to find: the median of the five numbers 1, 5, 6, 10, 14 is 6. More generally, if there are n numbers and n is odd, the median corresponds to the number in position $\frac{1}{2}(n + 1)$ when the numbers are arranged in ascending (or descending) order.

For an *even* number of values the median is defined to be half-way between the middle two. This makes it half the sum of the middle two. For example, the median of 1, 5, 6, 10, 14, 20 is halfway between 6 and 10, which makes it $\frac{1}{2}(6 + 10) = 8$. More generally, if the number of values n is even, then the median is defined to be halfway between the $\frac{1}{2}n$th value and the $(\frac{1}{2}n + 1)$th. This can be written as the value in position $\frac{1}{2}(n + 1)$, provided this is read as the mean of the values in positions n and $n + 1$.

Sometimes the *same values* of the variable may occur several times, as in 1, 3, 6, 6, 8, 8, 8, 9, 11, 13, 13. Here there are 11 values and the middle one (the sixth) is 8, which is the median. The first five values are less than or equal to the sixth value. The last five values are greater than or equal to it. Thus instead of thinking of the median as a value such that half the values are greater than it and half are less, we think of it as a value such that half are greater than or equal to it, and half are less than or equal to it, with no item being counted twice.

As an example consider the data in Table 7.5, which is reproduced as the first two columns of Table 8.1. There are 226 items, all with values of x lying between 0 and 15 inclusive. The second column shows the frequencies with which different values occur. The third column shows the cumulative sums of the frequencies Chapter 3). For example, $37 = 0 + 5 + 12 + 20$ is the number of items with an x of 3 or fewer.

Since there is an even number of values, the median is defined to be halfway between the values of item 113 (which is 7) and item 114 (which is also 7). The median is therefore 7.

This means that half the values are less than or equal to 7, and half are greater than or equal to 7. One way of being quite clear about this is to draw a cumulative frequency graph (Figure 8.1), as decribed in Chapter 3. If we then draw a

Table 8.1

x	f	Cumulative frequency
0	0	0
1	5	5
2	12	17
3	20	37
4	18	55
5	15	70
6	21	91
7	30	121 contains items 113 and 114
8	23	144
9	14	158
10	11	169
11	13	182
12	17	199
13	16	215
14	9	224
15	2	226

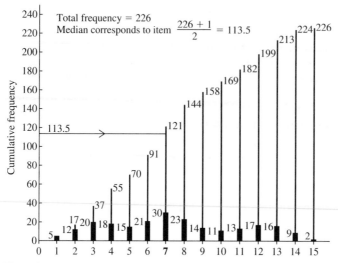

Figure 8.1

horizontal line at the level of cumulative frequency $\frac{1}{2}(n+1) = 113.5$ the median will be the value of the variable where this horizontal line cuts a cumulative frequency line.

Box 8.1

How to ... **find the median for ungrouped discrete data**

1. Arrange the data in a column in ascending order, with the lowest value of the variable coming first.

2. Construct a cumulative frequency table (Box 3.2). The last entry will be n, the total frequency.

3. Locate the cumulative frequency corresponding to item or individual $\frac{1}{2}(n+1)$, which gives the median value of x.

4. If n is even, the median value is halfway between the values for the two items on either side of $\frac{1}{2}(n+1)$.

In this example we have supposed that our values of x are precisely as listed, so they are *discrete*. For example, x may be the number of times that a person has arrived late for work during a month, which must be an integer.

But if x is *continuous* the tabulated data are necessarily rounded, which implies they are grouped, as we shall shortly see. We turn now to grouped data more generally.

8.3 The median for grouped data

When we have grouped data, the basic procedure is just the same, but as usual we have to risk inaccuracy if we seek precision. Consider again the data in Table 6.2, about the ages of Steven's prospective neighbours. We use it as part of Table 8.2. As there are 301 people, we have to find the age of person $\frac{1}{2}(301 + 1) = 151$. We see from the cumulative frequency column that

There are 115 persons aged up to almost 30
There are 38 persons aged between 30 and almost 40

Person 151 is in this second group, aged 30–39; that is all we can say with certainty. However, if we *assume* that these 38 people have ages evenly spread between 30 and almost 40, we can argue that person 151 is the $151 - 115 = 36$th person in this group, and that his or her age is given by

$$30 + \left[\frac{151 - 115}{38} \times (40 - 30) \right] = 30 + 9.474 = 39.474 \text{ years}$$

or, more credibly, 39 years.

Sometimes another assumption may be more plausible. In every case the result must be treated with caution, and it is important to consider whether there is point or meaning in an estimated average that is more precise than the original data.

The result 39.474 is too precise as an estimate of a median age. It implies correctness to a thousandth part of a year – a third of a day – which is more precise than many people can state their own ages. Even writing the median as 39.47 years, correct to about 3 days, is too precise. After all, we have assumed

Table 8.2

Age, x	Frequency, f	Cumulative frequency	
0–4	20	20	
5–9	22	42	
10–14	20	62	
15–19	18	80	
20–29	35	115	
30–39	38	153	person 151 in this group
40–49	44	197	
50–59	48	245	
60–64	20	265	
65–69	16	281	
70–79	16	297	
80+	4	301	

that the 38 persons in a group 10 years wide have ages evenly spaced over that period, which means that the 38 ages are 96 days apart from each other, so we cannot meaningfully make estimates more precise than about one-quarter of a year. The best we can do with 39.47 is to print it as $39\frac{1}{2}$, which conventionally does not have the same implications of precision as 39.5.

Box 8.2

How to ...   **find the median for grouped data**

1. Arrange the data in a column in ascending order, starting with the lowest value of the variable.

2. Construct a cumulative frequency table (Box 3.2). The last entry will be n, the total frequency.

3. Locate the group containing the median entry, which corresponds to item $\frac{1}{2}(n+1)$. If n is even, the median value is halfway between the values for the two items on either side of $\frac{1}{2}(n+1)$.

4. Denote the lower bound of the group containing the median by L, the upper bound by U (which will also be the lower bound of the next highest group).

5. Let the cumulative frequency up to the lower bound L of the median group (i.e. the total frequency in groups lower than the median group) be l, and the cumulative frequency up to the upper bound U be u, so the median group contains a frequency of $u - l$.

6. Assume that the $u - l$ items in the median group are evenly spread across it at intervals of $(U - L)/(u - l)$.

7. The median item, in position $\frac{1}{2}(n+1)$ measured from the first entry in the table, will be in position $\frac{1}{2}(n+1) - l$ in this median group.

8. The median value is therefore estimated at

$$L + \left[\tfrac{1}{2}(n+1) - l\right](U - L)/(u - l)$$

8.4   **Rounded continuous variables**

When we have a continuous variable presented as a discrete variable, such as ages rounded down to complete years, or masses rounded to the nearest kilogram, the normal practice is to assume that the individual items have values that are evenly spread over the interval of rounding. If the variable represents age, we assume that 30 people aged 7 have ages spread between exactly 7 and almost 8 at intervals of one-thirtieth of a year. If it represents mass rounded to the nearest kilogram, we assume the items have masses evenly spread between 6.50 and 7.49 kg. We can then apply the method used for finding the median of grouped data, with interval width $U - L = 1$.

For example, we know that for Table 8.1 with a total frequency of 226 the median corresponds to the value in position $\frac{1}{2}(226 + 1) = 113.5$. That locates the median as 7, but if the data relate to ages then, unless we have more exact information, we should think of the 30 people listed with an $x = 7$ as having ages that are spread out evenly over the year. The median age will be that of person $\frac{1}{2}(226 + 1) =$ person 113.5. We can find this from the formula or from first principles. Using the formula, we have

$$\text{Median} = L + \left[\tfrac{1}{2}(n + 1) - l\right](U - L)/(u - l)$$
$$= 7 + [113.5 - 92](1)/(122 - 92) = 7.717$$

We discuss this answer shortly, but first we derive it from first principles. We think of the age of person 113.5 as being halfway between the ages of person 113 and person 114, and we know there are 91 persons aged less than 7. Let person 92 be aged exactly 7. Similarly let person 122 be aged exactly 8 (Table 8.3). Then, as there are 30 people, we divide the year into 30 equal intervals, so person 113.5 will have an age of

$$7 + (113.5 - 92)/(122 - 92) = 7 + 21\tfrac{1}{2}/30 \text{ years}$$

which is defined to be the median age. Writing this as 7.717 years implies correctness to a thousandth part of a year.

An answer of 7.7 years, implying correctness to within 36 days, is more easily defended. It implies that we believe that persons 113 and 114 have an arithmetic mean age lying between 7.65 and 7.74, which is not unreasonable as there are 30 people numbered between 92 and 121 with ages between 7.00 and 7.99.

Box 8.3 explains how the median can be found (a) by calculation and (b) graphically.

Table 8.3

Person	Assumed exact age
92	7
93	$7 + (93 - 92)/(122 - 92) = 7 + 1/30$ years
94	$7 + (94 - 92)/(122 - 92) = 7 + 2/30$ years
95	$7 + (95 - 92)/(122 - 92) = 7 + 3/30$ years
⋮	⋮
113	$7 + (113 - 92)/(122 - 92) = 7 + 21/30$ years $= 7.700$ years
114	$7 + (114 - 92)/(122 - 92) = 7 + 22/30$ years $= 7.733$
⋮	⋮
120	$7 + (120 - 92)/(122 - 92) = 7 + 28/30$ years
121	$7 + (121 - 92)/(122 - 92) = 7 + 29/30$ years
122	$7 + (122 - 92)/(122 - 92) = 7 + 30/30$ years $= 8$ years

---

Box 8.3

How to ... **find the median for rounded continuous data**

By calculation

1. Locate the median item as for discrete data (Box 8.1).

2. Interpret the given value of x corresponding to this item as indicating an interval whose values have been rounded to x. Use your knowledge of the variable to decide whether the rounding has been down (as with age), to the nearest integer, or up (as with masses, perhaps). This locates the lower and upper bounds of the interval – probably 1 unit apart.

3. Assume that all items given as having this value of x in fact have values that are evenly spread through the interval and proceed as for grouped data (Box 8.2).

By drawing

1. Construct a cumulative frequency curve (or ogive) (Boxes 3.2 and 3.6).

2. Locate the value of half the total frequency on the vertical axis.

3. Draw a horizontal line through this value and continue it to intersect the cumulative frequency curve.

4. Drop a vertical from the point of intersection to the horizontal axis. The value of the variable is the median.

Consolidation exercise 8A

1. Give a simple verbal definition of the median. How must that definition be modified to take account of the fact that some values of the variable may occur several times?

2. Give an algebraic definition of the median that is applicable to both an even and an odd number of variables.

3. What assumptions are usually made in calculating the median for grouped data?

4. What are L, n, l, U and u in the following formula for the median?

$$\text{Median} = L + \left[\tfrac{1}{2}(n+1) - l\right](U - L)/(u - l)$$

5. Find the median of (a) the data for family size in Example 6.3 and (b) the age distribution given in Example 6.4.

8.5 Relationships between arithmetic mean, median and mode

If we draw a histogram or a frequency distribution curve (Figure 8.2), the mode will be the value of x corresponding to the highest value of the graph. We can also draw a vertical line dividing this graph into *two equal areas*. This line will cut the x-axis at the *median* value of x.

A vertical line through the *arithmetic mean* will pass through the *centre of gravity* of the area. (This will take account not only of the areas on each side but also of the shapes.)

If the graph is unimodal, the median will lie between the mode and the arithmetic mean. This is illustrated in Figure 8.2.

We have seen in Chapter 3 that if a unimodal histogram or frequency curve is not symmetrical, it will have one 'tail' longer than the other. If the longer tail is to the right of the mode, the distribution is 'skewed to the right', or **positively skewed**. If the longer tail is to the left, the distribution is 'skewed to the left', or **negatively skewed**.

There are various measures of skewness (one involving the median) which we shall meet when we study spread, or dispersion. However, we can appreciate the meaning of 'moderately skewed' without bothering too much about them. For moderately skewed frequency curves the empirical relationship

$$\text{Mean} - \text{mode} = 3(\text{mean} - \text{median}) \tag{8.1}$$

usually provides a good approximation that enables any one of these averages to be estimated from a knowledge of the other two.

Moreover, the mean tends to be on the same side of the mode as the longer tail of the distribution. This implies that if the mean is greater than the mode (which probably implies greater than the median), the two terms in equation [8.1] have a positive sign, indicating a positive skewness. If the mean is less than the mode (and the median), the sign indicates negative skewness. In Chapter 10 we describe a numerical measure of skewness based partly on this relationship. More advanced work uses a different definition and measure of skewness (Notebox 12.2).

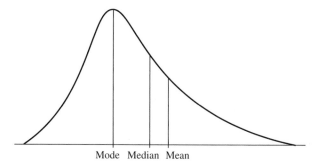
Mode Median Mean

Figure 8.2

We may now refer to the three measures of location (or average) that interest Lady Agatha in her concern about seat heights. We have found that

Arithmetic mean height = 72.6 cm (or 72.55 to 2 d.p.)

modal height   = 73 cm (data to nearest centimetre)

Median height   = 72.8 cm

These three averages are so close together that they are barely distinguishable from each other. The arithmetic mean is irrelevant to Lady Agatha's decision. What is encouraging is that the mode and median are so close. Seats 73 cm high will suit more people than any other height, and also ensure that complaints more or less cancel out. She is lucky.

We may also note that, as we would expect, the median lies between the mode and the mean. However in this example the empirical relationship, mean − mode = 3(mean − median), does not hold. If there had been more data, with higher frequencies, it might have been possible to make a meaningful calculation of the mode to one or two decimal places. But as it is, the estimate of the mode is too coarse to provide a check on the relationship.

There is further reference to the median when we study quartiles (of which the median is the second) in Chapter 10.

Consolidation exercise 8B

1. What can be said about vertical lines drawn through (a) the mode, (b) the median and (c) the arithmetic mean?

2. For what kind of distribution is the approximate relationship (mean − mode) = 3(mean − median) likely to hold ?

3. If the mean is greater than the mode, what can be said about skewness?

Summary

1. Another average unaffected by extreme values is the median. If there is an odd number of values, the median is the middle value when they are are arranged in order of magnitude. If there is an even number, it is taken as being halfway between the middle two. It has to be carefully interpreted if it has a value that occurs more than once. Use of a cumulative frequency graph clarifies this.

2. For continuous data and for grouped data, we have to make assumptions and our attempts at precision run the risk of inaccuracy. The calculations normally involve cumulative frequency tables (Boxes 8.1, 8.2 and 8.3).

3. The median can also be located as the value of the variable corresponding to an ordinate of 50% on the percentage ogive, which is also the value that divides the area under the distribution curve into halves.

4. Equation [8.1] is an empirical relationship between the mean, the mode and the median values; it is valid for moderately skewed frequency curves.

5. The median is the second quartile (Chapter 10).

6. Several non-parametric tests involve the median.

Other means

9.1 Introduction

Example 9.1 Tommy drove for three hours at 40 mph then for three hours at 60 mph. What was his average speed ?

Example 9.2 Eric drove 120 miles at 40 mph then another 120 miles at 60 mph. What was his average speed ?

Example 9.3 Harry has noticed that, in two successive months, petrol cost 50p per litre and 60p per litre. In each of these months he has bought the same quantities of petrol. He wants to know the average price he has been paying per litre.

Example 9.4 Stephanie discusses the prices with him and says that in her own case the price increase has caused her to reduce her travel, since every month she puts aside the same amount of money for petrol, and spends exactly that. She, too, wonders what average price she has been paying in the last two months.

Example 9.5 Between 1993 and 1994 the profits of Random Breweries rose by 20%. Over the next year they rose by 40%. What was the average annual percentage rise?

Example 9.6 Three squares have sides of 1, 2 and 3 cm. How long is the side of a square whose area is equal to the arithmetic mean of their three areas?

Example 9.7 The headmistress of Statingham Comprehensive finds that in the end-of-year assessment some people who do quite badly in mathematics have a high overall average, which leads to their preferment over people whose mathematics is better. She wants to find an average that will favour mathematicians.

Example 9.8 The retail prices of certain vegetables available in Statingham market on 1 July 1990 and 1 July 1992 are shown in Table 9.1. What happened to the average price of these vegetables?

Table 9.1

Retail prices for vegetables

	1990	1992
Medium-sized cauliflower	60p	70p
New potatoes (per kg)	40p	48p
Carrots (per kg)	40p	36p
Onions (per kg)	50p	48p

The arithmetic mean, the mode and the median are all averages, but only the arithmetic mean depends on the precise values of every item. However, in all except two of these examples the ordinary arithmetic mean is inappropriate. Some require the **harmonic mean**, some the **geometric mean**, and one the **root mean square**. Like the arithmetic mean, they depend on the precise values of every item. Example 9.7 introduces the idea of **weighting**, which we apply to the arithmetic, harmonic and geometric means. Example 9.8 is about index numbers, which involve the ratio of two weighted sums and are used to make comparisons between years or countries, for example. We merely touch upon a large subject.

The main sections of this chapter can be read in any order, and in a first reading the whole chapter can be omitted.

9.2 The harmonic mean

Speed (measured in distance per unit time) and price (measured in money per unit quantity) are examples of quotients – one thing divided by another. If you have to find the mean of two (or more) numbers that are quotients, ratios, percentages or growth rates, the arithmetic mean is usually not the one to use, as we now see.

Example 9.1 Remember that

$$\text{Speed} = \frac{\text{distance}}{\text{time}}$$

Tommy travels for 3 hours at 40 mph and then for another 3 hours at 60 mph. Note that he is travelling at two different speeds for equal *times*, and that *time* is on the *bottom* of the quotient. He covers 120 miles in the first 2 hours and another 180 miles in the next 2 hours. Adding up, he travels a total of 300 miles in a total of 6 hours, so his average speed is $300/6 = 50$ mph. We could have got this by taking the arithmetic mean of 40 and 60. There is no problem here; *time* is in the *denominator* (the bottom) of the definition of speed, and we averaged two speeds that were for the *same time*.

⟹ **Example 2** The definition of speed can be rewritten as

$$\text{Time} = \frac{\text{distance}}{\text{speed}}$$

Eric travels for 120 miles at 40 mph then for another 120 miles at 60 mph. He travels at two different speeds for equal *distances*, and *distance* is on the *top* of the quotient. The arithmetic mean of the two speeds is $\frac{1}{2}(40 + 60) = 50$ mph, but is it the right average? Note that, although time is on the bottom of the definition, we have averaged two speeds over the *same distance* (which is on the top). To check on the answer, go back to

$$\text{Time} = \frac{\text{distance}}{\text{speed}}$$

To cover the first 120 miles at 40 mph he takes 3 hours (distance/speed = 120 miles/40 mph). To cover the next 120 miles at 60 mph he takes 2 hours. Thus he travels the total distance of 240 miles in a total of 5 hours and his mean speed is therefore $240/5 = 48$ mph, which is *not* the arithmetic mean of the two speeds (40 and 60). Taking the arithmetic mean of speeds over the same distance does not give the right answer.

Examples 9.3 and 9.4 relate to prices. Keep in mind that

$$\text{Price} = \frac{\text{amount of money}}{\text{quantity of item}}$$

⟹ In **Example 9.3** the price of petrol rises from 50p per litre to 60p per litre. Harry buys *equal quantities* (say 20 litres) at both prices. Note that *quantity* of petrol is at the *bottom* of the quotient. The total amount spent $(20 \times 50\text{p} + 20 \times 60\text{p} = £22)$ divided by the total quantity (40 litres) will give an average price of 55p per litre, which is the arithmetic mean of the two prices (50 and 60). No problem. We kept the bottom item (quantity of petrol) the same.

⟹ **Example 9.4** is different. Stephanie spends *equal amounts of money* (say £12) at the two prices. Note that amount of money is at the top of the quotient. The total amount spent $(2 \times £12 = £24)$ divided by the total quantity $(£12 \div 50\text{p}$ plus $£12 \div 60\text{p} = 24 + 20 = 44$ litres) gives an average price of $(£24 \div 44) = 54.55\text{p}$ per litre, which is *not* the arithmetic mean of the two prices (50 and 60). We kept the top item (amount of money) the same, and by doing so ran into error.

Thus, putting it informally, we can use the arithmetic mean if we keep the item named in the bottom of the quotient the same, but we have to use a different method if we keep the item named in the top the same. It can be shown that when the top is kept the same, the correct mean to use is the **harmonic mean**.

Remembering that the reciprocal of something is 1 divided by that something, we may define the harmonic mean H of a set of numbers to be

the reciprocal of
   the arithmetic mean of
      the reciprocals of the numbers

Thus the harmonic mean of the three numbers 4, 6 and 9 is H, where

$$\frac{1}{H} = \frac{1}{3}\left[\frac{1}{4} + \frac{1}{6} + \frac{1}{9}\right]$$

$$= \frac{1}{3}[0.52778]$$

$$= 0.1759$$

$$H = 5.648$$

More generally the harmonic mean of n values of x is H, where

$$\frac{1}{H} = \frac{1}{n}\sum\frac{1}{x_i} \qquad\qquad [9.1]$$

Some generally remember this more easily if it is inverted

$$H = \frac{n}{\sum\dfrac{1}{x_i}} \qquad\qquad [9.2]$$

For example, the harmonic mean of the three numbers 1, 2 and 4 is

$$\frac{3}{1 + \frac{1}{2} + \frac{1}{4}} = \frac{3}{1\frac{3}{4}} = 3 \div \frac{7}{4} = 3 \times \frac{4}{7} = \frac{12}{7} = 1.71$$

Our results may be summarised in Table 9.2.

Box 9.1

How to ... **find the harmonic mean**

1. Often a variable is expressed as something *per* something else: miles per litre, pence per gram, grams per penny, dollars per franc. In such cases the variable is a quotient. If the entities in the top of the quotient are being kept constant over the process of averaging (as illustrated in Table 9.1), you should use the harmonic mean. If the entities in the bottom are being held constant, you should use the arithmetic mean.

2. To calculate the harmonic mean, first calculate the reciprocal of every value of the variable, thus getting n values of $1/x$.

3. Sum these values to obtain $\sum 1/x$.

4. Divide n by this sum. This gives the harmonic mean.

Table 9.2

Person	Quotient	Kept the same	Average to use
Eric	$\dfrac{\text{distance}}{\text{time}}$	distance 20 miles each speed	harmonic mean
Tommy	$\dfrac{\text{distance}}{\text{time}}$	time (30 min each speed)	arithmetic mean
Harry	$\dfrac{\text{cost}}{\text{quantity}}$	quantity (200 litres every month)	arithmetic mean
Stephanie	$\dfrac{\text{cost}}{\text{quantity}}$	cost (£80 every month)	harmonic mean

The rule that emerges is

	$\dfrac{\text{top}}{\text{bottom}}$	top	harmonic mean
	$\dfrac{\text{top}}{\text{bottom}}$	bottom	arithmetic mean

Consolidation exercise 9A

1. When is an average a mean?

2. What do speed and price have in common?

3. If we are dealing with ratios, when is the use of the arithmetic mean (a) acceptable and (b) unacceptable?

4. How is the harmonic mean defined? When must it be used?

5. What is the first step in calculating the harmonic mean? How do you then proceed?

6. Show that if you travel to Land's End at an average speed of 40 mph and then travel back at an average speed of 50 mph along the same route, your average speed will be given by the harmonic mean of these two speeds.

7. What is the average speed if you travel for 3 hours at 40 mph and for another 3 hours at 50 mph?

8. A delivery of 240 dolls was packed into boxes, each containing 20 dolls. Another delivery of 240 dolls was packed into boxes each containing 30 dolls. What is the correct mean to use for the mean content of a box of dolls? Why? Obtain the correct mean by two different methods.

9.3 The geometric mean

Many people would answer the question in Example 9.5 about Random Breweries by saying that the profits rose by an average annual percentage of $(20 + 40)/2 = 30\%$. This would be wrong. The correct answer is provided by the geometric mean, which is the subject of this section.

 Suppose in **Example 9.5** the brewery made a profit of £1 million in 1993 and £1.2 million in 1994. This represents an increase of 20%. Next year's profits rose by 40% *of what they were in 1994*, which means that the profits in 1995 came to $1.40 \times £1.2$ million $= £1.68$ million compared with £1 million in 1993. Thus the profits rose from £1million to £1.68 million, so they rose by 68% over the two years put together. They did this by rising 20% in the first year and 40% in the next year.

We may be tempted to say that this increase of 68% over two years averages 34% a year. But it does not. Suppose profits had risen 34% in the first year, turning £1 million into £1.34 million. If they now rise by this same percentage rate in the second year we get a figure of $1.34 \times £1.34$ million $= £1.796$ million, which is far too high.

What we want is a rate of growth that gives the right amount at the end of the complete period if applied every year. It can be shown that this is obtained not from dividing 68 by 2 but by taking the square root of 1.68, to get 1.296. This represents a growth of 29.6% each year in the profits of Random Brewery.

We can check this. Let us start with profits of £1 million in 1993. If they grow at 29.6% then in 1994 they will be £1.296 million. If this grows again by 29.6% in the next year, the 1995 profits come to $1.296 \times £1.296$ million $= £168$ million, which is what we got by having a rise of 20% in one year and 40% in the next.

To get a rule, note first that a growth rate of 20% was written as a decimal, 0.20. But it does not stop there. In the arithmetic mean we concentrated on the *levels* of profit, not the additional profit represented by the growth (of 0.20). The level of profit rose from £1 million to £1.20 million. Instead of saying 'profit rose by 20%', we say 'profit rose *by a factor of 1.20*'.

In working out the average profit over a few periods, we work with growth factors, not percentages. Accordingly, 20% becomes 1.20, 40% becomes 1.40 and the end result of our calculation, after taking a square root, is the growth factor of 1.296, which corresponds to a percentage growth of 29.6%.

Thus, if something grows by $r\%$ over one period then by $s\%$ over the next period (of equal duration), the average growth rate over each period is $\sqrt{(1 + r)(1 + s)}$ where r and s have been converted to decimals. If something grows by 5% one year and 20% the next year, the mean annual rate of growth is $\sqrt{1.05 \times 1.20} = \sqrt{1.26} = 1.122$ or 12.2%. This growth rate in each year will produce the same end result as 5% in the first year and 20% in the second year.

The mean we have just calculated is called the **geometric mean** of the two growth rates. It is the square root of their products after the percentage growth rates have been expressed as growth factors.

Be careful not to make the mistake of saying that if something grows by 5% in one year and 20% in the next, the geometric mean is $\sqrt{5 \times 20}$. This would be 10%, and something growing at that rate in years 1 and 2 would not achieve the right total growth. We must write a growth rate of 5% as 1.05.

If something grows by 5% in one year, 20% in the next year and 18% in the third year, the geometric mean of the three growth rates is obtained from the cube root of the product of these three growth factors. It is $\sqrt[3]{1.05 \times 1.20 \times 1.18} = \sqrt[3]{1.4868} = 1.1413$, indicating a mean growth rate of just over 14.1%. Something growing at this rate in each of three years will reach the same level as if it grew 5% in one year, 20% the next and 18% the next.

Box 9.2

How to ... **find the geometric mean**

1. The geometric mean is used mainly to average those ratios that reflect growth, and sometimes with other ratios. It cannot be used with data on the interval or lower scales (Noteboxes 9.2 and 33.3).

2. If you have growth rates of r_1, r_2, ..., express them as decimals and convert them into ratios $1 + r_1$, $1 + r_2$, ... (For example, 3% is written as .03 and converted into the ratio 1.03).

3. If there are only two values of the ratio, their geometric mean is obtained by multiplying them together then taking the square root of the product.

4. If there are n values of the ratio multiply them all together and take the nth root of the product by using a calculator, as described in Box 9.3 (or by taking logs and noting that the log of the geometric mean is the arithmetic mean of the logs of the original data).

Box 9.3

How to ... **find the nth root where n is any positive number**

1. Most good calculators have a button marked $[y^x]$. Some also have a second function marked $[\sqrt[y]{\ }]$.

2. If you have $[\sqrt[y]{\ }]$ then to find the 5th root of 32, written as $\sqrt[5]{32}$, follow this step:
   (i) Enter 5, $[\sqrt[y]{\ }]$ (probably preceded by [second function]) then 32, and if necessary [=]. Check that you answer is 2.
   If you have have to use $[y^x]$ then to find the 5th root of 32, follow these steps:
   (i) Express 1/5 as a decimal (0.2).
   (ii) Enter 32, press $[y^x]$, then press 0.2, and if necessary [=]. Check that your answer is 2.

3. Since different calculators have slightly different procedures, you may have to modify the steps outlined here.

Consolidation exercise 9B

1. When should the geometric mean be used? What is the geometric mean of the two numbers 3 and 12?

2. How is a growth of 3% written for use in calculating a geometric mean? How is a growth of 12% written? What is the geometric mean of (a) the two growth rates 3% and 12%? (b) 3, 8 and 9?

9.4 The root mean square

We now look at another mean that is used in later chapters.

Example 9.6 is not difficult. Three squares have sides of 1, 2 and 3 cm. Thus their areas are 1, 4 and 9 cm². The total area is therefore 14 cm², so the arithmetic mean of the three areas is $14/3 = 4.667$ cm². This is illustrated in Figure 9.1. A square of this area will have a side of $\sqrt{4.667} = 2.16$ cm. This value of 2.16 can be described as

> the square **root** of
> > the (arithmetic) **mean** of
> > > the **squares** of
> > > > the numbers 1, 2 and 3.

We therefore say that 2.16 is the **root mean square** of 1, 2 and 3. This form of average is used a great deal in some sciences and in statistical calculations.

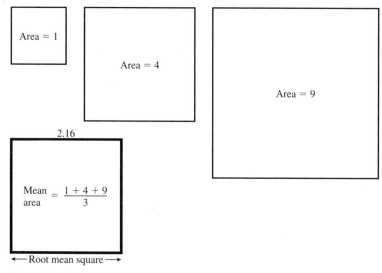

Area = 1

Area = 4

Area = 9

2.16

$\text{Mean area} = \dfrac{1 + 4 + 9}{3}$

←— Root mean square —→

Figure 9.1

Box 9.4

How to ... **find the root mean square**

1. This note does *not* describe how to find the root mean square *deviation*. This is the subject of a later note, which is related to this one.

2. To find the root mean square of a set of numbers, follow these steps:
   (i)   Square all the numbers – the results will all be positive or zero.
   (ii) Add the squares.
   (iii) Divide the total by the number of squares, remembering to count zero if this appears.
   (iv) Take the square root of the answer to (iii).

Consolidation exercise 9C

1. What is the root mean square? Find it for (a) 2, 3, 4, 5; (b) −4 , −3, −2, 1, 3, 5; (c) −4, −3, −2, 0, 1, 3, 5.

9.5 **The weighted arithmetic mean**

Example 9.7 introduces a common problem that is easily solved. To illustrate it in a simple case, suppose that three pupils have marks in four subjects as shown in Table 9.3. The three ordinary arithmetic means place Alec as top of the three. His low mark in mathematics has been compensated by his high marks in art and geography. On the other hand, Cheryl's high mark in mathematics has been insufficient to make up for her poor performance in art. The headmistress decides to attach twice as much importance to mathematics as to other subjects, and she does this by multiplying all the mathematics marks by 2 (Table 9.4). Then, when it comes to working out the mean, she has to remember that mathematics has been

Table 9.3

	Alec	Ben	Cheryl
Maths	30	60	80
English	60	50	60
Art	90	60	40
Geography	80	50	60
Total	260	220	240
Total/4	65	55	60

introduced as a double subject, so she has to divide by 5 instead of by 4. The result shown in Table 9.3 is that Cheryl has the highest 'weighted' arithmetic mean. The original information has been adjusted by giving mathematics an importance, or **weight**, of 2.

Table 9.4

	Alec	Ben	Cheryl
Maths × 2	60	120	160
English	60	50	60
Art	90	60	40
Geography	80	50	60
Total	290	280	320
Total/5	58	56	64

When the geography teacher sees what has been done, she says that if mathematics is being give a weight of 2, then geography should have a weight of $1\frac{1}{2}$. The headmistress agrees, and as the calculation is becoming complicated, she sets it out a little differently (Table 9.5). The entries in the weighted mark column are all obtained by multiplying the mark by the subject weight. The total of the weights represents the new 'number of subjects' with one subject being counted twice and another $1\frac{1}{2}$ times. The weighted mean is obtained by dividing the total of the weighted marks by the total of the weights. We see that Cheryl still has the highest weighted arithmetic mean.

Here, as with other weighted means, the difficult part of the procedure is deciding on what weights to use. It is usually simplest to give the least important value a weight of 1 and to work upwards (Box 9.4). However, it is equally permissible for you to give any weights, more or less than 1, as long as they represent your idea of the relevant importances. For example, mathematics, English, art and geography could be given weights of 2, 1.5, 0.8, and 1.3.

Table 9.5

	Weight	Alec		Ben		Cheryl	
		Mark	Weighted mark	Mark	Weighted mark	Mark	Weighted mark
Maths	2	30	60	60	120	80	160
English	1	60	60	50	50	60	60
Art	1	90	90	60	60	40	40
Geography	1.5	80	120	50	75	60	90
Total	5.5		330		305		350
Total/5.5			60		55.5		63.6

Box 9.5

How to ... **find a weighted arithmetic mean**

1. We want the arithmetic mean of several values of x but we do not want to give all the values the same importance (or weight) in the calculation.
2. Giving the least important value of x a weight of 1, choose what weights (denoted by various values of w equal to or greater than 1) to give the other values.
3. Multiply every x by its weight w, so now we have for every x a product wx.
4. Sum these products.
5. Add all the values of w, not forgetting to include 1 when it occurs.
6. Divide the total in step 4 by the total in step 5.

9.6 Other weighted means

The idea of weighting is not restricted to the arithmetic mean. We give two examples.

The harmonic mean of 4, 6 and 9

The harmonic mean of 4, 6 and 9 was worked out earlier. It is 5.684. Suppose we want to give the value 9 a weight of 3 but leave the other values with unit weight. We have to find H where

$$\frac{1}{H} = \frac{1}{5}\left[\frac{1}{4}+\frac{1}{6}+\frac{1}{9}+\frac{1}{9}+\frac{1}{9}\right] = \frac{1}{5}\left[\frac{1}{4}+\frac{1}{6}+\frac{3}{9}\right]$$

giving $H = 6.667$.

Note the use of 5 outside the bracket (as there are effectively 5 terms inside it, even when three of them are combined, as in the second bracket of the solution).

If we had wanted to give 9 a weight of 3.5, we would have had

$$\frac{1}{H} = \frac{1}{5.5}\left[\frac{1}{4}+\frac{1}{6}+\frac{3.5}{9}\right]$$

giving $H = 6.828$.

Box 9.6

How to ... **find a weighted harmonic mean**

1. The basic procedure is given in Box 9.1.
2. If there are n values and one of them, a, is to be given a weight of w then the sum of the reciprocals contains w/a instead of $1/a$, and instead of n we use $n + w - 1$.
3. If two (or more) values are weighted, we proceed similarly, and instead of n we use $n + (w_1 - 1) + (w_2 - 1) + \ldots$

The geometric mean of 3, 5, 6, 7

The geometric mean of the four numbers 3, 5, 6, 7 is

$$\sqrt[4]{3 \times 5 \times 6 \times 7} = \sqrt[4]{630} = 5.01$$

This has been worked out by using Box 9.3. If we want to give 6 a weight of 4, we have to work out the product

$$3 \times 5 \times 6 \times 6 \times 6 \times 6 \times 7 = 3 \times 5 \times 6^4 \times 7 = 136\,080$$

Then, since we have multiplied 7 numbers to get this, we have to take the seventh root and get

$$\sqrt[7]{136\,080} = 5.41$$

Note that in this calculation $7 = 1 + 1 + 4 + 1$, which is the sum of the weights.

We can do a rough check on the result. The unweighted mean gives all the numbers 3, 5, 6, 7 the same importance. It turns out to be 5.01. We have decided to give more importance to 6, which is bigger than this. We would therefore expect the mean to be pulled up above 5.01. On the other hand, we would not expect it to be as high as 6, because although there are many 6s, their product is pulled down more by the low values of 3 and 5 than it is pulled up by the single high value of 7. The calculated mean of 5.41 looks sensible, being above 5.01 but less than 6.

If we also wish to weight other values of x, we treat them similarly but we take a 'higher' root. Note that it is not necessary for the weights to be integers.

Box 9.7

How to ... **find a weighted geometric mean**

1. The basic procedure is as in Box 9.2.

2. If there are n values and one of them, a, is to be given a weight of w, the product of values contains a^w instead of a, and we take the $(n + w - 1)$th root instead of the nth root.

3. If two (or more) values are weighted, we proceed similarly but replace n with $n + (w_1 - 1) + (w_2 - 1) + \ldots$.

Consolidation exercise 9D

1. What is a weighted mean? What do the weights reflect?

2. What is the unweighted arithmetic mean of 2, 3, 4, 5, 6?

3. Say, without calculating it, whether the weighted arithmetic mean of 2, 3 (with a weight of 1.7), 4, 5 and 6 would be higher or lower than the mean you calculated in Question 2.

4. What is the harmonic mean of 1, 2 and 3 (a) unweighted, (b) if 3 has a weight of 4? What are the geometric means for (a) and (b)?

9.7 Index numbers

Index numbers form the subject of many books. Here we just introduce the idea, illustrating it with price indices, before making a general comment.

Price indices

 In **Example 9.8** the 'average price of these vegetables' is of greater interest if it reflects the different importances of the individual vegetables to the shoppers of Statingham. Suppose that the total weekly purchases of potatoes divided by the number of shoppers is 8 kg. We shall say that the 'average Statingham shopper' buys this amount of potatoes.

Suppose also that the prices and amounts purchased in a week in 1990, and in a week in 1992, by the average Statingham shopper are shown in column 2 of Table 9.6. We can see that the total cost of a basket of vegetables bought in 1990 was £6.29 whereas the cost of a basket bought in 1992 was £7.01. This represents an increase of 11.4%.

However, the basket bought in 1992 had a different mix of vegetables. It had more potatoes and onions, but fewer cauliflowers and carrots. In comparing £7.01 with £6.29, we are not comparing like with like. The totals combine the effects of price changes and of quantity changes. If our interest is in what has happened to prices, we need to see how much *the same basket* of vegetables would cost in the two years.

We do this by performing the calculations summarised in the last column. Here the 1990 quantities are multiplied by the 1992 prices. Thus in column 3 the 1990 prices are weighted by the 1990 quantities, whereas in column 7 the same 1990 weights are applied to the 1992 prices. The column total of £6.84 is the cost of the 1990 basket at 1992 prices. It is 8.7% higher than the 1990 cost. ⟴

Table 9.6

	1990			1992			1992 cost of 1990 quantities (pence)
	Price	Quantity	Actual cost (pence)	Price	Quantity	Actual cost (pence)	
	p_{90} (1)	q_{90} (2)	$p_{90}q_{90}$ (3)	p_{92} (4)	q_{92} (5)	$p_{92}q_{92}$ (6)	$p_{92}q_{90}$ (7)
Cauliflower	60	1.2	72	70	1.1	77	84
Potatoes	40	8.0	320	48	9.0	432	384
Carrots	40	4.8	192	36	4.0	144	172.8
Onions	50	0.9	45	48	1.0	48	43.2
Total expenditure			629			701	684
			$(\sum p_{90}q_{90})$			$(\sum p_{92}q_{92})$	$(\sum p_{92}q_{90})$
Percentage of base year (1990)			100			111.4	108.7

> ## Table 9.7
>
> . . . .
>
> | | Percentage of 1990 cost |
> | --- | --- |
> | Cost of 1990 basket at 1990 prices £6.29 | 100 |
> | Cost of 1990 basket at 1991 prices £6.62 | 105.2 |
> | Cost of 1990 basket at 1992 prices £6.84 | 108.7 |
> | Cost of 1990 basket at 1993 prices £7.07 | 112.4 |

Suppose that we also have price data for 1991 and 1993, and suppose these data enable us to work out the 1991 and 1993 costs of the 1990 basket of vegetables. The results of the calculation can be presented in Table 9.7.

In the last column of this table, we express the cost in each year as a percentage of the cost in 1990. We call these percentages 'values of the **price index**' in that they show how prices of the same basket have moved over the years. The year for which the index stands at 100 is called the **base year**; all other values are calculated as percentages of the base year. The weights used in this index are the **base year quantities**.

If we denote the base year by the subscript 0, we can denote the prices in that year by p_0, there being as many prices as there are items in the basket (four in our example). The quantities bought in that year can be denoted by q_0. The total cost of the basket at base year prices will be $\sum p_0 q_0$ (which may be compared with $\sum p_{90} q_{90}$ in Table 9.6).

The cost of the same basket in some other year denoted by the subscript i will be $\sum p_i q_0$ (which may be compared with $\sum p_{92} q_{90}$ in Table 9.6).

The ratio of these two values, multiplied by 100, gives the value of the price index in the year i with base year weights:

$$\frac{\sum p_i q_0}{\sum p_0 q_0} \times 100$$

Notice that we could use a similar approach to compare the costs of buying the same basket of goods in different countries, rather than in different years.

General comment

Index numbers are not restricted to prices. There are well-known indices of wage rates, earnings, productivity, traffic and crime to give just a few examples. They all use basically the same formulae.

Almost all governments publish official price and other indices, and there are many non-government indices. Their usefulness depends on how the compilers of the indices have collected (i) price or other data connected with the subject of the index, and (ii) data about spending or other matters on which the weights have been based. These are matters that are likely to vary between countries, and within a country from time to time. If any index is being used as an important part of an argument, it is advisable to check carefully on its compilation.

There are many difficult questions about the appropriateness of the weights and the coverage of an index.

Consolidation exercise 9E

1. Of what is an index number a ratio?

2. What is meant by the base year for an index? What does a base year weighted price index show? What does a current year weighted price index show? What are the formulae for these indices?

Notebox 9.1

The average may not matter

The average is not the only factor in decision making, and it may even be irrelevant. For example, a purchaser of metal beams may be very concerned to know the guaranteed lowest breaking load that a beam can support, rather than the average breaking load.

Notebox 9.2

A mean caution

We have defined the geometric mean for two ratios as the square root of their product. It can be calculated for any set of numbers in the ratio scale (Notebox 33.3), but not for numbers in the interval or lower scales. Even when this is kept in mind, it is easy to slip into error. Rather than rely on rules that may be too complicated or too simple, it is better to go back to first principles, as the following example illustrates:

Example 9.9 The ratio of average house prices to average household earnings of purchasers in north Statingham is 4:1, while in south Statingham it is 5:1. What is the ratio for the whole of Statingham?

The ratio involves two averages. The first is

$$\text{Average house price} = \frac{\text{Total paid for houses}}{\text{Number of houses bought}} = \frac{\text{TP}}{\text{NB}}$$

The second average is

$$\text{Average earnings} = \frac{\text{Total household incomes of purchasers}}{\text{Number of purchasing households}} = \frac{\text{TI}}{\text{NH}}$$

If we make the reasonable assumption that the number of purchasing households equals the number of houses bought, these two quantities cancel out when we divide them, and their ratio is

$$\frac{\text{TP}}{\text{NB}} \div \frac{\text{TI}}{\text{NH}} = \frac{\text{TP}}{\text{TI}} = \frac{\text{Total paid for houses}}{\text{Total household income of purchasers}}$$

⮕

We are told that in the north this ratio is 4:1, whereas in the south it is 5:1.

To get the ratio for the whole of Statingham, we need these two totals for the whole city. Suppose that, on making enquiries, we get the information in the first two columns of Table 9.8.

If we add the entries for the first two rows, we get that for the town as a whole the total paid is 28 (as in the last column) and the total income is 6. Dividing them gives the ratio of 14:3 (or 4.67). You can verify that for 4:1 and 5:1 the arithmetic mean is 4.5, the harmonic mean is 4.444 and the geometric mean is 4.472. Thus the average we have obtained from first principles is not given by any of these three means. (It is in fact a weighted arithmetic mean. What are the weights?)

Table 9.8

	North	South	Town
Total paid for houses (£m)	8	20	28
Total income of purchasers (£m)	2	4	6
Ratio	4:1	5:1	14:3

Notebox 9.3

Laspeyre and Paasche

Base year weighted price index numbers are known as Laspeyre price indices. Another index, known as the Paasche index, compares the cost of buying the current basket (with year i prices and quantities) with what the current basket (with year i quantities) would have cost in the base year at base year prices. This is given by

$$\frac{\sum p_i q_i}{\sum p_0 q_i} \times 100$$

There are other forms of price index.

Summary

1. When we want the average value of a variable that is defined as the quotient of two other variables, we may run into problems if we use the arithmetic mean. We may need to use the harmonic mean, defined by equation [9.1]. In Box 9.1 we summarise when and how to calculate it. Table 9.1 provides a useful quick reference.

2. Averaging ratios is sometimes tricky, as demonstrated in Notebox 9.3. it is always best to work from first principles, unless you are certain of what to do. Ratios derived from growth rates (but not the growth rates themselves) should be averaged by use of the geometric mean. The procedure is summarised in Box 9.2, and Box 9.3 describes a calculator technique that assists with its calculation.

3. Sometimes the root mean square is used. The procedure is summarised in Box 9.4. It is the square root of the arithmetic mean of the squares of the values being considered.

4. When we want to attach more importance to one item than to another, we weight the average in the direction of that item. More generally, we may attach different weights to the individual items, so that each item influences the mean proportionally to its weight. Boxes 9.5, 9.6 and 9.7 summarise the procedures for finding weighted arithmetic, harmonic and geometric means.

5. Index numbers are ratios of two weighted sums that have a common element.

6. A price index shows how much a basket of items costs in year i compared with its cost in year 0, which is called the base year.

7. The quantities in the basket (which are the weights) may be held constant at all times, so changes in the index I_i are due entirely to changes in prices:

$$I_i = \frac{\sum p_i q_0}{\sum p_0 q_0} \times 100$$

or the quantities may vary from year to year, as in

$$I_i = \frac{\sum p_i q_i}{\sum p_0 q_i} \times 100$$

so that in every year it compares the cost of the current basket with what the current basket would have cost in the base year.

8. Similar remarks apply to other indices. The basis of comparison can be spatial rather than temporal. We comment on some problems.

9. Noteboxes deal with (1) whether the average always matters, (2) the need for care when dealing with ratios and (3) Laspeyre and Paasche index numbers.

10

What about the width?

Introduction

Example 10.1 Twice a month Harry drives a small van to a secret location from which he collects bottles of Statisticians' Strength. He knows two routes quite well and has found that route A takes anything between 150 and 165 minutes with an arithmetic mean of 159 minutes. Route B takes between 148 and 188 minutes with an arithmetic mean of 155 minutes. He wants to choose one route. Which should it be?

Example 10.2 Harry also stocks imported alcohol-free beer, but only one brand at a time. When he stocks Kamela the daily sales average 14 bottles and vary between 12 and 18. When he stocks Pantha the daily sales average 17 bottles, ranging between 16 and 25. Which should he choose?

Example 10.3 Lady Agatha knows that the degree of discomfort arising from sitting on a seat of the wrong height depends on how wrong it is. She recognises that the requirements of the Ladies of Upper Sternum and the Working Women of Statingham could just possibly have the same mean, median and mode. Yet they could still be very different because one set of requirements would be more spread out than the other, and so imply a greater total discomfort. She wonders how to measure this spread.

Example 10.4 Steve wants to summarise his information about the ages of his neighbours in a way that will take account of its variety.

In the last few chapters we have been considering averages, which are measures of location. Now we look at measures of dispersion, which tell us how the values are spread. These are needed both for description of the data and for decision making. Here we look at some of the more important measures and show their relevance to a few problems. The most important – the standard deviation – is introduced in the next chapter.

10.2 **The range**

The simplest measure of spread, called the **range**, is the difference between the highest and the lowest value. For example, the range of the numbers 3, 9, 12, 10,

15, 22 is $22 - 3 = 19$. This is also the range of 3, 3, 4, 17, 22, 22 and of 7120, 7138, 7139, 7139, 7139, 7139

Although the range gives some idea of spread, it is based simply on the two extreme values and gives no idea of what is happening in between. Some idea is available if we also know an average, and the two together tell us more, but even then important points may be unrevealed. For example, the first two sets of numbers in the previous paragraph happen to have identical arithmetic means, as well as ranges, but they are very different.

In looking at Examples 10.1 and 10.2, we try to make the best use of the data, rather than restrict ourselves simply to the mean and the range. Although the range adds to our knowledge, we cannot make a good decision without asking other questions, and perhaps obtaining more information.

⮕ In **Example 10.1** the information available to Harry about travel times shows that route A has a higher mean travel time than route B (159 minutes compared with 155 minutes, a difference of 4 minutes), but a smaller range (15 minutes compared with 40 minutes). Thus the time taken on route A is *more predictable* than the time on route B.

Harry also compares the *worst that can happen* on route A with that worst that can happen on route B. Route A has never taken more than 165 minutes, whereas route B has taken 188 minutes at least once. True, the fastest route A trip took 2 minutes more than the fastest route B trip, but Harry looks upon that as trivial. So is an average saving of 4 minutes in a trip that lasts between $2\frac{1}{2}$ and 3 hours.

He decides in favour of route A. Here Harry is influenced not only by the differences in ranges, which indicate different degrees of predictability, but also by the fact that the longest route B trip took so much more time than the longest route A trip.

⮕ **Example 10.2** is a little more complicated. Making a choice between the two brands should involve much more information than is given. Harry knows that he needs to place his orders, and to arrange his shelf space, on the basis of the information he has, which includes a possible daily variation of as many as $18 - 12 = 6$ bottles of Kamela and $25 - 16 = 9$ bottles of Pantha.

He needs to consider prices, profit margins, suppliers' packaging, delivery and many other matters, including shelf space. He always tries to have enough stock on his behind-counter shelves to last for the current day. He will need to set aside less space if he sells Kamela. Also, at the end of the day, he is not as likely to have as many bottles left over as he would when selling Pantha.

But that means having lower average sales of alcohol-free beer. Possibly some of his 'lost' customers will choose some other non-alcoholic drink. He also has a little more space on his shelves that, if wisely used, may increase his sales of something else. Of course, if he always had plenty of space, these considerations would not arise.

Thus in both examples a knowledge of the range is useful, but it needs to be supplemented by other information.

10.3   Quartiles and the quartile deviation

Quartiles: ungrouped data

We have already seen that if the values of the variable are arranged in ascending or descending order, the middle value is called the median. It divides the total frequency into halves. We can also divide it into quarters, which helps us to see how the values are spread around the median.

The values of the variable that do this are called quartiles. The **first quartile** (Q_1) has one quarter of the total frequency less than or equal to it. The second quartile has half the frequency less than or equal to it, and is therefore the median. The **third quartile** (Q_3) has three quarters of the total frequency less than or equal to it.

If the number of items n is small, it may be dificult to decide which values divide them into quarters. But for larger numbers we can make the following definitions:

Q_1 is defined to be the $\frac{1}{4}(n+1)$th value

Q_3 is defined to be the $\frac{3}{4}(n+1)$th value

Thus for 39 values the first quartile is the $\frac{1}{4}(39+1)$th $=$ 10th value and the third quartile is the 30th value. More often than not the formula results in fractions, presenting the kind of minor problem that Winnie and Lady Agatha face below.

> **Example 10.3** uses the data in Table 3.2 about 60 seat heights. The first quartile will correspond to position $\frac{1}{4}(60+1) = 15\frac{1}{4}$ when the data are arranged in ascending order. At this stage some statisticians take the 15th value as the first quartile. Others, including ourselves, define the quartile to be one-quarter of the way between the 15th and the 16th values.
>
> Counting along the values, we find that the 15th value is $68 + 2.8 = 70.8$. The 16th value is $68 + 3.1 = 71.1$. The first quartile is one-quarter of the way between them, which is 70.9 to one decimal place.
>
> Similarly the third quartile will be the number corresponding to position $\frac{3}{4}(60+1) = 45\frac{3}{4}$. Some take this to be the number in the 46th position (which is 74.5). Others take it to be three-quarters of the way between the 45th number (which is 74.4) and the 46th (which is 74.5). These numbers are so close that, to one decimal place, both approaches give the same answer as the third quartile.
>
> Thus, although the procedure is not entirely satisfactory, we can at least say that one-quarter of the ladies need seat heights less than or equal to 70.9 cm and three-quarters less than or equal to 74.5 cm, which also means that one-quarter want seats as high as or higher than this 74.5 cm.
>
> Putting it another way, the middle half of the ladies want seats between 70.9 and 74.5 cm high. We are beginning to look at how the desirable heights are spread out.

Box 10.1

How to ... **find the first and third quartiles for ungrouped data**

1. Arrange the values in ascending order.

2. If there are n items, the first quartile will be the value in position $\frac{1}{4}(n+1)$ and the third quartile the value in position $\frac{3}{4}(n+1)$.

3. If the calculated position given by step 2 involves a fraction, the quartile lies between two consecutive values, a (which will be in the position just below the calculated postion) and b (which will be in the position just above it). Depending on the fraction, the quartile can be calculated as the value lying one-quarter, one-half or three-quarters of the way between a and b. Some statisticians round up or down to the nearer of a and b.

Quartiles: grouped data

The definitions are just the same for grouped data. Now, however, the fractions present no problem and enable us to be more precise, provided we accept an assumption. We show how to proceed by finding the first and third quartile ages of Steve's potential neighbours.

We return to Example 6.4, showing the information again in Table 10.1, where there is also a column showing the cumulative frequency. It is also shown as a histogram in Figure 10.1. Since there are 301 people, the **first quartile age** corresponds to the age of the person in position $\frac{1}{4}(301+1) = 75\frac{1}{2}$. The **third quartile age** corresponds to person $226\frac{1}{2}$.

From the cumulative frequency column we see there are 62 persons aged up to almost 15. In the next group are 18 persons aged 15–19. Taking age to be a continuous variable whose values are normally rounded down, so that somebody recorded as 19 may be almost 20, we *assume* that the 18 persons aged 15–19 have ages that are evenly spread over this five-year span at intervals of 5/18 years. Thus the ages of the 18 persons will be as shown in Table 10.2.

We have argued that Q_1 corresponds to person $75\frac{1}{2}$. This person will have an age midway between those of the 75th and 76th persons. We can calculate it by considering the Table 10.2 or by using the formula

$$Q_1 = \begin{bmatrix} \text{Lower bound of group} \\ \text{with } \frac{1}{4}(n+1) \end{bmatrix} + \text{Group width} \times \begin{bmatrix} \text{Position of } \frac{1}{4}(n+1)\text{th} \\ \text{item in group} \\ \hline \text{Number of items in group} \end{bmatrix}$$

$$= \qquad 15 \qquad + \quad 5 \quad \times \quad (75\tfrac{1}{2} - 62)/18$$

$$= 18.75 \text{ years}$$

Thus one-quarter of Steve's neighbours are aged under 18.75 years. A similar calculation shows that one-quarter of them are aged over 56.15 years, which locates Q_3. (In checking this, note that the group containing Q_3 is 10 years wide.) ⬭

Table 10.1

Age, x	Frequency, f	Cumulative frequency	
0–4	20	20	
5–9	22	42	
10–14	20	62	aged up to almost 15
15–19	18	80	person $75\frac{1}{2}$ is in this group
20–29	35	115	
30–39	38	153	
40–49	44	197	aged up to almost 50
50–59	48	245	person $226\frac{1}{2}$ is in this group
60–64	20	265	
65–69	16	281	
70–79	16	297	
80+	4	301	

Table 10.2

Person	63	64	65	66	...	79	80
Position in group	1	2	3	4	...	17	18
Age	$15\frac{5}{18}$	$15\frac{10}{18}$	$15\frac{15}{18}$	$16\frac{2}{18}$	...	$19\frac{13}{18}$	20

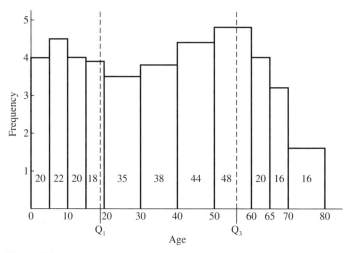

Figure 10.1

It follows that half of Steve's neighbours are aged between 18.75 and 56.15. This is illustrated in Figure 10.1, where half the area of the histogram lies between these values, indicated by Q_1 and Q_3.

Obviously the precision of this answer, far greater than the precision of the data, depends on the accuracy of our assumption that ages are evenly spread within a group. If the number of people in the group is large, this is more likely to be true than if it is small, but even then the truth may be quite different. Moreover, as in calculating the median, we need to think about what answers are possible, which can be a tricky question. The important point is that calculated quartiles should always be treated with caution, even when it is possible to check on the plausibility of the assumption.

Note that if, instead of ages, our data had been lengths of rods measured to the nearest centimetre, the lower bound of the group labelled 15–19 would not have been 15, it would have been 14.5. The calculation would have to be modified accordingly.

The quartile deviation or semi-interquartile range

We have seen that knowledge of the quartiles gives some idea of spread. The difference between the first and third quartiles is used as a measure of dispersion. The **semi-interquartile range** is defined to be half this difference and is given by

$$Q = \frac{Q_3 - Q_1}{2}$$ [10.1]

Since, by definition, half the total frequency must lie between the first and third medians, the semi-interquartile range is half the 'distance' that contains the middle half of the whole frequency. For Table 10.1 it is $\frac{1}{2}(56.15 - 18.75) = 18.7$ years.

Sometimes this measure is called the **quartile deviation.** It can be looked upon as the mean of the deviations of the first and third quartiles from the median. Other statisticians use the **interquartile range**, which is simply the difference between the first and third quartiles, without division by 2. We shall use the term *quartile deviation* because it is shorter and well understood. Our procedures are summarised in Box 10.2.

Box 10.2

How to ...   **find the quartiles and the quartile deviation for grouped data**

1. Arrange the data as for the median.

2. If there are n values, choose the $\frac{1}{4}(n+1)$th value as the first quartile.

3. Find the group that contains the first quartile.

4. Let L be the lowest possible value in this group.

5. Denote the 'width' (class interval) of the group by c (5 in the example). ⬤

6. Let the total frequency be n (301 in the example).

7. Let the total of all the frequencies in groups lower than the quartile group be f_l (62 in the example).

8. Let the frequency in the quartile group be f_q (18 in the example).

9. The first quartile is

$$Q_1 = L + \frac{c\left[\frac{1}{4}(n+1) - f_l\right]}{f_q} \qquad\qquad [10.2]$$

10. The third quartile Q_3 is found in exactly the same way, with the constants in the equation defined relative to the third quartile rather than the first. If there are n items, the third quartile corresponds to item $\frac{3}{4}(n+1)$.

11. Note that the first and third quartiles can also be found graphically as decribed in Notebox 10.1.

12. To obtain the quartile deviation, first calculate Q_1 and Q_3 as above then take half their difference.

Consolidation exercise 10A

1. What is the range of (a) 3, 3, 85, 89; (b) 3, 8, 6, −5, 4?

2. How is the first quartile defined?

3. What is another name for (a) the second quartile and (b) the quartile deviation? What is the formula for the quartile deviation?

10.4 **Box and whisker diagrams**

Sometimes the spread of the data is summarised in a simple diagram that portrays the range (as the difference between upper and lower limits), the median and first and third quartiles. They are particularly useful when comparing the spread of two sets of data. In that case they are drawn to the same scale and positioned along, or above and below, a common axis. Figure 10.2 shows the information in Table 10.3.

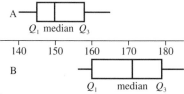

Figure 10.2

Table 10.3

Heights (cm) of pupils in two classes

	Class A	Class B	
Lower limit (l)	140	156	at end of left whisker
First quartile	144	160	at beginning of box
Median	150	172	dividing the box
Third quartile	158	179	at end of box
Upper limit (U)	165	185	at end of right whisker

The information for A is represented by a box stretching from the first quartile to the third quartile. It is divided in two by a vertical line through the median. Two whiskers stretch left to the lowest value and right to the highest. Thus the distance between the remote ends of the whiskers is the range, and the length of the box shows the separation of the quartiles.

If we are comparing A and B, it is obviously much easier to use the diagram than the table when picking out the important differences between the average heights (as measured by medians) and the spread of heights around these averages.

These five values are sometimes presented as a **five-figure summary**, witten as (140, 144, 150, 158, 165) or (156, 160, 172, 179, 185) without further labelling.

Box 10.3

How to ... **draw a box and whisker diagram**

1. Draw a horizontal scale and label it from less than the lowest value to more than the highest value.

2. On the axis itself or a line parallel to it, mark the lowest and highest values, which will be the outermost ends of the whiskers.

3. Mark the first and third quartiles, and draw a box with ends at these points.

4. Divide the box in two by a vertical line through the median.

5. Draw two horizontal whiskers from the box to the lowest and highest values.

10.5 Deciles and percentiles

Quartiles divide the ordered data into four quarters. If we divide the data into tenths, the successive dividers are called the first **decile**, the second decile, and so on. If we divide the data into hundredths we call the dividers **percentiles**. The tenth percentile (denoted by P_{10}) is the first decile; the ninetieth percentile (P_{90}) is the ninth decile. Sometimes spread is described by a specified interpercentile range, such as the **10–90 interpercentile range**, defined to be $P_{90} - P_{10}$.

10.6 The mean deviation

A different kind of measure of spread, or dispersion, is the mean deviation; it is not used a great deal but should be known. It is the average of the amounts (all given a positive sign) by which individual items differ (or deviate) from their arithmetic mean. Calculating it therefore requires a knowledge of the arithmetic mean μ, unlike the measures discussed above.

More formally, the **mean deviation** is the arithmetic mean of the absolute differences between the individual items and their arithmetic mean. It is given by

$$MD = \frac{\sum |x - \mu|}{n}$$

where μ is the arithmetic mean of the values of x and all the (unsigned) values of $|x - \mu|$ – one for each x – are summed.

Box 10.4

How to ... find the mean deviation

1. Calculate the arithmetic mean, μ.

2. For each value of x calculate the *deviation*, which is the numerical value (without regard to sign) of $x - \mu$, denoted by $|x - \mu|$.

3. Sum these deviations, to get $\Sigma|x - \mu|$.

4. Then divide this by the number n of xs.

5. If the data are in a frequency distribution, with the value x_i occurring f_i times, multiply each deviation by its frequency before summing. The formula is then

$$MD = \frac{\sum_{i=1}^{n} f_i|x_i - \mu|}{n}$$ [10.4]

Unlike the other measures of spread we have discussed, the mean deviation is based on all the values of the variable. It is not much used, partly because the need to get around the minus signs complicates the calculation and makes the measure unsuitable for more advanced work. It is supplanted by the **standard deviation** or its square, called the **variance**, which is less easy to calculate and requires a little more explanation. One advantage of the standard deviation is that it gets around the problem of minus signs. But it also has many other advantages and lies at the heart of statistical analysis and hypothesis testing.

Consolidation exercise 10B

1. What are deciles? How many are there? What are other names for the 50th percentile, the 75th percentile and the 90th percentile?

2. What is shown in a box and whisker diagram?

3. A distribution of 1200 values is summarised by (48, 129, 240, 290, 321). What are the numbers in the bracket?

4. How is the mean deviation defined? Can it be negative?

Notebox 10.1

Graphical derivation of quartiles

1. The quartiles, including the median, can easily be obtained from a cumulative frequency curve (Figure 10.3). Note that in Figure 10.3 the ordinate at 40 has a height representing the total number of people with ages up to the fortieth birthday.
   (i) On the vertical axis mark the points corresponding to $\frac{1}{4}(n+1)$, $\frac{1}{2}(n+1)$ and $\frac{3}{4}(n+1)$.
   (ii) Draw horizontal lines through these points to cut the cumulative frequency curve. Label the points on the curve P_1, P_2 and P_3 respectively.
   (iii) Draw vertical lines downwards through P_1, P_2 and P_3 to cut the x-axis. The points where these lines cut the x-axis give the values of Q_1, Q_2 (= the median) and Q_3.

2. Quartiles can also be derived from a histogram by an obvious adaptation of the method described in Box 8.3 for deriving the median.

Notebox 10.2

A quartile measure of skewness

The quartiles (including the median, which is the second quartile) are used in one measure of skewness:

$$\text{Quartile coefficient of skewness} = \frac{(Q_3 - Q_2) - (Q_2 - Q_1)}{Q_3 - Q_1}$$

If $(Q_3 - Q_2)$ is large, the quarter of the frequency contained within it is stretched out; if $(Q_3 - Q_2)$ is small, the quarter of the frequency is tightly packed. The same goes for $(Q_2 - Q_1)$. The difference between $(Q_3 - Q_2)$ and $(Q_2 - Q_1)$ shows the imbalance of packing on the two sides of the median. Dividing it by $(Q_3 - Q_1)$ gives this as a fraction of a measure of spread.

A positive value for this coefficient means that the distribution is skewed to the right. Another way of writing it is

$$\frac{Q_3 - 2Q_2 + Q_1}{Q_3 - Q_1}$$

⟹

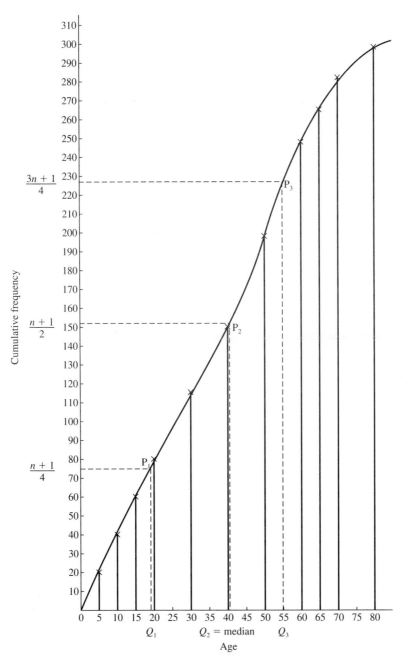

Figure 10.3

Sometimes people use the **percentile coefficient of skewness**, which can be obtained from either of the above expressions by writing P_{90} in place of Q_3, P_{50} in place of Q_2 and P_{10} in place of Q_1. Other measures of skewness are described in Notebox 12.2.

Summary

1. To describe a distribution we need to use a measure of its spread, or dispersion. Often a knowledge of this is essential to good decision making. We have introduced the simpler measures.

2. The range is the difference between the highest and lowest values (Box 10.1).

3. The lower (or first) and upper (or third) quartiles are values of the variable that respectively cut off the lowest quarter of the the population (or the total number of items), and the highest quarter. The second quartile is the median. Techniques for finding them are discussed in the text and summarised in Boxes 10.1 (ungrouped) and 10.2 (grouped).

4. Half the difference between the first and third quartiles defines the quartile deviation (or semi-interquartile range). Sometimes the interquartile range (of the whole difference between the quartiles) is used.

5. The quartiles, median and range are often summarised graphically in a box and whisker diagram Box (10.3) or a five figure summary. Deciles and percentiles are similarly defined in terms of tenths or hundredths of the total population.

6. Calculation of the mean deviation, which is not widely used, is described in Box 10.4. Two other measures of spread, the standard deviation and the variance, are the subjects of later chapters.

7. Noteboxes consider (1) graphical derivation of the quartiles and (2) a quartile measure of skewness.

The standard deviation: part 1

11.1 Introduction

⟹ **Example 11.1** Steve looks at his information about the sizes of the 291 families who become his neighbours (Example 6.3) and decides to measure the spread of family size by calculating the standard deviation.

⟹ **Example 11.2** Steve still wants to know how the ages of his neighbours are dispersed. He knows there can be large changes in their values and pattern without any effect on the range and quartile deviation. One advantage of the mean deviation is that it depends on every value, but Steve sees problems in calculating and using it. He seeks a measure of dispersion that depends on all the values.

In the last chapter we discussed various measures of spread, ending with the mean deviation, which was complicated by the need to ignore negative signs. We now come to a measure that overcomes the sign problem, but its importance goes beyond that. Before describing the logic of the standard deviation and how to calculate it, we anticipate a little to indicate, albeit inadequately, one of its powerful uses. Then we show how to find it in the simpler cases. Labour-saving techniques are discussed in Appendix B, and the next chapter deals with more complicated cases.

11.2 A preview of the normal distribution

Suppose that people watching a certain TV programme are invited to guess as accurately as they can the weight of the presenter, and then to phone in their guesses. As we discuss more fully in Chapters 19 and 20, it is very likely that if many people respond, the various guesses will have a symmetrical bell-shaped distribution (Figure 11.1). Because the distribution is symmetrical, the arithmetic mean of the guesses, the median and the mode will all have the same value, denoted here by μ.

Other examples of its likely occurrence, at least approximately, are

- the carefully measured lengths of rods cut by one person (or one machine) and intended all to be the same;

- the numbers of heads obtained in each of many successive throws of 1000 identical unbiased coins;
- marks gained by a very large number of candidates all sitting the same well-conducted examination.

In all these cases we can expect the distribution to have the shape shown, but it is not just any bell-shaped curve. It is a very special one, called the Normal distribution.

As with all distributions, one way of measuring the spread is to calculate the standard deviation. This is a bit like the mean deviation but a trick gets rid of the minus sign. We define it shortly. It can be calculated by using a formula or by pressing the right buttons on a pocket calculator. Conventionally we denote the standard deviation for a population by the small Greek sigma, σ, but for sample data we denote it by s; it may also be denoted by s.d.

Some idea of the usefulness of the standard deviation is given by the fact that, in all of the examples just mentioned, the following statements are true, at least to a very good approximation. We discuss them more fully in Chapters 19 and 20.

- A span of width σ on either side of the mean will contain 68.26% of the total population. In other words, 68.26% – just over two-thirds – of all guessed masses, measured lengths, etc., will have values lying within σ of the arithmetic mean.
- A span of width 2σ on either side of the mean will contain 95.44% of the total population.
- A span of width 3σ on either side of the mean will contain 99.7% of the total population – almost all of it.

These statements are illustrated in Figure 11.1. Whenever there is a Normal distribution, if we know the mean and standard deviation, we can repeat these statements and many similar ones. If the distribution is exactly Normal, the statements are exactly true, but if (as is more likely) the distribution is only approximately Normal, these statements will still be approximately true.

These results allow us later to develop ideas that are central to a great deal of statistical theory and investigation, especially assessment of the reliability of estimates based on samples. We have used 'Normal' with an N to emphasise that the Normal distribution is a special distribution of wide occurrence, at least

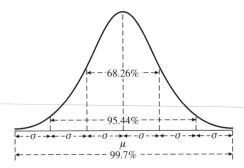

Figure 11.1

approximately. It would, however, be wrong to suggest that normally a distribution is Normal.

11.3 Defining the standard deviation

We must now consider the definition of the standard deviation, see the logic behind it, and learn how to calculate it. It is similar to the mean deviation but it is mathematically and practically more important. It makes use of the fact $-3 \times -3 = +9$, so that one way of getting rid of negative signs is to square the negative quantities.

Its calculation is helped by a formula given in equation [11.2], but we begin with a more cumbersome approach that helps to explain what it is all about.

The standard deviation of a set of numbers can be found by first calculating the arithmetic mean. Then we subtract this from every value of x to obtain a set of deviations from the mean. The calculation is finished by using Box 9.4 to find the root mean square of these deviations.

We now illustrate this, beginning with the set of numbers 1, 2, 4, 6 and 7, which have 4 as their arithmetic mean. They are repeated in Table 11.1 with an extra row. The entries in the second row have a total of 0 simply because they are all values of x that have had 4 subtracted from them, and 4 happens to be their arithmetic mean. Thus in row 2 every entry is a deviation from the mean, and the entries in row 3 are their squares.

We can call these squares the **squared deviations** (omitting 'from the mean' for brevity). We add them to get 26 then divide 26 by their number, 5. This gives the **mean squared deviation** of x. It is 5.2. The square root of 5.2 is 2.28, and 2.28 is the **root mean square deviation** of x. Another name for the root mean square deviation of x is the **standard deviation of** x.

The bigger the spread of the values of x, the bigger the squared deviations, hence the bigger the standard deviation. We can use the standard deviation to measure spread. Notebox 11.1 gives a diagrammatic interpretation.

The procedures just described for calculating the standard deviation can be summed up in a basic formula, which can be taken as its formal definition. The standard deviation is given by

$$\text{s.d.} = \sqrt{\frac{\sum (x - \mu)^2}{n}} \qquad\qquad [11.1]$$

Table 11.1

x	1	2	4	6	7	$(\sum x = 20)$
$x - 4$	-3	-2	0	2	3	$(\sum (x - 4) = 0)$
$(x - 4)^2$	9	4	0	4	9	$(\sum (x - 4)^2 = 26)$

In practice a formula derived from this and given shortly is simpler to use, but its meaning is less obvious.

Here there is an important point to be made. Remember that in these first chapters we are looking at populations, not samples. If we have information about 50 students then those 50 students form the population in which we are at present interested. We are not using them to make estimates for all students. This has to be stressed because some books introduce the standard deviation by talking about samples and define it with $n - 1$ where we have n. The formula we have given is right for the description of populations. We also use it for samples, but some other writers do not. The difference of opinion is summarised in Notebox 11.2 and considered more fully in Chapter 16, especially Notebox 16.2.

We take a simple numerical example of the use of equation [11.1]. Suppose we want the standard deviation of the six numbers in row 1. We begin by totalling them to get 84, then we divide 84 by 6 to get the arithmetic mean $\mu = 14$.

Table 11.2

x	10	12	14	15	16	17	$(\sum x = 84)$
$x - 14$	-4	-2	0	1	2	3	$(\sum (x - 14) = 0)$
$(x - 14)^2$	16	4	0	1	4	9	$(\sum (x - 14)^2 = 34)$

We use $\mu = 14$ in row 2 to get six values of $x - \mu$. As a check, we note that they add to zero (Table 11.2). We then square each value and write it in row 3. The total of these squares (all positive) is 34. Thus

$$\sum (x - \mu)^2 = 34$$

We use this in equation [11.1]. We also know that the number of items is $n = 6$. Thus from equation [11.1] we have

$$\text{s.d.} = \sigma = \sqrt{\frac{\sum (x - \mu)^2}{n}}$$

$$= \sqrt{34/6} = 2.38$$

Consolidation exercise 11A

1. How does the standard deviation differ from the mean deviation? How is it defined for a population?

2. Why is there a square root in the formula for the standard deviation?

3. We define the standard deviation for a sample in the same way as for a population. What different definition is used in some other books?

11.4 Calculating the standard deviation for ungrouped data

The trouble with using equation [11.1] is that if the mean value of x is not a round number, the calculations involve awkward figures. For example, if we had $\mu = 4.73$, then 4.73 would have to be taken from every value of x and the differences, each involving two decimals, would need to be squared, and so on. This is not a problem if you have a good calculator; all you have to do is to feed in the values of x then press a button that produces σ, as decribed in Appendix A. But even then it will be useful to know a few other versions of this formula, and a few tricks such as those in Appendix B.

Equation [11.1] can be written in a form that is easier for calculation:

$$\text{s.d.} = \sigma = \sqrt{\frac{\sum x^2}{n} - \left(\frac{\sum x}{n}\right)^2} \qquad [11.2]$$

If we use equation [11.2], Table 11.2 is replaced by Table 11.3, which has only two rows, one with the original data – values of x – and one with x^2. Inserting the totals in equation [11.2], along with $n = 6$, we get

$$\text{s.d.} = \sigma = \sqrt{(1210/6) - (84/6)^2} = 2.38$$

which agrees with our previous calculation.

Table 11.3

x	10	12	14	15	16	17	$(\sum x = 84)$
x^2	100	144	196	225	256	289	$(\sum x^2 = 1210)$

Coping with a frequency table

In Example 11.1 the data about family sizes appear as a frequency table. This contrasts with the example just considered, in which every value of x occurs just once.

We therefore have to calculate values of fx and of fx^2 as we show in Table 11.4. Then we have to sum them. In place of equation [11.2] we then have

$$\text{s.d.} = \sigma = \sqrt{\frac{\sum fx^2}{n} - \left(\frac{\sum fx}{n}\right)^2} \qquad [11.3]$$

where the summation terms are the sums of the columns containing values of fx and fx^2. The expression in parentheses is μ, the arithmetic mean of the family sizes.

Table 11.4

(1) Family size, x	(2) Number of families of this size, f	(3) = (1) × (2) Number of persons in these families, fx	(4) = (1) × (3) fx^2
1	30	30	30
2	65	130	260
3	40	120	360
4	55	220	880
5	50	250	1 250
6	30	180	1 080
7	12	84	588
8	5	40	320
9+	4	39	380*
Totals	$n = \sum f = 291$	$\sum fx = 1\,093$	$\sum fx^2 = 5\,148$

*The value 39 in column 3 implies that we have replaced 9+ in column 1 by 39/4 = 9.75. We have used this to calculate the entry in column 4 as 39 × 9.75.

The quantities to be calculated from these sums are

$$\left(\frac{\sum fx}{n}\right)^2 = \left(\frac{1093}{291}\right)^2 = 3.756^2 = 14.1076$$

$$\frac{\sum fx^2}{n} = \frac{5148}{291} = 17.6907$$

Inserting these values in equation [11.3] gives

$$\text{s.d.} = \sigma = \sqrt{17.6907 - 14.1076} = \sqrt{3.583} = 1.893$$

Note that $\sum fx / \sum f$ is the arithmetic mean of the family sizes, which is therefore 3.756 persons. The standard deviation of the family sizes is 1.893 persons.

We can see from the table (or from a histogram based on it) that the frequency distribution is bimodal and skewed to the left, so simple checks are not all that reliable. However, as a rough check, we may base some guesses on the results already stated for the normal distribution. These suggest that, for *most* hump-shaped distributions we are likely to come across, we can expect roughly 95% of the total frequency to lie within twice the standard deviation on either side of the mean. This is more likely to be true for unimodal distributions than for bimodal distributions, so it probably does not apply in our case. However, even here it seems to work. The check implies that, for a roughly normal distribution with mean 3.756 and standard deviation 1.893, we can expect 95% of the families to have sizes between a lower limit of 3.756 − 3.786 (which we can take as 0) and an upper limit of 3.756 + 3.786 which we can take as 7 or 8).

The table shows that almost all of the 291 families have a size between 0 and 7 persons, which conforms to this expectation even though the curve is not normal. Although this does not prove that our calculations are correct, almost all the frequency lies within a range of twice the standard deviation on either side of the mean, and this fact adds some credibility to them, as we shall see later.

Box 11.1

How to ... find the standard deviation for ungrouped data

If you are not using a calculator then consult Appendix B or proceed as follows. If you are using a calculator then see Appendix A.

1. Set out the data in columns, as in Table 11.4

2. For each x calculate fx then fx^2, and enter them in the appropriate columns.

3. Total the columns to get $n = \sum f$, $\sum fx$ and $\sum fx^2$.

4. Use them to calculate

$$\frac{\sum fx^2}{n} \quad \text{and} \quad \left(\frac{\sum fx}{n}\right)^2$$

5. Substitute these values in

$$\text{s.d.} = \sigma = \sqrt{\frac{\sum fx^2}{n} - \left(\frac{\sum fx}{n}\right)^2} \qquad [11.3]$$

We are not yet able to calculate the standard deviation of the ages of Steve's neighbours since the data about them is grouped. We use essentially the same method, but it has to be adapted for grouping, and, as we shall see in the next chapter, a special correction may be needed.

11.5 The variance

If you do not take the square root when calculating the standard deviation, you have a quanity called the **variance**. It is often important in advanced statistical work:

$$\text{var} = \sigma^2 = \frac{\sum f(x - \mu)^2}{n} \qquad [11.4]$$

Although the standard deviation is in the same units of measurement as the original data, the variance is in the square of those units.

We can compare the spread of two distributions using the standard deviation or the variance. However, in some calculations it is essential to use the variance rather than the standard deviation.

Consolidation exercise 11B

1. Give a formula for the standard deviation that does not involve calculating $(x - \mu)^2$.

2. How is this formula modified if the data are in an ungrouped frequency table? How do you further modify it if you have an assumed mean?

3. What is the square of the standard deviation called?

Notebox 11.1

A diagram for the standard deviation

The five values of x given in Table 11.1 are shown in Figure 11.2 by five points P, Q, R, S and T, that are at different heights just for convenience of drawing. The values of x are the distances of these points from the vertical line Oy. The vertical line AB is drawn through the arithmetic mean of the xs, μ.

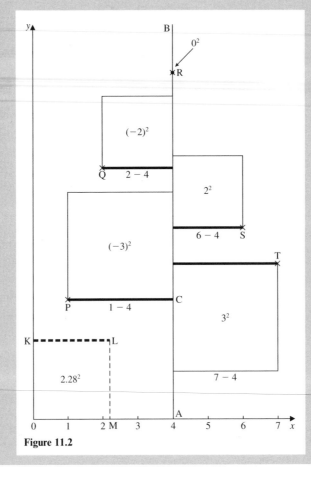

Figure 11.2

Draw five horizontal lines, one from each point to the line AB; line PC is an example. One of them is of zero length since the point R coincides with the line AB.

Now construct squares on each of these lines. These squares may be thought of as similar to the squares in Example 9.6. Each has an area given by the appropriate value of $(x - \mu)^2$. The square for $x = 4$ has zero area.

The areas total 26 square units. Their mean is calculated by dividing by 5 and getting 5.2, which is represented by the square KLMO. Its edge KL is the square root of its area, and is the standard deviation of the xs.

Notebox 11.2

Is it n or $n - 1$?

We often use sample material to estimate the mean and standard deviation for a population from which the sample is drawn. To explain why some books use n and others $n - 1$ in working out the standard deviation, we must mention a matter that is considered more fully in Chapter 16 and Notebox 16.2, which explains what is meant by a 'best' estimate. In this chapter we take the meaning on trust.

It can be shown that the best estimate of the population mean μ is the sample mean $\bar{x}$. But the best estimate of a population standard deviation σ is not the sample standard deviation s as defined by equation [11.1]. Instead, the best estimate of σ is

$$s\sqrt{n/(n-1)}$$

where s is given by the right-hand side of equation [11.1] and n is the sample size. By using this we quickly find that the best estimate works out to be

$$\sqrt{\frac{\sum (x - \bar{x})^2}{n - 1}} \qquad [11.5]$$

Because this provides a better estimate of the population standard deviation, this formula is sometimes used to define the standard deviation of the sample. We prefer to define it with n in the same way as we define the standard deviation of any set of numbers, using equation [11.1]; and to obtain the best estimate of σ (for the population) we say that it has to be multiplied by

$$\sqrt{n/(n-1)}$$

Some calculators are programmed to produce both σ (sometimes denoted by σ_n) and s (sometimes denoted by σ_{n-1}).

Notebox 11.3

Why do we take the square root?

It is important to understand the reason for taking the square root. Suppose we have data about the lengths of people's arms, in centimetres. When we square the deviations, we are suddenly working not in units of length but in units of area, since 3 cm becomes 9 cm^2. When we sum and divide by n we obtain an average area, which can hardly be used to describe the spread of measures of length. By taking the square root, we restore the unit of measurement.

Summary

1. A more useful measure of spread is the standard deviation. We illustrate one of its remarkable properties, in a preview of the Normal distribution.

2. The standard deviation is defined by equation [11.1]; some books use a slightly different formula, for reasons explained in Notebox 11.2.

3. We show in Box 11.1 how to calculate the standard deviation for ungrouped data. The essential idea is that for every value of the variable its deviation from the mean value is squared. The squared deviations are then summed and divided by the number of them, giving the average squared deviation. Its square root is the standard deviation. Labour can be saved by using an assumed mean (Appendix B).

4. The square of the standard deviation is called the variance [11.3].

5. Noteboxes deal with (1) a diagrammatic illustration of the standard deviation, (2) the n and $n - 1$ formulae and (3) the reason for taking the square root.

The standard deviation: part 2

Introduction

Example 12.1 Having found the standard deviation of the family sizes of his neighbours, Steve wants to find it for their ages, but his age distribution table is grouped.

Example 12.2 Steve tells his friends that his daughter Stella is brilliant in history, having been awarded 85% in the end-of-term examination. Eric announces that his daughter, Erica, who is of the same age but goes to a different school, has also been awarded 85%. Steven remarks that standards in the two schools are different and that Stella's mark of 85 is worth more than Erica's. How can this be checked?

Example 12.3 The annual sales figures for 10 men have a mean of £300 000 and a standard deviation of £30 000, whereas the figures for 20 women have the same mean but a standard deviation of £40 000. What is the standard deviation for all 30 people?

Example 12.1 continues our account of how to calculate standard deviations. If the data are grouped, assumptions are necessary and they may lead to an error for which there is a special correction. In tackling Example 12.2 we see how a knowledge of the means and standard deviations of two distributions can be useful in making comparisons, through the process of **standardisation**. The solution to Example 12.3 shows that we cannot add standard deviations but we can add variances.

12.2 The standard deviation for grouped data

Grouped data are necessarily imprecise. The basic assumption of the method for calculating their standard deviation is that the average value of x for all the items in a group is given by the value at the centre of the group. If we have 20 people with ages between 65 and almost 70 years, we assume their average age is 67.5 years. Then we take 67.5 as a precise value of x with a frequency of $f = 20$.

If we do this with the data in Table 6.2, then by using

$$\text{s.d.} = \sigma = \sqrt{\frac{\sum fx^2}{n} - \left(\frac{\sum fx}{n}\right)^2}$$

we can calculate the standard deviation as 21.89 years. This can be done using a calculator (Appendix A) or manually with a few tricks (Appendix B).

This answer looks reasonable. The mean is $\sum f/n = 38.09$ years. Twice the standard deviation (2×21.89) on either side of this contains virtually all the population. Three points should be noted:

(1) If all the intervals are of equal width, their midpoints will be evenly spaced. This enables effort to be saved (especially in manual calculations) by using new units (Appendix B). It may also be worthwhile to use new units if most (rather than all) of the intervals are of equal width, especially if the widths are related, such as some intervals of 5 and some of 10.

(2) There may be open-ended groups, as in Table 6.1. In this case the 'midpoint' has to be chosen as a figure that is both reasonable and convenient. There is very little error if the frequency is low. But if we are concerned about it, we can make slightly different choices and repeat the calculation to see what effect it may have.

(3) As we explain in Notebox 12.1, the basic assumption about the midpoint may lead to a systematic error, which is often corrected using Sheppard's correction. Notebox 12.1 should be read carefully.

Box 12.1

How to ... **find the standard deviation for grouped data**

1. Take the midpoints of the groups as the values of x.

2. Proceed as for ungrouped data in Box 11.1, supplemented with Appendix A if you are using a calculator, Appendix B if not.

3. In the light of Notebox 12.1, consider whether it is appropriate to use Sheppard's correction. If so, calculate it as $c^2/12$ where c is the width of the class interval, and subtract this from the calculated variance, as described in Notebox 12.1.

12.3 **Coefficient of variation**

A standard deviation of 3 cm when the arithmetic mean is 10 cm tells a different story from a standard deviation of 3 cm when the arithmetic mean is 1000 cm. Although the data are equally spread out, it may sometimes be wise to compare the spread with the mean. The usual way of making this comparison is to use the **coefficient of variation**, which is defined as

$$V = \sigma/\mu \qquad [12.1]$$

Sometimes we use the **quartile coefficient of variation:**

$$V_Q = \frac{Q_3 - Q_1}{Q_3 + Q_1} \qquad [12.2]$$

This takes the difference between the first and third quartiles as a measure of spread and compares it with their mean, which is taken as a measure of location.

 12.4 **Standardised scores**

The comparison of Stella's mark with Erica's in Example 12.2 depends on the availability of a little more information. Even then our method is not very reliable unless there are large numbers of examination candidates in both schools.

We assume this is so and we assume the examiners have the same ideas about what a good answer should say, but they do not mark with the same degree of strictness.

To illustrate the method, let us suppose that the arithmetic mean of the marks awarded by Stella's examiner is 55, with a standard deviation of 14, and that for Erica's classmates the average is 60 with a standard deviation of 10. Thus marks awarded in Stella's school have a lower average than those awarded in Erica's school, but they are more spread out.

This is an example of a frequently occurring kind of problem. To solve it we need to 'correct' each mark for (a) the difference in means and (b) the spread, as measured by the difference in the standard deviations. In effect, this means we have to shift one or both of the the distributions so their means coincide and then squeeze (or stretch) them so they have the same spread.

The *first basic assumption* is that if all the pupils had been examined by the same person, the average mark would have been the same in each school. The larger the number of pupils, the more likely this is to be true.

If we accept this, we can make progress by first seeing how the two marks compare with their averages:

Stella's mark $x = 85$	Erica's mark $x = 85$
School mean $\mu = 55$	School mean $\mu = 60$
Difference $x - \mu = 30$	Difference $x - \mu = 25$

Thus, in terms of the marks awarded in the two schools, Stella is more above average than Erica. Notice that if Clarissa, another girl in Stella's school, had scored a mark of 40 this calculation would give her a difference of -15. Our 'differences' are marks above or below average.

Our *second basic assumption* is that if all the pupils had been examined by the same person, the marks would have been spread out equally in both schools. The larger the number of pupils, the more likely this is to be true.

This means that if marks in one school have double the standard deviation of marks in the other, then a mark of 10 above average in the first school is going to be worth the same as a mark of 5 above average in the second school.

To allow for this sort of thing, we divide each difference by the relevant standard deviation, getting

Stella

$$\frac{\text{diff.}}{\text{s.d.}} = \frac{x - \mu}{\sigma} = \frac{30}{14}$$
$$= 2.14$$

Erica

$$\frac{\text{diff.}}{\text{s.d.}} = \frac{x - \mu}{\sigma} = \frac{25}{10}$$
$$= 2.50$$

Thus Stella has a mark that is above her class average by 2.14 standard deviations, whereas Erica has a mark that is above her class average by 2.50 standard deviations.

Since we have assumed the two schools would have had equal class averages and standard deviations if the same examiner had been used, this means that Erica has done better than Stella.

By using $x - \mu$ we have introduced a new variable with zero mean. Dividing by the standard deviation has further modified the variable to give it a unit standard deviation. It is as if the examiners agreed to give positive and negative marks in such a way that their mean was zero and standard deviation unity.

Scores (or other variables) that are adjusted in this way have been **standardised**. They are often used, both in statistical theory and testing, and in comparing empirical results. Note that these scores have no units of measurement. If instead of marks we had analysed prices expressed in pounds, we would have obtained the same standardised prices as if we had started with the same prices expressed in pence. The units cancel out because of the definition.

Box 12.2

How to ... obtain standardised scores (or variables)

1. Express each individual value of x as a signed deviation from the arithmetic mean μ and divide it by the standard deviation, giving

$$z = \frac{x - \mu}{\sigma} \qquad\qquad [12.3]$$

12.5   Adding variances

In Example 12.3 we want to find the standard deviation for 30 sales people when we know that for 10 men the mean is £300 000 and the standard deviation £30 000, whereas for 20 women the mean is the same but the standard deviation is £40 000.

Unfortunately, standard deviations cannot be added or averaged meaningfully. (This can easily be checked by working out the standard deviations for (i) 1, 2, 3, 4; (ii) 5, 7, 9, 10; (iii) 1, 2, 3, 4, 5, 7, 9, 10, and then seeing whether the answers to (i) and (ii) can produce the answer to (iii).)

However, variances (which are the squares of standard deviations) can be added and averaged, provided we allow for different sizes of populations or samples. This is easier if the means are the same, as in Example 12.4.

Suppose two sets of data have the same mean. Suppose that one set is of size n_1 and variance σ_1^2 and that the other set is of size n_2 and variance σ_2^2.

Suppose now that the two sets are combined into one larger set. It can be shown that the variance of the combined set is

$$n_1 \frac{\sigma_1^2 + n_2 \sigma_2^2}{n_1 + n_2}$$

which is the weighted arithmetic mean of the variances. This result assumes the values of the data in set 1 are not in any way influenced by the values in set 2. And remember that it is true only if the two sets have the same mean. If they do not, consult Notebox 12.4.

In Example 12.3 it will save some trouble if we work in units of £1000. We then have

$$n_1 = 10 \qquad \sigma_1 = 30 \qquad n_2 = 20 \qquad \sigma_2 = 40$$

Substituting these values in the formula, we have that the numerical value of the variance for the total sales force is

$$\frac{(10 \times 30^2) + (20 \times 40^2)}{10 + 20} = \frac{41\,000}{30} = 1366.667$$

so the standard deviation of the sales of all 30 people is the square root of this, which is 36.97 units (1 unit = £1000).

Notice that if we had tried to get the standard deviation for the combined set by taking a weighted sum, we would have obtained

$$\frac{(10 \times 30) + (20 \times 40)}{10 + 20} = \frac{1100}{30} = 36.66$$

In this example it happens to be close to the other result, but it is wrong. The only way to find the standard deviation for a combined group is to follow the correct procedure outlined above and presented more formally in Box 12.3.

Box 12.3

How to ... combine two standard deviations

1. In this set of instructions we assume the two means are the same. If they are not, see Notebox 12.3

2. Suppose that two sets of sizes n_1 and n_2, and with standard deviations σ_1 and σ_2 are combined into a new set of size $n_1 + n_2$.

3. Square the standard deviations to form variances σ_1^2 and σ_2^2.

4. The variance for the combined set is given by the weighted arithmetic mean of the separate variances:

$$\frac{n_1\sigma_1^2 + n_2\sigma_2^2}{n_1 + n_2}$$

5. The standard deviation for the new set is the square root of this.

12.6 Conclusion

We have now learned how to describe a statistical distribution by referring to its shape, its location (as measured by an average which may or may not be a mean) and its spread or dispersion. We know how to calculate the measures of location and dispersion, and we know some of their properties. Shortly we shall turn to questions about the reliability of measures of location and dispersion when we have only sample data, and to questions about the meaningfulness of differences between results for different samples. First, however, we discuss how to measure statistical relationships between two sets of data, such as a relationship between height and mass, or between some students' marks in mathematics and the same students' marks in physics. Later we will also consider the reliability of these measurements.

Consolidation exercise 12B

1. Why is it sometimes necessary to standardise marks or other values of a variable? What does standardisation do?

2. Tom earns £84 cleaning cars over the weekend. Many other people spend their weekends in the same way. If the arithmetic mean of their earnings is £54 and their standard deviation is £12, what are Tom's standardised earnings? Before consulting the answers, carefully consider whether there is anything wrong with your answer as you have stated it.

3. What information do you need about the means before you combine two standard deviations? How do you combine two standard deviations?

Notebox 12.1

Sheppard's correction

Grouping and assuming that the midpoint of the group is the mean value of the data within it usually introduces error. If the variable is continuous and the

⬎

frequencies tail off gradually in both directions, Sheppard's correction is often applied. It can lead to overcorrection. There is a limit to the accuracy of calculations based on imprecise data.

After using the grouped data to calculate the variance (which would be $21.89^2 = 479.17$ in the text example about ages), a small correction is subtracted from it to give a corrected variance. The square root of this gives the corrected standard deviation. If all the class intervals are of width c, the correction is given by $c^2/12$, which is $5^2/12 = 2.083$ in the example. This leads to a corrected variance of 477.09, hence a standard deviation of 21.84 years, just slightly less than the uncorrected value of 21.89 years.

Notebox 12.2

More on skewness

The standard deviation occurs in some measures of **skewness.** Two use the fact that for skewed distributions the mean and the longer tail tend to lie on the same side of the mode. This allows the separation of the mean and the mode, scaled down by the standard deviation, to be used in **Pearson's first coefficient of skewness:**

$$\text{Skewness} = \frac{\text{mean} - \text{mode}}{\text{standard deviation}}$$

Since the mode is sometimes difficult to measure, this is sometimes replaced by **Pearson's second coeffficient of skewness:**

$$\frac{3(\text{mean} - \text{median})}{\text{standard deviation}}$$

Another measure uses the cube of the sum of the deviations from the mean, along with the cube of the standard deviation. There is no need to go into the reasons for this. The formula is

$$\text{Skewness} = \frac{1}{n} \sum \left(\frac{x - \mu}{\sigma} \right)^3$$

Some books give this with $n-1$, $\bar{x}$ and s. In this form the formula uses sample data to estimate population skewness. See also Notebox 10.2.

Notebox 12.3

Adding variances when means differ

If two sets of data do not have the same mean, the formula for combining their variances is altered to take account of the separation of the means.

The formula is a little complicated. It will look less frightening if we use a slightly different notation, writing V for variance (where $V = \sigma^2$). If the set of size

n_1 and variance V_1 has a mean of μ_1, and the other set of size n_2 and variance V_2 has a mean of μ_2, then the variance of the combined set is V^* where

$$(n_1 + n_2)V^* = n_1 V_1 + n_2 V_2 + \frac{n_1 n_2}{n_1 + n_2}(\mu_1 - \mu_2)^2$$

Notice that we do not need to know the means. All we need is knowledge of their difference. If this is zero, the formula reduces to the simpler formula in the main text and Box 12.3.

Summary

1. The standard deviation for grouped data is calculated by assuming that the items in a group have a mean value of the variable given by the value at the midpoint of the group, as described in Box 12.1. Appendix C shows how to save work.

2. The midpoint assumption usually results in overestimation of the standard deviation. If the variable is continuous and the frequencies tail off gradually in both directions, Sheppard's correction may be applied to the variance, and the square root of the corrected varaiance gives the corrected standard deviation. It may be overcorrected (Box 12.2).

3. The standard deviation is used in the coefficient of variation [12.1] and the quartile coefficient of variation [12.2]. It is also used in calculating standardised scores, which are unitless. Box 12.2 summarises the technique which expresses every score as a deviation (from the mean) divided by the standard deviation.

4. Standard deviations can be combined only by combining the variances and then taking the square root. If the two means are the same, the procedure of Box 12.3 should be used; otherwise see Notebox 12.3.

5. Noteboxes deal with (1) Sheppard's correction, (2) further measures of skewness and (3) adding variances when the means differ.

Bivariate analysis and scattergrams

Introduction

Example 13.1 Harry believes that a good landlord should be well informed about his customers' families. Knowing that 18 of his customers had daughters studying history, he was delighted to obtain the examination marks for 17 of them. The marks varied from 17 to 80. He wondered if there was any relationship between the daughters' marks and his customers' weekly intake of NonTox.

Example 13.2 In looking for a new house, Steve comes across an estate of 11 new houses, all built by the same builder. One of them is empty, and he decides to buy it. When discussing the price, he discovers that he will receive a PIE rebate – an unknown allowance for 'personalised interior embellishment'.

Wondering how much this might be, he asks the purchasers of the other 10 houses how much they have received. The sums vary between £489 and £602. He guesses that perhaps the builder allows more for embellishment if the house has been empty for a long time. To check this he also asks the purchasers how long their houses had been on the market before they bought them.

Until now we have been concerned with problems involving one variable, usually denoted by X. The examples just given involve two variables. In Example 13.1 we have pints of NonTox drunk (with values denoted by x) and daughters' history marks (denoted by y). This information is available for n (in this case 17) persons. In Example 13.2 Steve knows for each of $n = 10$ houses the number of weeks on the market (X) and the PIE rebate (Y).

More generally we have n observations of two variables X and Y, and we are wondering whether there is some kind of relationship between their values x and y. It is not an idle curiosity. If there is a relationship, we may be able to use it.

In Example 13.1 perhaps we can use the relationship to help us to estimate the missing history mark from a knowledge of the father's consumption of NonTox. Or we may be able to find how much of the variation in the marks can be explained by the variation in drinking, but as we shall see, we have to be especially careful here.

In this chapter and the next two chapters we consider these examples of **bivariate data**, the search for a **bivariate relationship** and the uses of a bivariate relationship.

Note that a statistical relationship does not prove anything about causation. A strong relationship may be due to the value of X affecting the value of Y or the other way around. It may be that both X and Y are affected by some other factor we are not considering. Or the relationship suggested by the figures may be quite inexplicable and possibly due to chance. We return to such matters later.

We must also remember that if we find a relationship between NonTox drinking by those 17 fathers and their daughters' history marks, it does not imply that such a relationship exists for other fathers and their daughters. Much later we will consider when and how we can extend a finding about one set of people to a larger set.

13.2 The scattergram

Two of the most important techniques for the exploration of bivariate data are the related techniques of **regression** and **correlation**, which we describe in the next two chapters. Calculators have made them very easy to use. Unfortunately, one essential step that may be undertaken by hand or by calculator is all too often omitted, and then a great deal of statistical analysis becomes incorrect. It is all too easy to misuse regression and correlation techniques, especially if there is too much reliance on calculators. Note that some well-known calculators are not completely reliable for regression (Chapter 14 and Appendix A).

The first step in the examination of bivariate data must always be the construction of a scatter diagram, or **scattergram**. This is simply a diagram on a piece of squared paper showing for each observation the pair of values in the usual rectangular coordinates, so the horizontal position of the point indicates the value x of variable X and the vertical position shows the value y of variable Y. There will be n points, one for each **observation**, as in Figure 13.1, which has been drawn for the data in Table 14.1 of the next chapter. We refer to other features of the diagram later on. Whether the techniques of regression and correlation analysis described in Chapters 14 and 15 can properly be used depends on the appearance of this scattergram.

13.3 Application of standard techniques

Loosely stated, the basic assumption underlying the techniques we are about to describe is that the points in the scattergram are on a straight line or are scattered around one in an elliptical (cigar) shape, as sketched freehand in Figure 13.1. (More rigorously stated, the condition is that both variables should be at least approximately normally distributed, which is a statement that will be more easily

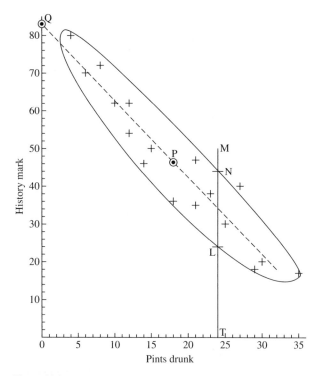

Figure 13.1

understood later.) This elliptical shape may be sloping upwards to the right, or downwards to the right. Normally we omit the words 'to the right'.

If the points are all exactly on a straight line, there is a perfect **linear relationship** between X and Y. The equation of the line will give the exact mathematical relationship between them (Notebox 13.1). If they are not on a line but scattered around one within an elliptical shape, then the equation of the line will give this relationship approximately. A task before us then is to find the straight line that gives the best approximation, which means knowing what we mean by 'best'.

If the points lie in an elliptical shape around a straight line, we can use standard calculations (Chapters 14 and 15) to find the slope and position of the straight line that allows us to make the 'best' estimate of y when we know the value of x, hence to estimate the missing mark. We shall see how to do this in Chapter 14, and later discuss how much confidence to put in such an estimate. For now all we need to note is that if the points do not lie in an elliptical shape around a straight line, the techniques of Chapters 14 and 15 cannot properly be applied.

If the points on the scattergram lie on a curved line, instead of a straight line (Figure 13.2), the relationship is described as **curvilinear.** In this case the standard techniques described in Chapters 14 and 15 are inapplicable. Their use will still lead to answers, but unless the data are transformed as described below, these answers will be meaningless and misleading.

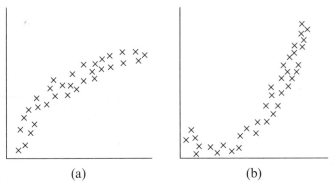

Figure 13.2

 Transforming the data

Sometimes it may be possible to convert the data into different units so that a linear relationship appears. Consider the following data:

x	1	2	3	3	4	5	5	6	7
y	1	3	10	8	15	27	23	40	45

They are are plotted in Figure 13.3(a) which shows a marked upward curve; this rules out the use of our techniques. However, we may notice that when $x = 3$, y has values around 9 (which is 3^2) and that when $x = 7$ the value of y is not far from 49 (which is 7^2). This prompts us to create a new variable named X' by squaring values of X. We then get the following table:

x	1	2	3	3	4	5	5	6	7
x'	1	4	9	9	16	25	25	36	49
y	1	3	10	8	15	27	23	40	45

The scattergram for values of X' and Y is shown in Figure 13.3(b); it reveals a close fit to a straight line. We can therefore legitimately use the techniques of regression and correlation described in Chapters 14 and 15, not for values of X and Y but for values of X' and Y. We have **transformed** x into x^2 (or, more generally, into x'). All subsequent calculations should involve x' rather than x.

 Structural breaks

Another advantage of the scattergram is that it reveals whether there is a **structural break** as in Figures 13.4 and 13.5.

Suppose we want to know whether there is a relationship between temperature

ORIGINAL DATA

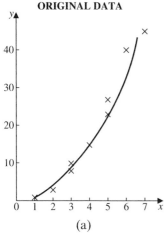

(a)

TRANSFORMED DATA

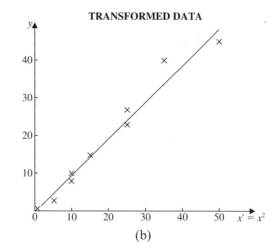

(b)

Figure 13.3

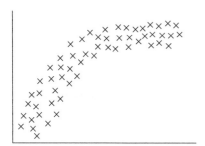

Figure 13.4

and the domestic consumption of electricity. With the help of power companies we get data for the domestic consumption of electricity in a town on the Mediterranean coast in each of the 36 months of the past three years.

Getting a monthly statistic for temperature is a little less straightforward. Daily weather records show the maximum and minimum temperatures. The weather station calls the mean of these two temperatures on any day the mean temperature for that day. If the daily mean temperatures are added over a month and then divided by the number of days in the month, we get a single statistic for each month, which we will call the mean monthly temperature.

In Figure 13.5 there are 36 points, one for each month of the past three years. The horizontal position of the point, corresponding to x, shows the mean temperature for that month. The vertical position (y) shows the electricity consumption.

If we wish to describe this scattergam as cigar-shaped, we have to admit that it is a singularly bent cigar. There is clearly not a linear relationship; but if we consider the diagram carefully, we may feel that two straight lines – one downward sloping for the lower temperatures, the other upward sloping for the higher temperatures – would provide a reasonable fit. Perhaps we should split our data and perform

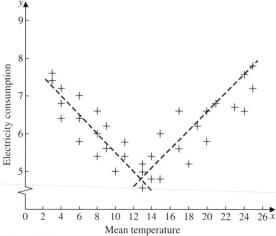

Figure 13.5

separate statistical analyses for the lower and higher temperatures. Is there something about the data that can justify this?

If we think of the domestic use of electricity, we can see that more electricity is used for heating in very cold weather. As the months become warmer this use declines, and the falling part of the scattergram seems to make sense. However, as the months become even warmer and then hot, more electricity will be used for cooling, for electric fans, air-conditioning, refrigerators, and so on. The upward part of the scattergram now makes sense. The bend is explained by the replacement of one relationship (based on the use of electricity for heating) with another (based on use for cooling).

Notice that we have explained the shape of the scattergram by introducing a causal relationship: changes in temperature cause changes in electricity consumption. But we may not always be able to do this quite so easily, and sometimes not at all.

It has been possible to speculate about these causal relationships, and to make some sense of the scattergram, only because we know something about the place to which the statistics relate and the kind of human behaviour that lies behind them. If we had just been given a scattergram showing domestic electricity consumption and temperature, we could not have been so positive in our interpretation. We would have needed to ask questions about where and how the temperature was measured, the area to which it related, and so on.

Without answers to such questions, we could have said little about causation. But we could still have noted the boomerang shape and the sharp bend in direction, and that would have directed our attention to some key questions:

- Why is it bending here?
- What happens when temperatures move from being a bit below k to being a bit above k?
- What happens when fuel consumption stops falling and starts to rise?
- Did the months when the scattergram was falling differ in some other way from the months when it was rising?

The shape of the scattergram is prompting inquiry into the underlying forces at work. Sometimes we may be unable to answer these questions; then we must be especially careful when we interpret any results obtained using techniques in later chapters.

 Outliers

Yet another use of a scattergram is the detection of **outliers** or 'odd men out'. An example is given in Figure 13.6.

There can be several reasons for an outlier. One should always check to see if there has been a mistake in plotting the point. If there has not been a plotting mistake there may still be a mistake in the recorded data, and if possible one should check for this.

If there is no mistake in the plotting or the recording, it is possible that the point corresponds to an individual who should not have been included. For example, if all the other points relate to regular customers of the Calculators' Arms, the inclusion of an occasional customer could result in an outlier. Yet another reason is that there is a genuine rare occurrence.

Sometimes it is fairly obvious that a point is an outlier but at other times the decision may be more difficult. Two techniques are commonly used to assist in the decision. We mention them now, even though they depend on ideas we have yet to describe in detail.

(1) We can calculate the equation of a line of best fit, taking account of all the data; then we repeat the calculation with the suspect point excluded. Afterwards we decide whether the exclusion has led to a line so different from the other that it really tells a very different story. If it does then we have to find out all we can about this point before we make our decision.

(2) Another technique is to draw a line of best fit and then two parallel lines, one

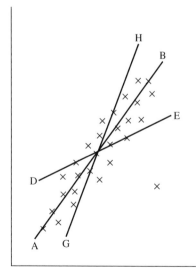

Figure 13.6

above it and one below, in such a way that say 95% of the points lie between the lines. Then we consider as suspect any point that lies well outside these lines. The meaning of 'well outside' is a matter for judgement.

If there is an outlier, it may be better to exclude it from the main statistical analysis, but be certain to mention it and try to find an explanation for it. From Figure 13.6 it seems better to suggest that the data are tightly bunched around the line AB, with one exception, than to say they are loosely bunched around DE or GH. The inclusion of the outlier in Figure 13.6 completely alters the story, pulling B down towards E or A across towards G.

Conclusion

Results will always emerge from the statistical calculations. If the scattergram does not warrant the use of the method, the result will be at best useless, but more often than not it will be misleading. A great deal of fruitless calculation and spurious argument can be avoided by first drawing the scattergram. In the next chapters we consider these matters further, beginning with the questions posed by Harry and Steve.

Consolidation exercise 13A

1. What is a scattergram? When and why is it necessary to draw one?

2. Draw scattergrams for the following two sets of data. Consider whether it is justified to use linear regression and correlation techniques.
   ....

   (a) | x | 3 | 7 | 9 | 2 | 5 | 11 | 4 | 1 | 6 | 8 | 5 |
   |---|---|---|---|---|---|---|---|---|---|---|---|
   | y | 7 | 13 | 18 | 5 | 9 | 20 | 10 | 3 | 11 | 17 | 11 |

   (b) | x | 3 | 7 | 9 | 2 | 5 | 11 | 4 | 1 | 6 | 8 | 5 |
   |---|---|---|---|---|---|---|---|---|---|---|---|
   | y | 8 | 45 | 85 | 3 | 20 | 3 | 17 | 12 | 5 | 40 | 36 |
   ....

3. Inspect the following sets of data and suggest in each case a transformation that is likely to result in an acceptable scattergram.
   ....

   (a) | x | 0 | 1 | 2 | 3 | 4 | 5 | 6 |
   |---|---|---|---|---|---|---|---|
   | y | 50 | 10 | 5 | 3 | 2 | 2 | 1.5 |

   (b) | x | 0 | 1 | 2 | 3 | 4 | 5 | 6 |
   |---|---|---|---|---|---|---|---|
   | y | 1 | 2 | 4 | 10 | 16 | 24 | 37 |

   (c) | x | 0 | 1 | 2 | 3 | 4 | 5 | 6 |
   |---|---|---|---|---|---|---|---|
   | y | 0 | 3 | 4.5 | 5.5 | 6 | 7 | 8 |
   ....

4. What is an outlier? How should you deal with one?

5. What is a structural break? How should you deal with one?

The equation of a straight line

Suppose a straight line goes through the point $(0, a)$ as shown in Figure 13.7 and has a slope b (so that for every 1 it goes across it goes up b). It can be shown that

(i) For all points on the line, the values of x and y (the coordinates) will satisfy the equation $y = a + bx$.
(ii) All points with coordinates satisfying this equation will lie on the line (or its extension in either direction).

We call this equation the **equation of the straight line.** It can also be written in other forms.

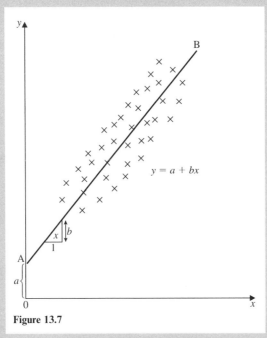

Figure 13.7

Finding a linear transformation

A useful first step in considering the transformation of variables is to examine the data, as in the text example, to see whether one of the following transformations is likely to work:

(i) *Squaring one variable* (or taking the square root of the other). If the result is simply to reduce the curvature rather than to eliminate it, there may be point in cubing rather than squaring. ⇒

(ii) *Taking reciprocals.* This is always worth trying if the original scattergram looks like Figure 13.8(a). Use $1/x$ instead of x.

(iii) *Taking logs.* If the curve rises very steeply (or tends to flatten out very quickly) as in Figures 13.8(b) and (c), try either $y' = \log y$ or $x' = \log x$.

Finding a suitable transformation often requires great experience (or great luck), and may not always be possible.

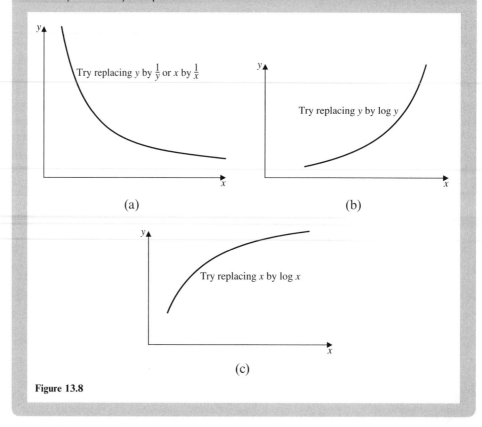

Figure 13.8

Summary

1. If there are several pairs of two variables, there may be a relationship between their values.

2. Most elementary techniques for the analysis of any such relationship assume that if it exists it is linear. It is important to begin by drawing a scattergram. Apart from being informative in its own right, this enables us to see whether the assumption of a linear relationship is valid. If it is not then the techniques of regression and correlation described in the next few chapters are inapplicable, and their use will lead to erroneous conclusions.

3. When a non-linear relationship appears, it is sometimes possible to find a transformation of one or both of the variables so that linearity occurs.

4. The scattergram may also help us by identifying 'an odd one out' or a structural break. With an odd one out, consider why it has arisen and whether it is better to exclude this piece of data from the analysis. With a structural break it may be advisable to split the data into two sets, each showing its own linear relationship.

5. Noteboxes discuss the equation of a straight line and techniques for finding a linear transformation.

14

Regression

14.1 Introduction

⇒ **Example 14.1** Harry tabulates his information about the amounts drunk by 17 customers and the marks awarded to their daughters, so that he can plot a scattergram as suggested in Chapter 13. Apart from the data in Table 14.1, another customer drinks 24 pints. He hopes he can use the scattergram or the table to estimate the history mark of the customer's daughter Tamandra.

⇒ **Example 14.2** Steve's enquiries (Example13.2) about time in the marketplace and the level of embellishment rebate give him the data in Table 14.2. His own house has been on the market for 17 weeks and he wants to estimate the rebate.

Harry and Steve make their estimates by using regression, which relates the values of one **dependent** variable to the values of one or more **independent** variables. This is important not only in estimating missing values but also for other purposes, including forecasting (which we mention briefly in Chapter 46). In this book we shall be concerned only with the simplest case in which there is only one independent variable.

Note carefully that in this chapter we assume Harry is interested only in those 17 customers (and Tamandra's father), their consumption of NonTox and the history marks of their daughters. He is not viewing his customers as a sample representing a larger number of drinkers, or the history marks as being in some way representative of other marks, except for the one 'missing' item. Similarly Steve is interested only in those 10 houses and the one he is thinking of buying.

Table 14.1

Pints	8	25	4	21	27	14	15	6	30	33	10	29	23	21	18	12	13
Marks	72	30	80	47	40	46	50	70	20	17	62	18	38	35	36	54	62

Table 14.2

Weeks	13	6	1	17	21	1	14	15	9	8
Rebate (£)	567	515	489	571	602	498	563	579	532	537

14.2 A subjective use of the scattergram

Harry's scattergram was shown in the last chapter (Figure 13.1). With one slight exception the points lie in a fairly compact cigar shape, which he sketches in. He thinks it is unlikely that the point corresponding to Tamandra and her father would lie outside this elliptical shape. He marks her father's drinking score of 24 on the x-scale and draws a vertical line TM. It cuts his ellipse at L and N. If Tamandra's point is somewhere within the cigar, it must lie between L and N; her mark must therefore lie between 24 and 44. He wonders whether his best bet may be halfway between, at 34.

He knows that if he had drawn the ellipse half an hour earlier, it might have been slightly different; then L and N would have indicated slightly different marks with a different midpoint. His technique is a bit too subjective.

Another approach is to accept there is a linear relationship and to try drawing the straight line that in some sense best fits the points. Then we can read off the value of y corresponding to an x of 24, and call it the best estimate.

This, too, is subjective. Lines drawn at different times or by different people would not all be the same. And what do we mean by 'best'?

14.3 The line of best fit for estimating y

Keep in mind that Harry's purpose is to use his information about 17 customers and their daughters. Based on what he knows about the drinking score for Tamandra's father, Harry attempts to estimate her history mark (y). This estimate will depend on how much Tamandra's father drinks (x). We say that Y is the **dependent variable** and X is the **independent variable**.

Notice that we are not saying that history marks depend on the amount drunk in any causal way. We are simply saying that our *estimate* of a history mark depends on our knowledge of the amount drunk. The dependency is statistical rather than causal.

Now 'best' has to be defined in this context. The 'best' line is the line that enables us to make a 'better' estimate of y on the basis of our knowledge of x than any other line would. We still have not defined 'best', any more than we have defined 'better', but we do know that it means 'best for estimating y from a knowledge of x'.

Consider Figure 14.1 overleaf. Each cross indicates a pair of values of X and Y. We draw a line AB that seems to summarise the approximate linear relationship between them. Then if we want to estimate the value of Y corresponding to the value x_1 of X, we can read it off from this line as being y_1.

As it happens, only one of the crosses lies exactly on this line – and even that is an accident. Some of them are above it, others below it. The **deviations** of two values of y from this line are shown by two vertical lines. They change as we alter the slope or position of this line. Statisticians usually define the 'best' line to be the one that minimises the root mean square (Section 9.4) of the deviations. In other words, the total area of all the squares similar to those drawn is lower than for any other line. The full name for the line that does this is the **least squares regression line of y on x**.

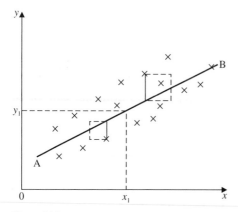

Figure 14.1

We saw in Notebox 13.1 that a straight line has an equation

$$y = a + bx$$

The slope of this line is given by b. The value of Y where the line cuts the axis Oy is a.

This definition of 'best' means that, to find the best straight line, we have to find the values of a and b that minimise the root mean square of the deviations. These values give us the regression line of Y on X – the line that is best for estimating y from a knowledge of x. We have found the line of best fit by the method of least squares.

This line will always pass through the point denoted by $(\bar{x}, \bar{y})$ for sample data but by (μ_x, μ_y) for population data. To avoid tedious repetition of these two versions of the same set of coordinates, we will temporarily refer to them as $(\bar{x}, \bar{y})$.

The constant a is the **regression constant**, and the coefficient b is the **regression coefficient**. Some books interchange a and b. The procedure for finding them is detailed in Box 14.1 and illustrated below.

In practice you will probably find the equation of this line by using a programmed calculator. If the observed values of x and y are tapped into the calculator as data input, it normally takes only a few seconds for a standard program to produce the values of a and b. Instructions for doing this are given in Appendix A. It is important to read this since the manuals and programs of some calculators are not completely reliable in this area.

 ## Using the original data to find the equation

The essentials of the method for calculating the equation of the regression line are best introduced with a simple example using only a few observations and easily managed numbers. With more realistic data the apparent burden of calculation can be a bit off-putting unless you have a suitable calculator or use some of the computational tricks (based on coded data) described in Appendix B. To keep these illustrative calculations short, we have used very few observations, therefore we will not be able to put much reliance on the result. (The reliability of regression estimates is considered in Chapter 32.)

Table 14.3 shows for five kennels the numbers of dogs being looked after and the weekly food costs. We know that another kennel houses 13 dogs and we want to estimate its food costs. The number of dogs is the independent variable (which we will denote by X) and we use it to estimate a value of food costs, which is therefore the dependent variable (denoted by Y).

A scattergram will show five points scattered in a cigar shape around a straight line. The calculation involves finding the sums of x, y and a few expressions involving x and y. This is done most easily by creating a new table (Table 14.4) that has the values of x and y in its first two columns. In the remaining columns we enter values of x^2, y^2 and xy. Check some of the entries in these columns to be certain you understand.

The columns are then summed, producing $\sum x = 69$, $\sum y = 139$, $\sum x^2 = 1045$, $\sum y^2 = 4215$, $\sum xy = 2094$ and $n = 5$.

We now have to use these results in a formula that gives us the values of a and b, hence the equation of the regression line. Notebox 14.1 gives a few versions of the formula, all leading to the same answers. Which one we use is up to us. Here, to make the explanation simpler, we shall begin by using the basic formula

$$\hat{y} = a + bx \qquad\qquad [14.1]$$

where $\hat{y}$ is the estimate of y based on a knowledge of x, and

$$a = \frac{(\sum y)(\sum x^2) - (\sum x)(\sum xy)}{n \sum x^2 - (\sum x)^2} \qquad\qquad [14.2]$$

$$b = \frac{n \sum xy - (\sum x)(\sum y)}{n \sum x^2 - (\sum x)^2}$$

Remember that the parentheses in expressions such as $(\sum x)(\sum y)$ mean that the two sums are multiplied together.

Table 14.3

Number of dogs, x	8	10	15	16	20
Weekly food cost (£), y	15	23	28	34	39

Table 14.4

x	8	10	15	16	20	$(\sum x = 69)$
y	15	23	28	34	39	$(\sum y = 139)$
x^2	64	100	225	256	400	$(\sum x^2 = 1045)$
y^2	225	529	784	1156	1521	$(\sum y^2 = 4215)$
xy	120	230	420	544	780	$(\sum xy = 2094)$

In the example we are considering

$$a = \frac{(\sum y)(\sum x^2) - (\sum x)(\sum xy)}{n \sum x^2 - (\sum x)^2}$$

$$= \frac{(139)(1045) - (69)(2094)}{5(1045) - 69^2}$$

$$= 769/464 = 1.657$$

and

$$b = \frac{n \sum xy - (\sum x)(\sum y)}{n \sum x^2 - (\sum x)^2}$$

$$= \frac{5(2094) - (69)(139)}{5(1045) - 69^2}$$

$$= 879/464 = 1.894$$

Thus the regression line for y on x is

$$y = 1.657 + 1.894x$$

The 'best' estimate of food costs (in £) for a kennel keeping 13 dogs is therefore

$$1.657 + 1.894(13) = 26.28$$

This is more accurate than the original data. If we wish, we can look upon it as an average of many estimates, as we describe later. We can also round it to the nearest pound, getting £26. Or if we want to know how many pound coins we will need to pay for the food, we can round it up to £27.

The same answer can be obtained by using another formula given in Notebox 14.1. It uses the same table and the same sums.

Box 14.1

How to ...   **find the regression line of Y on X**

1. Draw a scattergram to make certain that the points lie within a cigar shape around a straight line. If necessary find a suitable transformed variable (Notebox 13.2) and use this in place of x or y in the following instructions.

2. Calculate $\sum x, \sum y, \sum x^2$ and $\sum xy$. This can be done on the calculator or by a tabular method illustrated below.

3. Calculate the means from

$$\frac{1}{n} \sum x \quad \text{and} \quad \frac{1}{n} \sum y$$

For convenience we denote them by $\bar{x}$ and $\bar{y}$, as explained in the text.   ⬭

4. Calculate

$$S_{xx} = \frac{1}{n} \sum x^2 - \bar{x}^2$$

$$S_{xy} = \frac{1}{n} = \sum xy - \bar{x}\bar{y}$$

where n is the number of observations.

5. Calculate $b = s_{xy}/s_{xx}$ by dividing the results in step 4.

6. Use the values of b, $\bar{x}$ and $\bar{y}$ to find a from $a = \bar{y} - b\bar{x}$.

7. Then the regression equation of y on x is $y = a + bx$ with the values of a and b found in steps 5 and 6.

Use of the procedure just illustrated, or a programmed calculator, will show that in answer to **Example 14.1**, Harry's regression equation for estimating history marks is

$$y = 83.18 - 2.062x$$

Since Tamandra's father drinks 24 pints, this leads to a best estimate of her history mark of

$$y = 83.18 - 2.062 \times 24 = 33.69 \text{ (34 to the nearest mark)}$$

This should be compared with the estimates derived earlier by more subjective methods. We leave to later chapters the questions of reliability and of whether Harry can extend his finding to customers of other pubs.

The regression line just calculated goes through $(\bar{x}, \bar{y})$ (Notebox 14.2) and has a value of 83.18 when $x = 0$, going through the points labelled P and Q in Figure 13.1. The line does not slice the sketched cigar symmetrically down the middle. This is not because Harry's freehand sketch is not very good. There is a good reason why the line should not slice it symmetrically, as we see shortly.

In **Example 14.2** Steve's scattergram (Figure 14.2), slopes upwards to the right, indicating that y increases as x increases. Once again a linear relationship exists. The regression equation can be found by using a calculator or by using the above equations. In Appendix B we find it more economically by using assumed means. The equation is

$$y = 487.84 + 5.472x$$

Thus, since Steve's house has been empty for 17 weeks, his best estimate of the PIE rebate is

$$£487.84 + 5.47 \times 17 = £581 \text{ (to the nearest pound)}$$

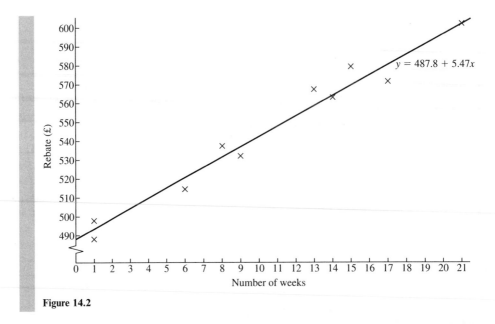

Figure 14.2

Consolidation exercise 14A

1. What do regression techniques do?

2. How can a scattergram provide a rough estimate of a missing value?

3. What is (a) a dependent variable and (b) a regression line?

4. What do we mean by 'best' when applied to a regression line?

5. What is minimised by the regression line of Y on X?

6. Through what point must a regression line always pass?

7. In the regression equation $y = 17.2 - 3.4x$ what is (a) the regression constant and (b) the regression coefficient? Does the line slope upwards or downwards?

 Reliability of the best estimate

Although we have estimated that a house standing empty for 17 weeks will have a PIE rebate of £581, we have little idea of the reliability of this estimate. There are several sources of possible error. One is that possibly the neighbours have shown a systematic bias in answering Steve's enquiries. Perhaps some of them have understated the size of their rebate because they want to create the impression that they paid a lot for the house. Perhaps some have deliberately overstated the rebate, for other reasons: but there is no reason why misstatements should balance out. Although we believe there is likely to be an overall bias, we have no way of knowing its sign. Even if the data are accurate, there are other sources of possible error in the estimate of the rebate. We look at a few.

The regression equation that we have calculated is completely accurate for the 10 houses for which we have data (provided the data are accurate). When we use it to estimate the unknown value of y corresponding to $x = 17$, we are assuming that if we had had complete information for 11 houses instead of 10, we could have calculated an identical regression line.

Moreover, the 10 points in the scattergram are spread around this regression line, rather than lying precisely on it. It would be remarkable if the point corresponding to Steve's house were to lie exactly on it. In other words, it would be remarkable if his rebate were to be exactly the value of y read from the regression line corresponding to $x = 17$. We can accept that £581, or even £580.87, is a 'best estimate' of the rebate but how reliable is this? What are the chances of Steve having a rebate as high as £600 or as low as £500?

We cannot answer this question now, but it is an example of questions about **confidence**. In Chapter 32 we shall see how to attach a confidence interval to an estimate of y. This is a range within which the correct value of y has a specified chance of lying, say 95%, assuming that our regression equation is correct.

This leads us to another source of possible error. Steve's information is about 10 houses of particular importance to him. Suppose, however, that we have data about X and Y for each of 10 houses sampled from those on a very large estate. We may wish to use this information about the sample to estimate the value of Y for *any* other house on that estate for which we know X. If we had chosen a different sample of 10 houses, we might have had somewhat different data, so our calculations would have led to a different regression equation. How good, then, is this regression equation based on sample data as a proxy for the true regression equation for all houses? How confident can we be in the values of the regression coefficients a and b derived from the sample data as estimates of the correct a and b for the whole large collection (or population) of houses?

Going to an extreme, is it possible that if we had data for all the houses, we would find a randomly scattered set of points with a calculated regression coefficient of virtually zero, so that a knowledge of x is of no use at all in estimating y?

Although we have drawn attention to these questions now, we cannot answer them until we have developed some other important ideas in later chapters.

14.6 The other regression line

The regression line of Y on X and the associated formulae are to be used *only* to estimate y from x. They are *not* to be used to estimate x from y. If we want the best least squares estimate of x, we must minimise *not* the sum of the squares drawn on vertical lines showing the deviations of y from the line, *but* the sum of squares drawn on horizontal lines showing the deviations of x from the line.

Box 14.2

How to ... find the regression line of X on Y

1. Go to the instructions in Box 14.1 for finding the regression line of Y on X. Write x wherever there is y, and y wherever there is x.

2. Use these amended instructions to obtain values of a and b, which we will call a' and b'.

3. Then the regression equation of X on Y is

    $$x = a' + b'y$$

4. Alternatively note that $b' = s_{xy}/s_{yy}$ and the line goes through the point $(\bar{x}, \bar{y})$.

Figure 14.3 shows the two regression lines for the data given in Table 14.5. The regression line of Y on X is

$$y = -0.1 + 1.16x$$

It minimises the sum of the squares of the deviations such as **PA**. The regression line of X on Y is

$$x = 3.18 + 0.61y$$

This minimises the sum of squares of deviations such as **PB**. Each line can be used to obtain a best estimate of one variable from a knowledge of the value of the other variable. They demonstrate two points.

We cannot expect a regression line to slice the ellipse down the middle, since they cannot *both* do this, and there is no reason to give preference to one over the other.

The second point is of greater practical importance. Suppose we are told that $x = 15$ and we want the best estimate of y. It is given by the regression of y on x and is $y = 19.5$, as at **K**. But if we are instead told that $y = 19.5$, the best estimate of x is given by the other regression line; it is not 15 but 16.8.

If British is the best estimate for the nationality of a man seen in the Sahara wearing a bowler hat and carrying a furled umbrella, it does not follow that the best estimate of what a British person will be doing in the Sahara is wearing a bowler hat and carrying a furled umbrella.

Unfortunately the manuals to well-known calculators have ignored this point. As explained in Appendix A, they correctly calculate the values of a and b for the regression of y on x, and correctly use them in estimating the 'missing' value of Y associated with a specified value of X. But they also quite incorrectly adapt these values of a and b to estimate the missing value of X associated with a specified

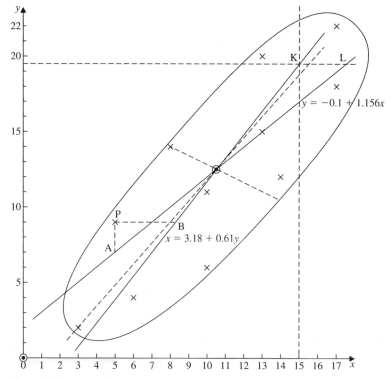

Figure 14.3

Table 14.5

x	3	5	6	8	10	10	13	13	14	17	17
y	2	9	4	14	6	11	15	20	12	18	22

value of Y. They should, instead, have reworked the regression as in Box 14.2; see Noteboxes 14.2 to 14.4 as well as Appendix A.

14.7 Non-linear regression and multiple regression

The method of least squares is also used to fit curves rather than straight lines to data, avoiding the need to transform the data. The calculations for **non-linear regression** are complicated and it is much easier to use appropriate computer programs. But be careful that you know the assumptions that should be valid for the application of the technique.

More complicated problems, in which we may try to relate the values of one variable Y (say building cost) to the values of two or more **explanatory** variables $X_1, X_2, X_3, \ldots$ (such as floor space, number of storeys and distance from a main road) can be tackled through the use of **multiple regression**, which is beyond the scope of this book.

In the simplest case, we assume that the independent variable is related to the explanatory variable by an equation such as

$$y = a + b_1 x_1 + b_2 x_2 + b_3 x_3 + \ldots + u$$

where u is an error or disturbance term assumed to be normally distributed around zero. We then calculate the values of a and the coefficients $b_1, b_2, \ldots$ that provide a least squares (or other) best fit. An important assumption is that the values of explanatory variables are independent of each other.

Consolidation exercise 14B

1. Write down a typical regression equation of (a) y on x and (b) x on y. How are the coefficients in (a) and (b) related?

2. What does the regression equation of X on Y minimise?

3. What are (a) non-linear regression and (b) multiple regression?

Notebox 14.1

Regression formulae

There are a few versions of the formula for the least squares regression of y on x. It can be shown that the values of a and b minimising the sum of squares of the y-deviations from the line are given by

$$a = \frac{(\sum y)(\sum x^2) - (\sum x)(\sum xy)}{n \sum x^2 - (\sum x)^2}$$

$$b = \frac{n \sum xy - (\sum x)(\sum y)}{n \sum x^2 - (\sum x)^2}$$

These formulae are not as complicated as they look. Note that the two denominators are the same and include only x and n. To use the formulae, we first calculate four sums: $\sum x, \sum y, \sum x^2$ and $\sum xy$. We then put these values, and n, in the above formulae.

Alternatively we can use formulae derived from equation [14.2] that make the calculations easier. These use our definitions of $\bar{x}$ and $\bar{y}$ as $\sum x/n$ and $\sum y/n$. First we define

$$s_{xx} = \frac{\sum x^2}{n} - \left(\frac{\sum x}{n}\right)^2 = \frac{1}{n}\sum x^2 - \bar{x}^2 \quad s_{yy} = \frac{1}{n}\sum y^2 - \bar{y}^2 \qquad [14.3a]$$

$$s_{xy} = \frac{\sum xy}{n} - \frac{(\sum x)(\sum y)}{n^2} = \frac{1}{n}\sum xy - \bar{x}\bar{y} \qquad [14.3b]$$

⇨

which make use of the calculated means. It then follows from equation [14.2] that

$$b = s_{xy}/s_{xx} \qquad\qquad\qquad [14.3c]$$

$$a = \bar{y} - b\bar{x} \qquad\qquad\qquad [14.3d]$$

which emphasises that the regression line goes through the point defined by the two means x (or μ_x) and $\bar{y}$ (or μ_y).

In the example of the text, we have

$$s_{xx} = \frac{\sum x^2}{n} - \left(\frac{\sum x}{n}\right) = (1045/5) - (69/5)^2 = 209 - 190.4 = 18.56$$

$$s_{xy} = \frac{\sum xy}{n} - \frac{(\sum x)(\sum y)}{n^2} = (2094/5) - (69 \times 139)/25$$

$$= 418.8 - 383.64 = 35.16$$

We now calculate first b from

$$b = s_{xy}/s_{xx} = 35.16/18.56 = 1.894(4)$$

and then a from

$$a = \bar{y} - b\bar{x} = 139/5 - 1.894(4)(69/5) = 1.657$$

Note for later use that s_{xy} is called the **covariance** of x and y, sometimes written $\text{cov}(x, y)$.

Some books use a capital S, which is defined slightly differently:

$$S_{xx} = \sum x^2 - \frac{(\sum x)^2}{n}$$

$$S_{xy} = \sum xy - \frac{(\sum x)(\sum y)}{n}$$

In this case $b = S_{xy}/S_{xx}$ and a follows as before. The version we have given seems easier to use.

Notebox 14.2

Checking the regression line

A regression line should go through $(\bar{x}, \bar{y})$. This means that in Example 14.2 the values of $\bar{x}$ and of $\bar{y}$ should satisfy the equation $y = 487.84 + 5.472x$.

Moreover, putting $x = 0$ in the equation $y = 487.84 + 5.472x$ gives $y = 487.84$. The regression line therefore passes through the point P on the y-axis.

The regression line is the line joining P and $(\bar{x}, \bar{y})$ as shown. It looks as if it may indeed be the line of best fit. If it did not, it would be time to check the calculation.

Notebook 14.3

b, b' and r²

If one of the two regression coefficients b and b' is known, the other can be found quickly by using the fact that their product equals the square of the correlation coefficient r defined in Chapter 15. Users of programmed calculators who want a' and b' are advised to proceed as follows:

(i)   Enter the data and use the appropriate output button to find r and b.
(ii) Square r.
(iii) Divide the answer by b. This gives b'.
(iv) Multiply b' by $\bar{y}$ (produced by another button).
(v)   Subtract the product in step (iv) from $\bar{x}$ (produced by another button). This gives a'.

For example, in Exercise 14.1 $b = -2.062$. With the same data input we can find, by using the right button, that $r = -0.9517$ (which we also calculate in Chapter 14). It follows that $b' = 0.9317^2 \div -2.062 = -0.439$. Use of $b' = s_{xy}/s_{yy}$ gives the same answer.

Summary

1. A scattergram may provide a visual suggestion of a linear relationship between two variables. Regression techniques are used to measure the extent to which values of one variable depend on values of another. This information is important both in its own right and in activities such as forecasting and estimating missing values. We use this second activity to explain the ideas and techniques.

2. The techniques described in this chapter are applicable only if the scattergram consists of points scattered over a roughly elliptical area. In such cases two straight regression lines may be drawn through the point corresponding to the arithmetic mean values of the two variables. Their slopes are such that one line provides the 'best' estimate of a 'missing' value of the dependent variable Y corresponding to a stated value of the independent variable X. This is called the (least squares) regression line of Y on X. The other provides the 'best' estimate of a 'missing' value of X corresponding to a stated value of Y, and so regresses X on Y. The two lines coincide only if all the points in the scattergram lie exactly on one straight line.

3. We define 'best' as minimising the root mean square of the deviations. The equation of the least squares regression line of Y on X is given by equation [14.1], in which the constant and the coefficient are calculated from equation [14.2]. In Notebook 14.1 we show that this leads to equations [14.3a] to [14.3d], which are easier to use. There are similar equations for the regression line of X on Y.

4. The procedures for calculating the two lines are summarised in Boxes 14.1 and 14.2. We emphasise the importance of using the right line, pointing out an error in the programs and instruction books of some well-known calculators. In particular, suppose that y_1 is the best estimate of y given that $x = x_1$. Then the best estimate of x given that $y = y_1$ will *not* be x_1 unless this happens to be the mean $\bar{x}$.

5. Questions about the reliability of estimates derived from the regression equations are answered in later chapters.

6. Other types of regression are non-linear regression and multiple regression.

7. Noteboxes discuss regression formulae, checking the calculation of the regression line, the 'other' regression line and the relationship between the regression and correlation coefficients.

Parametric correlation and a preview of significance

Introduction

Example 15.1 Looking at the data in Example 14.1, Harry wonders how much of the variation in history marks between one daughter and another can be explained by the variation in drinking by the fathers; and – to be open-minded about causation – how much of the variation in drinking can be explained by the history marks. He is not clear about what he means by 'variation'.

Example 15.2 Steve wants to know how well the period on the market and level of embellishment rebate (reported in Example 14.2) move together, and how much of the variation – a word picked up from Harry – in the rebate can be explained by variations in the period on the market.

In the last chapter we used linear regression to estimate the slope of the straight line that best fits the points of the scattergram of the data, first checking that the scattergram had the right shape. This slope, which tells us how much we can expect the dependent variable (usually y) to increase if the independent variable (usually x) increases by unity, is given by the regression coefficient. We saw that there are two regression lines, and two regression coefficients, depending on which variable is being treated as dependent.

The correlation coefficient, which we meet shortly, treats both variables as independent. It measures how closely the two variables move together. Its square tells us how much of the variance of one variable 'can be explained by' the variance of the other, but variance is not the same as 'variation'. Possibly Harry and Steve are being a little optimistic.

In the last chapter we assumed that we were interested only in the data we had, except perhaps for a single missing value. The same assumption (but with no missing values) is made in this chapter, until very close to the end when we take a preliminary look at the problem of extending the relevance of a correlation coefficient.

Note also the conditions of validity for the techniques we are about to describe:

(1) The data must be on interval or ratio scales, such as scores, marks, heights, numbers of pints, temperatures, and so on, as explained in Chapter 33 and

Notebox 33.1. The techniques cannot be applied to ranks; Chapter 38 describes an appropriate measure for them.

(2) The technique of correlation analysis developed in this chapter, more properly called linear parametric correlation, is valid only if linear regression analysis is valid, as described in Chapter 13. The scattergram has to be cigar-shaped.

15.2 A diagrammatic approach

We have seen that there are two regression lines – the regression line of Y on X, and the regression line of X on Y. This is illustrated in Figure 15.1

If these two lines are perpendicular to each other, there is no tendency at all for the values of X and Y to move together (or against each other). We say that there is **no (linear) correlation** between X and Y. Reflecting this, the correlation coefficient we define shortly has a value of zero. Figures 15.2(a) and (b) give examples of zero correlation.

If the two regression lines coincide, so that the angle between them is zero, a knowledge of either x or y allows the other to be calculated exactly and without any doubt. We say that there is **perfect correlation** between X and Y. The correlation coefficient is defined so that in this case its numerical value is unity. This is illustrated in Figure 15.3(a). The one important exception is that if the line is parallel to either the x-axis or the y axis, then Y or X is constant and the question of correlation does not arise.

In between the extreme cases of perfect correlation and no correlation, we may have a scattergram with a roughly elliptical shape and two regression lines cutting at an angle, as in Figure 15.1. The degree of correlation increases from zero through low and high to perfect as the angle between the lines decreases from 90° to 0°.

If there is a tendency for the values of X and Y to increase (or decrease) together, we say there is a **positive correlation**, and the calculated coefficient will

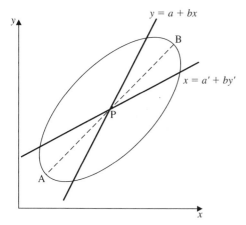

Figure 15.1

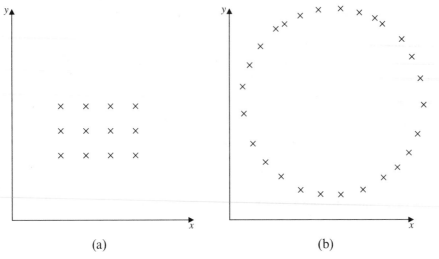

(a)   (b)

Figure 15.2

have a positive sign, as in Figure 15.3(b). If one variable tends to decrease as the other increases, we describe the correlation as **negative**, as in Figure 15.3(c).

15.3 **The product-moment correlation coefficient**

The most widely used measure is the (Pearson) product-moment correlation coefficient defined as

$$r = \frac{\sum (x - \bar{x})(y - \bar{y})}{[\sum (x - \bar{x})^2 \sum (y - \bar{y})^2]^{\frac{1}{2}}} \qquad [15.1]$$

This can take values ranging from +1 to −1. The numerical value of the coefficient measures the tightness with which the points cluster around the major axis – the longest straight line that can be drawn within the ellipse – as APB in Figure 15.1. The sign indicates whether the values of X and Y tend to move in the same or opposite directions.

Use of equation [15.1] is cumbersome. A formula derived from it and much easier to use is

$$r = \frac{S_{xy}}{\sqrt{S_{xx}S_{yy}}} \qquad [15.2]$$

where

$$S_{xx} = \frac{1}{n}\sum x^2 - \bar{x}^2 \qquad S_{yy} = \frac{1}{n}\sum y^2 - \bar{y}^2 \qquad S_{xy} = \frac{1}{n}\sum xy - \bar{x}\bar{y}$$

The layout for the calculation is similar to Table 14.4. Many calculators are programmed to give r immediately once the values of x and y have been tapped in

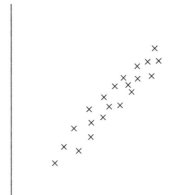

(a) Perfect positive correlation   (b) High positive correlation

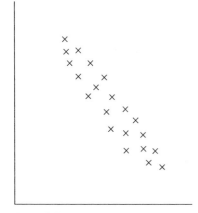
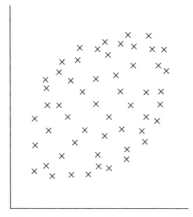

(c) High negative correlation   (d) Low positive correlation

Figure 15.3

(Appendix A), but be careful not to assume that the calculation is valid just because the calculator gives an answer. Always check first by drawing a scattergram. Box 15.1 gives the procedure in the absence of a programmed calculator.

Box 15.1

How to ... find the product-moment correlation between two variables

1. Draw a scattergram and use it to check the validity of applying the following procedure.

2. Tabulate the variables and find the sums of squares and products as for a regression calculation (Box 13.1).

3. Evaluate s_{xx}, s_{xy} and s_{yy} defined as

$$s_{xx} = \frac{1}{n}\sum x^2 - \bar{x}^2 \qquad s_{yy} = \frac{1}{n}\sum y^2 - \bar{y}^2 \qquad s_{xy} = \frac{1}{n}\sum xy - \bar{x}\bar{y}$$

4. Substitute these values in

$$r = \frac{s_{xy}}{\sqrt{s_{xx}s_{yy}}} \tag{15.2}$$

As examples we may take Harry's data in Example 14.1. Using the formulae given in Box 15.1 and the sums calculated in Chapter 14, we have

$$s_{xx} = (6869/17) - 18.1765^2 \qquad s_{xy} = (11\,541/17) - (18.1765 \times 45.706)$$
$$= 73.675 \qquad\qquad\qquad = -151.89$$

We also have

$$s_{yy} = \frac{1}{n}\sum y^2 - \bar{y}^2 = (41\,391/17) - 45.706^2 = 345.726$$

These values give

$$r = \frac{-151.87}{\sqrt{(73.675 \times 345.726)}} = -0.9517$$

which can also be obtained directly from programmed calculators. It means that the quantity drunk and the marks in history move in opposite directions (shown by the minus sign) and show a very close statistical relationship.

We have to be careful not to misinterpret this high negative correlation. It should not be taken as evidence of any causal connection.

Moroever, the conclusion, like the evidence, is restricted simply to those fathers in that pub, to pints of NonTox drunk by those fathers, and to the history marks of their daughters. We have no right to extend the result to other instances or to speculate on what might have been if there had been many other customers or many more pubs. Questions such as these can be answered only through the use of techniques developed for the analysis of sample data. We briefly touch on these in a moment, but we do not consider them in any depth until Chapters 31 and 32.

15.4   The size of the correlation coefficient

The value of a correlation coefficient depends on all the paired values of (x, y) corresponding to points on the scattergram. If these values are altered, the correlation coefficient will usually change. But there are important kinds of alteration to the values of (x, y) that leave the correlation coefficient unaffected:

(1) If an alteration to the values of (x, y) has the effect of shifting the whole scattergram vertically or horizontally (or both), without rotating it, then the correlation coefficient is unaltered. This means that if all the xs have an amount p added to (or subtracted from) them, and all the ys are similarly altered by an amount q, the correlation remains the same. This becomes important when we use assumed means, as described in Appendix B.

(2) If an alteration to the values of (x, y) affects the scattergram by stretching it (or compressing it) uniformly parallel to one or both of the axes, the coefficient is unchanged. Thus if all the xs are multiplied (or divided) by p and all the ys multiplied (or divided) by q, the correlation will be unaltered, provided that p and q are of the same sign. (If p and q differ in sign, then the sign of the correlation coefficient will be reversed.) This becomes important if we work in new units, as in Appendix B.

Any other alteration to the values of the variables will almost certainly change the correlation coefficient.

A value very close to zero has to be suspected. Try taking a set of 10 pairs (x, y) that are clearly unrelated and see how close to zero correlation you get by pure accident. We return to this subject in Chapter 31.

It can be shown that r^2 measures the fraction of the total variance in the values of y that can be attributed to or 'explained by', the variance of x. Thus a correlation coefficient of 0.5 means that one-quarter of the variance in y can be explained by the variance of x. The value r^2 is sometimes called the **coefficient of determination**.

It follows that, in a purely statistical sense, 0.939^2 (about 88%) of the variance in either of the variables of interest to Harry can be 'explained by' the variance in the other variable and the high correlation. 'Variance' and variation are not quite the same thing (Notebox 15.1), but it is as far as one can go in answer to the question.

Consolidation exercise 15A

1. What does a correlation coefficient measure?

2. Which is the dependent variable in the correlation of the data mentioned in Example 14.1?

3. What are the two main requirements of the data if we are to perform the kind of correlation analysis described in this chapter?

4. What is the angle between the regression lines if there is no correlation?

5. What is meant by a 'negative' correlation?

6. If all values of x are increased by 5 and all values of y are decreased by 3, what happens to the correlation coefficient?

7. If all values of x are doubled and all values of y are reduced by 20%, what happens to the correlation coefficient?

8. If the correlation coefficent between X and Y is 0.7, how much of the variance in Y can be explained by the variance of X? How much of the variance in X can be explained by the variance of Y?

15.5 A preview of significance

Whether the correlation indicated by a coefficient can be taken as evidence of a correlation among members of a much larger group (or, to use more technical language, whether a correlation coefficient based on sample data implies a correlation in the population from which the sample is drawn) depends on the size of the coefficient, the size of the (sample) group for which data was used, and how the sample has been taken. These points are considered more fully in later chapters but a preview may be useful.

Two six-sided dice labelled X and Y were rolled 10 times and the scores recorded. Their scattergram was acceptable and the calculated correlation coeffficient between these 10 pairs of values was 0.268. Can we take this as evidence of some kind of link between the results of throwing the die X and the results of throwing Y? If we had thrown the dice many thousands of times, would we get a similar correlation?

The experiment was repeated. The scores for this second set of 10 rolls of the dice gave a correlation coefficient of −0.440. This is a negative correlation, somewhat stronger than the positive correlation revealed by the first set of throws. Which, if either, of these coefficients can be taken as indicating what would happen if we threw the dice many times?

A further repetition of the experiment led to the third coefficient −0.136. These three results are so different that it would clearly be unwise to make a general assertion about these two dice on the basis of one set of ten throws.

Before trying anything else we might consider what the correlation of all 30 throws may be. By adding the totals for the three experiments we obtained totals that enabled the coefficient for this correlation to be calculated as −0.2778.

Instinctively we might feel that a result based on 30 throws is more reliable than one based on 10 throws, and also perhaps more reliable than a conclusion based on three sets of 10 throws taken separately. But is this so? In any case, when we consider the matter of throwing two properly balanced dice, surely we would expect no correlation between the results? Yet none of our answers is zero.

When we say that we would expect no correlation, we mean that if we took very many throws of the two dice and correlated the results, we would expect an answer of zero, or very close to zero. If that is so, what are the chances of getting a coefficient of −0.2778 from 30 throws by accident rather because of a real correlation?

The answers to such questions must wait. We have not yet considered problems of using sample data to estimate properties of the 'population' from which the sample is drawn. It is a big subject that we treat in detail in later chapters. We can, however, anticipate a little by making the following statements that here have to be taken on trust. They refer to the reliability of correlation coefficients derived from samples as estimates of the (true) correlation in a very large 'population' from which the sample has been drawn. We suppose that the act of sampling has not introduced any bias, a concept we discuss more fully later on.

It can be shown by methods described in Chapter 31 that if we have a large population in which the correlation between x and y is zero, there is a 10% chance that a properly drawn sample of 10 will produce a correlation coefficient of anything between -0.55 and $+0.55$ by pure chance. It can also be shown that there is a 5% chance of the sample producing a result as high (numerically) as 0.63.

Thus if we want to say 'there is a chance of 19 in 20 (95%) that there is a non-zero correlation', we would need to get a result numerically greater than 0.63 from a sample of 10.

If we take a sample of 30, there is a 10% chance of getting a result as high (numerically) as roughly 0.31 by pure accident. There is a 5% chance of getting one as high as 0.36.

Bringing all this and similar results together, we can produce Table 15.1. Let us see how to use this table. Suppose we have a *sample* of the size shown in the first row and that it gives us a correlation coefficient of size r. We know this may be accidental or it may indicate a real correlation in the population from which the sample has been drawn. If r is greater than the corresponding entry in the second row, we can be at least 95% certain there is a correlation in the population. If it is less than the value in the second row, there may still be a correlation in the population but we cannot be 95% certain of it.

Table 15.1

Sample size	3	6	10	20	30
Minimum value of correlation coefficient (5%)	0.997	0.811	0.632	0.444	0.361

Even if we are very confident there is a correlation in the population, we must be careful not to take this as evidence of causation. It remains a statistical association, and although it may prompt a search for some physical, biological or other relationship, there is no guarantee in the statistics that if we find one it will be the only one, or right; or indeed that one exists.

What we have just said is an oversimplification and it is better to wait until Chapter 31 before testing the meaning of correlation coefficients. This preview is part of a bridge between the study of populations and the study of samples, which we continue in the next chapter.

15.6 Multiple correlation and partial correlation

One property of the correlation coefficient r is that its square measures the amount of the variance in the values of the dependent variable Y that can 'be explained by' variations in the values of the independent variable X.

In many cases the values of a dependent variable Y may be related simultaneously to values of more than one independent variable, denoted perhaps by $X_1, X_2, X_3 \ldots$. **Multiple correlation** establishes the amount of the variance in Y that can be explained by simultaneous changes in the several independent variables. This can be decomposed through *partial correlation* coefficients into the parts played by these independent variables separately in explaining the variance of the dependent variable. Many theoretical and practical problems arise and the subject is beyond the scope of this book.

Notebox 15.1

Variance and variety

Variance is the square of a measure of spread. Insofar as variation from a mean value increases the spread, there may seem to be some justification for suggesting that in layperson's language it may be called a measure of variation or variety. But this is not always so. Consider two sets of numbers

(a) 3, 5, 7, 9, 10, 12, 14, 16
(b) 3, 3, 3, 3, 16, 16, 16, 16

The variance of (b) is greater than the variance of (a), but (a) shows the greater variety, or variation.

Summary

1. Although regression analysis is essentially concerned with using a value of an independent variable to estimate the value of a dependent variable, correlation analysis is concerned with examining how well the values of the two variables fit together; and these two variables are of equal status, neither being dependent.

2. The extent of correlation is measured by a correlation coefficient defined to have a value of zero when there is no correlation and a numerical value of unity when there is a perfect (positive or negative) correlation.

3. Here we are concerned only with the correlation of variables that are measured on the interval or ratio scales and produce an elliptical scattergram. We use the product-moment correlation coefficient defined by equation [15.1]. For purposes of manual calculation, equation [15.1] is usually rewritten as equation [15.2]. The procedure is summarised in Box 15.1. Appendix B contains some tricks.

4. Any alteration to the variables which simply moves the scattergram without altering its shape, or stretches or squeezes the scattergram uniformly parallel to one or both axes will leave the correlation coefficient unchanged.

5. A full discussion of the meaning of the size of the coefficient has to wait until Chapter 31. We anticipate ideas developed more properly there in order to point out that the chances of getting a coefficient of any given size by accident depend on the number of pairs of values being correlated. Table 15.1 shows that if there are only 6 pairs then a value as high as 0.811 is needed before we can be 95% confident that it is not a fluke result, but if there are 30 pairs any value as high as 0.361 passes this test.

6. Even correlation coefficients that pass the test mentioned in paragraph 5 should not be taken as proof of any causal or other relationship. There may be common factors at work, or the correlation may be quite inexplicable and possibly spurious. Common factors could also obscure a correlation, producing a value that is too low. Zero correlation has to be treated with suspicion.

7. The square of the correlation coefficient measures the fraction of the total variance in the values of one variable that can be attributed to, or 'explained by', the variance of the other variable. This square is also equal to the product of the slopes of the two regression lines, as explained in Notebox 14.3.

8. Two related ideas are multiple correlation and partial correlation.

9. A notebox shows why the variance is not a satisfactory measure of variety.

16

Linking sample and population

 16.1 **Introduction**

Example 16.1 A national sample survey conducted in 1991 by UK government statisticians showed that the average income of one-adult non-retired households, before deduction of tax and receipt of state benefits, was £11187. Many commentators remarked that this was far higher than the figure that would have been obtained if every household had been questioned. How reliable is £11187 as an estimate of the national average? (The published table gives the sample size as 888 and prints a 'standard error' of £497.)

Example 16.2 A public opinion survey conducted in Statingham on 10 May 1996 showed that 52% of those questioned opposed the proposal that the local cricket pitch should be sold to a developer, whereas 48% supported it. Councillor Brickman argued that a sample survey was useless. He said the truth could well be that if everybody had been questioned on that day, there would have been a majority in favour of the sale. Is he right?

Example 16.3 Harry has an idea that the average age of his customers' cars is greater than the average age of all cars still legally on the road in this country. Thinking about this, he wonders what the national average age of cars may be, and how he could find it.

Until near the end of the last chapter we were concerned only with populations. Then we asked whether a correlation in a sample had any implication for a population from which it was drawn. The questions in this chapter are of a similar kind: how useful and reliable are sample results as indicators of things to do with the population? twb = .10w > We investigate why a study of populations *must* sometimes be based on a study of samples; why samples are necessary; and what is expected from them. Then we take an informal overview of the reliability of sample results, paving the way for the rest of the book.

 16.2 **Why we need samples**

The government statisticians mentioned in Example 16.1 collect information for use by the government and those members of the public who wish to know about the social and economic well-being of the people. Decisions about what information they collect are taken more by politicians and high-powered advisory committees than by the statisticians themselves, who sometimes offer advice but are concerned mainly with collecting and processing information that will enable them to provide the required reliable statistics.

One well-known statistical enquiry is the population census, which is held in most countries. The main purpose of early censuses was to count the number of people as a basis for calculating potential military strength or tax revenue. Now this count also helps in calculations of housing, nutritional, medical and other needs. As society has become more sophisticated, the census enquiry, which is addressed to every household and aims at getting information about everybody, has grown to include questions about a wide range of topics, including employment, accommodation, education and car ownership. Government policies and actions, both national and local, are based on the answers.

In Great Britain this census is undertaken every tenth year. The huge scale of the tasks of organising it and collecting the data is sufficient to explain why it is not done more often. Another reason is the size of the task of converting the raw data collected on several millions of forms into reliable statistical tables, which normally takes a few years.

Society changes in the 10 years between censuses. Various other sources of data allow some updating from time to time, but during periods of rapid change this may not be adequate or sufficiently reliable for making up-to-date policy decisions. Another major enquiry may therefore be needed by midway between censuses.

This led in 1966 to the introduction of a 10% sample census midway between censuses. Questionnaires were issued to a carefully devised sample of 1 household in 10, and of 1 in 10 persons living in hotels and institutions. The aim was to make the sample as representative as possible of the whole population. Something like two million forms were completed and used as the basis of tabulations.

The introduction of this sample in place of a full census saved money and led to faster publication of information. Those are the main reasons for the use of samples. Another major reason is that sometimes making an enquiry changes, or even destroys, the item that is subject to it. For example, a manufacturer of ballpoint pens can know how many hours of continuous writing a pen will provide only by using it to exhaustion. Obviously the manufacturer has to test (and use up) only a sample of his or her output, and then make an estimate. How reliable will this estimate be, and how should the manufacturer take the sample so that it will be as reliable as possible?

Ideally, the manufacturer would like a reliable **point estimate** for the life of a pen, such as 843 hours. More realistically, the manufacturer will be content with an **interval estimate** such as 802–884 hours, within which he or she can expect the life to occur. The manufacturer also wants to know the chances of a pen not lasting as long as (say) 802 hours (so that he or she can have some idea of how often there will be a disgruntled customer); and of a pen lasting longer than (say) 884 hours (because in that case he or she may be understating the pen's merits and possibly sellng it too cheaply).

16.3   Relying on samples

So far we have been concerned almost entirely with populations. Admittedly we have often been concerned with fairly small numbers of items, ranging from 10 houses to 301 residents on an estate. But they have not been samples. Except for a few paragraphs in Chapters 14 and 15, our interest has been restricted to the set of individuals or other items about which we have had information: the 301 residents of that estate, the 60 people who were measured for seats, the 17 customers whose drinking habits were noted by Harry, and so on. Except to make a single estimate when illustrating a method, we have not attempted to use our information as the basis for statements of wider applicability. We did not use our data about the ages of the 301 residents of one estate to make estimates of the age distribution of all people in Statingham.

Yet in examples Examples 16.1 and 16.2 information has been gathered about a comparatively small number of people and used to support statements about the incomes or opinions of many more. The sample census questioned a much larger number of people, but they still amounted to only 10% of the total.

We all have experience of this being done, as in reports of public opinion polls. These polls are often based on interviews of just over 1000 people, yet they contain statements such as '35% of the country's electorate support greater grants for undergraduates.'

Now we must consider whether this kind of thing is justified. And if it is justified, why have we been so cautious about not using Lady Agatha's data to make a statement about seat heights for all offices?

To do so, we must use a few words in a rather special way. In Chapter 2 we described the complete collection of items in which we are interested as a *population*. In Example 16.3 Harry is interested in the ages of all cars still legally on the roads in this country. Those cars form his population, which has a size of several million.

We have seen how to describe a population on the basis of information about all its members. We have drawn frequency distributions, and calculated various averages, measures of spread, and regression and correlation coefficients. We could try to tackle Example 16.3 in the same way. With enough effort the information about car ages could be extracted from government files, but it would be an enormous job. There is obviously a temptation to go to the car records and take a *sample* of just some of them, and to find the age distribution of cars in that sample. Can we use this to make statements about the age distribution of all cars? Can we use the mean age of cars in the sample as a reliable *estimate* of the mean age of all cars?

16.4   Three factors to keep in mind

The answers to such questions depend on what we mean by 'reliable estimate' and 'safe', but they also depend on a few other factors. Two that are within our control concern the sample: (i) the nature of the sample and (ii) its size.

One other important factor concerns the population, and so is beyond our control. Perhaps rather surprisingly the size of the population does not matter, as long as it is very big compared with the sample and as long as the question is not too complicated. (If properly taken, a sample of 1000 from a population of 1 000 000 is as good as a sample of 1000 from a population of 50 000 000.) What does matter is (iii) the size of the population variance; in our example it is the extent to which car ages are spread out. We now take a brief look at these three factors.

The nature of the sample

If you derive your information from a sample, you can never be absolutely confident that your point estimate for the population is correct. But if you take the sample in the right way, you can be perhaps 95% confident about an interval estimate.

The key point is that the sample should be **representative** of the population, which is not always easy to achieve. A sample that is not representative is **biased**. If many samples are taken and they are all biased in the same way, the bias is **systematic**.

One method of sampling is called **random sampling**. There are a few different kinds of random sample. Whenever a *statistician* refers to a random sample, we may suppose it is a *simple random sample*, unless stated otherwise. A simple random sample is a sample such that every item in the population has exactly the same chance of being included in it; and whether one item is included is in no way affected by whether another item is included. If we take a random sample of 100 persons from all persons living in Statingham, the inclusion of Harry in no way affects the chance of his wife being included.

Unfortunately, many people use the phrase 'random sample' to refer to any convenient selection, which is likely to be far from random. Obtaining (or 'drawing') a random sample is not easy unless you know how. Some examples of how not to do it are given in Notebox 15.1. Random samples are not haphazard selections; they have to be very carefully chosen. In Appendix C we describe how to use a table (or a computer program) of random numbers, along with an appropriate 'sampling frame' to do this.

An important point discussed more fully in Appendix C is that the standard methods for taking random samples are strictly valid only if the sample is taken from an infinitely large population. In practice it is enough if the population is very large compared with the size of the sample. Use **sampling with replacement** for smaller populations, as described in Appendix C. This has the effect of converting the population into one that is infinitely large. Alternatively some of the formulae used to perform the calculations can be altered to take account of the known population size.

If the sample is random from a very large population, or has been taken with replacement, it is fairly easy to make correct statements about the reliability of results based on it. But this is more difficult for some other kinds of carefully controlled sample (such as mentioned in Appendix C). Sometimes a controlled sample may be more reliable than a simple random sample.

Random sampling is not guaranteed to produce a representative sample, but it is guaranteed not to produce a systematic bias. For example, suppose that 52% of all adults in the country are females. Let us take a random sample of 100 households. It could well be that in that sample only 47% are females. In that sense, it is a biased sample. However, if many random samples of 100 were taken in exactly the same way then half of them would have 52 or fewer females and half would have 52 or more females. Most of them would be biased one way or the other, but on average they would be correct. Random samples also have other properties that sometimes make their use essential.

The sample size

It can be shown that with a random sample the reliability of a sample result as an indicator of a population result depends on the square root of the size of the sample, so a sample that is nine times as big will treble the precision. As the costs of collecting and processing the data increase with the size of the sample, this usually means that costs increase a great deal faster than precision. In choosing a sample size, the gain in precision has to be weighed against a greater gain in costs.

The population variance

The reliability of a population estimate based on a sample also depends on the variance of the *population* data. A small variance means the data are bunched together, so a sample is not likely to be seriously unrepresentative; but a large variance means the data are so spread out that an unrepresentative sample may easily arise. Unfortunately, we are unlikely to know the population variance. In that case we may be able to assume a value for it (as illustrated in later chapters). More usually we fall back on information about the sample variance to make an estimate of the population variance.

Consolidation exercise 16A

1. How does a sample differ from a population? Is there a minimum size for a population ?
2. A bank examines the accounts of 20 customers to see (a) whether the customers should receive nasty letters and (b) whether the bank should change the rules about granting overdrafts. In which case (if any) should the 20 accounts be treated as a population, and in which a sample?
3. Sample data are used to make an estimate for a population. Name three factors on which the reliability of the estimate will depend.
4. What is the necessary relationship between sample size and population size?
5. Define a simple random sample. Is such a sample ever biased?

16.5 Using samples to make estimates for populations

We described our populations by calculating such things as means and standard deviations. More formally, we call these 'things' **parameters** of the population. They are numbers that help to describe the population distribution. Some other frequently used parameters are the mode, the median and the variance.

One use of a sample is to allow these population parameters to be *estimated*. We say 'estimated' because calculating them with certainty would be possible only if we had information about every member of the population.

Although we refer to the *population mean* as a parameter, we never call a *sample mean* a parameter. Staticians restrict the word *parameter* to populations. We call a sample mean a **statistic**. This is a word that already has a few different meanings (Chapter 2), and it seems to be a pity that some other word was not chosen, but it is too late to change it now. Every population parameter has its corresponding sample statistic.

Population parameters are conventionally indicated by Greek letters, and sample statistics by Roman letters. Here are the most important ones:

- The *population* arithmetic mean and standard deviation are *parameters* denoted by and μ, and σ.
- The *sample* arithmetic mean and standard deviation are *statistics* denoted by $\bar{x}$ and s.

The first step in estimating any population parameter is to calculate the corresponding sample statistic, but note the next paragraph.

The second, and essential, step is to ask whether the sample statistic (and especially the sample mean and the sample standard deviation) can be used immediately as an estimate of the population parameter, or whether some kind of modification of the sample statistic is necessary.

It can be shown that a sample mean provides a 'best' estimate of the population mean, where 'best' has the meaning described in Notebox 16.2. However, as explained in Notebox 11.2, a sample standard deviation does not provide a best estimate of the population standard deviation. It can be shown that on average the sample standard deviation defined by

$$s = \sqrt{\frac{\sum (x - \bar{x})^2}{n}}$$

will underestimate the population standard deviation σ. The best estimate of the population standard deviation σ is

$$\hat{\sigma} = \sqrt{\frac{\sum (x - \bar{x})^2}{n - 1}} = s\sqrt{\frac{n}{n - 1}} = s\sqrt{n/(n - 1)}$$

It is for this reason that some writers define the standard deviation of a sample to have $(n - 1)$ where we have n.

Reliability and probability

The estimates of average income in Example 16.1 and the proportion of people opposing the sale in Example 16.2 are point estimates. The questions relating to them can be expressed more formally:

- *Example 16.1.* How reliable is the statistic $\bar{x} = £11\,187$ as a point estimate of the population parameter μ?
- *Example 16.2.* How reliable is the sample statistic $p_{\text{sample}} = 0.52$ as a point estimate of the parameter p_{pop}?

These questions ask how much reliance, or **confidence**, we can put on the point estimates. Since a point estimate is highly unlikely to be absolutely correct (and is therefore highly likely to be wrong), a more useful question is: what is the *chance* of the estimate being wrong by more than a stated amount? Before we can make much sense of this, we have to learn a bit about chance, which means we must look at simple probability. Until we have done this, we cannot answer either Example 16.1 or Example 16.2. We consider these examples and Example 16.3 further in Chapter 21.

 Hypothesis testing

Another use of samples is to test ideas, theories, beliefs or hypotheses about the population. For example, we may have a theory that in this country daughters of fathers who drive many miles every week do better in school than daughters of fathers who drive very little.

One way of testing this is to take two random samples, one drawn from all fathers who drive a great deal and one from all fathers who drive very little. Suppose that we do this, difficult though it may be, and then look at the school performances of daughters of the two groups of fathers. We may conclude that, as far as the samples are concerned, daughters of heavy drivers do indeed perform better. Can we extend this conclusion to the whole country?

We know that if we repeat the sampling we will almost certainly come up with different people who would have different driving patterns and academic achievements. Possibly the second sample would not support the statement that daughters of heavy drivers do better. So how confident can we be about this statement? What are the chances of our conclusion based on one sample of heavy drivers and one of light drivers being wrong? Once again we are talking about confidence and chance – about probability.

In the next few chapters we look at all these matters in much more detail. To begin we have to familiarise ourselves with a few basic ideas about probability. More complicated applications of probability are not needed at this stage and are deferred until near the end of the book.

Consolidation exercise 16B

1. What is a parameter? Why do we estimate a parameter?
2. What are the the first two steps in estimating a parameter?
3. What do the following symbols mean:

    μ s $\bar{x}$ σ

4. What are (a) a point estimate and (b) the other kind of estimate? Which of these two kinds of estimate is more likely to be correct?
5. How do we try to measure confidence in an estimate?

Notebox 16.1

How *not* to take a random sample

Of 100 adults from a town that has 50 000 adults

1. Take every 500th person from an alphabetical list.
2. Stick pins in a telephone directory.
3. Stand outside a supermarket and stop people.

Of 20 streets from a town that has 500 streets

1. Take every 25th street from a street list printed on a town map.
2. Throw darts at a town map.
3. Drive around and note the street name when the mileage counter changes.

Notebox 16.2

Best estimates

When a sample statistic is to be used in estimating a population parameter we look for three properties:

- *Lack of bias.* If we use it many times as an estimator of the population parameter, will the estimate be right on average?
- *Consistency.* As the sample size increases, does the variance of the sample statistic become smaller and tend to zero?
- *Efficiency.* Do the many estimates obtained from this statistic for many samples have a smaller variance than the estimates obtained from some alternative statistic whose use might be contemplated?

A sample statistic is said to provide a *best estimate* of the population parameter if it is unbiased, consistent and efficient. The sample mean is a best estimate of the population mean, but the sample standard deviation is not – it needs modification.

Summary

1. The chapter is an introduction to the use of samples, which we sometimes need in order to make an enquiry cheaper, faster and even at times practicable.

2. We use samples to make estimates for populations. The reliability of the estimate depends on (i) how the sample has been taken (or drawn) from the population, (ii) the sample size and (iii) the population variance. It does not depend on the population size.

3. A random sample, such that every item in the population has exactly the same chance of being included in it, is widely used. It is guaranteed not to introduce a systematic bias, and has the advantage of facilitating the assessment of the reliability of estimates. It is not easy to obtain and has to be drawn very carefully.

4. The precision of an estimate based on the sample depends on the square root of the sample size. Doubling precision can be achieved by quadrupling the sampling size, which has cost implications.

5. The mean, standard deviation and other measures may be calculated for samples. They also exist for populations, and we use our sample results as the basis of estimates for the population values. We call the sample results statistics. The true population values are called parameters. Thus we use the sample mean as a statistic that provides an estimate of the true population mean, which is a parameter. A (sample) statistic may need modification before it can be used as an estimate of a (population) parameter. Parameters are usually denoted by Greek letters, while statistics are denoted by Roman letters.

6. Estimates may be point estimates, which are wrong unless they are precisely correct, or interval estimates, which are less precise.

7. The reliability of an estimate has to be stated in terms of probabilities. (What is *the chance* of an estimate being wrong by more than a stated amount?)

8. Samples are also used to test ideas, theories or hypotheses about a population.

9. Noteboxes illustrate how not to take a random sample and indicate the meaning of 'best estimate'.

17

Simple probability

Introduction

Example 17.1 The Member of Parliament (MP) for Statingham hears that, despite the survey mentioned in Example 16.1, the mean income of one-adult non-retired households is only £980. The MP wants to know the chance of this rumour being right.

Example 17.2 In the National Lottery 6 balls are chosen at random from 49 differently numbered balls. Entrants have tickets with their own choice of six numbers. What is the chance of one ticket having the six winning numbers? (See also Notebox 45.2.)

Example 17.3 The Council of Statingham University is buying a new car for the vice-chancellor. The chance of it being a Rolls-Royce is 1 in 50 and the chance of it being a Lada is 1 in 10. What is the probability that it will be (a) either a Lada or a Rolls-Royce, (b) neither of them?

Example 17.4 If I roll a well-balanced six-sided die five times, what is the chance of getting 4 five times?

Example 17.5 If a coin is tossed seven times and the result recorded as head or tail, how many different sequences of seven results can there be?

Example 17.6 A bus contains 30 people. What is the probability that at least two of them have a birthday in common?

The first example is about the reliability of an estimate based on a sample. It is the kind of question we met in the last chapter. Answering it involves using ideas about probability.

Examples 17.2 to 17.6 have no immediate link with populations and samples, but their solutions also depend on ideas used in the study of probability. In some ways they are simpler than Example 17.1, so we use them as an introduction to the language of probability and a few basic ideas.

In the next few chapters we apply these ideas to problems about the reliability of estimates and conclusions based on sample material, including the problem raised in Example 17.1. Other aspects of probability are considered much later.

17.2 Definitions and the language of probability

The activity of holding a draw in a raffle or a lottery is an example of an **experiment**. Statisticians do not restrict this word to its everyday use. Any action or process that results in a measurement or other observation (which may be simply a verbal statement) is called an experiment. According to this definition, all the following are experiments:

- asking people what they think of the prime minister;
- calculating quantities of materials needed for a building;
- asking if the post office is open;
- spinning a coin 100 times and counting the number of heads;
- rolling a six-sided die and noting whether it scores 5.

Thus there need not be anything 'experimental' in the common usage of that word.

Spinning a coin 100 times involves 100 repetitions; each repetition is called a **trial**. The result of an experiment (or of a trial) is called an **outcome**. It may be quantitative, as in spinning a coin or calculating amounts of material, or it may be non-quantitative, as in the other examples.

Experiments that interest students of probability have more than one possible outcome. In the simplest case, such as asking whether the post office is open, there may be only two (yes or no), but if the experiment is sufficiently complex or elaborate there may be many.

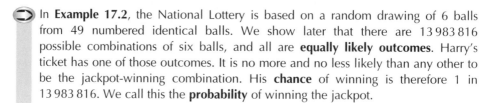

In **Example 17.2**, the National Lottery is based on a random drawing of 6 balls from 49 numbered identical balls. We show later that there are 13 983 816 possible combinations of six balls, and all are **equally likely outcomes**. Harry's ticket has one of those outcomes. It is no more and no less likely than any other to be the jackpot-winning combination. His **chance** of winning is therefore 1 in 13 983 816. We call this the **probability** of winning the jackpot.

We now apply the same idea to some much simpler experiments.

(1) Suppose we toss a perfectly balanced coin which cannot land on its edge. It will come down heads or tails. There are two outcomes, each equally likely. The chance of it coming down heads is 1 in 2, sometimes written as 1/2, 0.5 or 50%.

(2) Suppose we roll a six-sided die and get a prize if it stops with 5 uppermost. Since any one of 6 numbers may be uppermost, there are 6 possible outcomes. They are all equally likely. Only one of them will win the prize. The probability of winning in a single roll is 1 in 6, or 1/6.

(3) Suppose the prize for rolling the six-sided die is awarded if the uppermost number is divisible by 3. There are still 6 possible outcomes, one for each number that may be uppermost, all equally likely. Two of them (3 and 6) are winners. The chance of winning is therefore 2 in 6, which may be written 2/6 or 1/3.

In these experiments somebody has won something, and we have been concerned about the chance of winning. However, not all experiments involve winning. If we are interested in the probability of the outcome of an experiment having a specified characteristic, we call every outcome that has it a **success**. All other outcomes are **failures**. Thus, in the third experiment, getting a 3 or a 6 is a success; getting anything else is a failure. The fact that here success leads to a prize is of no interest to the statistician.

We define the **probability of a success** to be

$$P(S) = \frac{\text{number of outcomes that can be successes}}{\text{total number of equally likely possible outcomes}} \qquad [17.1a]$$

The phrase 'equally likely' is an essential part of the definition. Similarly the probability of a failure is

$$P(F) = \frac{\text{number of outcomes that can be failures}}{\text{total number of equally likely possible outcomes}} \qquad [17.1b]$$

In these expressions $P(S)$ and $P(F)$ are read as 'P S' and 'P F'. $P(S)$ is simply a shorthand form for 'the probability of a success'.

Defining probability in this way poses a few philosophical problems, at which we hint in Notebox 17.1, but for all practical purposes we can ignore them here.

Notice that the third experiment has 6 possible outcomes. Two of them are successes, so $P(S) = 2/6$ and $P(F) = 4/6$. The sum of these two probabilities is $2/6 + 4/6 = 1$.

Whenever a failure is defined to be the absence of success

$$P(S) + P(F) = 1 \qquad [17.2]$$

This is true because S and F are **complements.** One or the other must happen, but if one happens the other cannot. The two **events** S and F are **mutually exclusive.** The event F means 'not S'; and the event S means 'not F'.

The complement of an event denoted by A is often denoted by $\overline{A}$. Thus, as the complement of a success is a failure, we can write $\overline{S} = F$. This allows us to rewrite equation [17.2] as $P(S) + P(\overline{S}) = 1$, and more generally

$$P(A) + P(\overline{A}) = 1 \qquad [17.3]$$

where $P(A)$ denotes the probability of any event A, and $P(\overline{A})$ denotes the probability of the event 'not A'.

It follows from our definition that a probability is always a fraction lying between (or equal to) 0 and 1. If an event A has a probability 0 then it is certain not to happen; its happening is *impossible*. If it has probability 1 then it is *certain*.

Consolidation exercise 17A

1. Define an experiment. Which of the following is not an experiment?
   (a) asking the time
   (b) going for a walk
   (c) switching on the light
   (d) getting out of bed.

➡

2. What are (a) a trial and (b) an outcome? For all the experiments you have identified in answering Question 1, state whether the number of outcomes is countable on one hand.

3. If a properly balanced six-sided die is rolled, what are the probabilities of getting (a) 5 and (b) an even number?

4. What is the complement of getting 5 in Question 3? What is the probability of getting it?

5. What is meant by probabilities of (a) 50%, (b) 0.5, (c) 0, (d) 1 and (e) −1?

 17.3 Some probability theorems

Summing probabilities of mutually exclusive events

If two events are such that if one occurs the other cannot, then the events are called mutually exclusive. For example, getting a 5 on a single roll of a six-sided die means that I cannot get an even number on that roll. Moreover, getting an even number means that I cannot get a 5 on the same roll. Thus event A, defined as getting a 5, and event B, defined as getting an even number, are mutually exclusive events. Their probabilities in this example are $P(A) = 1/6$ and $P(B) = 3/6$.

The probability of getting A or B corresponds to the die showing 2, 4, 5 or 6, which means that 4 of the 6 possible outcomes are successes (in the sense that they are A or B). It follows that

$$P(A \text{ or } B) = 4/6 = 1/6 + 3/6 = P(A) + P(B)$$

Because of the + sign on the right-hand side of this expression, statisticians write $P(A \text{ or } B)$ as $P(A + B)$. This is a confusing use of notation, but it has become standard. Remember that $P(A + B)$ means the probability of event A *or* event B *or both*. If the events are mutually exclusive, they cannot occur together, then $P(A + B)$ means the probability of *A* or *B*. It does not mean the probability of *A* and *B*.

More generally, if A and B are mutually exclusive events, we have that

$$P(A \text{ or } B) = P(A + B) = P(A) + P(B) \qquad [17.4]$$

Example 17.3 can be solved by using equations [17.3] and [17.4]. The probability of the vice-chancellor getting a Rolls-Royce is given as 1 in 50 and the probability of the vice-chancellor getting a Lada is 1 in 10. Thus, with a little abbreviation and use of initials, we have

Probability of Rolls = $P(R) = 0.02$ (1 in 50)
Probability of Lada = $P(L) = 0.10$ (1 in 10)

Using equation [17.4] we find

Probability of Rolls or Lada = $P(R + L) = P(R) + P(L)$
$$= 0.02 + 0.10 = 0.12 \quad (12\%)$$

Using equation [17.3], which states that for any event A

$$P(A) + P(\overline{A}) = 1$$

we have

$$P(R + L) + \text{Probability of neither} = 1$$
$$\text{Probability of neither} = 1 - P(R + L)$$
$$= 1 - 0.12 = 0.88$$

There is thus a chance of 88% that the vice-chancellor's car will be neither a Rolls-Royce nor a Lada.

This approach can be extended to more than two mutually exclusive events. If there are three (or more) mutually exclusive events $A, B, C, \ldots$ with probabilities $P(A), P(B), P(C), \ldots$, then the probability of at least one of them occurring is the sum of the separate probabilities:

$$P(A + B + C + \ldots) = P(A) + P(B) + P(C) + \ldots \qquad [17.4a]$$

If these events are also **exhaustive** (in the sense that one of them must occur), then the sum of the probabilities on the right-hand side of equation [17.4a] equals 1, so

$$P(A + B + C + \ldots) = P(A) + P(B) + P(C) + \ldots = 1 \qquad [17.5]$$

Independent events

If two events are such that the probability of one happening is completely unaffected by whether the other happens, and this is true both ways round, and they are not influenced by a common factor, then the events are said to be **independent**. The failure of an electric light-bulb in my entrance hall, and the occurrence of a nine-letter word in the *Times* headline tomorrow are probably independent events, but not if they are due to the occurrence and reporting of a hurricane. Similarly my decision about whether to wear a raincoat does not affect, and is not affected by, Lady Agatha's decision about whether to plan to watch cricket, but both may be influenced by the morning's weather forecast, so they are not independent. Often it is difficult to detect the common influence, and events thought to be independent may not be.

We show in Chapter 45 that if A and B are two independent events then the probability of both events happening, denoted by $P(AB)$, is the product of the separate probabilities. Thus for two independent events, A and B, we have

$$P(AB) = P(\text{both } A \text{ and } B) = P(A) \times P(B) \qquad [17.6]$$

This can be demonstrated with a simple diagram. Figure 17.1 shows a square of unit area divided by a vertical line such that the part to the left of the line has an area equal to $P(A)$ and the part to the right of the line an area equal to $P(\overline{A})$. The two parts, like the two probabilities, add to unity.

The square is also divided by a horizontal such that the area above the line is equal to the probability $P(B)$ and the area beneath the line is equal to $P(\overline{B})$.

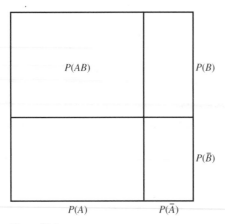

Figure 17.1

The rectangle in the top left corner corresponds to both A and B. Measured in terms of length × breadth, it is $P(A) \times P(B)$; equation [17.6] follows immediately.

If the probability of it raining in Statingham three months from next Thursday is 0.2 and the probability of there being a general election on the same day is 0.1, and these are two independent events, then the probability of that day seeing both rain and a general election is $(0.2) \times (0.1) = 0.02 = 1/50$.

The result embodied in equation [17.6] for two independent events can be easily generalised. If there are several independent events, the probability of their all happening is the product of their separate probabilities:

$$P(ABCD\ldots) = P(A) \times P(B) \times P(C) \times P(D) \times \ldots$$

We also show in Chapter 45 that if A and B are two independent events then the probability of *either* one *or* the other (or both) occurring is

$$P(A \text{ or } B \text{ or both}) = P(A + B) = P(A) + P(B) - P(A) \times P(B) \qquad [17.7]$$

In terms of Figure 17.1, the probability of A or B or both is the area to the left of the line, which is $P(A)$, plus the area above the line, which is $P(B)$, corrected for the fact that this sum includes the top left rectangle twice, so we have to reduce the sum by taking away (once) the area of this rectangle, which is $P(A) \times P(B)$.

Thus the probability of it *either* raining in three months time *or* of there being a general election that day *or* both is

$$0.2 + 0.1 - (0.2 \times 0.1) = 0.28$$

Notice that for mutually exclusive events we have equation [17.4]:

$$P(A + B) = P(A) + P(B)$$

but for independent events we have equation [17.7]:

$$P(A + B) = P(A) + P(B) - P(A) \times P(B) \qquad [17.7]$$

Consolidation exercise 17B

1. What are (a) mutually exclusive events and (b) independent events?

2. How do you write 'the probability of A or B' in the notation of probability?

3. Two events A and B have probabilities of 0.6 and 0.7. Can they be mutually exclusive? Why?

4. Two independent events C and D have probabilities of 0.6 and 0.2. What are the probabilities of (a) both happening and (b) neither happening?

Repeated occurrences

We consider two problems in which there is repetition.

In **Example 17.4** the chance of getting a 4 on a single roll of the die is 1/6. If I roll it twice there will be 36 ($=6^2$) possible outcomes: (1,1) (1,2), (1,3), ..., (1,6), (2,1), (2,2), ..., (5,5) (5,6), (6,1), (6,2), ..., (6,5), (6,6) where the first digit gives the score for the first roll and the second digit gives the score for the second roll. Only one of these 36 possible outcomes will be (4,4), so the probability of getting a 4 twice is $(1/6)^2 = 1/36$.

The same kind of argument will show that the probability of getting a 4 in each of three throws will be $(1/6)^3$, whereas the probability of getting five 4s will be $(1/6)^5 = 1/7776 = 0.0001286$, which answers Example 17.4.

More generally, if an event has the chance p of having a specified outcome A, then the chance of that outcome repeatedly occurring in *all* n repetitions of the event is p^n.

$$P(AAAA\ldots) = [P(A)]^n \qquad\qquad [17.8]$$

The next example does not involve probability but establishes a result that is used a great deal in the solution of problems about probabilities.

In **Example 17.5** the first toss of the coin can lead to two different results; H and T.

If the coin is tossed twice, four possible results are HH, HT, TH and TT, note that $4 = 2^2$.

If we toss the coin three times, then for every one of these four possible results it may come down H or T, resulting in the eight possible sequences HHH, HTH, THH, HHT, HTT, THT, TTH and TTT; note that $8 = 2^3$.

It should now be fairly clear that if we toss the coin four times, there will be $2^4 = 16$ possible sequences. Going one step further, if an event can have 2 different outcomes and it occurs r times, there can be 2^r different sequences of outcomes. Thus tossing a coin seven times can lead to $2^7 = 128$ different sequences, which answers Example 17.5.

Now suppose that, instead of tossing a coin, we roll a six-sided die, so that in the first roll there can be six outcomes. For each of these six outcomes there can be six outcomes in the second roll, so there are $6 \times 6 = 36\,(= 6^2)$ possible outcomes for the first two rolls: $(1,1)$, $(1,2)$, ..., $(1,6)$, $(2,1)$, $(2,2)$, ..., $(2,6)$, $(3,1)$, ..., $(6,6)$.

If we now roll the die again, for each of these 36 sequences of the first and second rolls, there can be 6 different outcomes of the third roll, making a total of $36 \times 6 = 216 = 6^3$ sequences. Going on in this way, we can establish an important general result:

> If an event can happen in r different ways and it is repeated n times to provide a sequence of results, there will be r^n different sequences.

 17.4 **Arranging and choosing**

This section is about a few important but simple ideas. We do not need to spend much effort on them, provided we know the results and remember what the symbols mean. We simply state and illustrate the results without any attempt to prove them

Arranging n different things in order

2 different things can be arranged in a row in $2 = 2 \times 1$ ways
3 different things can be arranged in a row in $6 = 3 \times 2 \times 1$ ways
4 different things can be arranged in a row in $24 = 4 \times 3 \times 2 \times 1$ ways

More generally, n different things may be arranged in a row (or in order) in

$n(n-1)(n-2)(n-3) \ldots 3 \times 2 \times 1$ ways

Instead of writing this out in full, we denote it by $n!$ and call it either **factorial n** or **n factorial**. It is important to remember this definition of $n!$. Many calculators have a button for it.

Arranging items in order in a row we sometimes call **permuting**. Do not confuse this meaning of 'permute' with its slightly different meaning used by addicts of football pools. In mathematics and statistics 'permuting' always means arranging in order in a row: first, second, third, etc. Arranging in order around a circular table, with no position called 'first' is not permuting.

Arranging r things out of n different things in order

If we are choosing first, second and third prizewinners out of 20 competitors, the number of ways of doing it is

$20 \times 19 \times 18 = 6840$

Notice that this is

$$\frac{20 \times 19 \times 18 \times 17 \times 16 \times 15 \times \ldots \times 2 \times 1}{17 \times 16 \times 15 \times \ldots \times 2 \times 1} = \frac{20!}{17!} = \frac{20!}{(20-3)!}$$

More generally, the number of ways of permuting r things out of n different things is

$$^{n}P_{r} = \frac{n!}{(n-r)!} \qquad [17.9]$$

The left-hand term is read as 'en pee ah'. Thus the number of ways of choosing 4 shops out of 30 to be given facelifts one after the other is $^{30}P_{4} = 30!/26! = 30 \times 29 \times 28 \times 27 = 657\,720.$

Choosing r things out of n, regardless of order

It can also be shown that the number of ways of choosing 3 things out of 8 different things, with no concern for order is

$$\frac{8!}{3! \times 5!} \quad \text{which is sometimes written} \quad \frac{8!}{3!5!}$$

This is often written as $^{8}C_{3}$ and called the number of ways of **combining** 3 things out of 8. This can be found by pressing the right button, or it can be worked out as

$$\frac{8!}{3! \times 5!} = \frac{8 \times 7 \times 6 \times 5 \times 4 \times 3 \times 2 \times 1}{(3 \times 2 \times 1) \times (5 \times 4 \times 3 \times 2 \times 1)}$$

$$= \frac{8 \times 7 \times 6}{3 \times 2 \times 1} = 8 \times 7 = 56$$

More generally, the number of ways that r things can be chosen out of n different things, with no concern for order, is

$$^{n}C_{r} = \frac{n!}{r!(n-r)!} \qquad [17.10]$$

For example, the number of ways of choosing 4 shops out of 30 to be closed down all at the same time is

$$^{30}C_{4} = \frac{30 \times 29 \times 28 \times 27}{4 \times 3 \times 2 \times 1} = 27\,405$$

Notice that when we use the word 'combined' we *always* mean chosen without concern for order. When we use the word 'chosen' we normally mean without concern for order, unless we specify otherwise.

The solution to Example 17.6 is surprisingly easy if we first think of a few versions of the same kind of problem with smaller numbers, which is often a good approach. We give the solution in Notebox 17.3; it is not essential reading. The important thing is to know the meaning of the symbols and the words 'permuting' and 'combining'. In common everyday use they often have different meanings.

Consolidation exercise 17C

1. The chance of winning any prize in a single entry in the National Lottery is (approximately) 1 in 54. (a) If you and a stranger have one ticket each, what is the chance of both of you winning prizes? (b) Why do we specify a stranger? (c) What is the chance of you and five strangers winning a prize each?

2. What is the value of 6!? Do this without a calculator and then check your answer with a calculator.

3. What does 'permute' mean? In how many ways can eight different letters be permuted?

4. What is the symbol for permuting four people out of seven? What is its value? Do this first without a calculator then with a calculator.

5. In how many different ways is it possible to arrange the letters of (a) RANGE, and (b) STRANGE? Try to find out in how many different ways it is possible to arrange the letters of (c) RANGER, (d) STRANGER and (e) ARRANGED, but do not spend a long time on them and do not be disheartened if you cannot do them without help from your tutor.

6. What does 'choose' mean in 'I am going to choose five people to send home in the next bus.'? Give your answer (a) as a sentence and (b) as a single word.

7. If an event can happen in four ways and it is repeated five times, how many different possible sequences of results will there be?

Notebox 17.1

Defining probability

Our definition of probability involves the word 'likely', which most people regard as synonymous with 'probable'. In advanced work a distinction is drawn between these words but some people argue that there is still some circularity in the definition. One other way of proceeding is illustrated by the following example.

Toss a coin and write H or T to indicate whether it comes down heads or tails. Do this many times and keep working out the ratio p of the number of heads to the total number of tosses.

Suppose that after 100 tosses there are 47 heads. Then $p = 0.47$. Now toss it again. If the coin comes down H then the new value of p will be $48/101 = 0.4752\ldots$. If it comes down T the new p will be $47/101 = 0.4653\ldots$.

As we go on tossing the coin this ratio will slowly change – sometimes upwards, sometimes downwards – but each change will be smaller than the one before, because we are always dividing by a bigger number. If, as we continue, the range of fluctuation narrows, and after 5000 tosses we are getting successive

⊃

values all between 0.4731 and 0.4730, we can suggest that the ratio p is settling down to say 0.47305 as a limit. We are getting closer and closer to it. We define the probability of a head to be that limit.

If the values of the ratio drift around with no tendency to converge, then we do not define a probability, even though we could possibly still argue that some values are more probable than others.

Notebox 17.2

0!

Using the definition that $n! = n(n-1)(n-2)\ldots 2 \times 1$, we have that

$$3! = 3 \times 2 \times 1$$
$$2! = 2 \times 1$$
$$1! = 1$$

Sometimes a calculation will involve 0!. Here the above definition breaks down. Instead we look upon 0! as meaning the number of ways in which 0 things can be arranged in a row, which is 1. In advanced mathematical work it is found that giving 0! this value leads to correct solutions of problems that can also be solved in other ways. Accordingly, 0! is defined to be 1.

Notebox 17.3

Birthdays in common

Example 17.3 asks about the probability that at least 2 out of 30 people have a birthday in common. We begin with two simpler problems.

What is the probability that two strangers will have their birthdays this year on the same day of the week?

Person A can have a birthday on any one of seven days; so can person B. The number of possible pairs of days is therefore $7 \times 7 = 49$. These are illustrated by the 49 pairs in Table 17.1. In this table we use the digits 1 to 7 to represent the

Table 17.1

11	12	13	14	15	16	17
21	22	23	24	25	26	27
31	32	33	34	35	36	37
41	42	43	44	45	46	47
51	52	53	54	55	56	57
61	62	63	64	65	66	67
71	72	73	74	75	76	77

↱

different days of the week; Sunday is day 1. The first of the numbers in each pair gives the day on which A's birthday falls, and the second the day on which B's birthday falls. Thus the number of possible pairs (49) takes order into account. We must therefore take order into account in the rest of the answer, or we will not be comparing like with like.

Seven pairs, lying along the downward-sloping diagonal, show A and B having birthdays on the same day of the week. The other 42 pairs show A and B having birthdays on different days, and they also take order into account. Their number is therefore given by $^7P_2$, which is the number of ways of selecting two different days out of seven, taking order into account.

Thus the $7^2(=49)$ pairs showing the total number of pairs of days are made up of $^7P_2(=42)$ pairs that have different days and the remainder (7, which is $7^2 - {}^7P_2$) that show both people having birthdays on the same day of the week.

The probability of two people having a birthday on the same day of the week is 7/49, which is derived as

$$\frac{7^2 - {}^7P_2}{7^2}$$

This is $1 - {}^7P_2/7^2$.

What is the probability that two (or more) strangers out of three will have their birthdays this year on the same day of the week?

We begin in the same way. There are $7^3 = 343$ possible triplets of days. This number takes order into account, so we must take order into account in the rest of the answer.

The number of ways in which we can select 3 different days of the week out of 7, taking order into account, is $^7P_3 = 7 \times 6 \times 5 = 210$. Thus, of the 343 possible triplets of days, 210 contain three different days. The probability of the three people having their birthdays on different weekdays is therefore $210/343 = 0.6122$. It follows that the probability of at least two people having a birthday on the same day is $1 - 210/343 = 0.3878$. This is $1 - {}^7P_3/7^3$.

We can now return to Example 17.3. In this we have 365 days instead of 7, and 30 people instead of 2 or 3. Apart from that, the problem is the same. The answer is therefore given by

$$1 - {}^{365}P_{30}/365^{30} = 1 - 0.29368$$
$$= 0.70632$$

Summary

1. To assess the reliability of sample results, we use ideas about chance and probability. This chapter deals with probability only insofar as it is needed for these purposes. Other important ideas in probability are developed in Chapters 39 to 43 and 45.

2. An experiment is any process or action resulting in one or more observations or measurements. A single performance of the experiment is a trial and the result is the outcome, which may or may not be quantitative.

3. Normally an experiment may have more than one possible outcome. We use event to describe the set of one or more outcomes in which we are interested. Events with a desired feature are successes and all other events are failures.

4. If there are n equally likely possible outcomes and m of them are successes, then we define the probability of a success to be m/n. An alternative definition is given in Notebox 17.1. A probability can never be less than zero or more than unity. Every possible event has a non-zero probability.

5. The non-occurrence of an event A is another event denoted by not-A or $\overline{A}$, with probability $P(\overline{A})$. Since either A or $\overline{A}$ must occur, $P(A) + P(\overline{A}) = 1$. $\overline{A}$ is called the complement of A. A and $\overline{A}$ are exhaustive, in the sense that since one or the other must occur, they exhaust all possibilities. They are also mutually exclusive: if one occurs the other cannot. (Some writers use A' instead of $\overline{A}$.)

6. If two events are mutually exclusive, the probability of one or the other occurring is the sum of their separate probabilities: $P(A \text{ or } B) = P(A + B) = P(A) + P(B)$. This result can be generalised. The probability of at least one of several mutually exclusive events occurring is the sum of their separate probabilities. This sum is unity if the events are also exhaustive.

7. Two events are independent if the probability of one event is quite unaffected by whether the other event occurs, and there is no common influence. If A and B are two independent events with probabilities $P(A)$ and $P(B)$, the probability of both of them happening, usually denoted by $P(AB)$, is the product of the two probabilities: $P(A \text{ and } B) = P(AB) = P(A)P(B)$. This result may be generalised for any number of independent events $P(ABC \ldots) = P(A)P(B)P(C) \ldots$.

8. If two events are not mutually exclusive, the probability of one or the other occurring is $P(A + B) = P(A) + P(B) - P(A)P(B)$.

9. If there are n repeated occurrences of an experiment, and every time the probability of success has the same value p, the probability of success in all n occurrences is p^n.

10. We introduce the factorial notation, with $n!$ defined to be the product $n(n-1)(n-2)\ldots 3 \times 2 \times 1$ for $n \geqslant 1$, and $0! = 1$. This is the number of ways of arranging n different things in order in a row.

11. We can arrange r out of n different things in order in a row in $n!/(n-r)!$ different ways. This is known as permuting r things out of n and is denoted by ^nP_r.

12. We can select r things out of n different things, regardless of order, in $n!/r!(n-r)!$ different ways. This is known as combining r things out of n and is denoted by ^nC_r.

13. If an event can happen in r different ways and it is repeated n times to provide a sequence of results, there will be r^n different sequences.

14. Noteboxes consider (1) an alternative definition of probability, (2) the meaning and value of 0! and (3) birthdays in common, a solution to Example 17.5.

18

Discrete, random and binomial

18.1 Introduction

⊃ **Example 18.1** Daphne de Bencha, manager of the Statingham branch of the Improvident Bank, has been asked to grant overdraft facilities to the local branch of the Organisation in Aid of Distressed Bookmakers. She agrees to do this, provided she gets exactly 3 heads when she tosses 7 brand new coins. What is the probability of this happening?

⊃ **Example 18.2** To keep control over his secret vice of drinking lots of cocoa, the Bishop plays Holy Rolo once a day. He rolls a ball down a sloping board studded with model angels. The ball bounces around and eventually comes to rest in one of three slots labelled 0, 1 and 2. The number on the label tells the Bishop how many mugs of cocoa he may drink that day. The probability of the ball coming to rest in slot 1 is 0.3, and the probability of it coming to rest in slot 2 is 0.1. How many mugs of cocoa is the Bishop likely to consume in 100 days?

⊃ **Example 18.3** Lady Agatha receives an antique urn containing a few thousand one-cup teabags. Sixty percent of them are Earl Grey tea. The rest are Cheepanas tea. When she entertains 20 Ladies of Upper Sternum she takes 20 tea bags out of the urn. What is the chance that exactly 5 of them will be Cheepanas tea?

The first two examples can be solved quite simply by use of the definition of probability. We use them to introduce a few more technical terms and the ideas of a **random variable** and its **probability distribution**. We then use some of this to tackle Example 18.3, which is typical of a frequently occurring type of problem and involves the discrete binomial probability distribution. This leads to the **binomial frequency distribution** and a statement of its mean and variance.

18.2 Discrete random variables

We saw in Chapter 17 that if an event can happen in r ways and it is repeated n times to provide a set of results then the number of different possible sets is r^n. Using this in Example 18.1 we have that if the 7 coins are tossed in order, there will be $2^7 = 128$ possible sequences. Those involving 3 heads and 4 tails are

198

tabulated in Notebox 18.1; there are 35 of them. (In practice we would calculate this from a simple formula given shortly, instead of writing them all out.) It follows that the probability of getting exactly 3 heads when 7 coins are tossed is

$$\frac{\text{Number of ways of getting 3 heads}}{\text{Total number of possible outcomes}} = \frac{35}{128} = 0.273$$

which is the probability of Daphne de Bencha having to grant overdraft facilities.

Daphne would have been safer if she had stipulated that she would grant facilities only if there were exactly 2 heads, since the probability of this is

$$\frac{\text{Number of ways of getting 2 heads}}{\text{Total number of possible outcomes}} = \frac{21}{128} = 0.164$$

Her decision is determined by the outcome (the number of heads) of the random process of tossing 7 coins; and some outcomes are more probable than others. Three heads are more probable than 2 heads, and there can be anything between 0 and 7 heads. The probabilities of the various possible numbers of heads arising are given in Table 18.1. Notice that the probabilities add to unity.

Table 18.1

Number of heads	0	1	2	3	4	5	6	7
Probability	1/128	7/128	21/128	35/128	35/128	21/128	7/128	1/128
	0.008	0.055	0.164	0.273	0.273	0.164	0.055	0.008

A discrete variable (such as the number of coins coming down heads) is called a **discrete random variable** if (i) it can take only specified values and (ii) each value has a fixed probability (which, as a check, add to unity).

The table of probabilities is called the **probability distribution.** Its entries are the same as those in a percentage frequency distribution, which is obtained by dividing the frequency with which a specified value occurs by the total frequency, as decribed in Chapter 3.

In theory it is possible to draw up a table like Table 18.1 for any discrete random variable. It may be difficult to find the probabilities, but in principle a table showing the probability distribution can be constructed. We shall see shortly that something similar can also be done for continuous random variables.

In **Example 18.2** there are three possibilities, indicated by the scores 0, 1 and 2. Nothing else is possible. The probabilities of 1 and 2 are given as 0.3 and 0.1, so the probability of the only remaining possibility (a score of 0) must be $1 - (0.3 + 0.1) = 0.6$.

Thus the score can take any of three values, each with a fixed probability. The score (or the permitted number of mugs of cocoa) is therefore a discrete random variable. Its probability distribution may be written as

$$P(x = 0) = 0.6 \qquad P(x = 1) = 0.3 \qquad P(x = 2) = 0.1$$

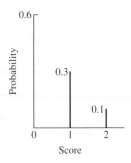

Figure 18.1

This probability distribution is shown by three vertical lines in Figure 18.1; their lengths add to unity.

There is an infinite number of possible probability distributions. In this chapter we begin a study of what they all have in common, and look at one of them more intensively. Other frequently occurring distributions are studied in later chapters. Remember that they all relate to random variables, and that the immediately following discussion relates to *discrete* random variables. Go back to the definitions and check that you understand them.

We now return to the probabilities listed in Table 18.1. They are graphed in Figure 18.2. Each line has a length proportional to the probability, so the total of the lengths is unity. Notice that the cumulated total of the lengths from the left extreme up to and including the line marked 5 shows the probability of getting 5 heads or fewer.

If Daphne had tossed 100 coins, instead of 7, there would be a similar diagram, with 101 lines. Its highest line (representing the probability of getting 50 heads

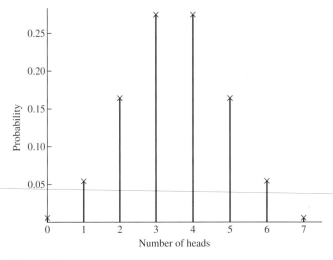

Figure 18.2

when you toss 100 coins) would have a height of only 0.0796. This may seem low, but the total probability of unity has to be split between 101 possible values, and none of them is going to be particularly high. With so many lines, each differing only slightly from its neighbours, it would be difficult not to think of a continuous curve joining their tops.

The cumulated total of the lengths of these lines up to and including line 35 would show the probability of getting 35 or fewer heads.

Consolidation exercise 18A

1. How many different sets of results can be obtained by rolling a six-sided die twice, and noting the results in order?

2. What is the probability of getting a total score of 10 when rolling two six-sided dice?

3. What conditions must hold for a variable to be a discrete random variable?

4. Are the following discrete random variables?

   (a) the number of passengers in a bus on route 19

   (b) the number of printing errors in this morning's *Times*

   (c) the number of letters arriving in my morning post

   (d) the number of words in a book of about 500 pages

5. What is a probability distribution?

6. What is shown by the height of a line in a discrete probability distribution?

7. What is shown by the total lengths of the lines to the left of, and including, the line for which the discrete random variable has a value of 7?

18.3 ## The binomial probability distribution

The diagram showing the probabilities of various numbers of heads when you toss 7 coins also shows the probabilities of getting these numbers of heads if you toss a single coin 7 times. This is an example of a **binomial experiment** with 7 trials. It is called a binomial experiment because (i) for each trial there are only two possible outcomes (H and T) and (ii) each trial is identical to all others, with a constant probability of success. Binomial experiments are sometimes called Bernoulli experiments.

If the coin is perfectly balanced, the probability of H is equal to the probability of T, each being 0.5. We can write $P(H) = P(T) = 0.5$, (or $p_H = p_T = 0.5$).

If the coin is **biased** the chance of a head may be greater than the chance of a tail, but both chances must still add up to unity since the two events are exhaustive. For example, if $p_H = 0.6$ then $p_T = 0.4$.

Now look at **Example 18.3**. The probability that a tea bag has Cheepanas tea is 0.4. We want to know the chance of a selection of 20 tea bags containing exactly 5 Cheepanas tea bags.

If we think of selecting a tea bag as an experiment there will be only two possible outcomes, Earl Grey and Cheepanas. It is an experiment we do 20 times, and we assume that the bags are well mixed up in this large urn, so the chance of getting a Cheepanas is constant (as long as the number of bags selected is very small compared with the number in the urn). This means that we have a binomial experiment.

Solving a binomial problem by using the formula

In Table 18.2 we compare the two problems. At the bottom of the table we state formulae that give the probability of getting 3 heads in 7 tosses and of getting 5 Cheepanas tea bags when selecting 20. They are derived from a more general formula given in the last column. This shows a binomial experiment in which there are n trials, each with a chance p of success, and the formula gives the probability of getting r successes.

It is easiest to state a general result first, and then to look at the two examples. In the binomial experiment of the third column there are n identical trials, each with a probability p that the outcome will be a success and of $q = 1 - p$ that it will be a failure. It can be shown that the probability there will be exactly r successes is

$$P(r) = {}^nC_r p^r (1 - p)^{n-r} = {}^nC_r p^r q^{n-r} \qquad [18.1]$$

where ${}^nC_r$ is defined in Chapter 17 and can quickly be found on many calculators. This allows us to draw up a table of probabilities like Table 18.1 by giving r all possible values from 0 to n. It is a table of the **binomial probability distribution**. We denote the distribution by $B(n, p)$.

This can be used to answer problems that would not normally be described as experiments in the everyday sense of the word. If any event may or may not occur, and it can happen in two (or more) places, occasions or ways, with the same

Table 18.2

Example 18.1		Example 18.2		More generally	
Number of coins	7	Number of bags	20	Number of trials	n
$P(H)$	0.5	P(Cheepanas)	0.4	P(success)	p
$P(F)$	0.5	P(Earl Grey)	0.6	P(failure)	$1 - p$
Interested in		Interested in		Interested in	
$P(H = 3)$		$P(C = 5)$		$P(r$ successes$)$	
(getting 3 heads)		(getting 5 Cheepanas)			
${}^7C_3\, 0.5^3\, 0.5^4$		${}^{20}C_5\, 0.4^5\, 0.6^{15}$		${}^nC_r p^r (1 - p)^{n-r}$	
$= 0.273 = 27.3\%$		$= 0.0746 = 7.46\%$			

probability, then the probabilities can be examined by using this result, as ilustrated in the following example.

⬭ **Example 18.4**   The Skew Distribution has two early-evening bar persons, Donna and Ben. The landlord estimates that each has a chance of only 2 in 7 of arriving for work on or before time. They act quite independently of each other, and travel to work by different means. What are the chances of (i) both being late, (ii) one and only one being there on time and (iii) both being there for opening time?

⬭ Let being on or before time be a success, and being late a failure. For each person the chance of a success is $p = 2/7$, while the chance of a failure is $q = 5/7$. We can now work out the probabilities from first principles, as in Chapter 17, or by using the formula we have just quoted.

From first principles

(i)   The probability of Donna being late is $P(D) = 5/7$.
  The probability of Ben being late is $P(B) = 5/7$.
  Therefore the probability of both being late is

$$P(D) \times P(B) = 5/7 \times 5/7 = 25/49 = 0.5102$$

(ii) Similarly the probability of Donna being late and Ben being on time is $5/7 \times 2/7 = 10/49$, whiich is also the probability of Donna being on time and Ben being late. Thus the probability of one and only one being on time is $10/49 + 10/49 = 20/49 = 0.4082$.

(iii) Finally the probability of both being on time is $2/7 \times 2/7 = 0.0816$

Using the formula

We can look upon each arrival of a member of staff as a trial in a binomial experiment. The trial is a success if the member arrives early or on time. Otherwise it is a failure. In our case there are 2 trials (so $n = 2$) with $p = 2/7$ and $q = 1 - p = 5/7$. With $n = 2$ the above formula gives us the following results:

(i)   The probability of both being late (0 successes) is

$$q^2 p^0 = (5/7)^2 (2/7)^0 = (25/49)(1) = 25/49 = 0.5102$$

(ii) The probability of only one being on or before time (1 success) is

$$^2C_1 \, p^1 q^1 = 2(5/7)(2/7) = 20/49 = 0.4082$$

(iii) The probability of both being on or before time (2 successes) is

$$q^0 p^2 = (5/7)^0 (2/7)^2 = (1)(4/49) = 0.0816$$

Note as a check that the probabilities add to unity.

This technique is summarised in Box 18.1.

> **Box 18.1**
>
> *How to ...* **find the probability of exactly *r* successes in *n* identical trials, each with a chance *p* of success**
>
> Either use the following procedure or use the tables described in Box 18.2.
>
> 1. By direct calculation (Chapter 17), by using tables or with a pocket calculator, find the value of
>
> $$^{n}C_{r} = \frac{n!}{r!(n-r)!}$$
>
> 2. Calculate p^{r}.
>
> 3. Calculate $(1-p)^{n-r}$.
>
> 4. Multiply the results of steps 2 and 3 then multiply this product by the result of step 1. This gives the value of
>
> $$^{n}C_{r}\,p^{r}q^{n-r} \qquad \text{(where } q = 1 - p)$$
>
> which is the probability of exactly *r* successes. Note that, under suitable conditions and for large values of *n*, a good approximation to the probability can be obtained by using Box 28.4 or Box 42.2.

We illustrate it by solving **Example 18.1**, using the formula to examine the probability of getting 3 heads out of 7 tosses of a coin, with $p = \frac{1}{2}$. It is

$$^{7}C_{3}\left(\tfrac{1}{2}\right)^{3}\left(1 - \tfrac{1}{2}\right)^{4} = \frac{7!}{3!4!}\left(\tfrac{1}{2}\right)^{3}\left(\tfrac{1}{2}\right)^{4}$$

$$= \frac{7 \times 6 \times 5}{3 \times 2 \times 1}\left(\tfrac{1}{2}\right)^{7} = 35 \times \left(\tfrac{1}{2}\right)^{7}$$

$$= 0.2734 \quad (27.34\%)$$

In **Example 18.3** the probability of getting exactly 5 Cheepanas tea bags in 20 when $p = 0.4$ is

$$^{20}C_{5}(0.4)^{5}(1 - 0.4)^{15} = 0.0746 \quad (7.46\%)$$

Using tables of cumulative binomial probabilities

Note that in Example 18.3 the probabilities of 0, 1, 2, 3, 4 and 5 Cheepanas bags would be shown by the lengths of the first six lines in a diagram similar to Figure 18.2. Thus the probability of getting 5 Cheepanas bags or fewer is the sum of the first six lengths.

With the help of computers it is easy to prepare tables showing such sums for the commonly occurring low values of *n* and *r*, and selected values of *p*. (For high values a different formula is usually used to give a very good approximation to the answer.) A table of cumulative binomial probabilities obtained in this way is given in Appendix D. As an example of how to use it, we return to the problem about tea bags.

⬭ In **Example 18.3** we have $n = 20$, $r = 5$ and we want the probability of exactly 5 Cheepanas tea bags. In statistical terms, this is the event in which we are interested and, despite Lady Agatha, we call getting a Cheepanas tea bag a success.

We identify the set of rows In Table D2 for which $n = 20$, and pick out the row in this set for which $r = 5$. We run along this row until we are in the column headed $p = 0.4$. The entry reads 0.1256. This is the probability of getting 5 or fewer Cheepanas tea bags, or successes:

$P(r \leqslant 5) = 0.1256$ (the total length of lines 0, 1, 2, 3, 4, 5)

Another row of the table shows the probability of 4 successes or fewer:

$P(r \leqslant 4) = 0.0510$ (the total length of lines 0, 1, 2, 3, 4)

It follows by subtraction that the probability of exactly 5 successes is

$P(r = 5) = 0.1256 - 0.0510 = 0.0746$ (the length of line 5)

which we have already obtained by calculation.

Thus, provided we have suitable values of n, r and p, we can use tables to get the result and completely eliminate calculation – except for a simple subtraction.

Note that the entries in the table are correct to four decimal places, which means that the difference between any two of them may be slightly out in the last digit.

Before summarising the use of these tables, we consider a few more examples.

⬭ **Example 18.5**   Statingham Magistrates' Court has a car park with 8 spaces. These spaces are reserved for use by 20 ticket holders, who are all magistrates. On average there are 6 magistrates using it at any lunchtime. What is the chance of magistrates turning up at lunchtime and being unable to find a space?

⬭ We assume that the magistrates act independently, and interpret this question as: what is the chance of more than 8 magistrates wanting to park there at lunchtime? We consider it as a binomial experiment, with $n = 20$ and $p = 6/20 = 0.3$ as the probability of a success, i.e. of a magistrate turning up. We want to know the probability of r or more successes.

A tedious way of finding this is to use Box 18.1 nine times (for 0, 1, 2, ... , 8 arrivals) and then sum the answers. This will give the probability of 8 or fewer magistrates arriving. If we subtract this from unity, we have the probability of 9 or more arriving, hence the probability of magistrates turning up and finding the car park full.

Alternatively we can use tables of the cumulative binomial probabilities. These show that for $n = 20$ and $p = 0.3$ the probability of 8 arrivals or fewer is 0.8867. There is thus a possibility of $1 - 0.8867 = 0.1133$, roughly 11%, that more than 8 will arrive, so 0.1133 is the probability that at least one of the magistrates will find the car park full.

Procedure when $p > 0.5$

If $p > 0.50$ we can still use Table D2 by looking at the column for $1 - p$, which gives the cumulative probabilities of various numbers of failures; from this we can infer the probability of the complementary number of successes. We illustrate this with Example 18.6.

Example 18.6 We spin 10 identical biased coins. What are the chances of getting (a) 3 heads or fewer, (b) exactly 3 heads, (c) more than 3 heads when $p = 0.6$ and $q = 0.4$?

Since our tables have no column for $p = 0.6$, we use the column for $p = 0.4$. This gives cumulative probabilities for tails rather than heads (or failures rather than successes). We use the fact that 3 heads or more means 7 tails or fewer. This immediately allows us to write

P (more than 3 heads, when $p = 0.6$)
$\quad = P$ (fewer than 7 tails, when $p = 0.4$) $= P$ (6 tails or fewer, when $p = 0.4$)

The cumulative table tells us, in the column headed 0.4, that this is 0.9452, which answers part (c) of the question.
We also have

P (exactly 3 heads $| p = 0.6$) $= P$ (exactly 7 tails $| p = 0.4$)
$\quad = P$ (7 tails or fewer $| p = 0.4$) $- P$ (6 tails or fewer $| p = 0.4$)

The column headed 0.4 gives this as $0.9877 - 0.9452 = 0.0425$, answering part (b).
Since part (a) is the complement of part (c), the answer to part (c) must be $1 - 0.9452 = 0.0548$.

We refer to this example in Box 18.2, which summarises the use of cumulative probability tables.

Box 18.2

How to ... **use tables of cumulative binomial probabilities to find the probability that the number of successes in n identical trials, each with a chance p of success, is (a) $\leqslant r$, (b) r, (c) $> r$**

If p appears at the head of a column in the table and there are rows corresponding to n and r

A1. Go down the p column and read the entry, call it P, in the row corresponding to n and and r. This gives the probability of r successes or fewer.

A2. Check that the entry above the entry used in step A1 gives the probability of $r - 1$ successess or fewer. Subtract this entry from P. This gives the probability of exactly r successes.

A3. Subtract P from 1, getting $1 - P$. This is the probability of getting more than r successes.

If $p < 0.50$ but there is no column for it

B1. Find two columns with values of p just above and just below the required value, call them p_a and p_b.

B2. Proceed as in step A1 for both p_a and p_b, getting P_a and P_b. The desired probability P will lie between these values. If the given value of p is closer to p_a than to p_b, then in most cases the desired P will be closer to P_a than to P_b.

B3. If necessary proceed in the same way for steps A2 and A3.

If $p > 0.50$ and there is no column for it

C1. Find $q = 1 - p$.

C2. Proceed as in Example 18.6, using the fact that the probability of r or fewer successes with a probability p is the probability of $n - r$ or more sucesses with a probability q.

If there is no set of rows for the given value of n

D1. Either do the best you can with higher and lower values of n, very similar to step B2, or use formulae instead of the table.

Consolidation exercise 18B

1. What is a binomial experiment?

2. If the probability of success in a binomial experiment is p, and it is performed n times, what is the probability of exactly r successes?

3. In your answer to Question 2, what does the term involving C mean?

4. What is the binomial probability distribution?

5. How do we use tables of cumulative binomial probabilities to find the probability of getting exactly 7 successes in 10 trials with a constant probability of success equal to 0.3?

18.4 **Expected frequencies in a binomial experiment**

If we spin 20 equally biased coins *once* then the *probability* of getting 5 heads (which we call a success) can be calculated from Box 18.1 as ${}^{20}C_5 p^5 q^{15}$, where p is the probability of any one coin being a head. Suppose that $p = 0.4$. Calculation (or the use of Table D2) will show that the probability of a success (getting 5 heads) is 0.0746.

This means that if we spin 20 such coins 1000 times and count the number of successes (the number of times on which we get 5 heads), we would have

$$
\begin{aligned}
\text{Expected number of successes} &= \text{number of trials} \times \text{probability of success} \\
&= 1000 \times 0.0746 \\
&= 74.6 \\
&= 75 \text{ (to nearest integer)}
\end{aligned}
$$

There is no guarantee there will be 5 heads 75 times, but it is more likely to be 75 than any other number of times.

More generally, if a binomial experiment with parameters n and p is conducted N times, each producing a number of successes lying between 0 and n, then as N increases, these numbers can be expected to have a frequency distribution that could be derived from the binomial probability distribution simply by multiplying it by N. This is called a **binomial frequency distribution**.

For example, Table 18.1 shows the probability distributions for various numbers of heads if we toss 7 coins. If we do this 1000 times, the frequencies with which we *expect* to get various numbers of heads are given by multiplying the probabilities by 1000.

18.5   Mean and variance of expected binomial scores

If we perform a binomial experiment many times, we expect some results more frequently than others. To keep the arithmetic manageable, we modify the tea bags example so that only 7 bags are selected. The probability of any bag being Cheepanas is still 0.4. If we pick up 7 tea bags 1000 times, we would expect to **score** anything between 0 and 7 successes each time. The frequencies with which we would expect these scores to arise are given by multiplying the probabilities by 1000, as in the third row of Table 18.3. The probabilities have been worked out by using Box 18.1 with $n = 7$ and $p = 0.4$, so $1 - p = 0.6$.

We call the mean of these **the expected score**. If the experiment is conducted many times, each time producing a score, we expect the actual average score to have this value.

The mean of scores 0–7 arising with these frequencies can be worked out by the methods of Boxes 6.1 and 11.1 but it is simpler to use a formula that gives the answer.

It can be shown that if a binomial experiment with a probability of success p is repeated N times, the mean of the expected numbers of successes is

Expected number of successes $= \bar{x} = Np$ [18.2]

Table 18.3

Score, x	0	1	2	3	4	5	6	7
Probability	0.0280	0.1306	0.2613	0.2903	0.1935	0.0774	0.0172	0.0016
Expected frequency in 1000 repetitions of experiment, f	28	131	261	290	194	77	17	2
Product, fx	0	131	522	870	776	385	102	14

$$\sum f = 1000 \quad \sum fx = 2800 \quad \bar{x} = 2800/1000 = 2.8$$

For example, in the case just examined, $n = 7$ and $p = 0.4$, and the expected number of successes is $7 \times 0.4 = 2.8$ as calculated.

It can also be shown that the variance of these numbers is

$$\text{Expected variance} = Np(1 - p) = Npq \qquad [18.3]$$

For example, the variance of the numbers of Cheepanas tea bags being chosen when the experiment of selecting 7 tea bags is conducted many times is $7(0.4)(0.6) = 1.68$, so the standard deviation of this number is $\sqrt{1.68}$, which is just under 1.3.

 Conclusion

This discussion of the discrete binomial distribution has hinted at what is to follow. We saw that when there are many coins the diagram suggests a continuous curve linking the tops of the discrete lines. In the next chapter we shall see how, as the number of coins increases, this discrete binomial distribution becomes indistinguishable from the highly important continuous normal distribution. The link between these two distributions helps us to solve binomial problems involving large numbers, but this is only one of many uses.

Consolidation exercise 18C

1. What is meant by the expected score?

2. An experiment has a probability of success equal to 0.36. If the experiment is repeated 900 times what is (a) the expected number of successes and (b) the expected variance?

Notebox 18.1

Possible permutations of 3 heads and 4 tails obtained by spinning 7 coins

Ending HHH:	T TTT HHH			
Ending THH:	T TTH THH	T THT THH	T HTT THH	H TTT THH
Ending HTH:	T TTH HTH	T THT HTH	T HTT HTH	H TTT HTH
Ending HHT:	T TTH HHT	T THT HHT	T HTT HHT	H TTT HHT
Ending TTH:	T THH TTH	T HTH TTH	H TTH TTH	T HHT TTH
	H THT TTH	H HTT TTH		
Ending THT:	T THH THT	T HTH THT	H TTH THT	T HHT THT
	H THT THT	H HTT THT		
Ending HTT:	T THH HTT	T HTH HTT	H TTH HTT	T HHT HTT
	H THT HTT	H HTT HTT		
Ending TTT:	T HHH TTT	H THH TTT	H HTH TTT	H HHT TTT

Summary

1. If a discrete variable can take only specifed values, each with a fixed probability of occurring, it is called a discrete random variable.

2. The distribution showing the probabilities of each value arising is a discrete random probability distribution. The graph will be a number of vertical lines.

3. An experiment consisting of n identical trials, each with the same chance p of success, is a binomial experiment. The chance of getting a stated number of successes in the n trials is a term in the binomial probability distribution. More precisely, if X is a discrete random variable, we say that it follows the binomial probability distribution if $P(X = r) = {}^nC_r p^r q^{n-r}$, where p is the constant probability of success in a single trial, $q = 1 - p$ and n is the number of trials. We write this as $X \sim B(n, p)$.

4. Box 18.1 summarises a procedure for finding the exact probability of r successes in n identical trials each with a constant probability p. This may also be found by using tables of cumulative binomial probabilities, as described in Box 18.2.

5. These tables may also be used to find the probabilities of the numbers of successes being (a) not more than and (b) more than the stated number.

6. Multiplication of the probabilities by the total frequency leads to a binomial frequency distribution. This has an expected value of np and a variance of npq.

7. A notebox lists the permutations of 3 heads and 4 tails in tossing 7 coins.

19

From binomial to standard normal

19.1   Introduction

Example 19.1   A well-balanced coin is spun 100 times. Harry wants to bet on there being fewer than 34 heads, but is put off by the arithmetic of working out the probability by using the binomial distribution. Is there some other way of finding it?

Example 19.2   Lady Agatha is meeting a train that is usually between 3 seconds and 4 minutes 11 seconds late. She says there is a 10% chance that it will be exactly 2 minutes 7 seconds late. Is this likely to be true?

Example 19.3   The heights of Statingham students are normally distributed with a mean of 180 cm and a standard deviation of 8 cm. Find lower and upper heights, symmetrically placed around the mean, that you expect to contain 95% of the heights.

These examples illustrate some of the questions we face in this chapter and the next. They all involve an idea that is best illustrated by the solution to Example 19.1 – the need to replace a discrete distribution by a continuous one. They also introduce other ideas developed more fully in later chapters. This chapter and the next are informal introductions to ideas that keep recurring in the rest of the book – continuous distributions and the normal distribution.

19.2   From discrete to continuous

Figure 19.1(a) shows the binomial probability distribution for tossing 10 well-balanced coins. It can be denoted by B(10, 0.5) as a special case of B(n, p) with $n = 10$ and $p = 0.5$. It has 11 lines, corresponding to the 11 possible numbers of heads, 0, 1, 2, ..., 9, 10. In Figure 19.1(b) we increase the number of coins to 20 and show the distribution B(20,0.5) with 21 lines. Comparison of the two diagrams shows that the two distributions are very much alike, but the one with the larger value of n is lower and more spread out.

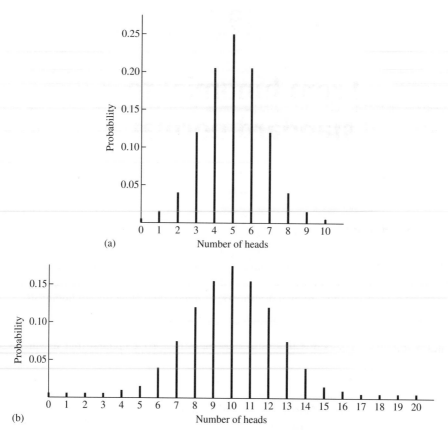

Figure 19.1

The heights of the distributions can be calculated as the probabilities represented by the longest lines. For $n = 10$ the longest line, corresponding to 5 heads, represents a probability of $^{10}C_5(0.5)^{10} = 0.246$. For $n = 20$ the longest line, corresponding to 11 heads, represents a probability of $^{20}C_{11}(0.5)^{20} = 0.176$.

We now turn to the widths, or spreads, of the two diagrams. One way of measuring this is to use the standard deviation. We know that the binomial distribution has a variance of npq, which here is $10(\frac{1}{2})(\frac{1}{2}) = 2.5$. The standard deviation is therefore $\sqrt{2.5} = 1.581$. For $n = 20$ the variance is $20(\frac{1}{2})(\frac{1}{2}) = 5$, and the standard deviation is 2.236.

If we go on increasing n, we get a series of distributions like those in Figure 19.1, all of basically the same shape, but flatter and more spread out. Some values of the maximum height (corresponding to the probability of the central value) and the spread (represented by the standard deviation) for selected values of n are shown in Table 19.1. We also show the products of the maximum height and the standard deviation. As we increase n the product also increases, but it does so less and less rapidly, and it seems to be converging to a value rather less than 0.4. It can be

Table 19.1

Height and spread of binomial distribution with $p = 0.5$

n	Maximum height	Standard deviation	Product
10	0.246	1.581	0.389
20	0.176	2.236	0.393(5)
30	0.144	2.739	0.394(4)
40	0.125	3.162	0.395(3)
60	0.1026	3.873	0.397
100	0.0796	5.000	0.398

shown that the product never quite reaches 0.39895, no matter how large n may be. We use this in Figures 19.2 and 19.3.

Figure 19.2 shows the frequency polygons that may be obtained by joining the tops of the lines for (a) the binomial probability distribution with $n = 10$ and

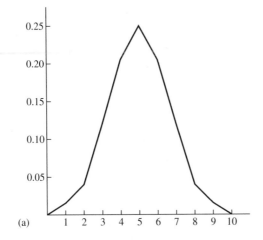

(a)

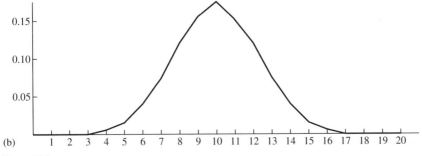

(b)

Figure 19.2

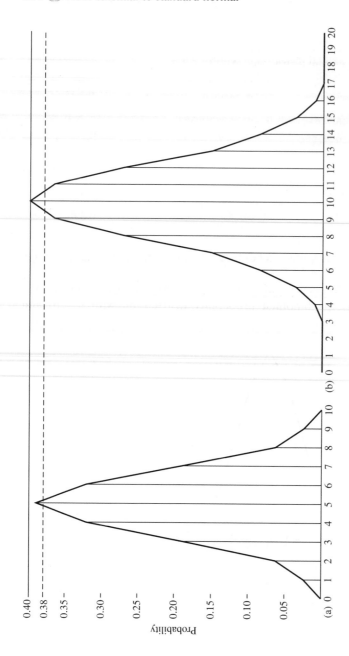

Figure 19.3

$p = 0.5$, and (b) the same distribution with $n = 20$ and $p = 0.5$. As we saw in Chapter 3, the straight lines forming the polygons have no meaning for discrete data, but here their corners represent probabilities. As long as we remember this, the polygons help us to see the shapes of the probability distributions.

The appearance of one of these distributions depends on what scale we choose. If we halve the horizontal scale, the vertical lines will be closer together and the polygon will be squeezed up. If we double the vertical scale, the graph will be more peaky.

In Figure 19.3(a) we have redrawn Figure 19.2(a) (for $n = 10$) with altered scales. It has been squeezed horizontally by dividing the horizontal scale by the standard deviation of 1.581 (from the first row in Table 19.1). This has the effect of producing a diagram of the same spread as one that has a standard deviation of unity.

It has also been stretched vertically by multiplying the vertical scale by the same amount. This means that its highest point is given by the entry at the end of the top row of the table, 0.389. This height no longer measures the probability; instead it measures the probability multiplied by the standard deviation. As we have multiplied the height by the same factor as we have divided the length, the area under the curve is unchanged.

Figure 19.3(b) shows the polygon for $n = 20$ treated in much the same way, but now the horizontal scale has been divided by 2.236 and the vertical scale multiplied by this amount. The greatest height is 0.394. Notice that, when their scales are adjusted in this way, one of these polygons is almost indistinguishable from the other, except for the number of corners. A polygon for $n = 100$, squeezed horizontally and stretched vertically in the same way, would be practically indistinguishable from a smooth curve with height 0.398.

Thus we can squeeze and stretch binomial polygons for discrete data so that they fit more or less on top of each other; and as n increases they fit more and more closely to the smooth continuous curve just mentioned. It can be shown by some hefty mathematics that this smooth curve is in fact the Normal curve introduced in Chapter 20. Before making use of this, we have to consider some implications of moving from discrete to continuous.

19.3 Continuous random variables

Suppose we choose a book and are told that its mass is more than 1.432 kg but less than 1.434 kg. Its mass will not have to be 1.433 kg. It could, for example, be 1.43279 kg, although we would probably not be able to measure as accurately as that. Exact mass does not change in steps. It is a continuous variable. Between any two values of mass there are infinitely many other values of mass.

Yet if the probability of getting exactly 50 heads is low when we toss 100 coins, think of how low must be the probability of a book having a mass exactly equal to 1.432 798 kg when this is just one of infinitely many possible weights; and how do we find it? Example 19.2 should be answered with this in mind, but see Chapter 21.

We get over these difficulties by using the following definition of a **continuous random variable** and its associated probability distribution (which, for reasons we shall soon see, we slightly rename). Read this definition carefully and read it several times.

The curves shown in Figure 19.4 all have the following properties:

- They move from left to right without doubling back.
- They are never below the *x*-axis.
- None of them shoots up indefinitely towards heaven.
- They all have a measurable area beneath them (which implies the previous property). Some extend indefinitely in a horizontal direction, but when this happens the curve approaches closer and closer to the horizontal axis, and eventually it becomes indistinguishable from it. The approach to the axis is fast enough for the total area to be measurable.

Now think of a continuous variable *X* that has the following properties:

- It can take any value on the part of the *x*-axis that is beneath the curve. This means it is between the values labelled A and B on the diagram.

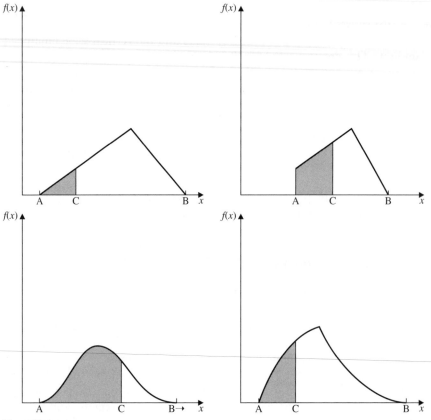

Figure 19.4

- It takes these values in such a way that the probability of X having the value less than or equal to C is given by

$$\frac{\text{The shaded area under the curve to the left of the vertical through C}}{\text{The total area under the curve}}$$

We call any such variable a continuous random variable. If the vertical scale is altered by dividing all the heights by the numerical value of the total area, the total area under the curve becomes unity. This means that the probability of X lying between A and C is given simply by the shaded area. We then say that the curve depicts (or is) the **probability density function** for a continuous random variable.

This curve shows not the probability of X having a specified precise value (which for a continuous distribution would be infinitesimal), but the density of the probabilities between one value and another. If X is more likely to have values between 1 and 2 than values between 2 and 3, then in terms of probability the values are more densely distributed between 1 and 2 than between 2 and 3.

Thus, unlike our procedure for a discrete probability distribution, we are not talking about *heights* representing individual probabilities. We are talking about *areas*, and in particular about the area to the left of the vertical line (which is equivalent to the total of the lengths of several lines in a discrete diagram such as Figure 19.1).

As an example, consider the probability distribution shown in Figure 19.5. The continuous variable X can take *any* value between 1 and 4 inclusive, but it cannot be less than 1 or more than 4. This distribution has an equation which is specified as

$$f(x) = \begin{cases} 2x/15 & \text{for } x \text{ between 1 and 4 inclusive} \\ 0 & \text{for all other values of } x. \end{cases}$$

It can easily be checked that the total area under this curve is unity, so this is a probability distribution of a continuous random variable.

The probability of x having a value of 2.45 or less is the hatched area up to and including $x = 2.45$, which is 0.3335. The probability of x having a value of 2.55 or less is the area (hatched and shaded) up to and including $x = 2.55$, which is 0.366 833. Thus the probability of x lying between 2.45 and 2.55 is $0.366\,833 - 0.333\,500 = 0.033\,333$.

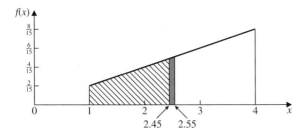

Figure 19.5

Similarly the probability of x lying between 2.495 and 2.505, two much closer numbers, is $0.351\,668\,33 - 0.348\,335\,00 = 0.003\,3333$, which is much smaller. The probability of it lying between the even closer numbers of 2.4995 and 2.5005 is very small indeed. For such reasons we do not define the probability of a precise value (such as 2.5) of a continuous random variable. Usually, instead, we calculate one of the following three probabilities:

- the probability of x lying within a specified narrow band around 2.5;
- the probability of x being equal to or less than 2.5;
- the probability of x being equal to or more than 2.5.

The mathematical function, or statement (even verbal), defining the continuous probability distribution is called the probability density function (p.d.f.).

Consolidation exercise 19A

1. How many vertical lines will there be in a binomial probability diagram when $n = 17$? What is represented by the lengths of the lines? What would show the probability of 4 or fewer successes?

2. What is a binomial probability polygon? What do the straight lines show?

3. What is the connection between the binomial polygon and the normal curve?

4. How is a continuous random variable defined? What is the area under a continuous probability distribution?

5. What is a probability density function?

 19.4   **The normal probability and frequency distributions**

We are particularly interested in the continuous distribution shown by Figure 11.1. If its total area is unity, it depicts the **normal probability distribution**. One definition is that it is the distribution to which the symmetrical binomial probability distribution tends more and more closely as the number of items increases indefinitely.

The **normal frequency distribution** can be derived from a normal probability distribution by multiplying its ordinates by the total frequency, along the same lines as in Chapter 18 and as illustrated shortly. In this case the vertical axis shows frequencies rather than probabilities, and the area under the curve to the left of a specified value will give the total frequency with which that value or less than that value occurs.

Here we are concerned mainly with the normal probability distribution. In more mathematical books this is defined in terms of a complicated formula that implies what we have just said. We do not need to know the formula but we do need to know two things about it. As we have seen, to describe the distribution, we need to know something about the spread and the maximum height. The formula for the curve therefore involves a measure of spread. This is σ, the standard deviation of the values of the variable whose probabilities concern us. As the area under the probability curve has to be unity, once we have defined its width (in terms of σ), we have also defined its height.

The formula also involves the mean μ, which locates the horizontal position of the highest point on the curve. We indicate a normal probability distribution with mean μ and standard deviation σ by $N(\mu, \sigma^2)$, where the parentheses contain the variance σ^2 rather than the standard deviation. If the continuous variable X has values x that are normally distributed in this way, we write $X \sim N(\mu, \sigma^2)$.

Thus there is a family of normal probability distribution curves, each defined precisely by its mean and standard deviation, all capable of fitting exactly on top of each other by suitable choice of mean and standard deviation. A few are shown in Figure 19.6. Curve B can be obtained from curve A by shifting the mean 10 to the right. Curve C can be obtained from curve A by reducing the standard deviation to 60% of A's and increasing the height by $1/0.60 = 1.67$, preserving the area under the curve.

The areas under the normal curve have been tabulated in great detail, allowing us to answer a surprising range of questions about the reliability of estimates, and similar matters.

For example in Example 19.3 we are told that the heights of undergraduates at Statingham are normally distributed with a mean of 180.00 cm and a standard deviation of 8 cm. As we shall see in a later chapter, this allows us to make two assertions with a high degree of confidence:

- Of the undergraduates, 95% will have heights within 1.96×8.00 cm of 180.00 cm, i.e. between 164.3 cm and 195.7 cm (which answers Example 19.3)

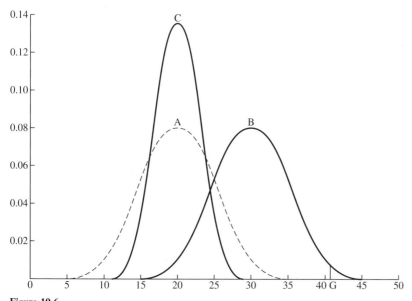

Figure 19.6

- Of the undergraduates, 99% will have heights within 2.58×8.00 cm of 180.00 cm, i.e. between 159.36 cm and 200.64 cm.

In these results the values of 1.96 and 2.58 are derived from a table for the areas under the normal curve.

As another example, we may be able to take measurements of ancient bones and say that there is a 98% probability they are between 300 and 400 years old and come from a young female.

One reason for studying probability and the normal curve is to make statements such as these. It enables us to squeeze the pips out of sample data.

19.5 Standardisation

In Chapter 12 we found it easier to compare two sets of marks by standardising them. We did this by shifting them so that their means coincided, and then squeezing them horizontally by dividing the variable by its standard deviation, so that both distributions had the same spread. This process is particularly useful when we have normal distributions.

For example, if we know that the history marks awarded in the two schools have normal distributions, we can draw Figure 19.7 to depict the problem of comparing the two marks of 85 scored in each of the schools. The marks for school A can be standardised by shifting the mean to 0 (as inset) and dividing the standard deviation by $\sqrt{196} = 14$, so the mark of 85 translates into 2.14 standard deviations above the mean. Similarly the marks for school B are standardised by shifting the mean of 60 to 0 and dividing by $\sqrt{100} = 10$, converting the mark of 85 into 2.5 standard deviations above the mean. All of that was done in Chapter 12, but because we are now asssuming there is a normal distribution, we shall shortly be able to use our knowledge of areas under the normal curve to calculate the probabilities of girls in the two schools getting marks of 85 or higher

This illustrates a highly important point. We have said that the normal distribution with mean μ and variance σ^2 is denoted by $N(\mu, \sigma^2)$. We can convert this into the **standard normal distribution** by subtracting μ from every value of x, providing a new mean 0 (so that we have $N(0, \sigma^2)$) then dividing every (new) value by the variance. This gives $N(0,1)$ which has zero mean and unit variance. The complicated mathematical equation of the curve ensures this horizontal compression is automatically accompanied by a vertical stretching that preserves the area. The curve is shown in Figure 19.8.

More generally, we can convert any normal distribution with mean μ and variance σ^2, i.e. $N(\mu, \sigma^2)$, into the standard normal distribution $N(0, 1)$ by defining a standardised variable Z such that any one of its values is derived from a value of X by using

$$z = (x - 20)/2 \hfill [19.1]$$

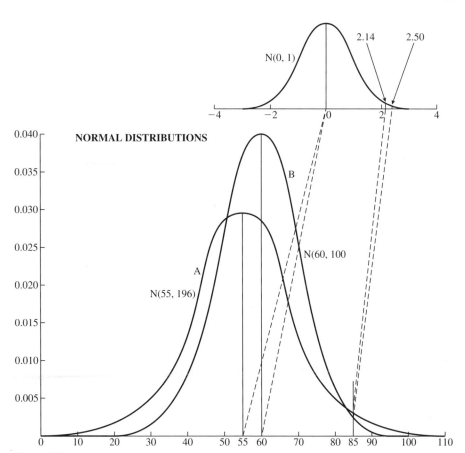

Figure 19.7

If this is done then the point G on curve B in Figure 19.6 is shifted to the point G in Figure 19.8, and the probability of X having a value corresponding to G or less on curve B is the same as the probability of Z having a value corresponding to G or less on the standard normal curve.

Box 19.1

How to ... **standardise a normally distributed variable**

1. Let the random variable X be normally distributed with mean μ and variance σ^2.

2. Define a new variable $Z = (X - \mu)/\sigma$. Be careful to have σ here, not σ^2.

3. Then Z is normally distributed with zero mean and unit variance.

4. The value z of Z corresponding to $X = x$ is $z = (x - \mu)/\sigma$.

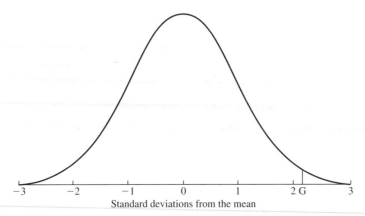

Standard deviations from the mean

Figure 19.8

 Looking forwards

We have completely answered only one of the three questions posed in the examples at the beginning of this chapter. The next few chapters answer the others.

Consolidation exercise 19B

1. How can the normal probability distribution be derived from the binomial distribution?

2. What is the symbolic notation for a normal probability distribution with mean μ and standard deviation σ?

3. What is meant by standardising a normal probability distribution? How is a standardised normal probability distribution represented symbolically? What is a shorter name for this distribution?

4. If X is distributed as in Question 2 and Z as in Question 3, write down an equation linking X and Z.

5. If $X \sim N(50, 16.00)$ find the values of X corresponding to standardised values of (a) -2 and (b) 1.96.

6. If $X \sim N(\mu, 25)$ and the standardised value corresponding to $x = 40$ is $z = -1$, what is μ?

7. If $X \sim N(30, \sigma^2)$ and the standardised value corresponding to $x = 50$ is $z = 2$, what is σ?

Summary

1. As we increase n the binomial probability distribution diagram has more lines. As their lengths represent probabilities that add to unity, the larger the number of lines,

the shorter they will be. The product of the maximum height and the standard deviation of the distribution approaches an upper limit of almost 4.

2. Joining the tops of the lines produces polygons. As we increase n these polygons can be made to fit over each other by dividing the horizontal scale by the standard deviation and multiplying the vertical scale by the same factor. As n increases indefinitely, this polygon approaches indefinitely close to the normal curve.

3. A continuous random variable can take all values on a specified part of the x-axis beneath a continuous rightwards-moving curve, and it takes these values in such a way that the probability of X having the value less than or equal to c is given by the area under the curve to the left of c divided by the total area under the curve. The curve depicts its probability density function.

4. When the normal curve has unit area, it depicts the normal probability distribution. It is the distribution to which the binomial probability distribution tends as the number of items increases indefinitely. It is defined mathematically by a complicated formula that involves the mean μ and the standard deviation σ. If a continuous random variable X has a normal probability distribution with mean μ and variance σ^2, we write $X \sim N(\mu, \sigma^2)$.

5. If the ordinates of the normal probability distribution are multiplied by the total frequency, we get a normal frequency distribution. Then 95% of the total frequency will lie within $\mu - 1.96\sigma$ to $\mu + 1.96\sigma$, and 99% within $\mu - 2.58\sigma$ to $\mu + 2.58\sigma$.

6. Any normal distribution $N(\mu, \sigma^2)$ can be converted into the standard normal distribution $N(0, 1)$ by defining a new variable Z such that $z = (x - \mu)/\sigma$.

The importance of being normal

20.1 Introduction

Example 20.1 Every lunchtime Anna Liszt, a statistics lecturer in Statingham University, observes the soup lady in the refectory. Mrs Dollop serves the students by pouring soup out of a large jug until she has counted six. She says that only 1 student in 40 gets short measure. Anna notes the number of jugs of soup used to serve 250 students and estimates that the mean serving is 240.2 ml. After careful observation she estimates that the standard deviation of the servings is 10.1 ml. She then ponders on what Mrs Dollop means by 'short measure'.

Example 20.2 Students assert that the chance of getting 272 ml (to the nearest millilitre) is just about equal to the chance of getting a smile with the soup. What is it?

Example 20.3 To ensure a regular supply of fresh meat, the refectory breeds rabbits. Every Saturday the 20 heaviest rabbits are picked out. If their total mass exceeds 90 kg, there will be rabbit pie on Tuesday. Nifty figuring enables Anna to estimate that last Saturday the mean mass of all 4837 rabbits was 2.87 kgs and the standard deviation of the masses was 0.80 kg. Will there be rabbit pie?

Answers to these questions depend on the population distributions: the distribution of the volumes of all servings of soup and the distribution of the masses of all 4837 rabbits. Experience over many decades has shown that the distributions of such variables are quite likely to be adequately represented by the normal distribution, which we are about to describe with the help of what we have already learned about the binomial distribution. If that is the case, these questions are easily answered. The exact occurrence of the normal distribution is less common than is sometimes asserted, but many variables have distributions that approximate to it very closely.

We begin by looking at the normal probability distribution (which we use in answering the first two questions), then go on to the normal frequency distribution (which we use in answering the third question). In later chapters we see that the normal probability distribution plays a remarkable role in estimation and hypothesis testing, even when the data come from populations that are not normal.

224

20.2 Areas under the normal curve

We know that the standard normal distribution has a mean of 0 and a standard deviation of 1. It has been intensively studied and tables of the areas to the left of an ordinate enable us to make the following important statements, which have already been illustrated in Figure 11.1:

- Approximately two-thirds of the area under the curve lies between -1 and $+1$. This means that the probability of z lying between -1 and $+1$ is approximately two-thirds. More accurately it is 0.6826.
- Approximately 95% of the area under the curve lies between -2 and $+2$. This means that the probability of z lying between -2 and $+2$ is approximately 95%. More accurately 95% of the area lies between -1.96 and $+1.96$.
- Almost all (more accurately 99.73%) of the area lies between -3 and $+3$. It is almost certain that z lies between -3 and $+3$.

Since the normal curve is symmetrical, the two tails defined by ordinates equidistant from the mean have equal areas. They are the shaded areas in Figure 20.1.

We now show how these results can be applied to *any* normal probability distribution through the use of standardisation. We can study any normally distributed variable, mean μ and variance σ^2, by first converting the distribution $N(\mu, \sigma^2)$ into the standard normal distribution $N(0, 1)$, by writing $z = (x - \mu)/\sigma$.

⬭ **Example 20.4** Figure 20.2 shows the distribution $N(20, 4)$ for a random variable X. Find two values, symmetric about the mean, such that there is a 95% probability that a randomly chosen value taken from this distribution lies between them.

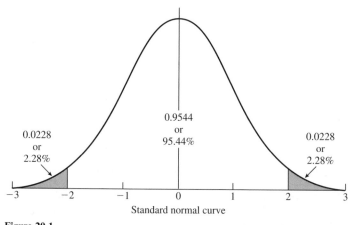

Figure 20.1

0.0228
or
2.28%

0.9544
or
95.44%

0.0228
or
2.28%

Standard normal curve

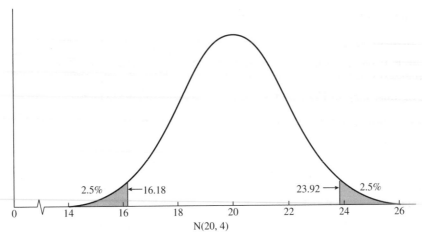

Figure 20.2

➡ The distribution N(20, 4) has a mean of 20 and a standard deviation of $\sqrt{4} = 2$, as shown in Figure 20.2. It can be standardised – converted into N(0, 1) – by writing

$$z = (x - \mu)/\sigma = (x - 20)/\sqrt{4}$$

We know that 95% of the area under N(0, 1) is contained between $z = -1.96$ and $z = +1.96$.

The value of x corresponding to $z = -1.96$ is such that

$$-1.96 = (x - 20)/2$$

Multiplying both sides by 2 gives

$$-3.92 = x - 20$$

$$x = 20 - 3.92 = 16.08$$

The value of x corresponding to $z = +1.96$ is such that

$$+1.96 = (x - 20)/2$$

which gives $x = 23.92$.

Thus if X has a normal probability distribution N(20, 4), there is a 95% probability that x lies between 16.08 and 23.92, as shown in Figure 20.2.

More generally

For N(μ, σ^2) 95% of the area lies between $\mu - 1.96\sigma$ and $\mu + 1.96\sigma$.

Thus if we know μ and σ, we can immediately write down the values of the variable that have between them 95% of the distribution, as we did in the last chapter for the heights of undergraduates.

⬭ **Example 20.5** Near the refectory till there is a collection box labelled 'For Beans'. On the third Saturday of May the Annual Baked Bean Binge is held. The person who consumes most baked beans in half an hour is given the contents of the box as a prize. Over the years it has been found that the value of this prize is a random sum with a normal probability distribution, a mean of £100 and a standard deviation of £8.30. Find a value (a) such that there is a 2.5% probability the next prize will be less than it, and (b) such that there is a 2.5% probability the next prize will be more than it.

⬭ We know that there is a 95% probability that the next prize will be between £100 − 1.96 × £8.30 and £100 + 1.96 × £8.30, which means between £83.73 and £116.27. Moreover, by symmetry, there is a 2.5% probability that the prize will exceed £116.27 and a 2.5% probability that it will be less than £83.73, which answers the questions.

The range between £83.73 and £116.27 is an example of a **symmetrical 95% probability range**. Other ranges, such as £84.39 to £117.05, can also be shown to contain a probability of 95%, but they do not have the mean at their centre. The values £83.73 and £116.27 are the symmetrical **lower and upper 2.5% probability bounds**. Usually we omit the word 'symmetrical', taking it for granted unless the contrary is specified.

There is no need to restrict ourselves to questions about 95% of the distribution. In Chapter 28 we shall see how to find the limits that contain any specified percentage, but for now we use Table 20.1 to summarise the relevant results for a normal probability distribution $N(\mu, \sigma^2)$. The first two rows show that there is a probability of one-half that the value of a normally distributed variable will lie

Table 20.1

z	Percentage of total probability within a range $z\sigma$ on either side of the mean	Percentage of total probability within each tail, P
0.6745	50	25.0
1.00	68.26	15.87
1.96*	95	2.50
2.00	95.44	2.28
2.33	98	1.00
2.58*	99	0.50
3.00*	99.73	0.135
3.018	99.75	0.125
3.09	99.80	0.10

*These rows are useful to note.

within a span of roughly $2\sigma/3$ on either side of the mean; and of roughly two-thirds that it will lie within a span of σ on either side. The other rows most usefully noted are marked with an asterisk. Some have already been illustrated in Figure 11.1.

At this introductory stage we shall confine our examples and exercises to the few values of z that are listed in this short table. The general procedure for using Table 20.1 to find the lower and upper values for a stated probability is summarised in Box 20.1.

Box 20.1

How to ... find the symmetrical lower and upper *P*% probability bounds for a random variable *X* distributed as $N(\mu, \sigma^2)$

1. Find the value of *P* in column 3 of Table 20.1 (or use a fuller table as described in Chapter 28).

2. Read off the value of *z* in column 1 corresponding to this value of *P*.

3. The lower and upper bounds, cutting off tails each with an area corresponding to a probability of *P*%, are given by $\mu - z\sigma$ and $\mu + z\sigma$, where *z* has the value obtained in step 2.

We can use these ideas to discover what Mrs Dollop means by 'short measure'. Anna assumes that the frequency distribution for the quantities received is normal. She knows it has a mean of 240.2 ml and a standard deviation of $\sigma = 10.1$ ml. This normal distribution is shown in Figure 20.3. The mean amount of soup is indicated by μ, and L and M are two points on either side of the mean, a distance of 1.96σ from it.

We know from Table 20.1 that a vertical line 1.96σ to the left of the mean cuts off a tail with area 2.5%. This represents the smallest 2.5% of the servings, and therefore must correspond to the servings that are 'short measure'.

Short measure must therefore equal or be less than

$$\text{Mean} - 1.96 \times \text{standard deviation} = 240.2 - 1.96 \times 10.1$$
$$= 240.2 - 19.8 = 220.4 \text{ ml}$$

Students getting amounts of soup indicated by the tail to the left of L are getting short measure.

As a further illustration, we consider Example 20.6.

Example 20.6 KlearBeam headlamps are made with a mean life of 10 000 driving miles and a standard deviation of 370 miles. KlearBeam want to advertise a guaranteed life, in terms of miles driven, and to offer free replacement if a bulb fails sooner than this. What should the guaranteed life be if they want the chance of having to replace the headlamp to be less than 1 in 200?

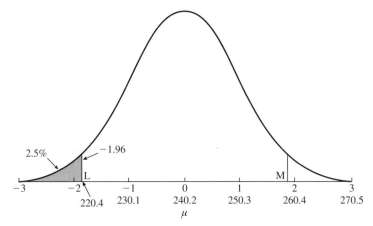

Figure 20.3

The given data are represented in Figure 20.4. We know from Table 20.1 that an ordinate a distance of 2.58σ from the mean will cut off a tail area of 0.5% of the total under the curve. There will be two such ordinates, cutting off lower and upper tails, at L and M. Lamps with lifetimes to the left of L will fail the guarantee.

But L is a distance of 2.58×370 to the left of the mean, which is 10 000, so L represents a lifetime of

$$10\,000 - 2.58 \times 370 = 9045.4$$

Thus the chance of a headlamp having a life of 9045.4 miles or less is just 0.5%. Playing safe, the manufacturer offers to replace any lamp that does not last 9000 miles.

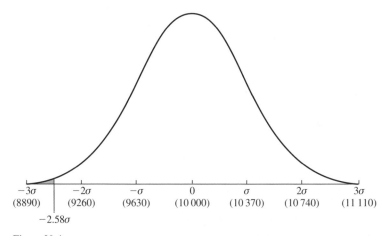

Figure 20.4

Consolidation exercise 20A

1. L and M are two points on the horizontal axis of a probability distribution diagram. What is represented by the area under the curve between L and M?

2. Approximately, what is the probability that a variable Z with standard normal distribution has a value lying between -1 and $+1$?

3. There is a probability of 0.95 that a standard normal variable has a value lying between $-k$ and $+k$. What is the numerical value of k?

4. A normally distributed variable has a mean μ and a variance σ^2. What is the chance that a randomly chosen value of it lies between $\mu - 3\sigma$ and $\mu + 3\sigma$?

5. A random variable is distributed as N (63.24, 16.00). Find the percentage probabilities that one of its values will be (a) more than 67.24, (b) more than 75.24, (c) more than 73.56, (d) less than 59.24, (e) less than 51.24, (f) less than 55.40, (g) between 59.24 and 67.24, (h) between 55.40 and 73.56.

6. If $X \sim N(\mu, 36)$ and $P(X > 40) = 0.1587$, what is μ?

7. If $X \sim N(20, \sigma^2)$ and $P(X < 16.08) = 0.025$, what is σ?

20.3 Correcting for discrete data

We may need to make a small correction because we are using a continuous distribution to examine values of a discrete variable.

⬭ **Example 20.2** involves amounts of soup measured correct to the nearest millilitre, so the recorded quantities are discrete. But the normal curve is continuous. Thus we have to take account of the fact that 272 to the nearest millilitre means anything between 271.5 and almost 272.5. Since we are interested in measures of soup that are 272 ml or more, we have to take account of all quantities equal to or greater than 271.5 ml, which is 31.3 more than the mean of 240.2.

  We standardise this difference of 31.3 by dividing by the standard deviation of 10.1. This gives us a difference from the mean of 3.099 standard units to the right. According to Table 20.1, the tail area cut off by 3.00 standard units is 0.0014. The area cut off by 3.099 units would be even smaller. Since 0.0014 is 14 in 10 000, there are few smiles. Serving soup must be a very solemn ritual.

Whether we apply this kind of correction is often a matter of judgement, depending on the accuracy of our data and on whether the correction is likely to lead to an important difference.

20.4 Examining a normal frequency distribution

We saw in Chapter 19 that if the probabilities in a normal probability distribution are multiplied by the total frequency, then we have a normal frequency distribution. The area will represent the *total frequency*. The area to the left of any value c will be the frequency with which values of c or less arise.

Sometimes a problem about frequencies can be solved simply by noting that Table 20.1 about probabilities associated with various spans also presents the fraction of total frequency contained within the spans. This means that 'probability' in the column headings can be replaced by 'frequency' to form a new table whose numerical content will be unchanged. We can use this to answer the next example.

⬭ In **Example 20.3** Anna assumes that the masses of the rabbits are normally distributed. She knows that the mean is 2.87 kg and that the standard deviation of the masses is 0.80 kg. She wants to know the weights of the heaviest 20 rabbits, which is 0.5% of the total of 4000. These rabbits therefore have the masses represented by the right-hand tail of the frequency distribution, corresponding to an area that is 0.5% of the total area (Figure 20.5). We know from the amended Table 20.1 that this corresponds to a distance of 2.58 standard deviations to the right of the mean. Thus, if the mass corresponding to this distance is m, we have

$$\frac{m - 2.87}{0.80} = 2.58$$

$$m = 2.58 \times 0.80 + 2.87 = 4.934 \, kg$$

The masses of the 20 heaviest rabbits are at least this and must therefore total at least $20 \times 4.934 = 98.68$ kg. The stated decision value is 90 kg; there will be rabbit pie.

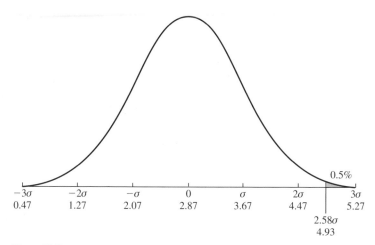

-3σ	-2σ	$-\sigma$	0	σ	2σ	3σ
0.47	1.27	2.07	2.87	3.67	4.47	5.27

0.5%

2.58σ
4.93

Figure 20.5

Box 20.2

How to ... find, for a normally distributed random variable, the symmetrical upper and lower bounds that contain between them *B*% of the total frequency

1. Replace 'probability' in the column headings of Table 20.1 by 'frequency'. ⬭

2. Find the value of B in column 2 of Table 20.1 (or use a fuller table as described in Chapter 28).

3. Read off the value of z in column 1 corresponding to this value of B.

4. The lower and upper values are given by $\mu - z\sigma$ and $\mu + z\sigma$, where z has the value obtained in step 3.

Consolidation exercise 20B

1. How can a normal probability distribution be translated into a normal frequency distribution?

2. Approximately what fraction of the total frequency of a normally distributed variable will lie within the standard deviation of the mean value?

3. Suppose 95% of the frequency of a normally distributed variable lies between 304 and 696. What is (a) the standard deviation of the distribution and (b) the mean?

4. A random variable is distributed as N(63.24, 16.00). State the fractions of the total frequency that will be (a) more than 67.24, (b) more than 75.24, (c) more than 73.56, (d) less than 59.24, (e) less than 51.24, (f) less than 55.40, (g) between 59.24 and 67.24, (h) between 55.40 and 73.56.

(20.5) Looking forwards

In the examples of this chapter we have been told, or had to assume, that the distributions are normal. In Chapter 44 we show how to test whether it is reasonable to describe a distribution in this way. But before then we look at a remarkable result that allows us to use the normal distribution to solve a vast variety of problems arising out of distributions that are very definitely not normal. That is one of the main reasons for being sure about the points made in this chapter. It means that being normal is not always important, but being familiar with normality is.

Summary

1. The normal probability distribution can be defined as the distribution to which the binomial probability distribution tends as the number of items increases indefinitely. The area beneath the curve is unity. The normal frequency distribution can be derived from it by multiplying each probability by the total frequency. The area represents the total frequency.

2. The standard normal distribution has zero mean and unit variance. It is written N(0, 1) and has a maximum height of just under 0.4.

3. Any normally distributed variable X with parameters μ and σ can be transformed into a standardised normal variable Z with parameters 0 and 1 by writing $Z = (X - \mu)/\sigma$. This allows properties of any normal distribution to be studied with the help of tables of the standard normal distribution.

4. As 95% of the area under $N(0, 1)$ lies between -1 and $+1$, then 95% of the area beneath $N(\mu, \sigma^2)$ lies between $\mu - 1.96\sigma$ and $\mu + 1.96\sigma$. The tail areas beyond these values contain the lowest and highest 2.5% of the area. Other lower and upper limits for various percentages of the area can be found with the help of Table 21.1 or the fuller version Table D4.

5. These results can be used to determine a range of values between which (or beyond which) there is a stated probability of a random variable falling. A correction may be necessary if the data are discrete rather than continuous. They can also be used to determine a range of values between which (or beyond which) a stated fraction of the total frequency will lie.

21

Estimating a mean with confidence

Introduction

⇒ **Example 21.1** Anna Liszt takes a random sample of 144 men students at Statingham University and measures their heights. She finds that the mean height is 192 cm. She knows nothing about the distribution of heights of all men in the university except, somehow, that their standard deviation (σ) is 9 cm. She says there is a 95% chance that the mean height of all men students lies between 190.5 cm and 193.5 cm. How does she get that?

⇒ **Example 21.2** The vice-chancellor was impressed but worried. To avoid allegations of genderism, there had to be a similar survey of female students.

⇒ **Example 21.3** Daphne de Bencha, the manager of the Statingham branch of the Improvident Bank, fears that her unauthorised overdraft customers may conspire and refuse to repay. She wants to know how much money might be involved, but her books are not in a good state. A total of 1601 customers have these overdrafts, and because there are so many, it would take Daphne a long time to check every single one. Fortunately, all the accounts are numbered, so she can use random number tables to take a simple random sample of 100 accounts. She finds that they have a mean overdraft of £346.43 and a standard deviation of £94.27. She wants to find a lower and upper limit such that there is a 99% chance the true mean for all 1601 accounts lies between them. Then she can estimate the total by multiplying by 1601.

⇒ **Example 21.4** The Nutrition Committee of Statingham Halls of Residence wants to estimate the fibre content of Stoj, a new cereal product packed in individual serving portions of 30 g. How many portions should it examine if it wishes to be 95% confident that the mean content of fibre per portion lies within 0.3 g of the estimated value?

We can solve such problems by using an approach described in this and the next few chapters. It is based on the **central limit theorem**, which allows us to use

234

random sample estimates to make statements about populations without having to assume that the populations have normal distributions, as described in Section 21.2. The exact procedure depends on the kind of information we have, as shown in Section 21.3 (which need not be fully mastered in a first reading). We also distinguish between two-tail and one-tail tests, and see how to modify the calculations if the sample comes from a smallish population.

In this and the next few chapters we *assume* that we have a **large sample**, which we define as a sample bigger than 30. Some more cautious people prefer to say bigger than 50. Small samples are discussed later, particularly in Chapter 29.

We end with an account of how to find the size of a large sample necessary to ensure a confidence interval not wider than a stated amount. Some of this can be omitted on a first reading.

21.2   The estimator's confidential friend

We now see how to make interval estimates in which we can have a stated confidence. (The precise meaning of this phrase appears shortly.)

We begin by assuming that *either* the population is many times bigger than the sample, *or* sampling is done with replacement, as indicated in Chapter 16 and Appendix C. In both cases this will allow us to use results that are strictly true only for infinitely large populations. We discuss sampling without replacement later in the chapter.

If we take many samples and calculate the mean for each sample, these means will have a distribution, called the **sampling distribution of the mean**, or the **distribution of the sample mean**. It can be shown that:

> If our sample is large then for any population, no matter what its distribution, the sample means will be approximately normally distributed. This normal distribution will be centred on the population mean μ with a variance of σ^2/n, where σ^2 is the population variance and n the size of the sample.

The approximation improves as the sample size increases. If the population itself has an exactly normal distribution, the sample means will have an exactly normal distribution, even for small samples. In all cases, however, the population must be many times larger than the sample, or sampling must be with replacement.

What we have just said is an informal statement of the **central limit theorem** applied to the mean. This is tremendously important. It means that for *any* very large population, *whatever its distribution*, if we have a large random sample we can (i) use its mean as a point estimate of the population mean, and then (ii) use what we know about the normal distribution to make statements about the reliability of this point estimate in terms of the population variance and the sample size.

It is easier to do this if we know the population variance. Shortly we shall see how to do it even without knowing the population variation. We now apply this theorem to the first example.

⬩ In **Example 21.1** Anna Liszt bases her argument on the central limit theorem:

(i) The best point estimate of the population mean height (μ) of all men in the university is the sample mean ($\bar{x}$) of 192 cm.

(ii) If she had taken many samples, their means would have been approximately normally distributed about μ with a variance of σ^2/n, and so have a standard deviation of

$$\sigma/\sqrt{n} = 9/\sqrt{144} = 9/12 = 0.75 \text{ cm}$$

The rest of her reasoning needs to be read carefully. Be certain you understand and agree with each sentence before you pass on to the next.

She knows that, for a normal distribution, 95% of the total frequency lies within $1.96 \times$ the standard deviation on either side of the mean. Statement (ii) therefore implies that approximately 95% of the sample means would lie within $1.96 \times 0.75 \text{ cm} = 1.47 \text{ cm}$ on one side or the other of the population mean μ. The general case, with σ instead of 0.75, is shown in Figure 21.1(a).

It follows that there is a 95% chance that our single sample mean ($\bar{x}$), which is 192 cm, lies within 1.47 cm of μ, so μ lies within 1.47 cm of the single sample mean ($\bar{x}$), as illustrated for the general case in Figure 21.1(b). Anna therefore argues like this:

- By (ii) there is a chance of approximately 95% that the population mean height of all men students in Statingham lies within 1.47 cm of 192 cm (which means in the interval 190.53 cm to 193.47 cm).
- By (i) 192 cm is the best point estimate of the population mean height.

She rounds the limits to 190.5 cm and 193.5 cm.

The standard deviation of the sample mean is called the **standard error** of the mean. In more mathematical books it may be denoted by $\sigma_{\bar{x}}$, but we shall usually write s.e.(mean).

Consolidation exercise 21A

1. What assumptions have been made for this chapter?

2. What is meant by the sampling distribution of the mean? What is its shape? What are its mean and variance?

3. What does the central limit theorem say? What about small samples?

4. A sample of size 100 taken from a very large population has a mean of 127 g. It is known that the population standard deviation is 13 g. (a) What is the best estimate of the population mean? (b) Between what values is there a 95% chance of the population mean lying?

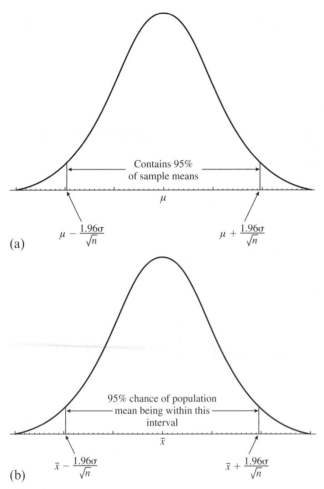

Contains 95%
of sample means

μ

$\mu - \dfrac{1.96\sigma}{\sqrt{n}}$ $\mu + \dfrac{1.96\sigma}{\sqrt{n}}$

(a)

95% chance of population
mean being within this
interval

$\bar{x}$

$\bar{x} - \dfrac{1.96\sigma}{\sqrt{n}}$ $\bar{x} + \dfrac{1.96\sigma}{\sqrt{n}}$

(b)

Figure 21.1

21.3   Confidence intervals for the sample mean

Known population variance

In the example just considered, we used information about the population standard deviation. We now look at it a little more formally. We know that 95% of the sample means will lie within 1.96 standard errors on either side of the population mean. Moreover, if we know the population variance to be σ^2 then we can say that the standard error of the sample mean is $\sigma/\sqrt{n}$. We can therefore write that 95% of the *sample means* will lie within

$$\mu - 1.96\sigma/\sqrt{n} \quad \text{and} \quad \mu + 1.96\sigma/\sqrt{n}$$

as shown in Figure 21.1(a).

This implies that if $\bar{x}$ is any one of the sample means, there is a 95% chance that the *population mean* will lie within

$$\bar{x} - 1.96\sigma/\sqrt{n} \quad \text{and} \quad \bar{x} + 1.96\sigma/\sqrt{n} \qquad [21.1]$$

as shown in Figure 21.1(b).

This statement is *approximately* true for *any* distribution if the sample is *large*, and *exactly* true even for *small* samples if the population distribution is *normal*. We call the range $\bar{x} - 1.96\sigma/\sqrt{n}$ to $\bar{x} + 1.96\sigma/\sqrt{n}$ the central 95% **confidence interval** for μ. The word 'central' is usually omitted. The two extreme values are called the **95% confidence limits** or the **95% fiducial limits**.

Beause of the symmetry of the normal curve, we can also say that 2.5% of the sample means would be in the upper (or right-hand) tail, with values exceeding $\mu + 1.96\sigma/\sqrt{n}$ and 2.5% would have values in the lower tail, less than $\mu - 1.96\sigma/\sqrt{n}$.

These results have been stated for the **95% confidence level**. Other confidence levels are frequently used, especially the 99% level, for which we have to replace 1.96 in the above account by 2.58. This is derived from the area under the normal curve, as described in Chapter 20. We consider another example.

➡ **Example 21.4** A sample of 100 cars owned by Statingham undergraduates shows a mean recorded mileage of 83 020 miles. It is known that the standard deviation of the mileage of all cars owned by Statingham undergraduates is 2500 miles. (a) Calculate an interval such that we can be 95% confident the mean mileage of all cars owned by Statingham undergraduates lies within it. (b) Within what range can we expect 95% of the mileages to lie?

➡ We assume there is a very large number of cars.

(a) The best point estimate of the population mean is the sample mean of 83 020 miles. The standard error of the mean is $\sigma/\sqrt{n}$ where $\sigma = 2500$ and $n = 100$. Therefore s.e.(mean) = 2500/10 = 250 miles. The 95% confidence interval for the population mean (in miles) is therefore

$$83\,020 - 1.96 \times 250 \quad \text{to} \quad 83\,020 + 1.96 \times 250$$
$$83\,020 - 490 \quad \text{to} \quad 83\,020 + 490$$
$$82\,530 \quad \text{to} \quad 83\,510$$

This is usually written as (82 530, 83 510) miles.

(b) We expect 95% of the mileages to be within $\bar{x} \pm 1.96\sigma$, which is $82\,020 \pm 1.96 \times 2500 = 77\,120$ to 86 920 miles.

In this example the factor 1.96 arises because we have been concerned with a 95% confidence level; $z = 1.96$ helps to define an area under the normal curve equal to 95%, as in Figure 21.1. More generally, if we are concerned with a c% confidence level, instead of 1.96 we have the value of z that helps to define an area of c%. Sometimes this is denoted by z_c. This is called the **c% confidence coefficient**. It

Table 21.1

One tail			Two tails	
Area of 1 tail (%)	Rest of area = c confidence level	z_c	Area of 2 tails (%)	Rest of area = c% confidence level
0.135	99.865	3.000	0.270	99.730
0.500	99.500	2.575	1.000	99.000
1.000	99.000	2.326	2.000	98.000
2.000	98.000	2.054	4.000	96.000
2.275	97.725	2.000	4.550	95.450
2.500	97.500	1.960	5.000	95.000
5.000	95.000	1.645	10.000	90.000
10.000	90.000	1.281	20.000	80.000
15.866	84.134	1.000	31.732	68.268
25.000	75.000	0.6745	50.000	50.000

corresponds to the values of the standardised variable that define an area under the curve equal to the level of confidence c. Some of them are shown in Table 21.1, which refers to *any* normally distributed variable (not just to standard errors of a sample mean) and is easily derived from Table 20.1.

For example, the values at the two-tail 99% level (or $c = 0.99$) are ± 2.575. The interval $\bar{x} \pm 2.575$ s.e.(mean) will contain 99% of the area under the curve, leaving each tail with an area of 0.5%. Another way of putting this is $z_{0.99} = \pm 2.575$ (or ± 2.58).

Because the excluded area is divided between the two tails of the curve, we refer to confidence intervals of this kind as **two-tailed confidence intervals**. We look at one-tailed intervals later on.

Box 21.1

How to ... find a two-tailed confidence interval for the mean (with known population variance): large sample

1. Specify the desired level of confidence c.

2. Use Table 21.1 (or Table D4) to determine the value of z_c corresponding to the chosen c (in the last column).

3. Determine the sample mean $\bar{x}$ and the population standard deviation σ and the sample size n.

4. The two-tailed confidence interval is $(\bar{x} - z_c\sigma/\sqrt{n}, \bar{x} + z_c\sigma/\sqrt{n})$.

The 50% confidence coefficient is 0.6745. The quantity $0.6745(\sigma/\sqrt{n})$ is sometimes called the **probable error** of the estimate. It defines the range on either side of the sample mean that has a 50% chance of containing the population mean.

Consolidation exercise 21B

1. What is a two-tailed 99% confidence interval?

2. A sample of size n has a mean of $\bar{x}$. What other information do we need before we can write down a 95% confidence interval for the mean?

3. Is the confidence interval for the sample mean or the population mean?

4. What is the width of a two-tailed 95% confidence interval for the mean?

5. Write down (a) the two-tailed 95% confidence intervals, and (b) the 99% two-tailed confidence intervals for the mean when you have the following information:

   (i) $n = 100$, $\bar{x} = 30$, $\sigma^2 = 64$

   (ii) $\bar{x} = -0.3$, $n = 81$, $\sigma = 3$

6. What is a confidence coefficient? How is the 90% confidence coefficient written? Using Table D4, what is its value?

Unknown population variance

All that is splendid – if you know the population variance, which you usually do not. However, it is suprisingly easy to get a good estimate of it, and sometimes you do not need even that. We look at four cases. One of them also illustrates how we have to look carefully at the nature of our data, and when and how to adjust for sampling without replacement.

Known standard error

In Chapter 16 we quoted from a government survey; it reported a figure of £11 187 for the mean sample income of one-adult non-retired households in 1991. The sample was a complicated stratified sample. The government statisticians calculated that if it had been a simple random sample, the standard error would have been £497, which they published with the mean. The more complicated sample would have produced a smaller standard error, but we use the one they have given. This enables us to calculate a pessimistic version of the 95% confidence interval as

$$\pounds 11\,187 \pm 1.96 \times 497 = 11\,187 \pm 974 \text{ (to the nearest £)}$$

$$= (\pounds 10\,213,\ \pounds 12\,161)$$

There is a chance of only 5% that the population mean lies outside this interval. Now check on the guess you were asked to make in Chapter 16.

Known population range

If the distribution is approximately normal, or at least is mound-shaped, then an approximate value of the standard deviation is one-quarter of the range (between the highest and lowest values), as explained in Notebox 21.2. If we know the *population* range, we can therefore estimate the population variance and use this as in Section 21.1. We are more likely to know the *sample* range (see later).

Known sample variance
We stated in Chapter 16 that if we do not know the population standard deviation, the best estimate we can make is

$$\hat{\sigma} = s\sqrt{n/(n-1)} \qquad [21.2]$$

where s is the standard deviation for a sample of size n.

If n is very large then $n/(n-1)$ is very close to unity, so as a very close approximation, we have $\hat{\sigma} = s$. Otherwise, using s as an estimate of σ will introduce bias; on average the values of s will understate the value of σ. In any case the procedure we have described should not be used if the sample size is under about 30, when small-sample techniques should be used (Chapter 29).

As an example, we see how the vice-chancellor's concern was met.

⬤ In **Example 21.2**, not knowing the standard deviation of the heights of female students, Anna Liszt took a sample of 312 students and used the sample standard deviation as the basis of an estimate. Remembering that this is a biased estimate of the population standard deviation, she obtained her best estimate of the population standard deviation from the formula discussed in Chapter 16.

The mean height was 174.5 cm, and the sample standard deviation was 7.2 cm. She used this to estimate the population standard deviation as

$$\hat{\sigma} = 7.2 \times \sqrt{312/311}$$

Her estimate of the 95% confidence interval for the mean height of all female students was therefore

$$174.5 \pm 1.96(\hat{\sigma}/\sqrt{312}) = 174.5 \pm 1.96(7.2/\sqrt{312})$$
$$= 174.5 \pm 0.800 \quad \text{or} \quad (173.7, 175.3)\,\text{cm}$$

Unknown population variance but known sample range
If we know the sample range, we can take one-quarter of it as an approximation to the sample standard deviation, but this is a biased estimate of the population standard deviation, which we need for the confidence interval. We therefore have to use the procedure for known sample variance.

Box 21.2

How to ...   **find a two-tailed confidence interval for the mean (with unknown population variance): large sample**

1. Specify the desired level of confidence c.

2. Use Table 21.1 (or Table D4) to determine the value of z_c corresponding to the chosen c (in the last column). ⬤

3. Note the sample size n and find the sample mean $\bar{x}$.

4. Estimate the sample standard deviation s by using one of the following: a known standard error, the population range, the sample variance, the sample range.

5. Either
    (i) Estimate the population standard deviation $\hat{\sigma}$ by multiplying s by $\sqrt{n/(n-1)}$.
    (ii) Calculate the two-tailed confidence interval as $(\bar{x} - z_c\hat{\sigma}/\sqrt{n},\ \bar{x} + z_c\hat{\sigma}/\sqrt{n})$.

    Or
    Calculate the two-tailed confidence interval as $(\bar{x} - z_c s/\sqrt{n-1},\ \bar{x} + z_c s/\sqrt{(n-1)})$.

We can help Lady Agatha in Example 19.2 by finding a 10% confidence interval for the lateness of the train, centred on 127 seconds. The standard deviation can be estimated from the range, which is 248 seconds. The value of z_c cannot be found until after reading Chapter 28.

Consolidation exercise 21C

1. Mention four ways of finding a confidence interval for the population mean when the population variance is not known. Why does $n-1$ appear?

2. A random sample of 101 students of Statingham University were asked to say how many times they had eaten meat in the last six months. The mean number was 147.38 and the standard deviation was 27.21. Calculate a 95% confidence interval for the population mean. What does this tell you, in plain English?

Allowing for smallish population size

We now consider the slightly more complicated problem facing Daphne de Bencha.

⬭ In **Example 21.3** Ms de Bencha wants a 99% confidence interval. She argues that the best estimate of the mean μ is $\bar{x} = £346.43$ and estimates the population standard deviation as

$$\hat{\sigma} = s\sqrt{n/(n-1)} = 94.27 \times \sqrt{100/99} = 94.7449$$

As $n = 100$ she then estimates the standard error of the sample mean as

$$\hat{\sigma}/\sqrt{n} = £94.7449/10 = £9.4745$$

(Later she realises that she could have found this more quickly by calculating $s/\sqrt{n-1}$.) ⬭

There is therefore a 95% chance that the population mean is contained within the confidence interval

£346.43 $\pm$ (1.96 × 9.47)

i.e. between £327.87 and £364.99, and a 99% chance that it will be contained within the confidence interval

£346.43 $\pm$ (2.58 × 9.47)

i.e. between £322.00 and £370.86.

Over her mid-morning break Daphne discusses the width of the confidence interval with her office boy, who points out that her sample comes from a population that is not all that large, and in theory it should be infinitely large. Therefore Ms de Bencha should have sampled with replacement, or revise her calculation. (This is explained in Chapter 16, and the procedure is described in Box 21.3.)

Billy explains that if a simple random sample of n items is taken from a population of size n_p, the correct standard error of the mean involves a correction factor of

$$\sqrt{\frac{n_p - n}{n_p - 1}}$$

This will always be less than unity, because the bottom is bigger than the top, and so by not applying it Ms de Bencha has overestimated the width of the confidence interval.

Jubilant, she tells Billy to work out the proper answer. He says the correction factor is

$$\sqrt{\frac{1601 - 100}{1601 - 1}} = 0.969$$

so the correct 99% confidence interval is given by

£346.43 $\pm$ (2.58 × 0.969 × 9.47)

i.e., between £322.75 and £370.11.

Dismayed that there is only 3% difference, Ms de Bencha sacks her office boy.

Box 21.3

How to ...　find the standard error of the mean for a large sample from (i) a very large population and (ii) a smallish population

1. If the population is very many times the size of the sample, or if sampling is done with replacement, the standard error of the sample mean is $\sigma/\sqrt{n}$ where σ is the population standard deviation and n the size of the sample. ⬭

2. If the population size n_p is not very many times the sample size, the standard error of the sample mean is

$$\frac{\sigma}{\sqrt{n}} \sqrt{\frac{n_p - n}{n_p - 1}}$$

Consolidation exercise 21D

1. Why was it necessary for Daphne de Bencha to allow for the size of the population, as well as the size of the sample?

2. Did this make the confidence interval wider or narrower?

3. Ms de Bencha allowed for the size of the population by using a formula. How else could she have allowed for it?

(21.4) **One-tailed confidence intervals**

Sometimes a **one-sided**, or **one-tailed**, confidence interval is used.

 Example 21.5 A few days after dismissing her office boy Ms de Bencha is told that she is going to be sued for wrongful dismissal. She is advised that the award will be based not on Billy's earnings but on the national average annual earnings of people of the same age in a variety of jobs. A sample of 144 such people reveals a mean income of £8420 per annum with a standard deviation of £530. She wants to know the greatest amount that she is at all likely to have to pay. Thus she wants an upper limit that is very unlikely to be exceeded. She defines 'very unlikely' to be a chance of one in a thousand – 0.001.

To answer this, she wants a confidence interval centred on £8240 with an upper tail area of 0.001, as shown in Figure 21.2. Since the confidence interval is symmetrical, the two tails will cut off an area of 0.002, so she needs a 0.998 (or 99.8%) confidence interval.

Tables show that this area is cut off by $z_{0.998} = 2.88$. Thus the two-tailed 99.8% interval is

$$£8240 \pm 2.88 \times (£530/\sqrt{144}) = £8240 \pm 127.20$$

Between them, the two tails cut off 0.2%, which means that each cuts off 0.1%. It follows that the upper one-tailed 99.9% interval is

$$(0, £8240 + 127.20)$$

Here the 0 indicates simply that the lowest value cannot be negative but the manager is quite uninterested in it. Thus the upper limit to a 99.8% two-tailed

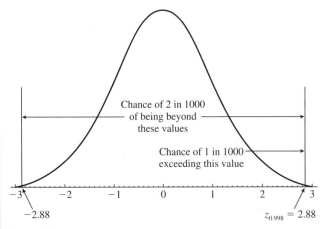

Chance of 2 in 1000 of being beyond these values

Chance of 1 in 1000 exceeding this value

-2.88 $z_{0.998} = 2.88$

Figure 21.2

confidence interval is also the upper limit of the 99.9% one-tailed interval. There is a chance of only 1 in 1000 that Ms de Bencha will have to pay more than £8367.20, which is all that concerns her.

On the other hand, Billy is interested in knowing how much is the least he can expect. Applying the same confidence limit of 1 in 1000, he calculates the lower one-tailed interval as (£8240 − 127.20, ∞) where the infinity sign indicates that he is not interested in the upper limit. Thus his answer is given by the extreme of the lower one-tailed interval, namely £8112.80.

Box 21.4

How to ... find a one-tailed *P*% confidence interval for the mean

1. Proceed as for a two-tailed interval, but when using Table 21.1, use the left-hand half. The *P*% one-tail value of z_c corresponds to the $(2P - 100)$% two-tail value.

Consolidation exercise 21E

1. A two-tailed 95% confidence interval is written as $\bar{x} \pm 1.96$ s.e.(mean). If a one-tailed confidence interval is written as $\bar{x} + 1.96$ s.e.(mean), what percentage level of confidence is to be attached to it?

2. What are (a) the upper and (b) the lower one-tailed 99% confidence intervals?

3. Determine an upper 95% one-tailed confidence interval for the mean, given the following information about a random sample drawn from a large population:

   $n = 51, \sum x = 1020, \sum x^2 = 21\,420.$

4. Suppose that in Question 4 the sample had been drawn without replacement from a population of 200. Calculate a lower 95% one-tailed confidence interval.

21.5 **Approximations and confidence intervals**

The formulae we have used express the width of the interval as a multiple of the population standard deviation. If we have only an estimate of the population standard deviation, then we can have only an estimate of the width of the interval. Thus when we say there is a 95% chance that the population mean lies within a certain interval, it is true only if our estimate of the population variance is correct.

We know that the adjusted sample variance produces an unbiased estimate of the population variance, so we have an unbiased estimate of the width of the confidence interval: it is as likely to be too high as too low. But when we use one-quarter of the range as an approximation, there is no such guarantee.

There can be other sources of inaccuracy. In particular, the sample may not be a random sample, and then the theory behind the central limit theorem is inapplicable.

Thus, if we have a value that the confidence interval just includes or just excludes, we should remember that we cannot be completely confident about our degree of confidence, especially if we have used the range or if the sample is suspect.

 21.6 **The sample size**

The width of the confidence interval also depends on the sample size. This means that if we know the population variance we can calculate the size of sample needed to provide an interval of a stated width. If we do not know the population variance we can still make progress as described below.

We know that the 95% confidence interval for the mean is

$$\mu \pm 1.96 \text{ s.e. (sample mean)} = \mu \pm 1.96 \times \sigma/\sqrt{n}$$

where the sample size is n and the population variance is σ^2. Thus the maximum error contained within the 95% confidence interval is $1.96\sigma/\sqrt{n}$, which is half the width of the interval.

If we wish to limit the error, with 95% confidence, to a maximum of B (regardless of sign), we should choose the sample size n so that $1.96\sigma/\sqrt{n}$ is no more than B. Thus we write

$$\frac{1.96\sigma}{\sqrt{n}} \leqslant B$$

which gives, on squaring and rearranging, $n \geqslant 3.8416\sigma^2/B^2$. Thus, to achieve an estimate that has a 95% confidence interval with a half-width B, we need a sample size of approximately 4 times the population variance divided by the square of the half-width. (If we use 4 rather than 3.8416, we overestimate the necessary sample size by about 4%.)

If we want to limit the error in one direction only, so that we can be 95% confident that our estimate is at most a quantity H too high, then we are thinking of a one-tailed confidence interval, and the 95% coefficient is 1.645 instead of 1.96. We need a sample size of $1.645^2\sigma^2/H^2 = 2.71\sigma^2/H^2$.

The difficulty is that probably we do not know the population variance; and until we take a sample, we do not know even the sample variance. The following account of how to deal with this can be omitted at a first reading.

When we do not know the variance, we sample in two stages. Briefly, we begin with a pilot sample of what seems to be a sensible size and use this to estimate the population variance. Then we use our estimate to work out the size of a supplementary sample. We take this and merge the results, using Box 12.3 (and possibly Notebox 12.3). The procedure is illustrated by solving Example 21.4 then summarised in Box 21.5.

 In Example 21.4 the population variance is unknown. The committee knew that the half-width of the 95% confidence interval derived from a sample of size n is $1.96\sigma/\sqrt{n}$. Since the desired maximum half-width is 0.3 g, this means that n should satisfy $1.96\sigma/\sqrt{n} \leqslant 0.3$, giving $n \geqslant 42.68\sigma^2$.

Unfortunately, the committee did not know σ^2. Accordingly, it chose a sample of 40 portions (a convenient number) of Stoj and found that the mean fibre content was 4.3 g with a sample standard deviation of 1.15 g. This implied a population standard deviation of

$$\sqrt{40/39} \times 1.15 = 1.1647$$

The 95% confidence interval was therefore

$$4.3 \pm 1.96 \times (1.1647/\sqrt{40}) \qquad \text{or} \quad \pm 1.96 \times (1.15/\sqrt{39})$$

$$= 4.3 \pm 0.3609$$

which meant that the half-width was too big. Denoting it by L, the committee took a supplementary sample, determining its size m in terms of the half-width L of the confidence interval based on the pilot sample, the desired half-width B, and the size of the pilot sample n, from the formula

$$m = n(L^2 - B^2)/B^2$$

$$m = 40(0.3609^2 - 0.3^2)/0.3^2 = 17.6$$

This suggested a supplementary sample size of 18, based on an estimated variance that derived from a not very large sample. To be safe it was decided to take a slightly bigger supplementary sample of 20.

The procedure is summarised in Box 21.5.

Box 21.5

How to ...   **find and use the minimum sample size if there has to be c% confidence that the error in an estimate of the mean does not exceed B**

1. Decide whether you want a one-tailed or a two-tailed confidence interval, and its level c. ⬭

2. Use Table 21.1 to find the value of z_c corresponding to the chosen confidence level c (in the last column).

3. If the population variance σ^2 is known, calculate the minimum sample size as $z_c^2 \times \sigma^2/B^2$.

4. If it is not known, but can be estimated (perhaps as in Box 11.2 or Box 21.2), then use the estimate in place of σ^2 in step 3 to obtain a provisional sample size n.

5. If the population variance is not known and no estimate is possible, guess a sample size, at least 30 and preferably at least 50; call it n.

6. Take a sample of size n (determined in step 4 or step 5) and use the sample data to calculate a sample variance s^2.

7. Use this to estimate a population variance from

$$\hat{\sigma}^2 = ns^2/(n-1)$$

8. Calculate the half-width of the confidence interval as $z_c \times \hat{\sigma}/\sqrt{n}$; call this L.

9. If $L > B$ then estimate m, the size of a supplementary sample, from $m = n(L^2 - B^2)/B^2$.

10. Take a supplementary sample of size m (or a little bigger if you want to play safe) and use this to obtain a supplementary sample variance. Combine this with the sample variance s^2 in step 6, according to Box 12.3 (and possibly Notebook 12.3), to obtain a new estimate of the population variance. Divide this by $(n + m)$ to obtain the standard error of the mean, and use this in the calculation of the confidence interval. If it is still too small, iterate.

Consolidation exercise 21F

1. In estimating a mean, what is the relationship between a confidence interval and sample size?

2. You are estimating a mean, you know the population variance and you know the maximum allowable error at a specified level of confidence. What two things must you do before you begin to calculate the necessary sample size?

3. In estimating a mean, how do you calculate the necessary sample size if you do not know the population variance?

4. A pilot sample of 100 items yields a mean of 18.0 and a variance of 64. A 95% confidence interval of half-width no more than 1.0 is required. What size of supplementary sample should be taken?

5. A supplementary sample is taken; its size is the same as determined in Question 4. Suppose it yields a mean of 17.8 and a variance of 60. Combine the samples to obtain new estimates of the sample mean, the population variance and a 95% confidence interval for the mean.

Notebox 21.1

A rough guide to the standard deviation

We saw in Table 21.1 that for a normal distribution a span of 2σ on either side of the mean contains 95.44% of the total frequency. Moreover, the span of the total frequency is the range. Thus, for a normal distribution, a span of 4σ approximates to the range. This is taken as justification for using one-quarter of the range as a rough estimate of the standard deviation for approximately normal, or at least mound-shaped, distributions. This is usual practice, but there seems to be no compelling reason for not using 3σ and so taking one-sixth of the range. Both approximations are rough unless the sample is very big.

Summary

1. According to the central limit theorem, for large samples ($n > 30$) or sampling with replacement, the means of many samples of the same size drawn from the same population will be approximately normally distributed, whatever the distribution of the population. The mean of this normal distribution will be the population mean. The variance of the sample means will be given by the variance of the population divided by the sample size, i.e. by σ^2/n. The square root of this quantity is called the standard error of the mean, sometimes written as s.e.(mean) or as $\sigma_{\bar{x}}$. The sample means therefore have a distribution that is approximately $N(\mu, \sigma^2/n)$.

2. The approximation to the normal distribution improves as the sample size increases. If the population has a normal distribution, the sample means have an exactly normal distribution, whatver the sample size.

3. The 95% (central) two-tailed confidence interval for the population mean is $(\bar{x} - 1.96\sigma/\sqrt{n}, \ \bar{x} + 1.96\sigma/\sqrt{n})$. This is an interval, symmetrical about the sample mean, which has a 95% probability of containing the population mean. There are similar expressions for other levels of confidence. A 50% confidence interval embraces twice the probable error. Box 21.1 summarises the procedure for finding a two-tailed confidence interval when the population variance is known.

4. Box 21.2 summarises the procedure when the population variance is not known. The procedures are considered more fully on pages 240–1. Most of them involve using a modification of the sample variance.

5. If the size of the population n_p is not very large compared with the sample size n, then either there must be sampling with replacement or the standard error of the mean has to be corrected by being multiplied by $\sqrt{(n_p - n)/(n_p - 1)}$. The procedure is summarised in Box 21.3.

6. One-tailed confidence intervals are such that there is a stated probability that the population mean is not beyond a single specified limit that provides an upper bound to it (or a lower bound). The upper limit of a 95% two-tailed confidence interval provides the only limit for a 97.5% one-tailed interval. The procedure is summarised in Box 21.4.

7. The accuracy of a confidence interval depends on the accuracy of any estimates we have had to make, and on the sampling.

8. If the population variance is known, the size of sample necessary to produce a confidence interval of half-width B at a stated level of confidence is given by $n = z_c^2(\sigma^2/B^2)$. If it is not known then a two-stage sampling procedure can be used, augmenting a pilot sample of guessed size with a supplementary sample of size determined by use of a formula given in Box 21.2. The results of the two samples are merged by adding the variances as in Box 12.3 and Notebox 12.3.

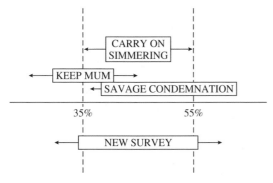

Confidence in proportions and other things

22.1 Introduction

Example 22.1 It is reported that 53% of a national sample of 1100 voters are going to vote for the Larger Pint Pots Party. Assuming that the sample is randomly drawn from the whole country, what is the 99% confidence interval for the proportion voting for Larger Pint Pots?

Example 22.2. It was reported to the Bishop of Greater Statingham that some of his enormous flock disapprove of his interest in dramatic societies. The editor of the *Statingham Declaimer* organised an opinion survey, which showed that 57 out of a sample of 120 churchgoers held this opinion.

Knowing that samples have a margin of error, the editor decides to calculate a 98% confidence interval and see where it lies in relation to two boundaries (of 35% and 55%) which he uses as guides for action (Figure 22.1). If the confidence interval includes the upper boundary of 55% but does not stretch as far as the lower boundary, he will argue that the balance of evidence is in favour of savage condemnation. If it stretches beyond the lower boundary but does not include the upper boundary, then his policy is to keep mum.

A third possibility is that it will be entirely between these two boundaries, touching neither. There is negligible chance of the truth being as high as 55% or

Figure 22.1

251

as low as 35%. In that case it is a matter of carry on simmering. The fourth possibility is that the confidence interval embraces both boundaries. The margin of error is so great that the truth could be more than 55% or less than 35%. In that case he will conduct a new and larger survey. What happens?

 Example 22.3   A random sample of 120 undergraduates are asked to estimate the height of the lower sill of the vice-chancellors's window. Their estimates have a mean of 20.3 metres and a standard deviation of 1.7 metres. Assuming that if all undergraduates had been asked, their estimates would have been approximately normally distributed, give confidence intervals for the mean and standard deviation of these estimates. Is the assumption necessary?

 Example 22.4   The statistics department at Statingham has been retained by the Larger Pint Pots Party to see if the public would like its leader to resign in favour of Mr Tuffee. It intends to take a national random sample, and wants to be 95% certain of producing an answer that is not more than 3% wrong. How large should the sample be?

The first two of these questions involve finding confidence intervals not for means but for proportions. After looking at how to do this, we extend the method to estimates of other population parameters, using Example 22.3 as an illustration. Example 22.4 is used to illustrate how to determine the necessary sample size.

(22.2)   A confidence interval for a proportion

The procedures described in Chapter 21 for calculating confidence intervals for means work because the sample mean is approximately normally distributed. This is also true of a sample proportion. If we can find the standard deviation of a distribution of sample proportions, the **standard error of a proportion**, we can therefore find confidence intervals.

We must begin with some terminology. We use a sample statistic to estimate a population parameter, and we are interested in finding an interval in which we can be, say, 95% confident that the population parameter lies. Thus our confidence interval is for the population parameter, rather than for the sample statistic. However, it has become common practice to refer to a confidence interval for a sample statistic, rather than for a population estimate based on a sample statistic.

In finding confidence intervals, we use the important result that if many large samples of size n are taken from a very large population, the sample proportions will be approximately normally distributed around the (true) population proportion (p) with a standard error given by

$$\text{s.e.} = \sqrt{pq/n} \qquad (q = 1 - p) \qquad\qquad [22.1]$$

But we do not know p; it is what we are trying to estimate. As with the bank manager's mean, the best estimate is the sample statistic.

Thus the procedure for finding a confidence interval for a proportion is just the same as for a mean, except that the standard error is given by equation [22.1]. In using this standard error we have easier calculations if we express the value of p as a decimal fraction rather than a percentage.

⬭ In **Example 22.1** we have $p = 0.53$, $q = 1 - p = 0.47$, and $n = 1100$. Thus a good approximation to the standard error of the sample proportion voting for Larger Pint Pots is

$$\sqrt{(0.53 \times 0.47)/1100} = 0.015\,05$$

Therefore the 99% confidence interval for the population proportion is

$$0.53 \pm 2.58 \times 0.04537 = 0.53 \pm 0.039$$

which means that, if opinions do not change before the election, there is a 99% chance that between (approximately) 49.1% and 56.9% will vote for Larger Pint Pots.

The general procedure is given in Box 22.1, which refers to Table 21.1. Two cautions are necessary:

(1) Equation [22.1] is valid only if $np > 5$ and $n(1 - p) > 5$. This is because it is based on using the normal distribution as an approximation to the binomial distribution.
(2) Equation [22.1] assumes that the sample is large and that the population is very large compared with the sample (or that sampling with replacement is used). If this is not so, a correction should be used (Chapter 21).

Box 22.1

How to ... find (i) a two-tailed and (ii) a one-tailed confidence interval for a proportion

1. Check that $np > 5$ and $n(1 - p) > 5$.

2. Specify the desired level of confidence c.

3. Use Table 21.1 (or Table D4) to determine the value of z_c corresponding to c, being careful to use the correct half of the table and noting Box 21.4.

4. Calculate the standard error of the proportion from

$$\text{s.e. (proportion } p) = \sqrt{pq/n} \qquad (q = 1 - p)$$

5. The $c\%$ confidence interval is

$$p - z_c\sqrt{pq/n},\ p + z_c\sqrt{pq/n}$$

We apply Box 22.1 to the next example.

⇨ In **Example 22.2** $np = 57$, so the condition in step 1 of Box 22.1 holds. The editor is going to use a two-tailed confidence interval such that there is a 98% chance that the true proportion lies within it; then he will use Figure 22.1.

He finds from Table 21.1 that the value of z_c which cuts off two tails, each of area 1%, leaving a confidence interval of 98%, is 2.326.

With $p = 0.475$, $q = 0.525$ and $n = 120$ the standard error of the sample proportion is

$$\sqrt{(0.475 \times 0.525)/120} = 0.0456 \text{ (to 3 d.p.)}$$

Using an argument similar to the one employed for the sample mean, the two-tailed confidence interval at level c for the proportion is

$$p \pm z_c \times \text{s.e. (sample proportion)} = 0.475 \pm z_c(0.0456)$$

Thus there is a 98% chance the proportion of the whole flock that is critical of the Bishop lies between

$$0.475 - 2.326 \times 0.0456 \quad \text{and} \quad 0.475 + 2.326 \times 0.0456$$

which gives the 98% confidence interval (0.369, 0.581).

This clearly does not extend as far as the lower boundary of 0.35, but it passes the upper boundary of 0.554. Savage condemnation is indicated.

Consolidation exercise 20A

1. What is meant by the standard error of a proportion? What is its value? (Say what your symbols stand for.)

2. What is the 95% confidence interval for a proportion?

3. What condition must apply to the size of p before equation [22.1] for the standard error of the proportion can be used?

4. What is denoted by z_c? The value of z_c for a two-tailed 98% confidence interval is 2.326. What one-tailed confidence interval has the same value?

 22.3 **A confidence interval for any normally distributed sample statistic S**

We can use the same procedure to find confidence intervals for sample statistics that have approximately normal sampling distributions. The most commonly occurring are the standard deviation, the variance, the median, quartiles, and differences between sample means and proportions (Chapter 23).

We continue to assume that our data come from a large sample taken from a population very many times larger than the sample, or from a smaller population

Table 22.1

Sample statistic	Symbol and standard error	Comment
Arithmetic mean	$\sigma_x = \sigma/\sqrt{n}$	Valid for any size n if population is normal; and for large $n (\geqslant 30)$ if non-normal
Proportion	$\sigma_p = \sqrt{pq/n}$	Same comments as for mean
Standard deviation	$\sigma_s = \sigma/\sqrt{2n}$	Valid only if the population is (approximately) normal and $n \geqslant 100$. For smaller n a different formula has to be used
Variance	$\sigma_s^2 = \sigma^2\sqrt{2/n}$	Same comment as for standard deviation
Median	$\sigma_{med} = \sigma\sqrt{\pi/2n}$ $= 1.253\sigma/\sqrt{n}$	Valid only if the population is approximately normal and $n \geqslant 30$
First and third quartiles	$\sigma_{Q1}, \sigma_{Q2} = 1.363\sigma\sqrt{n}$	Same comment as for median

Continued as Table 23.1

with replacement. If any approximately normally distributed sample statistic S is calculated from the data, it can be shown that the 95% two-tailed confidence interval at level c is given by

Mean value of $S \pm z_c \times$ standard error of S

where z_c is derived from tables of the area under the normal curve (when $c = 95\%$, $z_c = 1.96$).

However, the standard error of S may not be easy to find. We already know it for the mean and a binomial proportion, but for the other commonly occurring statistics there is a simple formula for the standard error only if the *population* is normal and certain sample size conditions are observed. These conditions and the appropriate standard errors are shown in Table 22.1. The procedure is summarised in Box 22.3.

Box 22.2

How to ... **find the confidence interval for a sample statistic S at level c if S is approximately normally distributed**

1, 2, 3. Follow Box 22.1.

4. Calculate the standard error by using the formulae given in Table 22.1, reading the comments carefully.

> **5.** Use Table 21.1 to find the value of z_c.
>
> **6.** The $c\%$ confidence interval is
>
> $s - z_c \times$ s.e., $\quad s + z_c \times$ s.e.

 In **Example 22.3** the confidence interval for the mean is found as in Chapter 21. Here we are concerned only with the confidence interval for the standard deviation. The data satisfy the conditions in column 3 on using the formula for the standard error of the standard deviation. This is therefore given by

$$\sigma_s = \sigma/\sqrt{2n} = 1.7/\sqrt{240} = 0.110\,\text{m (to 3 d.p.)}$$

It follows that a 95% two-tailed confidence interval for the standard deviation of the estimates is

$$(1.7 - 1.96 \times 0.11, \; 1.7 + 1.96 \times 0.11) \quad \text{or} \quad (1.48, 1.92) \text{ metres}$$

In other words, there is a 95% chance that if all undergraduates had been asked to estimate the height, the standard deviation of their estimates would have been between 1.48 m and 1.92 m.

The assumption stated in the question is necessary if we are going to use the formula in Table 22.1 for the standard error of the standard deviation.

Consolidation exercise 22B

1. In finding a confidence interval by the methods of this chapter, what is assumed about (a) the sample and (b) the population?

2. Under suitable assumptions, what is a 95% two-tailed confidence interval for a sample statistic S? What is a more correct way of describing this interval?

3. What is the formula for a 95% confidence interval for a proportion?

22.4 **The sample size**

In using proportions, we may want to find the size of sample needed to provide a confidence interval of stated width. The basic approach is the same as for the mean, but some of the detail is simpler. We assume throughout that we have a large sample, with $np > 5$ and $nq > 5$.

If we wish to limit the error, with 95% confidence, to a maximum of B (regardless of sign), we should choose the sample size n so the standard error is no more than B. Thus, at a 95% level, $1.96\sqrt{(p_s q_s/n)} \leqslant B$. Squaring both sides leads to

$$n \geqslant \frac{1.96^2 pq}{B^2} = 3.8416pq/B^2$$

In deriving this formula, we have assumed that, like p, B is expressed as a decimal. If we want a confidence interval that has a half-width of 3% – so that we are not more than 3% wrong – we must write $B = 0.03$.

If we want to limit the error in one direction only, so that we can be 95% confident that our estimate is at most a proportion H too high, the minimum sample size is $1.645^2 pq/H^2 = 2.71 pq/H^2$.

The difficulty is that we do not know the value of pq in the the population. Indeed, the purpose of our enquiry is to estimate p (and therefore q).

One trick is to assume the worst. The highest possible value for the product pq is obtained by putting $p = q = 0.5$, giving $pq = 0.25 = 1/4$. If we insert this in the formula for n, we get that the minimum size is

$$n = 3.8416/4B^2 = 0.96/B^2 \qquad \text{(approximately } 1/B^2) \qquad [22.2]$$

However, if the true value of pq is smaller than 0.25, this process means that we are overestimating the necessary sample size. We illustrate this by considering Mr Tuffee and his rival.

⬭ In **Example 22.4** the department wants to be 95% certain of producing an answer that is not more than 3% wrong.

We begin with equation [22.2], writing $B = 0.03$ gives

$$n = 0.96/0.03^2 = 1066.7$$

Thus the sample should be 1068 or larger.

This is the minimum that will guarantee, at 95% confidence, that the error is not more than 3%, on the worst possible assumption about the size of the product pq. Thus we are likely to be overestimating the minimum sample size.

If we wish to avoid this, we can use a pilot survey and a supplementary survey, more or less as in the case of the mean.

⬭ **Example 22.4** is now reconsidered with the further information that the proportion in favour of Mr Tuffee is expected to be around 20%. Remember that we want a maximum error of 3%, which has to be written as 0.03.

Use of $p = 0.20$ with a specified B of 0.03 gives

$$n = 1.96^2 pq/B^2 \qquad [22.3]$$
$$= 1.96^2(0.160/0.0009) = 683$$

as the size of the pilot sample.

Suppose that a survey of this size leads to a point estimate of 0.39. In this case the half-width of the confidence interval will be

$$L = 1.96\sqrt{p_s q_s/n} = 1.96\sqrt{(0.39 \times 0.61)/683} = 0.0366$$

which is too big. The stipulated maximum is 0.03.

Accordingly, we use the same formula as in the case of the mean to calculate the size of a supplementary sample as ⬭

$$m = n(L^2 - B^2)/B^2 \qquad\qquad [22.4]$$
$$= 683(0.0366^2 - 0.03^2)/0.03^2 = 334$$

Suppose that a sample of this size yields a value of $p = 0.38$. In that case we merge the supplementary sample result with the pilot sample result, by taking their arithmetic mean weighted by sample size to get an estimated proportion of

$$\frac{(683 \times 0.39) + (334 \times 0.38)}{683 + 334} = 0.3867$$

The width of the 95% confidence interval associated with this is

$$1.96\sqrt{p_s q_s/n} = 1.96\sqrt{(0.3867 \times 0.6133)/1017} = 0.0299$$

which is acceptable. Possibly it would not have been acceptable if we had assumed that our surveys had produced widely different sample proportions; iteration would then have been necessary.

The procedure for finding sample size is summarised in Box 22.2. Note that, although Example 22.4 has been based on a 95% confidence level, there is nothing to prevent us adopting different levels of confidence. More generally the sample size at a level $c\%$ is given by $n = (z_c/2B)^2$

Box 22.3

How to ...   **find and use the minimum sample size if there has to be $c\%$ confidence that the error in an estimate of a proportion does not exceed B**

1. Decide whether you want to ensure that the proportion is (a) not wrong, *regardless of sign*, by more than B, or (b) not wrong *in a given direction* by more than B.

2. Use Table 21.1 to find the value of z_c corresponding to the $c\%$ confidence level, using the two-tailed value if option (a) is chosen in step 1, but the one-tailed value if option (b) is chosen.

3. Be certain that B is expressed as a decimal.

4. The largest value that the minimum sample size can be is

$$n = (z_c/2B)^2$$

rounded up to the next integer.

5. If step 4 is rejected, proceed as follows:
  (i)   Guess a value for p and use it in equation [22.3] to find a sample size n.
  (ii)   Take a pilot sample of size n. Use the data to calculate first the sample proportion and from this calculate the width of the confidence interval using Box 22.1. If this is less than $2B$ there is no need to go further. ⬭

(iii) If the confidence interval in (ii) has a half-width of $L > B$, use equation [22.4] to calculate m and take a supplementary sample a little bigger than this.

(iv) Calculate a sample proportion from the supplementary survey data.

(v) Merge the proportions of the two samples, using their arithmetic mean weighted by sample size, to obtain a new estimated proportion. Use this to recalculate the standard error from $\sqrt{pq/(n+m)}$.

(vi) Multiply this by z_c to obtain the half-width of the confidence interval.

(vii) If necessary, repeat along the lines of (iii) to (v) with yet another supplementary sample.

Consolidation exercise 22A

1. What is meant by (a) 'necessary sample size' and (b) 'maximum permitted error'? How must the maximum permitted error be expressed?

2. When is it right to use (a) the one-tailed and (b) the two-tailed value of z_c? What are the one-tailed and two-tailed values of z_c for a 95% level of confidence?

3. What is the formula for determining the highest possible value of the (minimum) size of sample necessary to ensure that, at $c\%$ confidence, the error in a proportion does not exceed B? What is the objection to using this formula?

4. A sample of the residents of Wales is used to determine to within 2% the proportion who speak Welsh. Determination is made (a) with 95% confidence and (b) with 99% confidence. Assuming the proportion of Welsh-speakers is approximately 1/4, calculate the necessary minimum sample sizes.

5. If you cannot make the assumption specified in Question 4, what is the safest assumption to make in order not to end up with samples that are too small? What sample sizes does this assumption lead to?.

6. A pilot random sample of 100 households in Statingham shows that 35% of them have dogs. If an estimate for the whole town has to be correct to within 2% at a 95% confidence level, what size of supplementary sample should be taken?

7. Suppose the supplementary sample taken in Question 6 shows that 38% of the households have dogs. Combine the two sample results to obtain a new estimate for a 95% confidence interval for the population proportion.

Summary

1. If many large samples of size n are taken from a very large population, the sample proportions will be approximately normally distributed around the population proportion as mean with standard error $\sqrt{pq/n}$, provided that $np > 5$ and $nq > 5$.

2. As p is unlikely to be known, we use the sample value as a best estimate.

3. If the population is not very large, we have to use sampling with replacement or apply the correction indicated in Chapter 21.

4. Subject to the above, the $c\%$ two-tailed confidence interval for a proportion is $p \pm z_c \sqrt{pq/n}$.

5. More generally, for any approximately normally distributed statistic S, then subject to assumptions or conditions summarised in Table 22.1, the $c\%$ two-tailed confidence interval is $p \pm z_c$ s.e.(S), where s.e.(S) is the standard error of the statistic concerned. The standard errors of the more commonly used statistics are given in Table 22.1.

6. The necessary sample size for a given confidence interval for a proportion can be found in much the same way as for a mean. The detail is summarised in Box 22.3. A pessimistic estimate of the size can be found by putting $p = 0.5$, as in the first part of the solution to Example 22.4.

23

What's the real difference?

Introduction

Example 23.1 Daphne de Bencha of the Improvident Bank was told by her bosses that in the Upper Sternum branch the mean overdraft was only £230.18, compared with £346.43 in her branch. She felt that her own mean had been overestimated and the Upper Sternum mean underestimated. What confidence can she have in the alleged difference of £116.25?

Example 23.2 A random sample of 400 adults from Statingham had a mean height of 168 cm with a standard deviation of 10 cm. One of 900 from Upper Sternum had a mean of 170 cm with a standard deviation of 17 cm. What is the 99% confidence interval for the difference between the two population means?

Example 23.3 The Bishop is a fighter. The *Declaimer* shouts that a sample survey showed that 57 out of 120 members of his flock disapprove of him. In retaliation, he commissions a sample survey of opinions about the Editor's relationships with his numerous secretaries, and finds that 92 of the 196 people in the sample disapprove of the 'goings on'. He wonders whether there is really any difference between this and the extent of disapproval of his own behaviour.

Finding a confidence interval for the difference between two sample results is much the same as finding one for a mean or any other sample statistic. All we need to know is how to calculate the standard error of the difference, which is easy if we remember an important point about normal distributions. If two variables are normally distributed with means μ_1 and μ_2 and variances σ_1^2 and σ_2^2, their difference is also normally distributed with mean $\mu_1 - \mu_2$ and variance $\sigma_1^2 + \sigma_2^2$.

23.2 **The difference between two sample means**

Daphne de Bencha consults a book found in Billy's locker. As she fears, the best point estimate of the difference between the average of all overdrafts with her

261

branch (μ_a) and the average of all overdrafts with the other branch (μ_b) is the difference between the two sample means ($\overline{x}_a - \overline{x}_b$) which is indeed £116.25. Her hope lies in the confidence interval.

She knows that the two sample means are normally distributed, hence the same is true of their difference. She reads that the 95% confidence interval for $\mu_a - \mu_b$ is

$$\overline{x}_a - \overline{x}_b \pm 1.96 \text{ s.e.}(\overline{x}_a - \overline{x}_b)$$

She also reads that the easiest way to calculate the standard error of the difference between two sample means is first to calculate the standard errors for each of the two samples separately. She already knows that for her own branch the standard error of the sample mean is £9.47.

Her best estimate of the standard error of the sample mean for the Upper Sternum branch is obtained from $s/\sqrt{n-1}$, where s is the sample standard deviation for the branch and n is the sample size. Her bosses reveal that the survey in Upper Sternum used a sample of 50 accounts and the standard deviation of the overdrafts was £70.35. She uses this to calculate the standard error of the Upper Sternum sample mean as $£70.35/\sqrt{49} = £10.05$.

Peeping at Billy's book, she is surprised to read, but decides to believe, that the standard error of the difference between the two means is the same as the standard error of the sum of two means, and is obtained as

$$\sqrt{\left[\left(\begin{array}{c}\text{s.e. of sample mean} \\ \text{for Statingham branch}\end{array}\right)^2 + \left(\begin{array}{c}\text{s.e. of sample mean} \\ \text{for Upper Sternum branch}\end{array}\right)^2\right]}$$

Thus she gets

$$\text{s.e.}(\overline{x}_a - \overline{x}_b) = \sqrt{9.47^2 + 10.05^2} = \sqrt{190.68} = £13.81$$

The 95% confidence interval is therefore

$$\overline{x}_a - \overline{x}_b \pm 1.96 \text{ s.e.}(\overline{x}_a - \overline{x}_b) = 116.25 \pm 1.96 \times 13.81$$

or (£89.18, £143.32).

By any standards, this is a pretty big positive quantity. It cannot be used to refute the suggestion that the mean overdraft is lower in Upper Sternum.

She also calculates the 99% confidence interval, replacing 1.96 by 2.58. It does not make things much better. Then she tries correcting each of the standard errors because of sampling without replacement (Chapter 21), but that makes the confidence interval narrower.

Then she realises that she should have taken a one-tailed test, and really that makes it even worse. There is a chance of only 2.5% that her overdrafts exceed Upper Sternum's by as little as £89.18, or of 97.5% that they are bigger by at least that amount. She has to admit it: her overdrafts are on the high side.

The general procedure is summarised in Box 23.1.

Box 23.1

How to ... **find a confidence interval for the difference between two means**

1. Specify the desired level of confidence c.

2. Use Table 21.1 to determine the value of z_c corresponding to the chosen level of confidence.

3. Determine the sample means, sample sizes and (if possible) the population variances. If the last of these cannot be determined, then determine the sample variances.

4. The standard error of the difference between two means is given by

$$\sqrt{\left[\left(\begin{array}{c}\text{s.e. of sample mean}\\ \text{for sample a}\end{array}\right)^2 + \left(\begin{array}{c}\text{s.e. of sample mean}\\ \text{for sample b}\end{array}\right)^2\right]}.$$

$$= \sqrt{\sigma_{\bar{x}_a}^2 + \sigma_{\bar{x}_b}^2}$$

$$= \sqrt{(\sigma_a^2/n_a) + (\sigma_b^2/n_b)}$$

If the population standard deviations are not known, the best estimate of s.e.$(\bar{x}_a - \bar{x}_b)$ is obtained from

$$\sqrt{\frac{s_a^2}{n_a - 1} + \frac{s_b^2}{n_b - 1}}$$

where s denotes the sample standard deviation.

5. The confidence intervals are calculated as

$$(\bar{x}_a - \bar{x}_b) \pm z_c \text{ s.e.}(\bar{x}_a - \bar{x}_b)$$

We now use this procedure to find the 99% confidence interval for the difference between the population mean height of Statingham adult residents and the population mean height of Upper Sternum adult residents.

⊜ In **Example 23.2**, as the samples are large, we can ignore the $n/(n-1)$ correction and simply substitute s^2 for σ^2.

We have that the standard error of the difference between the two means is

$$\sigma_{\bar{x}_1 - \bar{x}_2} = \sqrt{\sigma_{\bar{x}_1}^2 + \sigma_{\bar{x}_2}^2}$$

$$= \sqrt{(\sigma_1^2/n_1) + (\sigma_2^2/n_2)}$$

$$= \sqrt{(10^2/400) + (17^2/900)}$$

$$= 0.7557 \text{ cm}$$

The 99% confidence interval is thus

$$\bar{x}_1 - \bar{x}_2 \pm 2.58\sqrt{(\sigma_1^2/n_1) + (\sigma_2^2/n_2)}$$
$$= 170 - 168 \pm 2.58 \times 0.7557 \text{ cm}$$

giving the interval

$$2.00 - 1.950 \quad \text{to} \quad 2.00 + 1.950 \quad \text{or} \quad (0.050, 3.950) \text{ cm}$$

This result is probably more accurate than the measurements. It is better to write it as 0.05 cm to 3.95 cm.

Consolidation exercise 23A

1. Refer if necessary to Chapter 22. If a sample of size n has a mean of $\bar{x}$ and the population standard deviation is σ, what is is the standard error of the sample mean?

2. Two sample means have standard errors of 3 and 4. What is the standard error of their difference?

3. Why do we need to find a confidence interval for the difference between two sample means? If the two means mentioned in Question 2 above have values of 20 and 22, what is the 95% confidence interval for their difference?

23.3 **The difference between two sample proportions**

⬤ The Bishop argues that his behaviour is no worse than the Editor's. Finding no help from the Book of Numbers he turns to his *Ecclesiastic Statistics*, where he reads:

> The 95% confidence interval for the difference between two proportions p_1 and p_2 is the difference $\pm 1.96 \times$ s.e.(difference between two proportions).

And he finds the formula

$$\text{s.e.(difference between two } proportions) = \sqrt{(p_1 q_1/n_1) + (p_2 q_2/n_2)}$$

He knows that in his own case $p_1 = 57/120 = 0.475$, so $q_1 = 0.525$ and $n = 120$. In the Editor's case $p_2 = 92/196 = 0.469$, $q_2 = 0.531$ and $n = 196$. The 95% confidence interval for the difference between these two proportions is therefore

$$(0.475 - 0.469) \pm 1.96\sqrt{(0.475 \times 0.525/120) + (0.469 \times 0.531/196)}$$

$$= 0.06 \pm 1.96 \times \sqrt{0.003\,349} = 0.06 \pm 0.113$$

$$= (-0.107, 0.119)$$

The Bishop is jubilant. There is a 95% chance that the true difference between the proportions lies between a proportion of 0.107 (or 10.7%), putting the Editor worse, and a proportion of 0.119 (or 11.9%), putting himself worse. With that range of difference, the Editor should chew his pen most carefully.

The general procedure is summarised in Box 23.2. Table 23.1 extends Table 22.1.

Box 23.2

How to ... **find a confidence interval for the difference between two proportions**

1. Check that for each sample $np > 5$ and $n(1 - p) > 5$.

2. Provided that step 1 is satisfied, proceed exactly as for the difference between two sample means (Box 23.1) except that in the formulae you replace σ_1^2 by $p_1 q_1$ and σ_2^2 by $p_2 q_2$.

Table 23.1

Sample statistic	Symbol standard error	Comment
Difference between two means	$\sigma_{x_1 - x_2} = \sqrt{\dfrac{\sigma_1^2}{n_1} + \dfrac{\sigma_2^2}{n_2}}$	Valid for any size n if population is normal; and for large $n\ (\geqslant 30)$ if non-normal
Difference between two proportions	$\sigma_{p_1 - p_2} = \sqrt{\dfrac{p_1 q_1}{n_1} + \dfrac{p_2 q_2}{n_2}}$	Same comments plus $pqn > 5n > 5$

Consolidation exercise 23B

1. Write down formulae for the standard error of (a) a sample proportion and (b) the difference between two sample proportions, saying what your symbols mean.

2. Two samples of sizes 101 and 65 have means of 30 and 40. The standard error of the difference between these means is 17.2. What is the standard error of the sum of these means?

Summary

1. We estimate confidence intervals for (a) the difference between two population means and (b) the difference between two population proportions. The estimates depend on the fact that if two variables are normally distributed with means μ_1 and μ_2 and variances σ_1^2 and σ_2^2, then their difference is also normally distributed with mean $\mu_1 - \mu_2$ and variance $\sigma_1^2 + \sigma_2^2$.

2. In both cases the confidence interval can be written as

$$\text{difference} \pm z_c \times \text{s.e.(difference)}$$

3. The standard error of a difference is the root of the sum of the squares of the two standard errors. Thus for two means

$$\text{s.e.(difference)} = \sqrt{(\sigma_a^2/n_a) + (\sigma_b^2/n_b)}$$

If the population variances σ^2 are not known, the sample variances are used with $n - 1$ instead of n (Box 23.1).

4. For two proportions

$$\text{s.e.(difference)} = \sqrt{(p_1 q_1/n_1) + (p_2 q_2/n_2)} \qquad \text{(Box 23.3)}$$

5. The conditions for the use of these results are given in Table 23.1.

Are they genuine ScrawlBalls?

24.1 Introduction

Example 24.1 Statingham's industrial estate houses the British branch of ScrawlBall Pens Inc. The pens are advertised as having an arithmetic mean life of 500 h of continuous writing, and the manufacturer knows that the variance of the lifetimes is 100 h².

Retailer A says that its own ScrawlBall pen supplied by its wholesaler had a life of only 485 h. Retailer B examines a sample of 81 pens that are supposed to be ScrawlBalls and finds that the mean of their lives is 497 h. They complain to the parent company. Is ScrawlBall Pens justified in asserting that the pens are almost certainly not genuine ScrawlBalls?

In the last few chapters we have been using samples to make estimates of population parameters. Now we begin to use them to test hypotheses. Both retailers have taken samples, one of 81 pens and one of only a single pen. We test whether the lives of the sampled pens are so low that the hypothesis that the pens are genuine is highly improbable. Before considering the tests we say a word about the use of statistics in testing hypotheses.

24.2 Statistics and hypothesis testing

The nature and testability of hypotheses

A **hypothesis** is a theory or statement whose truth has yet to be proven or disproven. It may have been put forward because it is believed to be true, or because it is to be used as the basis of an argument.

Some hypotheses can be tested easily. For example, a hypothesis that using metal containers in microwave ovens is unwise can be tested in several ways. A reliable but possibly expensive way is to try it and see. Other ways, that may involve some cost, are to ask other users, or to read cookery books or the manufacturer's instructions. The cost of the test and the cost of being wrong should always be considered.

Other hypotheses may be untestable. An example is that while she was drawing her last breath, Queen Victoria secretly began to feel amused. There is no way of obtaining evidence.

Yet other hypotheses may be incapable of being proven or disproven with complete certainty by use of the available information. But the evidence may allow us to make a statement about the chance of such a hypothesis being true. Examples are predictive hypotheses such as that it will rain over London tomorrow. The weather forecast states the chance of rain.

Another example is the statement that some bones found in a cave belonged to a young girl who died 4000 years ago. This cannot be proven with certainty but the experts may decide there is a 99% chance that the bones belonged to a young girl and a 95% chance that their age is between 3500 and 4500 years.

Using statistics to test hypotheses

Unless they are just guesswork, the examples we have given imply there has been some kind of statistical investigation, which has been used to assess the probability of the hypothesis being right. Hypotheses about tomorrow's weather over London and the age or sex of ancient human remains are not in themselves statistical. It is in attempting to ascertain the probability of their being correct that we have to use statistical methods.

When statistical methods are used in testing hypotheses, the result is always a statement of probability. Statistical methods will never prove or disprove a hypothesis with certainty, but sometimes it may be possible to attach a very high probability to one's conclusion.

We begin to use statistics in testing a hypothesis by replacing the non-statistical hypothesis (N) by an associated statistical hypothesis S. We try to choose S so that if N is right then S must be right; and preferably so that if S is right then N must be right. Doing this is often difficult and we may have to content ourselves with something a little less clear-cut.

For example, we may have the non-statistical hypothesis that young people living in London are brighter than young people living in Devon. An associated statistical hypothesis may be that the mean intelligence quotient (IQ) of school pupils in London is higher than the mean IQ of school pupils in Devon.

As we see below, we can test this statistical hypothesis about the mean IQ by statistical methods and make a statement about the probability of its being right. Whether this allows us to make a similar statement about the original non-statistical hypothesis depends on whether 'brightness' of 'young people' is adequately measured by the mean IQ of school pupils.

We need to take account of this in adding a paragraph to our note about ScrawlBall Pens Inc.

⬭ **Example 24.1 continued**   ScrawlBall Pens Inc. will accept a statistical hypothesis about the life of a single pen, or the mean life of a sample of pens, as a substitute for a non-statistical hypothesis about genuineness.

The plan of attack

It may be useful to summarise the argument that is to follow. There will always be two hypotheses, such as hypothesis *A* that the pens are genuine and hypothesis *B* that they are inferior imitations. The statistician will identify two numerical values associated with these hypotheses, and will then focus on the difference between them. This difference enables the value of a **test statistic** to be calculated. If this is higher than a predetermined **critical value**, it is highly improbable that hypothesis *A* is correct. In that case hypothesis *A* is rejected in favour of hypothesis *B*.

How to choose and calculate the test statistic and how to determine its critical value are parts of the technique about to be described, but once the test statistic has been chosen, its critical value must be determined *before* the actual value is calculated. There must be no possibility of the choice of critical value being influenced by knowledge of the actual value.

Types of error

There are two ways in which the result of a hypothesis test may be wrong. Suppose a man has been told by his wife that she believes the next train goes at 8.55. Rejecting this, he arrives at the station at 8.56 and finds the train has just gone. He has committed a **type I error**, having been wrong in rejecting a correct hypothesis.

On the other hand, suppose she tells him that she thinks it goes at 9.00. He believes her, and arrives at 8.56 only to find it has just gone. Now he has committed a **type II error**, having been wrong in accepting a wrong hypothesis.

Occasionally we shall use this terminology. It may help you to remember the two types of error if you note that in the above definition of a type I error 'wrong' appears once, whereas in the definition of a type II error 'wrong' appears twice.

Consolidation exercise 24A

1. What is meant by (a) a hypothesis and (b) testing a hypothesis?
2. Can statistical methods prove the truth of a hypothesis?
3. What is the first step in using statistics to test a hypothesis?
4. What is a test statistic?
5. Why do we speak of rejecting one hypothesis in favour of another?
6. What are (a) a type I error and (b) a type II error?

24.3 Testing the origin of a sample

We now test hypotheses about the genuineness of the pens, considering first the sample of 81 pens, helped by the fact that ScrawlBall Pens Inc. have already stated the variance of their own pens. The procedure is very similar to that of finding confidence intervals, but there is an important difference: we are not asking for a range within which a population mean probably lies. Instead we are asking whether the mean life of pens in the sample (or the actual life of the single pen) is so much less than the known (or hypothesised) population mean that it is improbable the sample (or the single pen) comes from the same population.

One often forgotten assumption must be stressed. We assume that the single pen can be looked upon as having arisen at random, and that the sample of 81 pens was a properly taken random sample, or can reasonably be treated as one.

The origin of a single pen

Step 1: specifying the hypotheses

We begin by specifying that the single pen comes from a population with mean μ and variance σ^2. Our hypothesis is that this is the population specified by ScrawlBall as having a mean of 500 and a variance of 100. In other words, the difference between the mean life of the population from which the pen comes and the mean life of the advertised population is zero. We therefore call this a **null hypothesis**, and denote it by H_0, writing

$$H_0: \mu = 500 \quad \text{or} \quad H_0: \mu - 500 = 0$$

The complainant is obviously suggesting that the pen comes from a population with a lower mean life than the advertised life of 500 h. This leads to the specification of an **alternative hypothesis**, which can be written

$$H_1: \mu < 500 \quad \text{or} \quad H_1: \mu - 500 < 0$$

The question now is whether the single pen with a life of 485 h has a life so much lower than 500 h that it is highly unlikely to come from a normal population with mean 500 and variance 100. If it has then we reject H_0 in favour of H_1.

Step 2: choosing the level of significance

We have to be more precise about what we mean by 'so much lower than 500 that it is highly unlikely to come from a normal population with mean 500 and variance of 100'. We may decide to define 'highly unlikely' as meaning 'at least 95% improbable'. In other words, if the difference between 485 and 500 stands a chance of more than 5% of arising by pure chance, we do not reject the hypothesis that the pen comes from a genuine batch of ScrawlBalls. We call 5% the **level of significance** and we are testing whether the difference is **significant at the 5%** level. (Notebox 24.1).

Step 3: choosing the test statistic

We know from Chapter 19 and Table D4 that the distribution of a normally distributed variable is best studied by standardising the variable. We therefore define a new variable $Z = (X - \mu)/\sigma$. To every value of X there is a corresponding value of Z. Instead of testing the value of X, we can therefore test the value of Z, which is easier.

For the standard normal distribution, we also know that 5% of the area under the curve lies to the left of $z = -1.64$. This value therefore defines a **critical region**. If the calculated z is less than -1.64 then the value of X is also improbably low and we reject the null hypothesis; Z is a test statistic.

Step 4: calculating the value of the test statistic
Under the null hypothesis, $\mu = 500$ and $\sigma^2 = 100$. This means that the test statistic is

$$z = (x - 500)/10$$

Thus as $x = 485$ for the single pen, $z = -15/10 = -1.5$.

Step 5: completing the test
As the calculated value of z lies above (or to the right of) the critical value of -1.64, the life is not so improbably lower than 500 that we have difficulty in accepting the null hypothesis. The chance of a single genuine ScrawlBall with a life as low as 485 h is more than 5%.

Testing the origin of a sample of 81 pens

The basic approach is much the same, but now we know the mean life for a sample of 81 pens. This affects the precise wording of the hypotheses and the definition of the test statistic. To explain it more easily we do not describe the test in exactly the same order as for a single pen.

Step 1: specifying the hypotheses
We begin with the hypothesis that the sample pens are genuine and so come from a population with $\mu = 500$ and $\sigma^2 = 100$. In other words there is no difference between the sample pens and the pens that are known to be genuine. Thus we have the null hypothesis $H_0: \mu = 500$ or $H_0: \mu - 500 = 0$.

The complaint made to ScrawlBall is that the pens in the sample had a lower mean life. This could be a consequence of sampling, or because the pens truly come from a population whose mean is less than 500. This leads to the alternative hypothesis $H_1: \mu < 500$ or $H_1: \mu - 500 < 0$.

Step 2: choosing the test statistic
We know from the central limit theorem (Chapter 21) that if many samples of size $n = 81$ are taken from the population of genuine pens, with $\mu = 500$ and $\sigma^2 = 100$, these means will have a normal distribution with a mean of 500 and a standard error of $\sigma/\sqrt{n} = 10/9$. This would not necessarily have been true had n been small. The following argument is not valid for small samples.

If we denote a sample mean from this population by $\bar{x}$, the standardised variable Z, defined by

$$Z = \frac{\bar{x} - \mu}{\sigma/\sqrt{n}} \qquad [24.1]$$

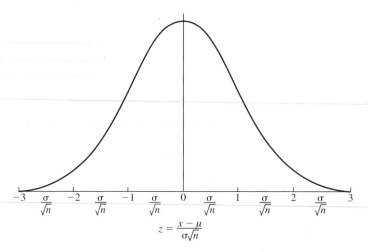

$$z = \frac{x - \mu}{\sigma\sqrt{n}}$$

Figure 24.1

will be normally distributed about zero mean with unit variance (Figure 24.1). Notice how this is just like the distribution for a single pen except the sample size comes into it in a way that reduces the size of Z.

We know that 90% of the area under this probability curve will lie between $z = -1.645$ and $z = +1.645$. Thus the chance of Z having a value of -1.645 or less is given by the area of the left-hand tail, which is 5%.

If we obtain a value of Z so low that the chance of reaching it is only 5% or less if the hypothesis is true, we have to choose between (i) accepting the hypothesis even though our evidence has led to a value of Z that is highly improbable and (ii) rejecting the hypothesis.

We can therefore use Z as a test statistic. For every value of the sample mean $\bar{x}$ there will be a unique value of Z, hence if $\bar{x}$ is very much below 500 it will lead to a Z that is *too negative* for us to accept the hypothesis H_0. We reject H_0 and accept instead that H_1 is more probable, in that if H_1 is true then such a negative value of Z is quite in order.

Step 3: specifying a level of significance

We have to decide on what we mean by 'too negative'. In the above illustration we suggested that we might reject H_0 if it led to a value of Z so low that there was a chance of only 5% (or less) that it would be reached. In such a case we say that we are rejecting H_0 at a *5% level of significance*. If the value of Z had not been so low, we would have said (choosing our words very carefully) that 'the evidence is not strong enough to justify rejection of H_0 at a 5% level of significance' (Notebox 24.2).

ScrawlBall can choose whatever level of significance it likes. It thinks that 1% would mean there is a high risk of deciding that the pens are not imitation when in fact they are; but 10% would mean that the firm may be involved in unnecessary costs such as in trying to trace the origin of imitations that are in fact not being made. It chooses a significance level of 5%.

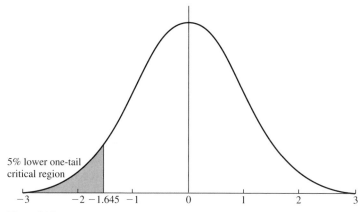

5% lower one-tail
critical region

Figure 24.2

Step 4: finding the critical value for the rejection zone

The decision about whether to accept the null hypothesis depends on the magnitude of z, defined in step 2, but we do not need to know this magnitude precisely. All we need to know is whether it is so big that it has less than a 5% chance of occurring if the null hypothesis is correct.

Recall that the probability of Z being equal to or less than a specified value z_1, is given by the area under the probability density curve to the left of z_1.

Thus if we can find the value of Z that has an area of 5% to its left (which we denote by z_c) then we have a *critical value*. Beyond this critical value z will be so highly negative that it lies in the *rejection zone*. If the negative value derived from our data equals or numerically exceeds this critical value, we should reject the null hypothesis.

We saw in our discussions of the normal curve and confidence intervals (Chapters 20 and 21) that a tail area of 5% is cut off by $z = 1.645$, which therefore becomes the numerical value of z_c. Since we are looking at the left tail, we attach a negative sign.

If we calculate the test statistic Z and find it is negative and numerically greater than 1.645, then we reject the hypothesis. Thus our rejection zone, or critical region, is defined by $z < -1.645$. (Figure 24.2). This critical region has been obtained before any calculations have been performed. It depends only on our choices of the test statistic and the significance level.

Step 5: calculating the value of the test statistic

We know that

$$Z = \frac{\bar{x} - \mu}{\sigma / \sqrt{n}}$$

Using our ScrawlBall data, we therefore have

$$Z = \frac{\bar{x} - 500}{10/9}$$

Since the sample mean $\bar{x} = 497$, our calculated test statistic is

$$z = \frac{497 - 500}{10/9} = \frac{-3}{1.111}$$

which is obviously negative and numerically not much less than 3. It is well beyond the critical value of -1.645 and is firmly in the rejection zone. There is no need to complete the calculation.

Step 6: conclusion
The value of z falls in the rejection zone at the 5% level of significance. The sample mean has less than a 5% chance of arising if H_0 is true. We therefore reject H_0 in favour of H_1. ScrawlBall Pens Inc. is probably justified in asserting that the pens are inferior imitations.

24.4 Second thoughts

Having decided to reject the null hypothesis if z falls in the 5% rejection zone, we should not now change our minds about the level of significance. If we do that, we may soon be reaching only decisions we want to reach. However, it is instructive to see how different the calculation and conclusion would have been (i) if we had adopted a different level of significance and (ii) if the sample mean had been a little higher.

If we had adopted a 1% level of significance
The analysis would have been just the same except that we would have used $z_c = -2.33$, which is the value that cuts off a tail of 1%.

The calculated z would still have been in the rejection region, still leading to rejection of the null hypothesis, despite the greater strictness of the test.

If the sample mean had been 498.5 (instead of 497)
We will adopt a 5% level of significance and proceed exactly as before. The critical value of z is again -1.645. The calculated value of z is

$$z = \frac{498.5 - 500}{10/9} = -1.423$$

which is closer to 0 than the critical value, so we must now reach a different conclusion.

Step 6: conclusion in altered example
The calculated z has a chance of more than 5% of arising if the null hypothesis is true. The evidence is therefore not strong enough to warrant rejection of H_0. We cannot assert with 95% confidence that the pens are imitations. Note carefully that we have not said the pens are genuine. There could be some other test that would reveal them to be imitation, perhaps a test of their lifetimes, their physical widths, or the chemical composition of their ink or material of manufacture. To establish genuineness could entail a major investigation.

24.5 Looking ahead

The ScrawlBall problem arises because the sample of pens has a mean life that is *less than* the official life of genuine ScrawlBalls. We have been seeing only if the life is significantly *less than* the official life. It is a one-sided question.

If we have a different problem – perhaps the number of non-working pens in a batch of 1000 is higher than the officially declared number of defectives – then we might want to see whether the actual number is significantly *greater than* the hypothesised number. This, too, is one-sided.

Another possibility is that the specifications and materials used in making ScrawlBall pens give them a mean mass with a 95% confidence interval of (4.65, 5.35) g. Pens of mass either less than 4.65 g or more than 5.35 g are likely to be imitations. In this case the statistical interest would be in whether the sample mean is significantly *different from* the hypothesised mean of 5.00 g, which is two-sided.

Solving problems such as these is not fundamentally different from what we have just done, but a few details have to be changed. We consider this in Chapter 25, where Box 25.1 describes the general procedure applicable to the ScrawlBall problems as well as to others.

Consolidation exercises 24B

1. State in order the the six steps in the testing of a statistical hypothesis.

2. What is (i) a null hypothesis and (ii) an alternative hypothesis?

3. What is the formula for the test statistic used to test a hypothesis involving a large-sample mean? How is this test statistic distributed?

4. What is meant by a 5% level of significance? Why should a level of significance be chosen before calculations are begun?

5. Which is the higher, a 5% level of significance or a 1% level?

6. What is meant by (a) critical value and (b) rejection zone?

Notebox 24.1

The level of significance

Conventionally, two levels of significance have become important: the 5% level and the 1% level. The 5% level means that we reject our null hypothesis H_0 if our sample leads to a value of the test statistic that we would expect to reach or to exceed only 5 times (but no more) in a hundred if H_0 is true. By adopting this level of significance, we are taking a 5% risk of rejecting H_0 (in favour of H_1) when in fact H_0 is right – a type I error.

If we adopt a 1% level, we are refusing to reject H_0 unless the evidence has led to a result that is likely to occur less frequently than once in a hundred times if H_0 is true. ⊃

For values between 5% and 1% we tend to reject the hypothesis with increasing degrees of conviction. Notice that if we adopt a 1% level of significance, we are less likely to reject a wrong hypothesis than if we adopt a 5% level. Note that a 1% level of significance is 'higher than' a 5% level of significance. See also Chapter 25.

Notebox 24.2

Evidence and verdict

Courts of law in Scotland may be presented with one of three verdicts: guilty, not guilty and not proven. In the court of statistical testing, the third of these has to be kept well in mind.

Summary

1. Statistical methods cannot prove or disprove a hypothesis. But if a hypothesis can be acceptably represented by a statistical hypothesis, it may be possible to make a probabilistic statement about it.

2. Errors of two types may arise when we test hypotheses: type I is when we wrongly reject a correct hypothesis, and type II is when we wrongly accept a wrong hypothesis.

3. We begin by formulating a null hypothesis H_0 (that there is no difference between a sample statistic and a hypothesised population parameter) and an alternative hypothesis H_1 (which in the case of the example in this chapter is that the sample statistic is lower than the population parameter).

4. This allows a *test statistic* to be calculated. In the case of a single item, this is the standardised variable $Z = (x - \mu)/\sigma$. For a large sample mean this is replaced by

$$z = \frac{\bar{x} - \mu}{\sigma/\sqrt{n}}$$

which is also distributed as $N(0,1)$. Every observed sample mean $\bar{x}$ leads to a unique value of Z. The more remote this is from zero, the less likely is H_0 to be true.

5. A level of significance has to be specified before calculations are begun. This is a low level of probability such that H_0 is rejected if the calculated value of Z has that (or lower) probability of arising.

6. Choice of test statistic and level of significance allow the use of tables to determine a critical value of Z, such that if the value of X leads to a value of Z lying beyond it then H_0 is rejected. Such values are in the rejection zone.

7. Substitution of the values for x (the single value) or $\bar{x}$ (the sample mean), μ (the hypothesised population mean), σ (the population standard deviation) and n (the sample size) in equation [24.1] allows a value of Z to be calculated. If this falls in the rejection zone, H_0 is rejected at the chosen level of significance. Otherwise we say the evidence we have used is not strong enough to warrant rejection; but there could well be some other evidence or test that would lead to rejection (Notebox 23.2).

8. The interest has been in whether (a) a single item or (b) a sample mean was so much lower than a hypothesised value that a hypothesis about the authenticity of the pen or the sample of pens should be rejected. Other examples, when we are interested in whether an observed value is (i) so much higher than and (ii) so different from a hypothesised value are discussed in Chapter 25, where several important points arise.

25

A tale of two tails

25.1 Introduction

Example 25.1 Members of the Distinguished Dachshunds' Diet Club (DDDC) state that the daily costs of keeping a dachshund are normally distributed with a mean of 98p and a variance of 30. Lady Agatha asserts that she spends £1.17 per day feeding hers. Test the hypothesis that she adheres to the DDDC's diet sheet.

Example 25.2 The *Proceedings of the Society for Promoting More Moral Behaviour in Statingham* lists the seven deadly sins. A special article says that having sinful thoughts is also sinning, and that on average adults in that city confess to sinning 170 times a month, with a standard deviation of 13 times a month. The *Statingham Declaimer* engages Anna Liszt to test the truth of this statement. She questions a random sample of 441 adults and finds that, on average, they sin 171.1 times a month. (a) Does this sample result imply that, on average, adults from Statingham are even worse than the *Proceedings* suggest? (b) Does it imply that either Anna's sample is biased, or the Society is wrong?

Example 25.3 A random sample of 81 male students from Statingham revealed that, on average, they spent 19.7 hours a week with their girlfriends, with a standard deviation of 4.6 hours. (a) Can this be taken as refuting the assertion that the average time spent together is 21 hours? (b) Can it be taken as confirmation of the assertion that the average time spent together is less than 21 hours?

Example 25.4 KuriousKarparts have found that there is only a 2.5% chance of one of their windscreen wipers providing as few as 190 000 satisfactory wipes. They advertise that their wipers have an average life of 210 000 wipes.

They buy their wipers in batches. Knowing that if they sell wipers with a shorter life than 210 000 wipes they will lose customers, they reject a batch if the mean sample life is so much lower than 210 000 wipes that the chance of it arising is less than 10% if the batch mean is 210 000.

If the life is too long, their satisfied customers will take a long time to come back, so they also reject a batch if the mean sample life is so much higher than 210 000 wipes that there is only a 1% chance of it arising if the batch mean is as stipulated.

They buy large batches of wipers from various suppliers, first taking random samples of 50 and testing their lives. What are the lowest and highest acceptable sample mean lives?

278

The problems facing us in these examples are essentially the same as the ScrawlBall Pens problem discussed in the last chapter. All of them contain information from a sample to be used for reaching a decision about a population. The main difference is that the ScrawlBall problem involved asking whether a sample statistic (the mean) was *so much lower than* a hypothesised population parameter that the difference was unlikely to arise by chance. But here we ask whether the sample statistic is (a) *so much higher than* the population parameter or (b) *so different from* the population parameter (regardless of sign) that a highly unlikely result has been obtained.

Answering these questions is almost exactly the same as answering the ScrawlBall questions, but involves the very important question of whether to apply a one-tailed or a two-tailed test.

We begin with problems in which the population variance is known. Then we consider examples in which it is not known.

The first example is a simple case of testing whether a single item comes from a stated population. In the next three examples we use the mean as our sample statistic. In all these examples our purpose is to test a hypothesis, and we do this by applying a **significance test** to the (standardised) sample mean. We therefore sometimes say that we are **testing the significance of the mean.** We are seeing whether the sample mean is so different from the hypothesised mean that it signifies the hypothesis is to be rejected.

Similar questions about sample proportions and other sample statistics are considered in the next few chapters.

25.2 The origin of a single item

 In **Example 25.1** we ask whether Lady Agatha is feeding her dog according to the diet sheet. If so, the daily cost of £1.17 should belong to the normal distribution with a mean of 98p and a variance of 30p². We set up the *null hypothesis* that this is so.

She could fail to observe the diet sheet by spending either more or less than the amount it indicates. This means that we are concerned with seeing whether £1.17 is so *different from* 98p that it is unlikely her cost does belong to a population with this mean. The *alternative hypothesis* is that Lady Agatha's cost belongs to a distribution whose mean is not 98.

If we adopt a 95% significance level, we are asking whether £1.17 (or 117p) is so different from 98p that it is unlikely to arise in a normal distribution with a mean of 98 and a standard deviation of $\sqrt{30}$ ($= 5.477$). The easiest way of testing this is to form a standardised variable, converting cost (X) into a new variable Z such that

$$Z = \frac{X - \mu}{\sigma} = \frac{X - 98}{\sqrt{30}}$$

We use this as our test statistic. On its distribution (Figure 25.1) we have marked two values of Z such that the combined chance of being lower than A or higher

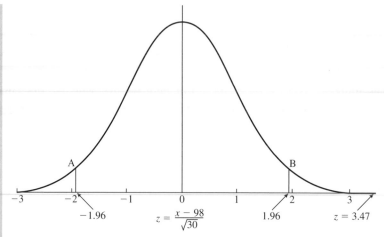

Figure 25.1

than B is 5% or less if Lady Agatha is observing the diet sheet. We have a two-tailed test. We know from Table 21.1 that the critical values of Z at A and B are -1.96 and $+1.96$. Inserting $x = 117$ we have a calculated z of

$$z = (117 - 98)/5.477 = 3.47$$

This is clearly outside the range between A and B. The fact that it is in the upper tail rather than the lower tail is irrelevant to our test, which is whether 117 is significantly *different* from 98.

At our chosen level of 5% significance, we reject the hypothesis that Lady Agatha observes the diet sheet.

25.3 Significance of sample means with known population variance

Upper tail

 Example 25.2 concerns moral behaviour in Statingham. Before starting to test any hypotheses, we ask whether we have continuous or discrete data. This important question is sometimes difficult to answer. We will treat the data as continuous for reasons given in Notebox 25.1.

Part (a) considers whether the sample result implies that, on average, adults from Statingham are even worse than the *Proceedings* suggest. Our concern is not with whether a single value belongs to a specified distribution, but with whether the mean of a sample of values belongs to it. This affects our hypotheses and introduces the sample size into the test statistic. We go through the same steps as we employed in the ScrawlBall example.

Step 1: specifying the hypotheses

We hypothesise that the mean number of sinning times of all adults in Statingham is as reported in the *Proceedings* and is 170.0. Thus, if we denote the population mean number of times by μ, we may write the *null hypothesis* as $H_0: \mu = 170.0$ or as $H_0: \mu - 170.0 = 0$, emphasising that the difference between the population mean μ and the sample mean is nil.

Once we have stated the null hypothesis, there are only three possible alternative hypotheses, one of them embracing the other two. The obvious alternative hypothesis is that (1) the population mean number of sinning times is *not* 170.0. If so, then one of two other more specific statements will be true: either (2) the mean number is *more than* 170.0 or (3) the mean number is *less than* 170.0.

The question asks whether the sample result implies that the mean sinning rate is even worse than the proceedings suggests. This requires a yes/no answer and the answer should be yes only if the sample result is so much *bigger than* the hypothesised mean of 170.0 that it is unlikely to have arisen if the null hypothesis is true.

This means that our *alternative hypothesis* should be the second of the three possible alternatives – the mean number of sinning times for the population is *more than* 170.0. Thus $H_1: \mu > 170$ or $H_1: \mu - 170.0 > 0$.

Step 2: choosing the test statistic

To test a hypothesis about a large-sample mean, we use our knowledge that the sample mean has a normal distribution centred on μ and with a variance of σ^2/n. We convert this into the standard normal variable, writing

$$Z = \frac{X - \mu}{\sigma/\sqrt{n}}$$

and then take as our test statistic the value of this corresponding to the sample mean:

$$z = \frac{\bar{x} - \mu}{\sigma/\sqrt{n}}$$

where n is the sample size and σ is the population standard deviation. The probability of $\bar{x}$ being different from μ by as much as the observed amount is the same as the probability of z being different from 0, and this can be obtained from tables of the normal distribution, as we see below.

Step 3: specifying a level of significance

We will adopt a 5% level of significance. This means that we reject H_0 if the calculated value z is so big that there is a chance of no more than 5% of its being reached by an accident of sampling. It is one-sided.

Step 4: finding the critical value for the rejection zone

We need as our critical value of z the positive value that cuts off a tail with an area of 5% of the area under the normal curve. This is 1.645, so we define the critical region to be $z > 1.645$ (Figure 25.2).

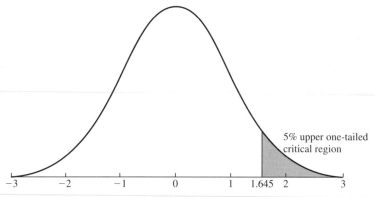

Figure 25.2

Step 5: calculating the value of the test-statistic
We know that $\sigma = 13$ and that $n = 441$, giving $\sqrt{n} = 21$. The formula therefore gives

$$z = \frac{\overline{x} - 170}{13/21} = \frac{171.1 - 170}{13/21} = 1.777$$

Step 6: conclusion
Since our calculated z is greater than 1.645, we reject H_0 at the 5% level of significance, in favour of H_1: that the mean of population from which the sample was taken is greater than 171.1.

Two tails

We now consider version (b) of the sinning problem, where the sample result is the same as in (a) but we are interested in a somewhat different question.

◯ We have asked whether the sample result implies that either Anna's sample is biased, or that the Society is wrong. Another form of this question is to ask whether it allows us to assert with a prespecified confidence that the sample does not come from a population with $\mu = 170.0$ and $\sigma = 13$.
   In the ScrawlBall example we asked whether a sample came from a population whose mean was *less than* 500 hours, and used the *lower tail* of the normal distribution of sample means.
   In answering Example 25.2(a) we asked whether the sample came from a population whose average rate of sinning was *higher than* 170 times a month, and used the *upper tail* of the distribution.
   Now we are asking whether a sample comes from a population whose mean is *different from* 170.0 cm. Here 'different from' means 'either less than or more than'. We have to use *two tails*, and our critical region has to be differently defined, giving z_c a *different* value, as we now see. ◯

Step 1: specifying the hypotheses

We hypothesise as before that the mean rate of sinning for all adults in Statingham is 170.0 times a month. Thus, if we denote the mean rate by μ, we may write H_0: $\mu - 170 = 0$.

Our alternative hypothesis is that the mean rate is not (or is different from) 170.0 times a month, H_1: $\mu - 170.0 \neq 0$.

Step 2: choosing the test statistic

We use the same test statistic as in the one-tailed test, using z defined by

$$z = \frac{\bar{x} - \mu}{\sigma/\sqrt{n}}$$

where n is the sample size and σ is the population standard deviation.

Step 3: specifying a level of significance

We will adopt a 5% level of significance.

Step 4: finding the critical value for the rejection zone

The null hypothesis is to be rejected if the probability of the calculated value of z being so different from 0 is less than 5%.

Since the normal probability curve is symmetrical about a mean of 0, this means that we have to divide the 5% probability of *being different from* into a 2.5% probability of *being more than* (in the upper tail), and a 2.5% probability of being less than (in the lower tail).

This is illustrated in Figure 25.3. The numerical value of z_c that cuts off two tails and evenly shares a total probability of 5% (in a two-tailed test at 5% significance), is the same as the value that cuts off one tail of 2.5% for use with a one-tailed test at 2.5% significance. It is 1.96, so we define our rejection region as

$$z < -1.96 \quad \text{and} \quad z > 1.96 \qquad (|z| > 1.96)$$

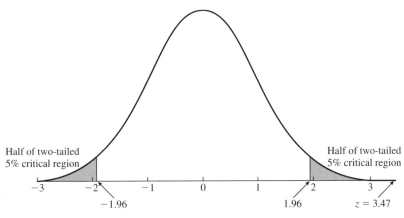

Half of two-tailed 5% critical region

Half of two-tailed 5% critical region

−1.96

1.96

$z = 3.47$

Figure 25.3

Step 5: calculating the value of the test statistic
We know that $\sigma = 13$ and that $n = 441$, giving $\sqrt{n} = 21$.
The test statistic is

$$z = \frac{\bar{x} - 170}{13/21} = \frac{171.1 - 170}{13/21} = 1.777$$

Step 6: conclusion
Since our calculated z is less than 1.96, the evidence is not strong enough for rejection of the null hypothesis that the sampled people come from a population with a mean sinning rate of 170 times a month. The result is compatible (at a 5% level of significance) with the Society's assertion that the average rate is 170.

Consolidation exercise 25A

1. What are we doing when we test the significance of the mean?

2. Stating the hypotheses is the first step in testing a significance. What is the second step?

3. The test statistic, denoted by z, involves four quantities. One of them is n, the sample size. What are the other three?

4. With a one-tailed test at a 5% significance level, what is the numerical value of the critical value for the rejection zone?

25.4   **Significance of sample means with unknown population variance**

Using the mean to test whether a large sample comes from a specified population is much the same whether or not we know the population variance. If we know it, we proceed as above; if not, we estimate it. We saw in Chapter 16 that the best estimate is a slightly modified version of the sample variance. More precisely, our best estimate of the population variance is

$$\hat{\sigma}^2 = ns^2/(n-1) \tag{25.1}$$

where the sample size is n and the sample variance is s^2.

We then use this estimated value $\hat{\sigma}^2$ in place of σ^2. As we have seen, a little algebra quickly shows that, if we do this, the standard error becomes

$$\sigma_{\bar{x}} = \hat{\sigma}/\sqrt{n} = s\sqrt{n/(n-1)} \times 1/\sqrt{n} = s/\sqrt{n-1}$$

where s is the sample standard deviation. The solution of the problem is therefore just the same as if we had known the population variance, except that in step 3 we use this amended version of the standard error.

Example 25.3 asks two questions. We now summarise their solutions, with a few comments. We follow the six steps used in solving Example 25.2 but we set them out more concisely.

Part (a)

The null hypothesis is that there is no difference between the mean time spent with their girlfriends by all male students and the asserted mean of 21 hours. Thus H_0: $\mu - 21 = 0$ and H_1: $\mu - 21 \neq 0$, implying a two-tailed test.

The test statistic is

$$z = \frac{\bar{x} - \mu}{\hat{\sigma}_{\bar{x}}} \quad \text{where} \quad \hat{\sigma}_{\bar{x}} = s/\sqrt{n-1}$$

and we choose a significance level of 1%. This means two tails, each with an area of 0.5%, and Table D4 gives the rejection zone as $-2.575 < z < +2.575$.

With $n = 81$, $s = 4.6$ and $\bar{x} = 19.7$ the calculated value of z is -2.528. Numerically this just falls short of the critical value of 2.575, so we do not reject the null hypothesis; the evidence is not (quite) strong enough for this.

Part (b)

The null hypothesis is as in part (a) but H_1: $\mu - 21 < 0$. The test statistic is the same as in part (a) but the critical level for the one-tailed test is 2.326. The calculated value is again -2.528, which is now numerically greater than the calculated value. We have therefore obtained a value that is less than 1% probable if the null hypothesis is correct. Since we have chosen this level of significance, we reject the null hypothesis in favour of the alternative hypothesis that the average time spent together is less than 21 hours.

25.5   Choice of tail and levels of significance

How to formulate the alternative hypothesis, with its implications for the tail characteristics of the test, is sometimes a difficult question. Significance tests are normally carried out to assist in a decision. As a broad rule, if it is clear that *any* big enough *difference* (regardless of sign) matters so far as that decision is concerned, then we need a two-tailed test. A common example of this is when we are wondering whether items in a sample come from a stated population whose mean is known. Any significant difference, positive or negative, in the mean will cast doubt on the origin of the sample.

Other cases concern only differences of one sign, then a one-tailed test is appropriate. Suppliers of items sold to wholesalers often state that the items in a delivery have a certain mean mass. The wholesaler will be interested only in whether the deliveries he or she receives have a mean mass of less than the stated amount. If the wholesaler wishes to check by sampling, he or she will apply a one-tailed test.

On the other hand, the supplier will not want any deliveries to have a mean mass that is too low, in case it should lead to prosecution, or one that is too high, which will erode profits. For the supplier, a two-tailed test is appropriate. If the supplier's loading (or filling) machine is not functioning properly, he or she needs to know about it.

Sometimes a person may be interested in both tails separately, for different reasons, and possibly with different ideas of significance. We consider KuriousKarparts.

⬭ In **Example 25.4** the first paragraph tells us that, if we can assume the lifetimes are normally distributed, the population mean life is $\mu = 210\,000$ wipes and the standard deviation σ is such there is only a 2.5% chance that the lifetime is less than 190 000 wipes. We use this to find the value of σ. With a normal distribution, the lower 2.5% tail corresponds to a value of -1.96σ (Figure 25.3). It follows that $1.96\sigma = 210\,000 - 190\,000$ hence $\sigma = 20\,000/1.96$. This allows us to calculate the standard error of the large sample mean, as $\sigma/\sqrt{n}$, which is

$$\frac{20\,000}{1.96 \times \sqrt{50}} = 1443 \text{ wipes}$$

The wipers are rejected if the mean life is so low there is only a 10% chance of getting it. Thus we have to calculate the 10% lower critical value. Now that we know the standard error, we can write this lower critical value as

$$\mu - 1.282 \times \text{s.e.} = 210\,000 - 1.282 \times 1443 = 208\,150 \text{ wipes}$$

A sample mean as low as this has only a 10% chance of arising if the batch mean is 210 000. Wipers from a batch with a mean less than this would have too high a risk of being short-lived.

The 1% upper critical value for the sample mean (needed because wipers that last too long are rejected) is

$$\mu + 2.326 \times \text{s.e.} = 210\,000 + 2.326 \times 1443 = 213\,356 \text{ wipes}$$

A sample mean as high as this has only a 1% chance of arising if the batch mean is 210 000. Wipers from a batch with a mean more than this would have too high a risk of being too durable.

Thus only batches with sample mean lives greater than 208 150 wipes and less than 213 356 wipes should be accepted.

Box 25.1

How to ... **test the significance of a large-sample mean**

1. Set up the null hypothesis H_0: $\bar{x} = \mu$ where μ is the hypothesised population mean.

2. State the alternative hypothesis H_1, normally
   (a) $\bar{x} \neq \mu$, requiring a two-tailed test
   (b) $\bar{x} > \mu$ or $\bar{x} < \mu$, each requiring a one-tailed test.

3. Specify the level of significance for the test.

4. Define the test statistic

$$z = \frac{\bar{x} - \mu}{\sigma/\sqrt{n}} \qquad \text{(if } \sigma \text{ is known)}$$

$$\frac{\bar{x} - \mu}{s/\sqrt{n-1}} \qquad \text{(if only } s \text{ is known)}$$

where σ is the population standard deviation and s is the sample standard deviation.

5. Find from tables the critical value of the test statistic at the chosen significance level, thus defining the rejection zone. In most cases it will be as follows

(a) For the two-tailed test. At 5% significance: the areas beyond 1.96 units from the central value in either direction. At 1% significance: the areas beyond 2.58 units.

(b) For the one-tailed test. At 5% significance: the area beyond 1.645 units from the central value in the appropriate direction. At 1% significance: the area beyond 2.33 units.

6. Calculate the value of z for the data being investigated and compare it with the critical value. Reject the null hypothesis if the numerical value of z exceeds the numerical critical value.

25.6 General procedure

The test procedure just described is valid for all large-sample tests of significance of normally distributed statistics. Although we have restricted this chapter to tests of the significance of a large-sample mean, the basic approach is applicable to many other statistics, such as a proportion, the difference between two large-sample means, and so on. We consider some of them in later chapters.

Consolidation exercise 25B

1. How do you test the significance of the sample mean if the population variance is unknown? What do you use instead of $\sigma_{\bar{x}}$? How else does the test differ when you do not know the population variance?

2. When is it necessary to use (i) a one-tailed test and (ii) a two-tailed test?

3. How is the critical value in a one-tailed test related to the value in a two-tailed test?

Notebox 25.1

Does 'an average rate of 171.1 times a month' imply continuous?

The number of times that any one person sins in a month must be an integer. So must the total number of times for 441 persons. When we multiply the average rate of 171.1 by 441 we obtain 75 455.1, which is an impossible total of integers. This shows that 171.1 is a rounded average. Its lowest value could be 171.05 and corresponds to a total of 75 433.05, which also is impossible. A total of 75 433 would not lead to a rounded average as high as 171.1, so we have to interpret this as 75 434. In other words, 75 434 is the lowest total compatible with a rounded mean of 171.1. A similar argument shows that the highest total compatible with this rounded mean is 75 477.

The actual total can therefore be any integer between 75 434 and 75 477 inclusive. Taking a few consecutive possible values around the centre of this interval, just for illustration, we have 75 454, 75 455, 75 456, 75 457. These would lead to means of 171.0975, 171.0998, 171.1020, 171.1043 expressed to 4 d.p. The separation between these consecutive possible values is so much smaller than the error of rounding that we are justified in treating the variable as continuous. (Note that the separation of the possible values is approximately 0.0023. More accurately it is 0.002 267 57 . . ., which is 1/441, the reciprocal of the sample size.)

Summary

1. The one-tailed test of Chapter 24 is extended to (a) the upper tail instead of the lower tail, and (b) to a two-tailed test.

2. The one-tailed test of whether an observed value is greater than a given value is identical to the test described in Chapter 24, except that the critical value lies at the beginning of the upper tail of the distribution. Both one-tailed tests are summarised in Box 25.1.

3. The two-tailed test of whether the observed value is different from a given value divides the critical region into two equal parts, one at each end of the distribution. For a 5% two-tailed test, each tail contains 2.5% of the distribution. Thus a two-tailed 5% test has two critical values, $\pm z_c$, each the same as the numerical value of the one-tailed 2.5% test. There are similar results at other levels of significance.

4. The wording of a problem and its precise nature should be carefully considered before deciding whether to use a one-tailed test or a two-tailed test.

5. The general procedure is given in Box 25.1.

6. Sometimes a problem may involve two tails at different levels of significance; then it becomes two one-tailed problems.

7. As indicated in Notebox 25.1, when data are reported as discrete measurements, it is necessary to consider whether they may be treated as continuous (using the methods in this chapter) or whether a continuity correction may be necessary (Chapter 26).

73 out of 100 adults drink Demons' Downfall

26.1 Introduction

Example 26.1 The Bishop of Statingham says that 73% of the adults in his diocese were drinkers of Demons' Downfall. Anna Liszt took a random sample of 81 adults and found that 58 of them confessed to this vice. Does this support the Bishop's opinion of his flock?

Example 26.2. The Mayor of Statingham is proud of his city's equal opportunities programme. He says that half the employees of the council and its agencies are female and half male. A random sample of 500 employees has 272 females. Does this cast doubt on his statement?

Example 26.3 According to the manufacturer's advertising, 8 out of 10 households use Wyshwash. A random sample of 100 households in Statingham shows that 74 households use it. If the manufacturer's claim is right at a national level, does the sample result suggest that the use of Wyshwash by Statingham householders is below the national level?

We have seen how to test the significance of a difference between a sample mean and a population mean. We use the same approach to test the difference between a sample proportion and a population proportion. The formula for the test statistic is slightly different from the formula in Chapter 25, and we have to use a continuity correction.

26.2 Testing the significance of a sample proportion

The questions in this chapter are examples of a more general question:

A sample of size n reveals a proportion p_s of successes. Does this differ significantly from the population proportion of successes p?

It can be shown (Chapter 22) that for large samples the sample proportion p_s is normally distributed about the population mean p and with a variance pq/n.

This normal distribution can be standardised by defining a new variable Z such that

$$Z = \frac{p_s - p}{\text{s.e. of proportion}} \qquad [26.1]$$

where the standard error of the proportion is (by definition) the standard deviation of the distribution of the sample proportions. We have just seen that this is $\sqrt{pq/n}$.

Thus the probability of the sample proportion differing from the population proportion by any given amount is the same as the probability of Z differing from 0 by the corresponding value z of the standardised variable. As in the case of the mean, we can use z as a test statistic. We need to find the value of

$$z = \frac{p_s - p}{\sqrt{pq/n}} \qquad [26.2]$$

which will be distributed as N(0,1).

In considering Example 26.1, we come across a few matters of more general application.

Example 26.1 suggests that we are concerned with the compatibility of Anna's sample result (58 out of 81) and the stated population result (73%). This indicates a null hypothesis H_0 that the difference between the p_s and the stated population proportion of 0.73 is zero. $H_0: p_s - p = 0$

If we were asking whether the Bishop's statement is an exaggeration, we would use a one-tailed test, but we are not. We will reject the stated null hypothesis if there is a big enough difference, regardless of sign; this means a two-tailed test. Thus the alternative hypothesis is that the difference is not zero, $H_1: p_s - p \neq 0$

We choose a 5% level of significance, which means that our critical value of z for a two-tailed test is 1.96.

Before we apply this test, we must note that, in the case of a proportion, we almost always need to apply a **continuity correction**, which is usually small but may be very important.

In **Example 26.1** the proportion drinking Demons' Downfall is $58/81 = 0.7160$. This differs from the value of 0.73 specified by the Bishop. We are concerned with whether the difference between them is bigger than the critical value.

The proportion of people drinking Demons' Downfall has been derived from the number drinking it in a not very big sample. This number (58) can alter only by one at a time; it is discrete. As we are using the continuous normal distribution to analyse it, we have to allow for this.

In continuous terms, 58 has to mean anything between 57.5 and almost 58.5. When divided by 81 this leads to a proportion that can be as high (or as close to the population proportion) as $58.5/81 = 0.72222$. This is the value we should use for p_s in performing the test, since it leads to the smallest possible value of the separation of the sample and population proportions compatible with there being 58 imbibers of Demons' Downfall.

We must also ask what the Bishop means by 73%. We may come up with two interpretations:

(a) We should take him at his word, and interpret it as meaning $p = 0.73000$.
(b) Even the Bishop is human and he means anything between 72.5% and almost 73.5%.

To illustrate the procedures, we consider these possibilities in turn with the help of Figure 26.1.

The Bishop means that $p = 0.73000$

As $p = 0.73$ we have that $q = 1 - p = 0.27$. As we have seen, the sample value of the proportion could be as high as $58.5/81$ (= 0.72222). We are testing whether this value is likely to come from a normal population with a mean of 0.73 and a variance of $pq/n = (0.73)(0.27)/81 = 0.0024333$ (and a standard deviation of $\sqrt{pq/n} = 0.049329$).

We set up a null hypothesis H_0: $p_s - 0.73 = 0$. Since we are asking whether the sample could have come from this population, regardless of whether the sample value is higher or lower than 0.73, we have a two-tailed test and the alternative hypothesis is H_1: $p_s - p \neq 0$. We decide to use a 5% level of significance.

The test statistic is essentially the same as for the test of a mean, except that the divisor is $\sqrt{pq/n}$. Thus

$$z = \frac{p_s - p}{\sqrt{pq/n}} = \frac{p_s - 0.73}{0.049329}$$

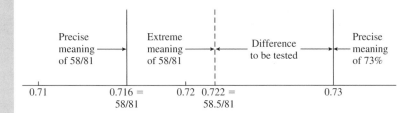

(a)

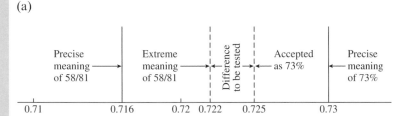

(b)

Figure 26.1

We know that the 5% critical value for the two-tailed test is 1.96. Inserting our corrected value of p_s, this formula for the test statistic gives a calculated value of

$$\frac{0.722\,22 - 0.730\,00}{0.049\,329} = -0.157\,72$$

Thus the calculated value of z lies between zero and the critical value of -1.96, so it does not fall in the rejection zone. The evidence of the sample does not justify rejection of the Bishop's statement. Notice that the continuity correction in this example decreases the separation of the means by $1/2n$ where n is the sample size. This is a general result. We now consider the same problem but with the Bishop's remark interpreted less precisely.

The Bishop means that the population proportion is between 72.5% and almost 73.5%

Now a sample value will be significantly different from the stated population value only if it is either significantly lower than 0.725 or significantly higher than 0.735.

Since our sample value is lower than the population proportion, we have to consider whether its highest possible value (of 0.722 22) is significantly lower than 0.725. This is shown in Figure 26.1(b). It means that the top of the expression for the test statistic becomes $-0.002\,78$ instead of $-0.007\,78$. The numerical value of z is obviously less than when we can take the Bishop at his word, so the difference remains non-significant.

We now consider the next example, with fewer digressions and numbering the steps of our argument.

⬭ 1. In **Example 26.2** we adopt a null hypothesis that the equal opportunities programme is working: that the sample imbalance is not great enough to cast doubt on this assertion. Symbolically, $H_0: p_s - 0.50 = 0$. We are interpreting 'half' as exactly 0.5.

2. Again we have a two-tailed test. We are wondering whether there is an imbalance, in one direction or the other. The alternative hypothesis is that the sample proportion is more different from 0.5 than can be ascribed to chance, $H_1: p_s - 0.50 \neq 0$.

3. We decide that the evidence casts doubt on the mayor's statement if there is less than a 5% chance of getting the recorded difference between the sample proportion of females and the hypothetical proportion of 50%.

4. The test statistic is

$$z = \frac{p_s - p}{\sqrt{pq/n}}$$

and there is a continuity correction which decreases the top of this expression by $1/2n$. As $n = 500$ this is very small and may seem to be unimportant, but wait and see! ⬭

5. The 5% critical level (two tails) is 1.96.

6. We can now calculate n using $p_s = 272/500 = 0.544$, $p = q = 0.500$ and the continuity correction of 0.001. We have

$$z = \frac{0.544 - 0.500 - 0.001}{\sqrt{(0.50)(0.50)/500}} = 0.043/0.022\,36 = 1.923$$

which is less than the critical level of 1.96.

(Note that without the continuity correction we would have had $z = 0.044/0.02236 = 1.9678$, which just exceeds the critical level of 1.960. In this example the conclusion is changed by the use of the small continuity correction. However small the correction may be, it is wise to include it, unless it is clear that the critical value is going to be well avoided.

7. Since the calculated value of the test statistic z is less than the critical value, the chance of the null hypothesis being correct and the observed difference arising accidentally is greater than 5%. Therefore we do not reject the null hypothesis in favour of the alternative hypothesis that the proportion of women is not equal to 0.5. The evidence does not justify the assertion that the equal opportunities programme is not working.

The procedure for testing the significance of a proportion is set out in Box 26.1

Box 26.1

How to ... **test the significance of a large-sample proportion**

1. Set up the null hypothesis $H_0: p_s = p$ where p is the hypothesised population proportion.

2. State the alternative hypothesis H_1 which will normally be
   (a) $p_s \neq p$, requiring a two-tailed test
   (b) $p_s > p$ or $p_s < p$, each requiring a one-tailed test

3. Specify the level of significance for the test.

4. Define the test statistic

$$z = \frac{p_s - p}{\sqrt{pq/n}}$$

5. Find from tables the critical value of the test statistic at the chosen significance level, as in step 5 of Box 25.1, thus defining the rejection zone.

6. Calculate the value of z for the data under study and compare it with the critical value. The numerical value of the numerator in the expression for z has to be decreased by a continuity correction of $1/2n$, where n is the sample size.

7. Reject the null hypothesis under the conditions described in step 6 of Box 25.1.

We consider another example, involving a one-tailed test.

⬭ In **Example 26.5** we are interested in whether a sample proportion of 74/100 (or 0.74) is so much lower than the national proportion of 8 out of 10 (or 0.8) that it is unlikely to be a chance effect of sampling.

We set up the null hypothesis that the sample of people from Statingham comes from a population with a proportion (denoted by p) of 0.8, $H_0: p = 0.8$ (it is used by 0.8 of Statingham households). And the alternative hypothesis is $H_1: p < 0.8$ (it is used by fewer than 0.8 of them).

Under the null hypothesis, the sample proportion is normally distributed with mean $p = 0.8$ and variance $pq/n = (0.8)(0.2)/100 = 0.0016$. The test statistic is

$$z = \frac{p - 0.8}{\sqrt{0.0016}} = \frac{p - 0.8}{0.04}$$

We decide to use a 5% one-tailed test and to reject H_0 if the calculated z is numerically greater than 1.645.

At first sight, we may now insert the sample value $p = 0.74$ and obtain $z = -1.5$, which is clearly less (numerically) than 1.645. However, when we are dealing with proportions, we have to be particularly careful to introduce a continuity correction.

In the sample of 100 people, 74 have said they use Wyshwash. The smallest amount by which this can be wrong is one person, so the proportion can take only values such as ... , 0.73, 0.74, 0.75, ... To allow for this, we take 0.74 to mean any value between 0.735 and 0.745. Thus we have to acknowledge that the proportion in Statingham could be as high as 0.745, and this extreme has to be used in calculating our test statistic, which becomes

$$z = \frac{p - 0.8}{0.04} = \frac{0.745 - 0.8}{0.04} = -1.375$$

This does not lie within the lower tail, as it is numerically less than 1.645. We have no grounds for rejecting H_0.

In this case the continuity correction has not changed the conclusion. Once again the continuity correction has decreased the difference between the observed proportion and the hypothesised proportion in the expression for z by $1/2n$, where n is the sample size. We have decreased it by 0.005, which is the same as $1/200$.

Consolidation exercise 26A

1. What is being tested when we test the significance of a proportion?

2. How are sample proportions distributed? What, in words, are (i) the variance of the distribution and (ii) the standard error of a sample proportion? ⬭

3. What is the formula for the standard error of a sample proportion?
4. What is the formula for the test statistic used to test a sample proportion?
5. Why is a continuity correction necessary when testing the significance of a proportion? What is its usual value?

Notebox 26.1

A notation for proportions

In Chapter 16 we said that normally we use Greek letters for population parameters, such as σ for the population equivalent of s for the standard deviation. Some authors use π for the population proportion and p for the sample proportion. To avoid confusion with the constant associated with the circle, we follow a common practice of using p for the population mean, using p_s for the sample mean when necessary.

Notebox 26.2

Continuity corrections

In testing proportions it is always necessary to use a continuity correction. The magnitude of this is $1/2n$ where n is the sample size. Give it the sign that decreases the numerical value of the test statistic. As the borderline between discrete and continuous is often blurred by the precision of measurement, the same kind of correction (with the same value of $1/2n$) sometimes has to be used when testing the mean or other statistics.

Notebox 26.3

An effect of imprecision

A further correction is sometimes introduced when solving problems like Example 26.1 using the second interpretation of the Bishop's 73%. If we are seeing whether the sample value is significantly less than 0.725 rather than 0.730, this is the value of p that should be used in the denominator, and q becomes 0.275 instead of 0.27. These changes lead to a slight reduction in the value of z.

Summary

1. To test the significance of a proportion, we use the fact that for large samples the sample proportion p_s is normally distributed about the population mean p with a variance pq/n. This normal distribution can be standardised by defining a new variable Z such that

$$Z = \frac{p_s - p}{\text{s.e. of proportion}} \qquad [26.1]$$

where the standard error (s.e.) of the proportion is (by definition) the standard deviation of the distribution of the sample proportions, $\sqrt{pq/n}$.

2. Accordingly, we adopt the test statistic

$$z = \frac{p_s - p}{\sqrt{pq/n}} \qquad [26.2]$$

which will be normally distributed with zero mean and unit variance.

3. A continuity correction of $1/2n$ has to be applied in a way that decreases the numerical value of the numerator of the test statistic.

4. The general procedure for one-tailed and two-tailed tests is very similar to the procedure for means and is given in Box 26.1.

5. It may be necessary to consider whether any imprecision in the statement of p requires some modification of its value in the test statistic. Any such modification should appear in both the numerator and the denominator.

6. Noteboxes concern a notation for proportions, continuity corrections and an effect of imprecision.

27

Ours are bigger than theirs

27.1 Introduction

Example 27.1 The *Celtic Fringe Property Gazette* reports that the average advertised selling price for a random sample of 100 Welsh houses is £40 000, but the average for 144 Irish houses is £38 000. Can we infer that, on average, Welsh house prices are (i) higher than Irish house prices and (ii) at least £1000 higher than Irish house prices?

Example 27.2 The mean number of hours per week spent attending lectures by a sample of 100 Statingham science undergraduates was 7.7. A sample of 100 arts undergraduates had a mean of 8.0 hours. In both cases the standard deviation was 1 hour. Is the assertion that the arts undergraduates at Statingham spend more hours at lectures than the science undergraduates justified?

Example 27.3 Apart from the samples mentioned in Example 27.2, a sample of 64 students of social science showed a mean attendance time of 7.9 hours with a variance of 1.21 hours². Does this warrant the assertion that the time spent at lectures by social science students was somewhere between the times spent by arts students and science students?

Example 27.4 The landlord of the Calculators' Arms found that 23 out of a random sample of 86 customers were female. In the Skew Distribution it was 19 out of 78. Is the landlord of the Calculators' Arms justified in saying that the fraction of his customers who are female is not the same as for the Skew Distribution? What assumptions are involved in answering this?

In Chapters 25 and 26 we asked whether a sample mean or proportion differed from a known or hypothesised population value by more than can be explained by chance. The examples in this chapter ask a different question, Do the sample results differ from *each other* by more than can be attributed to chance?

This is answered by essentially the same technique as before, but now we focus on the difference between two samples. In most cases we are concerned with the difference between two sample means, but in Example 27.4 it is between two sample proportions. Under certain assumptions these differences will be normally distributed. This allows us to use a procedure that is applicable to any normally distributed large-sample statistic. We begin with an account of this.

27.2 Testing the significance of any normally distributed large-sample statistic

Hypotheses about any large-sample statistic can be tested by essentially the same procedure as we used in the last two chapters, as long as the statistic is at least approximately normally distributed. The one thing we need to know, or be able to work out, is the standard error of the relevant sample statistic, since this forms the denominator of the test statistic. Apart from that, the procedure is just as for the mean and proportion. We summarise it in Box 27.1 and will be using it later. Some of the more frequently used standard errors are given in Tables 22.1 and 23.1.

Box 27.1

 How to ... test the significance of any normally distributed large-sample statistic y_s

1. Set up the null hypothesis $H_0: y_s = y$ where y is the hypothesised population statistic.

2. State the alternative hypothesis, H_1, which is usually
   (a) $y_s \neq y$, requiring a two-tailed test
   (b) $y_s > y$ or $y_s < y$, each requiring a one-tailed test

3. Specify the level of significance for the test.

4. Define the test statistic

$$z = \frac{y_s - y}{\text{standard error of } y}$$

5. Find from tables the critical value of the test statistic, as in step 5 of Box 25.1, thus defining the rejection zone.

6. Find, possibly from Table 22.1 or 23.1, the formula for the standard error of the sample statistic.

7. Using step 4 and the results of steps 5 and 6, calculate the value of z for the data being investigated and compare it with the critical value. If the data are discrete, the numerical value of the numerator has to be decreased by a continuity correction, which should be determined carefully.

8. Reject the hypothesis if the numerical value of z exceeds the critical value, as in step 6 of Box 25.1.

27.3   Testing the difference between two large-sample means

We illustrate the application of this procedure by considering Example 27.1, which contains two questions. As the example stands, the answers to both questions must be negative. Before we can perform any calculations about the significance of the difference between the two sample means, we need to know something about the variances. But so far we know nothing about them.

The questions can be answered with greatest assurance if we know the two population variances, i.e. the variances of advertised house prices for all houses in Wales and in Ireland. It is highly unlikely that we will know these, but the test that we would then apply has something to tell us.

It relies on the fact that if two statistics are normally distributed then so is their difference. We know that for large samples the sample means are normally distributed, and so is their difference. This means that we can test the difference between two means by using the procedure summarised in Box 26.1, provided we can find the appropriate standard error.

We begin with Example 27.1(i). It can be shown that the sampling distribution of the difference between two means is normal with a mean equal to the difference between the two population means, which is

$$\text{mean of } (\bar{x}_1 - \bar{x}_2) = (\mu_1 - \mu_2) \qquad\qquad [27.1a]$$

and a standard error that depends on the two population variances and the sample sizes, being

$$\text{s.e. of } (\bar{x}_1 - \bar{x}_2) = \sqrt{(\sigma_1^2/n_1) + (\sigma_2^2/n_2)} \qquad\qquad [27.1b]$$

This involves the sum of the two population variances. If we do not know them, we estimate them from the sample variances (as explained below), and apply basically the same test but with a little less confidence (in the ordinary sense of that word). The calculations are simplified if the population (or sample) variances are equal. We consider a few different possibilities.

The two population variances σ_1^2 and σ_2^2 are known (large samples, $n_1 > 30$, $n_2 > 30$)

Example 27.1 This case has more information. We are also told that the two population standard deviations are $\sigma_1 = £8000$ (for Wales) and $\sigma_2 = £9000$ (for Ireland). We ask if we can now infer that mean Welsh house prices are higher than Irish house prices.

Step 1: setting up the hypotheses

We have two populations: one of all Welsh house prices and the other of all Irish house prices. We ask whether the difference between the sample means is so big that it requires us to reject the hypothesis that the average of all Welsh house prices (μ_1) equals the average of all Irish prices (μ_2). We therefore set up the null hypothesis that there is no difference, writing $H_0: \mu_1 - \mu_2 = 0$. Does the evidence justify the statement that Welsh house prices are higher than Irish prices? This implies a one-tailed test, and the alternative hypothesis $H_1: \mu_1 - \mu_2 > 0$.

Step 2: specifying a significance level

We will use a 5% significance level.

Step 3: choosing the test statistic

We know that the difference between the two sample means is distributed normally with a mean given by equation [27.1a] and a standard error given by equation [27.1b]. Our test statistic is derived by standardising this normal distribution. It is

$$z = \frac{\text{difference in sample means} - \text{difference in population means}}{\text{standard error of difference}}$$

giving

$$z = \frac{(\bar{x}_1 - \bar{x}_2) - (\mu_1 - \mu_2)}{\sqrt{(\sigma_1^2/n_1) + (\sigma_2^2/n_2)}} \qquad [27.2a]$$

Step 4: finding the critical value for the rejection zone

The 5% one-tailed critical value is 1.645. Therefore the rejection zone is $z > 1.645$.

Step 5: calculating the value of the test statistic

We work in units of £1000 and use the following values:

$$n_1 = 100, \; \bar{x}_1 = 40, \; \sigma_1^2 = 8^2 \qquad n_2 = 144, \; \bar{x}_2 = 38, \; \sigma_2^2 = 9^2$$

and if H_0 is true then $\mu_1 - \mu_2 = 0$, giving

$$z = \frac{(40 - 38) - 0}{\sqrt{(64/100) + (81/144)}} = 1.824$$

Step 6: conclusion

Our value of z is higher than we would expect by chance with a probability of 5% if the null hypothesis is true. At a 5% significance level, we reject the null hypothesis that there is no difference between the mean house prices in favour of the alternative hypothesis that Welsh house prices are higher.

There is a known common population variance, σ (large samples, $n_1 > 30$, $n_2 > 30$)

In the case we have just considered, the two populations had different variances, σ_1 and σ_2, both of which we knew. Now we examine a case in which the populations have the same variance and we know its value. The test statistic is obtained from equation [27.1] by putting $\sigma_1 = \sigma_2 = \sigma$, which gives

$$z = \frac{(\bar{x}_1 - \bar{x}_2) - (\mu_1 - \mu_2)}{\sigma\sqrt{(1/n_1) + (1/n_2)}}$$ [27.2b]

If we want to answer a question like this, we need a new version of the further information.

⬭ **Example 27.1 continued** Suppose we are told that the two population standard deviations equal £9500.

⬭ The solution now proceeds as before except that the test statistic is given by equation [27.2b]. The 5% critical value is still 1.645 and the calculated value of z becomes

$$z = \frac{(40 - 38) - 0}{9.5\sqrt{(1/100) + (1/144)}} = 1.619$$

Step 6: conclusion

Our value of z does not fall in the rejection region (defined by > 1.645). The probability of getting 1.619 if the null hypothesis is true is more than 5%. Our evidence does not warrant rejection (at 5% significance) of the null hypothesis that there is no difference between the mean house prices in the two countries.

Possibly the two population variances differ but neither of them is known (large samples, $n_1 > 30$, $n_2 > 30$)

Here we know nothing about the variances. Our only guide to their magnitudes is the evidence of the samples. We use them to make the best estimates we can of the population variances, as explained in Chapter 16.

The best available estimates for the population variances are

$$\hat{\sigma}_1^2 = n_1 s_1^2/(n_1 - 1) \quad \text{and} \quad \hat{\sigma}_2^2 = n_2 s_2^2/(n_2 - 1)$$

where s_1^2 and s_2^2 are the sample variances. These estimates are substituted for σ_1^2 and σ_2^2 in equation [27.2a] for the test statistic. They give

$$z = \frac{(\bar{x}_1 - \bar{x}_2) - (\mu_1 - \mu_2)}{\sqrt{[s_1^2/(n_1 - 1) + s_2^2/(n_2 - 1)]}} \qquad [27.2c]$$

If the samples are very large, $n/(n - 1)$ will be so close to unity that the sample variance can be taken as the population variance. (If this is done and the resulting value of the test statistic is very close to the critical value, then it is prudent to repeat the calculation properly.) We now answer a version of Example 27.1 augmented by some further information.

⬭ **Example 27.1 continued** We are told that possibly the two population standard deviations differ, and that the sample standard deviations are $s_1 = £8000$ and $s_2 = £9000$.

⬭ The solution now proceeds as before except that the test statistic is given by equation [27.2c]. The 5% critical value is still 1.645. Calculations along these lines show that $z = 1.811$, which exceeds the critical value. We therefore reject the null hypothesis that there is no difference between the mean house prices in the two countries in favour of the alternative hypothesis that mean Welsh prices are higher.

There is an unknown common population variance (large samples, $n_1 > 30$, $n_2 > 30$)

Finally in this examination of the first question raised in Example 27.1, we look at the case in which it is known (or believed) that the two populations have the same variance but we do not know what it is. (Possibly we have applied a test to be described in Chapter 30 and decided that the sample variances are not significantly different from each other, so we accept the hypothesis that the two populations have a common variance.)

Once again we use our data about sample variances and sample sizes to estimate the population variance. The best estimate is obtained by pooling the sample variances, weighted by the sample sizes, but with the total of the sample sizes reduced by 2. Thus

$$\hat{\sigma}^2 = \frac{n_1 s_1^2 + n_2 s_2^2}{n_1 + n_2 - 2}$$

The value of $\hat{\sigma}$, given by the square root of this is then substituted for σ in equation [27.2], yielding

$$z = \frac{(\bar{x}_1 - \bar{x}_2) - (\mu_1 - \mu_2)}{\sqrt{\left[\dfrac{(n_1 s_1^2 + n_2 s_2^2)(n_1 + n_2)}{(n_1 + n_2 - 2)(n_1 n_2)}\right]}} \qquad [27.2d]$$

If the samples are very large, the denominator can be simplified by cancelling out the two almost equal brackets $(n_1 + n_2)$ and $(n_1 + n_2 - 2)$. We now consider our example, augmented by some new data.

Example 27.1 continued We are told that the two populations have a common variance, and that the sample standard deviations are $s_1 = £8000$ and $s_2 = £9000$.

The solution now proceeds as before except that the test statistic is given by equation [27.2d]. The 5% critical value is still 1.645 and the calculated value of z becomes 1.778. This exceeds the 5% critical value, so it is higher than we would expect by chance with a probability of 5% if the null hypothesis is true. We reject the null hypothesis that there is no difference between the house prices in the two countries in favour of the alternative hypothesis that Welsh house prices are higher.

As an exercise, explore the data further by altering the values of the standard deviations given in the four cases. The procedure is summarised in Box 27.2.

Box 27.2

How to ... **test the difference between two large-sample means**

1. Set up the null hypothesis, usually $H_0: \mu_1 - \mu_2 = 0$, but see Box 27.3.

2. State the alternative hypothesis H_1, which will usually be

   (a) $\mu_1 - \mu_2 \neq 0$, requiring a two-tailed test
   (b) $\mu_1 - \mu_2 > 0$ or $\mu_1 - \mu_2 < 0$, each requiring a one-tailed test

3. Specify a level of significance for the test.

4. Decide whether the two population variances are

   (a) known and diferent
   (b) known and the same
   (c) unknown but different
   (d) unknown but the same

5. Choose the the appropriate test statistic by using equation [27.2a], [27.2b], [27.2c] or [27.2d] according to the case selected in step 2. If the hypothesis is that the two samples come from populations with the same mean, then $\mu_1 - \mu_2$ in the chosen formula becomes zero. (If the hypothesis is that there is a specified difference then see Box 27.3.)

6. Follow steps 5 and 6 of Box 25.1.

Example 27.2 is more straightforward than Example 27.1. Here we have been given the two sample variances, but we do not know whether the populations have different variances. Except for one thing, it would therefore be appropriate to perform two tests, one using equation [27.2c] and one using equation [27.2d]. However, these tests are simplified because the samples have the same size and the same standard deviation. This means that in equations [27.2c] and [27.d] the denominator is simplified to $s\sqrt{2/(n-1)}$. Thus only one test is necessary. If two samples of the same size have the same standard deviation, the best assumption is that the two populations also have the same standard deviation. The solution is left as an exercise.

Example 27.3 is more complicated. It involves two one-tailed tests. If the time spent by social science students is 'somewhere between' the times spent by arts and science students, and the sample time for arts students is greater than for science students, we have to perform two tests that can be summarised as follows:

(i)   Test whether social science time is less than arts time, which has to be rephrased in terms of a null hypothesis and an alternative hypothesis.

(ii) Test whether social science time is greater than science time, which also has to be rephrased in the same way.

In each instance it is necessary to use both equation [27.2c] and equation [27.2d]. Once again the calculations are left as an exercise. Do not confuse the testing of these two hypotheses with a more complicated test of how the variance between the three sample means compares with the variance within the samples, a test that we do not describe.

Consolidation exercise 27B

1. Two large samples of sizes n_1 and n_2 have means $\bar{x}_1$ and $\bar{x}_2$ and variances σ_1 and σ_2. How (in words) is the difference between the means distributed? What are (a) the mean and (b) the variance of this distribution? What condition(s) must hold for your answer to be correct?

2. What (in symbols) is the null hypothesis used when testing the difference between two means? What (in words) is the test statistic then used?

27.4   Testing whether a difference is greater than a stated amount

In Example 27.1(i) we asked whether we could infer that Welsh house prices are higher than Irish house prices. In Example 27.1(ii) we asked if we could infer that Welsh house prices are at least £1000 higher than Irish prices. We know that we need further information about variances, so to illustrate the procedure we rewrite our example as follows.

Example 27.1 continued A random sample of the advertised selling prices of houses in Wales has 100 prices which average £40 000, with a standard deviation of £8000. A similar sample of 144 Irish house prices has a mean of £38 000 and a standard deviation of £9000. Can we infer from this that Welsh house prices are on average more than £1000 higher than Irish? Possibly the two populations have different variances.

The procedure here is almost exactly the same as in Section 27.2 but we have to alter our hypotheses.

Step 1: setting up the hypotheses

We set up the null hypothesis that Welsh prices are on average £1000 higher than Irish prices. Strictly, our hypothesis is that they are *at most* £1000 higher, but if we take 'equal to £1000' and then find this is almost certainly too low as a statement of the difference, then anything less than this must also be too low. The wording implies a one-tailed test. Using units of £1000, we write $H_0: \mu_1 - \mu_2 = 1$ and the alternative hypothesis $H_1: \mu_1 - \mu_2 > 1$

Step 2: specifying a significance level

We will use a 5% significance level.

Step 3: choosing the test statistic

As before, we use

$$z = \frac{\text{difference in sample means} - \text{difference in population means}}{\text{standard error of difference}}$$

accompanied by our knowledge of the sampling variances and that possibly the two populations have different variances. We define z as in equation [27.2c].

Step 4: finding the critical value for the rejection zone

The rejection zone for the 5% one-tailed test is $z > 1.645$

Step 5: calculating the value of the test statistic

Again we work in units of £1000, using the following values

$$n_1 = 100, \bar{x}_1 = 40, \sigma_1 = 8 \qquad n_2 = 144, \bar{x}_2 = 38, \sigma_2 = 9$$

but now in the numerator of equation [27.2c] we have $\mu_1 - \mu_2 = 1$. This gives $z = 1/\sqrt{1.2025}$, clearly less than unity and therefore less than 1.645, which defines the rejection zone.

Step 6: conclusion

Our value of z is lower than we would expect by chance with a probability of 5% if the null hypothesis is true. We cannot on this evidence reject (at the 5% level) the null hypothesis that the average Welsh price is £1000 higher than the average Irish price. The evidence is not strong enough to support the alternative hypothesis that Welsh prices are on average more than £1000 higher.

The general procedure is summarised in Box 27.3

Box 27.3

How to ...   test whether the difference between two large-sample means is equal to (or larger or less than) *k*

1. Set up the null hypothesis, usually $H_0: \mu_1 - \mu_2 = k$.

2. State the alternative hypothesis H_1, usually
   (a) $\mu_1 - \mu_2 \neq k$, requiring a two-tailed test
   (b) $\mu_1 - \mu_2 > k$ or $\mu_1 - \mu_2 < k$, each requiring a one-tailed test

3 and 4. As steps 3 and 4 of Box 27.2.

5. Choose the appropriate test statistic by using equation [27.2a], [27.2b], [27.2c] or [27.2d] according to the case selected in step 4.

6. Find the critical value for the rejection zone from Table 21.1.

7. Using $\mu_1 - \mu_2 = k$, calculate the value of the test statistic.

8. Compare the calculated value with the critical value and reject the null hypothesis if the calculated value is numerically greater than the critical value.

27.5   **Testing the difference between two proportions**

Proportions almost always involve discrete data. For the moment, however, we are going to ignore this, in order to concentrate on the basic method for solving a problem such as Example 27.4.

Example 27.4 is a test of alertness. Our samples are supposed to come from very large populations, or to involve sampling with replacement. In this example we cannot assume that the population of customers of the Calculators' Arms is many times larger than the sample of 86, nor do we know the size of the population of all customers. We may also have doubts about the alleged randomness of the sample. The tests we have been describing cannot properly be applied unless we assume that the sampling was properly conducted and involved replacement (as described in Appendix C).

Suppose we have two random samples of sizes n_1 and n_2. Let the proportions of successes in these two samples be p_{s1} and p_{s2} (where the subscript s refers to the sample). And suppose that in the two populations from which these samples are drawn the proportions of successes are p_1 and p_2.

We know that if the sample sizes are large, the two sample proportions are normally distributed such that

$$p_{s1} \sim N(p_1, p_1 q_1/n_1)$$

(where $q = 1 - p$)

$$p_{s2} \sim N(p_2, p_2 q_2/n_1)$$

It follows that the difference between these two proportions will also be normally distributed with a mean equal to the difference between the means $p_1 - p_2$ and a variance equal to the sum of the variances

$$(p_1 q_1/n_1) + (p_2 q_2/n_2) \qquad [27.5]$$

This normal distribution is the sampling distribution of the difference between two proportions. Its standard deviation, which is the square root of equation [27.5], is the **standard error of the difference**, which leads to the denominator of our test statistic used below.

When we test the significance of the difference between two sample proportions, the question is usually about two samples from populations that have the same population proportion p. In this case the above normal distribution has a mean of zero. We incorporate this in our test statistic, which becomes

$$\frac{\text{The difference between the sample proportions}}{\text{Standard error of the difference}}$$

Two cases concern us:

(a) where the population proportion p is known
(b) where the population proportion p is not known

Case (a)
Using that $p_1 - p_2 = 0$, our test statistic becomes

$$z = \frac{p_{s1} - p_{s2}}{\sqrt{pq[(1/n_1) + (1/n_2)]}}$$

and we see whether the calculated value z is significantly different from 0.

Case (b)
A more common case is illustrated by Example 27.4. In this we do not know the population proportion. We estimate it as a weighted average of the sample proportions, from

$$\hat{p} = \frac{n_1 p_{s1} + n_2 p_{s2}}{n_1 + n_2} \quad \text{and} \quad \hat{q} = 1 - \hat{p}$$

We use these estimated values in place of p and q in the denominator of the test statistic.

Example 27.4 is easily solved if we assume that there has been sampling with replacement. For the Calculators' Arms $n_1 = 86$ and the proportion of females is $p_{s1} = 23/86 = 0.267\,44$ whereas for the Skew Distribution $n_2 = 78$ and $p_{s2} = 19/78 = 0.243\,59$. Thus the best estimate of the assumed common population proportion is

$$\hat{p} = \frac{n_1 p_{s1} + n_2 p_{s2}}{n_1 + n_2} = \frac{23 + 19}{86 + 78} = 0.256\,10$$

We now set up the null hypothesis that there is no significant difference between the proportions, $H_0: p_1 = p_2 = p$. The alternative hypothesis, that there is a significant difference, is $H_1: p_1 \neq p_2$.

For a two-tailed test at the 5% level, we reject the null hypothesis if z is numerically greater than 1.96. Step 4 in Box 27.4 with $k = 0$ gives

$$z = \frac{0.267\,44 - 0.243\,59 - 0}{\sqrt{(0.256\,10)(0.743\,90)(1/86 + 1/78)}} = 0.349$$

which is much less than the critical value. There are no grounds for rejecting the null hypothesis that the proportions are the same in both pubs.

The difference between the calculated z and the critical value is so great that the matter of a continuity correction is academic. We should, however, consider the procedure.

Suppose we are interested in the difference D between two quantities A and B, with $B > A$, so that $D = B - A > 0$. A has a margin error of a on either side, so that for A we have to read 'between $A - a$ and $A + a$', as illustrated in Figure 27.1(a). Similarly, for B we have to read 'between $B - b$ and $B + b$'.

The diagram shows that the difference D has to be viewed as anything between $D - (a + b)$ and $D + (a + b)$. The margin of error d of the difference D is the sum of the separate margins of error a and b.

In our example the sample proportion of 23/86 is bigger than the sample proportion of 19/78, so the positive difference D is $23/86 - 19/78$. However, the value of 23/86 has a margin of error on each side equal to $b = 1/2n = 1/(2 \times 86)$. Similarly 19/78 has a margin of error equal to $a = 1/(2 \times 78)$. Thus the difference D has a margin of error d of $(a + b) = 1/(2 \times 86) + 1/(2 \times 78) = 0.012\,22$.

This gives the value of the continuity correction, but we still have to see how to apply it, as shown in Figure 27.1(b). Suppose there is a nominal difference of D, smaller than a given quantity k, and we wish to test whether it is significantly smaller. We have to see whether $D + d$ is significantly smaller than k. We narrow the gap.

Similarly, if D is bigger than k and we wish to test whether it is significantly bigger, we have to see if $D - d$ is significantly bigger. Again we narrow the gap. The correction reduces the numerical value of the calculated test statistic.

In our example the gap is already smaller than the critical value, so it is not significant. Narrowing it would make it even less significant.

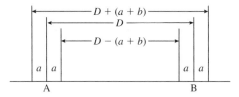

A   B

(a)

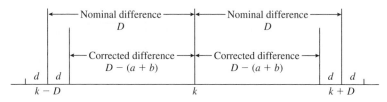

(b)

Figure 27.1

Box 27.4

How to ... **test the difference between two proportions**

1. Set up the null hypothesis $H_0: p_1 - p_2 = k$ where k is the hypothesised difference between the two proportions and may be zero.

2. Set up the alternative hypothesis H_1:

   (a) $p_1 - p_2 \neq k$, requiring a two-tailed test
   (b) $p_1 - p_2 > k$ or $p_1 - p_2 < k$, each requiring a one-tailed test

3. Specify a level of significance.

4. If both samples come from a population with a known common proportion p, then use the normally distributed test statistic

$$z = \frac{p_{s1} - p_{s2} - k}{\sqrt{pq[(1/n_1) + (1/n_2)]}}$$

5. If the population proportion is not known, use the same test statistic but replace p and q (in the denominator) by the estimated values

$$\hat{p} = \frac{n_1 p_{s1} + n_2 p_{s2}}{n_1 + n_2} \quad \text{and} \quad \hat{q} = 1 - \hat{p}$$

6. Use Table 21.1 or tables of the normal distribution to determine the critical value of z appropriate to the level of significance and the number of tails. This defines the rejection zone.

7. Reduce the numerical value of the numerator in the expression for z by adding or subtracting the continuity correction $(1/2n_1) + (1/2n_2)$.

8. Calculate the corrected value of z from step 4, taking account of steps 5 and 6 as needs be.

9. Compare the calculated z with the critical value. If the calculated value is further from 0 than the critical value, then reject the null hypothesis.

Consolidation exercise 27B

1. Two large samples of sizes n_1 and n_2 have means $\bar{x}_1$ and $\bar{x}_2$ and variances σ_1 and σ_2. You wish to test whether the population means differ by more than 100. What (in symbols) are (a) the null hypothesis and (b) the denominator of the test statistic?

2. What (in words) is the test statistic used when testing the difference between two proportions?

3. If two large samples of sizes n_1 and n_2 reveal proportions p_{s1} and p_{s2}, what is the best estimate of the population proportion? What assumption is made about the population in testing the difference between two proportions?

Summary

1. Hypotheses about any large-sample statistic can be tested by essentially the same procedure as used in the last two chapters, provided that the statistic is at least approximately normally distributed.

2. Applying the test requires knowledge of the standard error of the sample statistic concerned, since this forms the denominator of the test statistic. Apart from that, the procedure is just as for the mean and proportion. We summarise it in Box 27.1. Some of the more frequently used standard errors are given in Tables 22.1 and 23.1.

3. The difference between two sample means can be tested using equation [27.1b] to calculate the standard error. Four cases may be considered, depending on how much is known about the population variance(s). Equations [27.2a] to [27.2d] are used, and the procedure is summarised in Box 27.2. Box 27.3 extends it to testing whether the difference exceeds a stated value.

4. Discrete measurements need to be handled differently from continuous measurements. A continuity correction needs to be applied when testing the difference between two proportions; the basic formula is equation [27.6]. The procedure is summarised in Box 27.4. The continuity correction for a difference is the sum $(1/2n_1) + (1/2n_2)$, which should be applied so that it reduces the numerical value of the calculated test statistic.

28

Probability, precise significance and the normal approximation to the binomial

28.1 Introduction

▷ **Example 28.1** Sacks of potatoes packed by a local farmer have normally distributed masses with $\mu = 20.02$ kg and $\sigma = 1.69$ kg. What is the probability that a sack will have a mass (a) greater than 22.00 kg, (b) less than 19.00 kg, (c) between 19.00 and 22.00 kg? If the farmer packs 800 sacks during a day, how many sacks can he or she expect to have a mass greater than 23.00 kg?

▷ **Example 28.2** Letters are delivered to the Calculators' Arms at times that are normally distributed around 10.20 with a standard deviation of 8 minutes. What is the chance of them arriving (a) later than 10.30, (b) before 10.15 and (c) between 10.15 and 10.30?

▷ **Example 28.3** Marks awarded to 600 candidates for a Diploma in Work Avoidance were normally distributed with $\mu = 45$ and $\sigma = 14$. (a) If the pass mark was 40, how many diplomas were awarded? (b) Distinctions were awarded to the top 4% of the candidates; what was the lowest mark earning a distinction? (c) What was the semi-interquartile range of the 600 marks?

▷ **Example 28.4** An enquiry into the leases of all shops in Upper Sternum showed that, on 31st December last, the mean time before the next rent review was 3.98 years. A sample enquiry involving 400 shops in Statingham showed a mean time of 3.90 years, with $\sigma = 0.47$ years. How significant is the difference between the Statingham sample mean and the Upper Sternum population mean?

▷ **Example 28.5** In the city of Statingham the probability that a resident is female is 0.55. If 200 residents are selected at random, what is the probability that (a) 120 or more, (b) fewer than 90, (c) exactly 100 are female?

311

In the last few chapters our interest has been in testing hypotheses, using a method based on the fact that for large samples certain sample statistics (including the sample mean) are at least approximately normally distributed. We have introduced a test statistic and described a value that has only a 5% chance of being reached or passed as critical. Then we have said that if our calculated value of z is (numerically) higher than this critical value, our data are significant at the 5% level. We have not been interested in the precise probability of getting the calculated value of z, only in whether the calculated value is less probable than the chosen level of significance.

Now we widen our view. There are many tests of hypotheses that do not involve the normal distribution, as we will see in later chapters. And sometimes we do not particularly want to test a hypothesis but we are interested in knowing the precise significance (which we define shortly) of getting the calculated value of a test statistic, be it based on the normal distribution or not. Doing this for normally distributed statistics involves knowing more about the area under the normal curve than we learned from Chapters 19 and 20. This will also enable us to use the normal distribution to provide very good approximations to the solutions of binomial problems when n is large.

 ## 28.2 The area under the normal curve

In Chapter 21 we had a table showing the areas cut off under the standard normal curve by various values of z. They were presented both as tail areas and as areas for a symmetric range bounded by $-z$ and $+z$. For example, we had that for $z = +1.96$ the right-hand tail area is 0.025 (or 2.5%) of the total area; and a range from -1.96 to $+1.96$ contains 0.95 (or 95%) of the total area. This is illustrated in Figure 28.1, which also shows that the total area under the curve to the left of 1.96 is 0.975 (or 97.5%).

Figure 28.2 shows the standard normal curve (with unit area) with a superimposed cumulative probability curve. For any value of z the ordinate of this

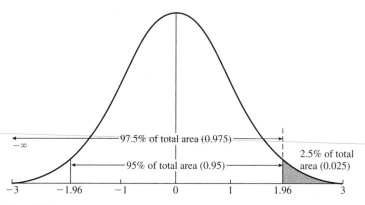

Figure 28.1

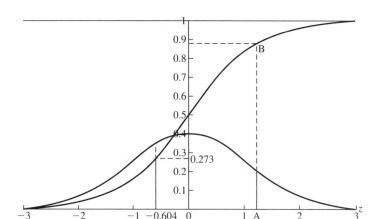

Figure 28.2

cumulative curve shows the area under the normal curve to the left of this ordinate. Thus the area under the normal curve to the left of A is given by the height AB.

The areas given by ordinates of this curve have been tabulated, as in Table D4. The table itself lists these areas for positive values of z. Values for negative z can be quickly derived by using the symmetry of the curve (Notebox 28.1). Sometimes the table is published with values of the area between z and the mean, rather than the total area to the left of z. The diffference in values is 0.5. We now illustrate some points about the use of this table.

In **Example 28.1** we have a normally distributed variable (X, the mass of a sack of potatoes) with $\mu = 20.02$ kg and $\sigma = 1.69$ kg. First we define a new variable Z whose values z are the standardised values of x:

$$Z = \frac{X - \mu}{\sigma} = \frac{X - 20.02}{1.69}$$

For any value of X there is a corresponding value of Z given by this equation. Figure 28.3 shows the normal curve with the original values of X and their corresponding values of Z. The probability of X having a specified value is the same as the probability of Z having the corresponding value. We now look at the various parts of the question.

(a) The probability of x being greater than 22.00 (which is the area to the right of $x = 22.00$) is the probability of z being greater than the corresponding value, which is

$$z = \frac{22.00 - 20.02}{1.69} = 1.17$$

The table of the area under the normal curve (Figure 28.2) shows that the area to the left of 1.17 is 0.8790. This has a greater degree of accuracy than 1.17, so we

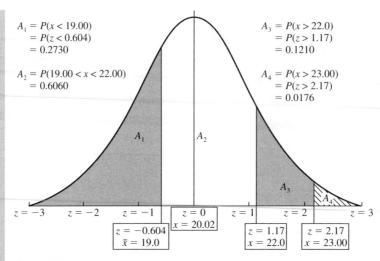

$A_1 = P(x < 19.00)$
$\quad = P(z < 0.604)$
$\quad = 0.2730$

$A_2 = P(19.00 < x < 22.00)$
$\quad = 0.6060$

$A_3 = P(x > 22.0)$
$\quad = P(z > 1.17)$
$\quad = 0.1210$

$A_4 = P(x > 23.00)$
$\quad = P(z > 2.17)$
$\quad = 0.0176$

Figure 28.3

shall take it simply as 0.879. We can then take the area to the right of $z = 1.17$ as $1 - 0.879 = 0.121$. This, the probability that z is greater than 1.17, is also the probability that x is greater than 22.00, so it is the probability we are seeking.

Notice our argument. By the process of standardisation, we have

$$Z = \frac{X - 20.02}{1.69}$$

and so

$$P(X > 22.00) = P\left(Z > \frac{22.00 - 20.02}{1.69}\right) = P(Z > 1.17)$$

$$= \text{area to right of } z = 1.17$$

$$= 1 - (\text{area to left of } z = 1.17)$$

$$= 1 - 0.879 = 0.121$$

This procedure will be used frequently.

(b) By a similar argument, the probability of x being less than 19.00 kg is the probability of z being less than the value corresponding to $x = 19.00$, so

$$P(X < 19.00) = P\left(Z < \frac{19.00 - 20.02}{1.69}\right) = P(Z < -0.604)$$

$$= \text{area to left of } z = -0.604$$

$$= \text{area to right of } z = +0.604 \qquad \text{(by symmetry)}$$

$$= 1 - \text{area to left of } z = +0.604 \qquad \text{(Notebox 28.1)}$$

$$= 1 - 0.727 = 0.273$$

which can be checked by referring to Figure 28.2.

(c) The requirements of (a) and (b) have cut off two fairly large tails, one for masses greater than 22.00 kg and the other with masses less than 19.00 kg. The probability of having a mass between these values is given by the area under the remaining (or central) part of the curve. Since the total area is unity, this must be

$$P(19.00 < x < 22.00) = 1 - P(x < 19.00) - P(x > 22.00)$$
$$= 1 - 0.121 - 0.273 = 0.606$$

as shown in Figure 28.3.

The next part of the question involves sacks heavier than 23.00 kg. As in part (a), we have that the probability of a sack weighing more than this is

$$P(X > 23.00) = P\left(Z > \frac{23.00 - 20.02}{1.69}\right) = P(Z > 1.76)$$

$$= \text{area to right of } z = 1.76$$

$$= 1 - (\text{area to left of } z = 1.76)$$

$$= 1 - 0.9608 = 0.0392$$

which is shown by the right-hand tail in Figure 28.3.

If there are 800 sacks each with this probability of being very heavy, then the number of sacks expected to be heavier than 23.00 kg is 800×0.0392, which is $31.36 = 32$ sacks.

The basic techniques used in solving these questions are summarised in Box 28.1

Box 28.1

How to ... **find the probability that a normally distributed continuous variable X (a) is less than x_1, (b) exceeds x_1**

1. Calculate

$$z_1 = \frac{x_1 - \mu}{\sigma}$$

where μ and σ are parameters of the normal distribution of X. The variable Z will have a standard normal distribution.

2. If $z_1 > 0$ use Table D4 (or the cumulative curve in Figure 28.2) to find the area to the left of z_1. This gives the probability (a) that X is less than x_1.

3. Subtract from unity the area found in step 2. This gives the right-hand tail area, which answers (b).

4. If $z_1 < 0$ then give it a $+$ sign and use steps 2 and 3 to find the right-hand tail area. By symmetry this is also the left-hand tail area for values less than z_1, which answers (a). Subtraction from unity gives (b).

The table can also be used in reverse. For example, in solving Example 19.2, as in Chapter 21, we want to find a 10% confidence interval, which means finding a value of z_c that will contain 5% of the area on each side of the mean. We search the body of the table for 5%, written as 0.05 if the areas are measured from the mean, or as 0.55 if they are measured from the left extreme. This corresponds to $z = 0.125$, which is the value we need in calculating the 10% confidence interval.

⊃ **Example 28.2** involves *discrete measurements*. Time is a continuous variable, but it seems reasonable to infer from the question that it is measured in minutes, and that 10.30 means any time between $10.29\frac{1}{2}$ and $10.30\frac{1}{2}$.

(a) To avoid having to cope with hours, minutes and seconds, we measure all time in minutes (and fractions of minutes) after 10.00, so that 10.05 becomes 5. Denoting delivery time by X, we have that it is distributed as $N(20,8^2)$. We define a standardised variable

$$Z = \frac{X - 20}{8}$$

The letters are recorded as 'arriving later than 10.30' if they come after $10.30\frac{1}{2}$, which means that X has a value greater than 30.5. We therefore use this in our calculation of probability, getting

$$P(\text{arriving later than } 10.30) = P(X > 30.5)$$

$$= P\left(Z > \frac{30.5 - 20}{8}\right) = P(Z > 1.3125)$$

$$= 1 - (\text{area to left of } z = 1.3125)$$

$$= 0.0947$$

(b) We have to interpret 'before 10.15' as meaning before $10.14\frac{1}{2}$, so our variable X has a value less than 14.5. This leads to

$$P(\text{arriving before } 10.15) = P(X < 14.5)$$

$$= P\left(Z < \frac{14.5 - 20}{8}\right) = P(Z < -0.6875)$$

$$= \text{area to left of } z = -0.6875$$

$$= 1 - (\text{area to left of } z = +0.6875) \qquad \text{(Notebox 28.1)}$$

$$= 0.2459$$

(c) Here 'between 10.15 and 10.30' is interpreted as not before $10.14\frac{1}{2}$ (when $X = 14.5$) but before $10.30\frac{1}{2}$ (when $X = 30.5$). The probability of this is easily obtained as

$$P(14.5 < X < 30.5) = 1 - P(X < 14.5) - P(X \geqslant 30.5)$$

$$= 1 - 0.2459 - 0.0947 = 0.6594$$

⟹ **Example 28.3** about the diploma in work avoidance involves the same basic ideas about probability and the area under the normal curve, but it is a little less straightforward.

As always, we begin by standardising the variable, defining

$$Z = \frac{X - 45}{14}$$

(a) To know how many diplomas were awarded, we work out the probability of a candidate getting a diploma and then multiply this by 600, the number of candidates. Since the pass mark is 40, we want the probability of X being at least 40. We do not know whether candidates are awarded fractional marks, or whether (if they are) the marks are rounded up, down or to the nearest integer. We shall suppose that the pass mark of 40 is so rigidly observed that even 39.99 would fail. In that case we have

$$P(X \geqslant 40) = P\left(Z \geqslant \frac{40 - 45}{14}\right) = P(Z \geqslant -0.3571)$$

$$= \text{area to right of } z = -0.3571$$

$$= \text{area to left of } z = +0.3571$$

$$= 0.6394$$

This is the probability of passing when the pass mark is strictly observed. Since there are 600 candidates, the expected number passing is $600 \times 0.6394 = 383.64 = 383$ for a tough board.

(b) Distinctions were awarded to the top 4% of the candidates; there were 4% of $600 = 24$ distinctions. A more useful way of looking at it is that the probability of getting a distinction is 0.04, so the corresponding value of z is the value that cuts off a tail with an area of 0.04, or has 0.96 to its left.

To find this value, we have to find $1 - 0.04 = 0.96$ in the body of the table, then read off the value of z. The value .9599 appears at $z = 1.75$. From the right-hand part of the table, we see that 1.751 leads to the addition of 1 to the last digit, giving an area to the left of 0.96.

Thus distinctions correspond to $z \geqslant 1.751$. From the definition of Z this gives

$$\frac{x - 45}{14} \geqslant 1.751 \quad \text{so} \quad x \geqslant 69.51$$

which we have to interpret as $x \geqslant 70$ for the distinction mark. The lowest mark earning a distinction is 70.

(c) To obtain the semi-interquartile range, we must usually find the first and third quartiles. However, with a symmetrical curve it is simply the separation of either of these quartiles from the mean, as can be seen from Figure 28.1. If we find the value of z that has 0.75 to its left, we have the z corresponding to the semi-interquartile range; tables show it is 0.674. ⬭

We have to be clear about what this is. A deviation of 0.674 unit standard deviations from the mean will cut off a tail area of 0.25. With the units of the original variable, this means that the deviation from the mean has to be $0.674 \times 14.0 = 9.44$. A range of this magnitude on either side of the mean will contain 50% of the total frequency, so 9.44 is the semi-interquartile range.

Consolidation exercise 28A

1. For the standard normal curve what values of the variable will cut off tails with areas of 1.96% of the area under the curve?

2. If the curve mentioned in Question 1 is a probability distribution, what will the upper tail area represent?

3. If the curve mentioned in Question 1 is a frequency curve, what will the lower tail area represent?

4. Using Table 28.1, what is the area under the standard normal probability distribution (a) to the left of $z = 2.043$, (b) to the right of $z = 2.043$, (c) to the left of $z = -2.043$?

Table 28.1

Number of heads	0	1	2	3	4	5
Probability	0.0060	0.0404	0.1209	0.2150	0.2508	0.2007

Number of heads	6	7	8	9	10
Probability	0.1114	0.0425	0.0106	0.0016	0.0001

5. In finding the probability that a normally distributed continuous variable exceeds a specified value, how do you begin? What formula do you use for this? How differently do you proceed if the variable is discrete?

28.3 Precise significance or significance probability

For normally distributed variables

We now apply our knowledge of areas under the normal curve to questions about significance.

⬭ **Example 28.4** asks for the significance of the difference between a *sample* mean of 3.90 years for Statingham and the Upper Sternum *population* mean of 3.98 years.

To answer this, we need to know or assume something about the population standard deviation of lease lives in Statingham. We assume it is the same as the population standard deviation for Upper Sternum, which we can estimate by modifying the sample standard deviation, reported as 0.47.

So far, when we have met problems like this, we have been testing the hypothesis that the sample could have come from the named population, and compared the calculated value of a test statistic with a critical value that depends on the chosen significance level. But there is another approach which focuses less on hypothesis testing and does not specify either an alternative hypothesis or a level of significance. There is therefore no critical value of the test statistic or a rejection zone.

Instead, if the normal distribution is being used, the value of Z for the experimental data is calculated. Then a table of the area under the normal curve is used to find the precise probability of getting a value of Z at least as high as this if the null hypothesis is true. Thus we are not asking whether the probability is less than a critical amount. Instead we are calculating the precise value of the probability of getting the observed result or one that is even less likely. This is called the **significance probability (S.P.)**. We are not seeing if it is more or less than some critical value. If we wish to use our new knowledge to make a judgement, that is a different matter.

 Example 28.4 asks us to measure the probability of getting a sample result of 3.90 if the sample of size 400 comes from a population of mean 3.98. We assume that as the sample standard deviation s is 0.47, the best estimate of the population standard deviation is given by

$$s\sqrt{n/(n-1)} = 0.47 \times \sqrt{400/399} = 0.4706$$

We have used $n-1$ instead of n to emphasise that this is the correct formula, but with $n = 400$ the correction is trivial. Note that this estimate is the best estimate of the standard deviation for the Statingham population. We are asuming, in the absence of other information, that this is also the best estimate for Upper Sternum.

Since the sample mean is normally distributed about the population mean, finding the probability of it deviating from the population mean by a given amount involves the standardised variable

$$Z = \frac{\bar{x} - \mu}{\sigma_{\bar{x}}} \quad \text{where} \quad \sigma_{\bar{x}} = \frac{\sigma}{\sqrt{n}}$$

We find the value of this by using the best estimate of σ, which is 0.47, and $n = 400$. This gives us

$$\sigma_{\bar{x}} = 0.47/20 = 0.0235$$

$$\text{so} \quad Z = \frac{3.90 - 3.98}{0.0235} = -3.40$$

A value of Z as high as this (numerically) is exceedingly unlikely if in fact there is no difference between the mean remaining lease lives in Statingham and Upper Sternum. As the tables show, it cuts off a single tail area of 0.034%. There is a chance of about 1 in 3000 of getting a mean lease life as low as this if things are the same in Statingham as in Upper Sternum; and a chance of about 2 in 3000 of the Statingham life being so different from (rather than so much less than) the Upper Sternum life. The difference, regardless of sign, is significant at the 0.068% level. This statement is sometimes stated as 'the difference has a **significance probability** of 0.068'.

How we interpret this is up to us. If we wish, we can therefore reject the hypothesis that in this respect there is no difference between the two towns. Or we can say that if in fact there is no difference in lease life between the two towns, our sample has produced a highly unlikely result, but we say no more than that. Perhaps such a finding will prompt further enquiry, or caution about a proposed action or policy, but that is not the statistician's concern.

Summing up, we have measured the probability of getting the obtained result or something no more likely when knowledge or hypothesis has suggested something else. We have called this the significance probability of the observed result. One piece of knowledge used by us in this example is that the result in which we are interested (the sample mean) is normally distributed. We have used the area of the tail of the curve as a measure of the probability of getting this result or one even less likely.

Box 28.2

How to ... **find the significance probability of a normally distributed large-sample statistic**

1. Set up the hypothesis that the sample statistic y is normally distributed around a specified value y_0 with a standard deviation σ.

2. Find or estimate σ by any appropriate method.

3. Find the standard error of the statistic by using your knowledge of σ and n and one of the formulae in Tables 22.1 and 23.1.

4. Calculate

$$z = \frac{y - y_0}{\text{s.e.}(y)}$$

5. From the table of areas under the normal curve, find the area A_z between the central ordinate and the ordinate having the value z.

6. The significance probability of the sample statistic is then calculated as follows:
   (a) $2 \times (0.500 - A_z)$ if you are interested in its difference from y_0, or by
   (b) $(0.500 - A_z)$ if you are interested only in a difference in one direction

---

For other variables

We have just examined the probability of getting some observed result when some knowledge and/or hypothesis has led us to expect a different result. The observed result was a sample mean, and because of the central limit theorem we were able to measure the probability of getting that result (or worse) by using the normal distribution. We now look at a case that does not use the normal distribution. The example we choose uses the binomial distribution, but there are other possibilities. The point is that if we know the distribution, we can find the significance probability. Notice that the following example does not involve a large sample.

Suppose we toss $n = 10$ coins, believed to be all equally biased with the chance of a head given by $p = 0.4$. We would expect $np = 4$ heads. However, we get 2 heads. As a measure of the significance of this result, we calculate the total probability of getting it or getting even less likely results.

By using the formula in Chapter 18, or Table D2, we obtain Table 28.1. This gives the probability of 2 heads as 0.1209. By adding the probabilities of 0 heads and 1 head to this, we find that the probability of 2 heads or fewer is 0.1673.

However, there are other possible events with probabilities no more than the probability of getting 2 heads. These have listed probabilities of 0.1114, 0.0425, 0.0106, 0.0016 and 0.0001, corresponding to 6 or more heads, and totalling 0.1662.

Thus the probability of getting 2 heads or an even less likely result *in the same (or obtained)* direction away from the expected result is 0.1673; and the probability of getting equally or even less likely results *in the opposite* direction is 0.1662.

Adding them gives 0.3335, which measures the total probability of getting the observed result or one even less likely.

This is the **significance probability (SP)**. The complement of the SP, $1 - 0.3335 = 0.6665$, gives the probability of getting something more likely than the obtained result.

Thus we have calculated the SP by considering all possible events, listing their probabilities, and adding the probabilities equal to or less than the probability of the obtained result.

This is the basis of a general procedure. If the variable is discrete, the various possibilities can be identified and the smaller probabilities separately calculated and totalled. If it is continuous then some formula for the area under part of the curve (or integration) will have to be used.

Box 28.3

How to ... **find the significance probability of sample data**

1. State the hypothesis.

2. Decide upon a suitable test statistic and its distribution when the hypothesis is true.

3. Use your data to calculate the 'observed value' of the test statistic.

4. Identify all the other values of the test statistic that are no more likely to arise than the observed value, and list their probabilities. ⬤

5. The values identified in step 4 can be put into two groups: those corresponding to values higher than the hypothesis would suggest and those corresponding to lower values.

6. One of these groups will contain the observed value. The sum of the probabilities for that group is called the **significance probability in the obtained direction**.

7. The sum of the probabilities for the other group is called the **significance probability in the opposite direction**.

8. The total of the probabilities in steps 6 and 7 is the **total significance probability**.

Consolidation exercise 28B

1. What is measured by significance probability?

2. When a normally distributed large-sample statistic is standardised, it produces $z = 1.3$. Use Table D4 to find (a) the SP in the obtained direction, and (b) the total SP.

(28.4) **The normal approximation to the binomial**

Knowledge of the tables of areas under the normal curve allows us access to a very useful way of handling binomial problems when large numbers are involved, as in Example 28.5. We use an important result that links the discrete (and, as we shall see, non-parametric) binomial distribution to the (parametric) continuous normal distribution. There will be frequent references to it in later chapters.

In **Example 28.5** the probability that a resident is female is 0.55. If 200 residents are selected at random, what is the probability that (a) 120 or more, (b) fewer than 90, (c) exactly 100 are female?

This is a binomial problem, with the probability p of one person selected at random being female given as 0.55, and with $q = 1 - p = 0.45$ and $n = 200$.

The problem is easily understood. By the method of Chapter 18, the probability of there being exactly 120 females is given by the term containing $p^{120}q^{80}$. The numerical coefficient of this term is the value of $^{200}C_{120}$, so the probability of there being exactly 120 females is

$$^{200}C_{120}p^{120}q^{80} = \frac{200!}{120!80!}(0.55)^{120}(0.45)^{80}$$

which is an unattractive thing to work out. To obtain the probability of 120 or more, we would need to do similar calculations for 121, 122, ..., 200 and then add the answers.

We can get round this by using the relationship between the discrete binomial distribution and the continuous normal distribution (Chapter 19). It can be shown that with a sufficiently large n and a p not too far from 0.5, a good approximation to the binomial distribution $B(n, p)$ is given by the normal distribution summarised as $N(np, npq)$. The larger is n and the closer is p to 0.5, the better the approximation. (Box 28.4 gives a rough rule.)

We illustrate the use of this approximation, and the need to remember a continuity correction, by solving Example 28.5. Instead of performing laborious exact calculations of the binomial probabilities, we can argue as follows.

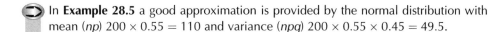

In **Example 28.5** a good approximation is provided by the normal distribution with mean (np) $200 \times 0.55 = 110$ and variance (npq) $200 \times 0.55 \times 0.45 = 49.5$.

The probability of there being 120 or more females

Noting that 120, as a value of a discrete distribution, corresponds to the interval 119.5–120.5 of a continuous distribution, we have that the probability of there being 120 or more is given by the area of the right-hand tail of the normal distribution beyond 119.5.

In the usual way we find this area by first standardising the variable. Denoting the number of females by X, we define a new variable Z such that

$$Z = \frac{X - 110}{\sqrt{49.5}} = \frac{X - 110}{7.0356}$$

where we have gone to four decimal places just in case the last figure or two affects the answer importantly. We now have

$$P(x \geqslant 120) = P\left(z > \frac{119.5 - 110}{7.0356}\right)$$

$$= P(z > 1.3503) = 1 - P(z \leqslant 1.3503)$$

which Table D4 shows to be 0.0885. Thus there is a probability of almost 9% of getting 120 or more females.

The probability of getting fewer than 90 females

This can be calculated from the normal distribution as the area to the left of 89.5. Arguing as above, we have

$$P(x < 90) = P\left(z < \frac{89.5 - 110}{7.0356}\right) = P(z < -2.9138)$$

which is the same as $P(z > 2.9138)$; Table D4 gives this as $1 - 0.9982 = 0.0018$, a chance of fewer than 2 in 1000.

The probability of there being exactly 100 females

This is given by the binomial term $^{200}C_{100}(0.55)^{100}(0.45)^{100}$, which involves such large numbers in $^{200}C_{100}$ and such small ones in $(0.55)^{100}(0.45)^{100}$ that most calculators give up. ⬭

To use the normal approximation, we have to interpret the discrete variable with a value of 100 as a continuous variable with a value between 99.5 and 100.5. The probability of x being between these values is given by the area under the graph between these two values. This is the area to the left of 100.5 *minus* the area to the left of 99.5:

$$P(99.5 \leqslant x \leqslant 100.5) = P(x \leqslant 100.5) - P(x \leqslant 99.5)$$

This is translated into standardised form by using

$$Z = \frac{X - 110}{\sqrt{49.5}} = \frac{X - 110}{7.0356}$$

which gives us

$$P(x \leqslant 100.5) = P\left(z < \frac{100.5 - 110}{7.0356}\right) = P(z < -1.350)$$

$$= P(z > +1.350) = 0.0885$$

and

$$P(x \leqslant 99.5) = P\left(z < \frac{99.5 - 110}{7.0356}\right) = P(z < -1.492)$$

$$= P(z > +1.492) = 0.0678$$

Thus

$$P(99.5 \leqslant x \leqslant 100.5) = P(x \leqslant 100.5) - P(x \geqslant 99.5)$$

$$= 0.0885 - 0.0678 = 0.0207$$

There is thus a probability of just over 2% of having exactly 100 females in the sample.

The general procedure for using the normal distribution as an approximation to the binomial distribution is given in Box 28.3. Note carefully the procedure for deciding how to deal with the continuity correction of 0.5.

Box 28.4

How to … **solve a binomial problem by using the normal distribution**

---

1. When n is large enough and p is not too different from 0.5, a good approximation to the binomial distribution B(n,p) is provided by the normal distribution N(np, npq). As a rough rule, the approximation is acceptable for $n > 10$ if p lies between 0.4 and 0.6, and for $n > 30$ if p lies between 0.1 and 0.9. ⬭

2. To convert a binomial problem into a normal distribution problem, use the following table:

Binomial problem $B(n, p)$	Normal problem $N(np, npq)$
$P(X < x_1)$	$P(X < x_1 - 0.5)$
$P(X \leqslant x_1)$	$P(X < x_1 + 0.5)$
$P(X \geqslant x_1)$	$P(X > x_1 - 0.5)$
$P(X > x_1)$	$P(X > x_1 + 0.5)$
$P(X = x_1)$	$P(x_1 - 0.5 < X < x_1 + 0.5)$

3. Solve the normal problem by using Box 28.1 with the problem specified in terms of a normal distribution with mean np and variance npq.

The procedure described in Box 28.4 provides an easy solution to Example 19.1.

28.5 Looking forwards

This chapter has already begun to get away from large samples and indicated a link between parametric and non-parametric tests. In the next few chapters we deal mainly with small samples, before going on to study some non-parametric tests and other probability distributions. The ideas about areas under the normal curve, significant probability and the normal approximation to the binomial distribution will help us.

Consolidation exercise 28C

1. When is it appropriate to use the normal approximation to the binomial?

2. A random variable X has the probability distribution $B(40, 0.4)$. What are the mean and variance of the normal approximation to it?

3. We are interested in the following quantities for the variable in Question 2: (a) $P(x < 15)$, (b) $P(x \leqslant 15)$, (c) $P(x = 15)$. What probabilities do we seek when using the normal approximation?

Notebox 28.1

Tables of areas under the standard normal curve

We illustrate the use of Table D4 by picking out the row beginning 1.9, along with the next row (beginning 2.0), and the heading in the upper margin. They are shown in Table 28.2. The figures in the main part of the upper margin represent the second decimal place in the value of z. For example, to find the area to the

Table 28.2

z	0	1	2	3	4	5	6	7	8	9	1 2 3 4 5 6 7 8 9 ADD
1.9	.9713	.9719	.9726	.9732	.9738	.9744	.9750	.9756	.9761	.9767	1 1 2 2 3 4 4 5 5
2.0	.9772	.9778	.9783	.9788	.9793	.9798	.9803	.9808	.9812	.9817	0 1 1 2 2 3 3 4 4

left of $z = 1.90$, we enter the row 1.9 and go along it until we are under 0 (which is the first entry), getting an area of 0.9713 (97.13%). Similarly the area to the left of $z = 1.95$ is 0.9744 ('0' is omitted to save space).

The figures in the upper margin above the right-hand part of the table represent the third decimal place in the value of z, and indicate how much has to be added to the last digit. For example, the area to the left of $z = 1.951$ is obtained by first finding the area to the left of 1.950 (which is the same as 1.95). We have just seen that this is 0.9744. We add to the last digit (4) the entry under 1 in the same row in the right-hand part of the table (which is 1). Thus the area to the left of 1.951 is 0.9745. To the nearest decimal place, this is also the area to the left of $z = 1.952$, but the area to the left of 1.973 has 2 (under 3) added to the last digit of 0.9744 to give 0.9746. Similarly the area to the left of $z = 1.959$ has 5 (under 9) added to 0.9744, giving 0.9749. The area under $z = 1.960$ is the next entry (under 6) in the main part of the table.

Check that the area to the left of z=2.054 is 0.98. By symmetry the area to the left of $z = -2.054$ is the same as the area beyond $z = +2.054$, which is $1 - 0.98 = 0.02$. More generally,

Area to left of $z = -k = 1 - $ (area to left of $+k$)

Summary

1. The area to the left of any ordinate of the standard normal curve can be found from Table D4 or Figure 28.2. This enables us to solve problems about any normal curve, as indicated in Box 28.1

2. Sometimes we want the probability of a random variable having a specified value or one that is just as unlikely or more so. If the variable is normally distributed, this is the area between the appropriate ordinate of the normal curve and the nearest tail (in that direction, or double this if both directions are considered). It is called the significance probability of that value. Box 28.2 restates Box 28.1 with an emphasis on this.

3. Finding the significance probability for random variables that are not normally distributed may involve separate calculations of the probability of getting the obtained result, and then of all equally likely or less likely results. These are then summed to provide the significance probability. The sum of the results that differ

from the expected result in the same direction as the obtained result is called the significance probability (SP) in the obtained direction. The sum of the other results is the SP in the opposite direction (Box 28.3).

4. Tables of the area under the normal curve can also be used to provide good approximations to binomial problems, as detailed in Box 28.4. This contains a statement of the conditions that must apply and of how to apply the necessary continuity correction.

5. A notebox explains how to use the tables of areas under the standard normal curve.

Small samples, the *t*-test and freedom

29.1 Introduction

Example 29.1 The Bishop of Statingham is critical of the literary standards of the *Statingham Declaimer*. To prove his point, he takes advice from Anna on how to draw a sample. By using random numbers, he chooses five weekday dates from last year and one page from each of the five issues published on those days. Then he persuades the Professor of Proper English to count the grammatical errors. The professor's research assistant finds a total of 85 errors, averaging 17 a day, with a standard deviation of 3. The Bishop wants a 95% confidence interval for the mean daily number of errors per page.

Example 29.2 Bricks used in the construction of Lady Agatha's dog kennel are supposed to be best Statingham Steam-Stained. They are known to have a mean mass of 4.13 kg, but Lady Agatha finds that a sample of 10 bricks from the latest delivery has a mean weight 3.98 kg, with a standard deviation of 0.14 kg. Is she justified in asserting that the delivered bricks are not what she ordered?

Example 29.3 A machine filling bags of Growler dry dog food works within well-defined limits on either side of a nominal weight that can be set at will. A sample of 8 bags filled on Monday has a mean mass of 19.40 kg and a standard deviation of 0.10 kg. A sample of 10 bags filled on Tuesday has a mean weight of 19.70 kg and a standard deviation of 0.08 kg. Has the setting of the nominal weight been changed?

We have stressed that the test procedures described in Chapters 24 onwards are valid only for large samples. For sample sizes of between 30 and 50 some of them may be acceptable, but to use large-sample tests when a sample is less than 30 is to ask for trouble. Large-sample tests are based on good approximations that degenerate as the sample becomes smaller. In particular, many sampling distributions become less well approximated by the normal distribution, yet it is for this that the confidence coefficients and the critical values in large-sample significance tests are strictly valid. To tackle the examples in this chapter, involving small samples, we must therefore use other tests.

Unlike large-sample tests, the tests we are about to describe involve no approximation and can be used for samples of all sizes. They are slightly more complicated than large-sample tests, and if the sample is large the gain in accuracy does not compensate for the extra effort involved in using them. They are called small-sample tests not because their use with large samples is wrong – it is not – but because their use with small samples is right.

29.2 The *t*-test

In introducing the central limit theorem, we pointed out that if the parent population is normal with mean μ and standard deviation σ then, whether the sample is large or small, the means of samples of the same size are exactly normally distributed with mean μ and standard deviation $\sigma/\sqrt{n}$. This is approximately true for large samples from any distribution.

On this basis, when we described large-sample confidence intervals and tests, we defined the test statistic

$$Z = \frac{\bar{X} - \mu}{\sigma/\sqrt{n}} \qquad [29.1]$$

If the population standard deviation σ is unknown, we use the best estimated value

$$\hat{\sigma} = s\sqrt{n/(n-1)}$$

where s is the sample standard deviation and n is the sample size.

This gives an approximate value of Z as

$$z \approx \frac{\bar{x} - \mu}{s/\sqrt{n-1}} \approx \frac{\bar{x} - \mu}{s/\sqrt{n}} \qquad \text{(for large } n\text{)} \qquad [29.2]$$

where $\approx$ means 'is approximately equal to'. The lower the value of n, the less accurate this procedure becomes. But this is simply a caution about replacing $\sqrt{n-1}$ with $\sqrt{n}$. It does not question the validity of replacing σ by an estimate based on s, or of using the normal distribution for the test statistic.

It is this that breaks down for small samples. For n less than about 30, the test statistic Z defined above does not follow the normal distribution. We therefore use a different test statistic. It looks very like Z but has an important difference. Notice carefully what this is. As we have seen, Z is defined in terms of the *population* parameter and we have approximated it by using an estimate based on the sample standard deviation. But t is defined in terms of the *sample* standard deviation:

$$t = \frac{\bar{x} - \mu}{s/\sqrt{n-1}} \qquad [29.3]$$

This involves no approximation.

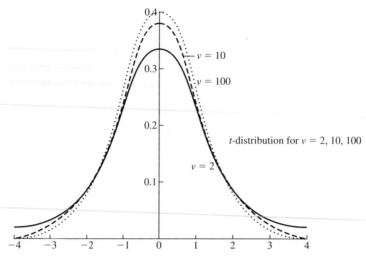

Figure 29.1

It can be shown that, provided the values of x in the population are at least approximately normally distributed, this statistic has a probability distribution with three properties:

- It is unimodal and symmetrical about a zero mean.
- It depends on the value of n in a way described later and illustrated in Figure 29.1.
- For large samples (at least 30) it is approximately normal.

It is this third point that justifies the use of the normal distribution in *large-sample* tests. For small samples the normal distribution is a bad approximation and this new distribution should be used; it is called the *t*-distribution. To use the *t*-distribution we have to know about degrees of freedom.

 29.3   Degrees of freedom

When the research assistant in Example 29.1 was asked for details of the grammatical errors, he discovered that he had mislaid his records. His first reaction was to guess five numbers: 23, 16, 11, 29 and 17. He was about to give them to his boss when he realised that they did not total 85. Economical with his energy, he decided to leave the first four figures as they were, totalling 79. Then, as the overall total had to be 85, he was forced to put the fifth number as $85 - 79 = 6$. He had no choice. He had lost a little freedom.

He could, of course, choose four other numbers, such as 17, 23, 12 and 9, but then he would have no choice in selecting his fifth number, which would have to be 24 to get the total right.

Thus, although he thought he could choose any five numbers with complete freedom, the requirement that they should total 85 had reduced the number of numbers he could choose. More technically, if there had been no such requirement, he would have had five **degrees of freedom**; now the restriction had **used up** one degree of freedom and so he had only four.

Now return to the *t*-statistic defined by equation [29.3]. Its value is based on *n* sample observations. Each observation (like each number guessed by the assistant) is potentially associated with one degree of freedom, but if we are using the data to calculate the test statistic *t*, some of the *n* potential degrees of freedom are used up before we can begin the calculations. We need to look further at this.

When we used a large-sample result to estimate a population mean, we decided that the best estimate was the sample mean and the 95% confidence interval was

$$(\bar{x} - 1.96\hat{\sigma}/\sqrt{n}, \quad \bar{x} + 1.96\hat{\sigma}/\sqrt{n})$$

where 1.96 was the value of *z* that cut off a tail area of 2.5% under the curve of the normal probability distribution.

For small samples the normal curve approximation does not apply, and instead of 1.96 we have to use a value that cuts off a tail area of 2.5% under a probability distribution curve whose precise shape depends on the value of *n*. However, what keeps appearing in the mathematics of this curve is not *n* but *n* − 1. This is because, in estimating $\hat{\sigma}$ from the usual formula for a standard deviation, we have had to assert that the *n* values of *x* have a mean equal to the calculated value of $\bar{x}$, just like we asserted that errors in five days had to total 85 (or to average 17). This reduces by 1 the number of degrees of freedom, usually denoted by *v* (the Greek letter *n*, pronounced 'new'). So, in using the *t*-statistic to examine *n* independent observations, we have *n* − 1 degrees of freedom; Notebox 29.2 puts this in different language. All we need to remember for now is this. The values of the *t*-distribution are tabulated for different degrees of freedom, so if we are using it in a simple case and we have *n* values of *x*, we use the column headed $v = n - 1$. Figure 29.1 shows the distributions for $v = 2, 10$ and 100.

29.4 Using the *t*-test

Just as the *t*-distribution curve depends on the number of degrees of freedom, so does the area under any part of it. Tables show the values of *t* such that the area between the left extreme of the curve and the ordinate equals one of several specified fractions of the total area, for various degrees of freedom. Consider Table D5; if there is 1 degree of freedom, we use the row labelled $v = 1$ (the top row). The value of $t = 6.31$ comes in the column headed $t_{0.95}$, which means that 95% of the total area lies to the left of the ordinate. But if there are 2 degrees of freedom, 95% of the total area lies to the left of the entry in row 2 under the column headed $t_{0.95}$, which is $t = 2.92$ (Figure 29.2). For 10 degrees of freedom it is 1.812 (Figure 29.3); for 120 degrees of freedom it is 1.658, very close to the value of 1.645 used in the large-sample test (which we know to be an approximation). As the sample size increases, the value of *t* falls.

Armed with this table, we can calculate confidence intervals for small samples. For example, if we wish to calculate the 95% confidence interval, we use the fact that 95% of the area under the curve lies between the two ordinates that cut off

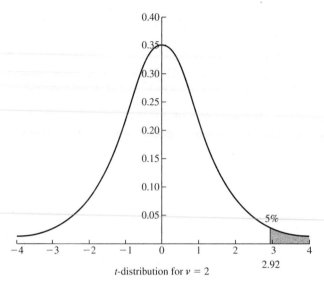

t-distribution for $\nu = 2$

Figure 29.2

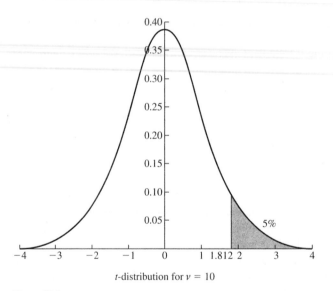

t-distribution for $\nu = 10$

Figure 29.3

2.5% at each tail. The ordinate $t_{0.975}$ (whose value will depend on the number of degrees of freedom) cuts 2.5% off the right-hand tail, and by symmetry the ordinate $-t_{0.975}$ cuts 2.5% off the left-hand tail.

It is easy to show from equation [29.2] that there is therefore a 95% probability that the *population mean* μ lies within the confidence interval

$$\bar{x} - t_{0.975}\, s/\sqrt{n-1} < \mu < \bar{x} + t_{0.975}\, s/\sqrt{n-1} \qquad [29.4]$$

Notice that the confidence interval narrows as *n* gets bigger. There are two reasons for this: the denominator increases, so we are dividing by a larger quantity, and the value of *t* gets smaller.

We illustrate the use of this formula shortly, but first we summarise the procedure we are about to use.

Box 29.1

How to ... **find the 95% confidence interval for the mean from a small sample of size *n***

1. Write the number of degrees of freedom $v = n - 1$.

2. From the *t*-table find the value of $t_{0.975}$ for this value of *v*.

3. Use that value of $t_{0.975}$, and your values of the sample mean $\bar{x}$, the sample size *n* and the sample standard error *s* in equation [29.4] to obtain the 95% confidence interval for the population mean μ.

Example 29.1 is now considered, first in the version stated at the head of the chapter, and then with slightly altered numbers to illustrate the effects of sample size and the number of degrees of freedom.

➲ We have a sample of $n = 5$ observations and use them to calculate the sample mean, which is $\bar{x} = 17.00$. The sample standard deviation is $s = 3.00$. We want to find the 95% confidence interval for the population mean.

➲ We assume that the numbers of errors in the population consisting of all pages of all issues published last year are approximately normally distributed.

As we seek a two-tailed 95% confidence interval, each tail has to cut off an area of 0.025. We therefore need to know $t_{0.975}$ for the right number of degrees of freedom, which is $5 - 1 = 4$. The 95% confidence interval is given by equation [29.4] with the above values:

$$17 - t_{0.975} \times 3/\sqrt{5 - 1} < \mu < 17 + t_{0.975} \times 3/\sqrt{5 - 1}$$

which is

$$17 - 1.5t_{0.975} < \mu < 17 + 1.5t_{0.975}$$

Tables give $t_{0.975} = 2.776$ for $v = 4$, so the confidence interval is

$$17 - 4.164 < \mu < 17 + 4.164 \quad \text{or} \quad 12.836 < \mu < 21.164$$

Thus the Bishop will be quite safe in proclaiming there is a 95% chance that last year the mean number of errors per page was between 12 and 22.

➲ Suppose instead that the sample size is 26 and the sample standard deviation is 7.5.

⬭ You can verify that in this case the 95% confidence interval will again be given by

$$17 - 1.5t_{0.975} < \mu < 17 + 1.5t_{0.975}$$

but now $v = 26 - 1 = 25$ so the value of $t_{0.975}$ is 2.060, which gives a confidence interval of $13.910 < \mu < 20.090$. Thus the larger sample produces a narrower confidence interval, even though the standard deviation happens to be greater.

⬭ Now suppose the larger sample of size 26 retains the standard deviation of the smaller sample, which is 3. Show that the confidence interval is 17 ± 1.236.

Consolidation exercise 29A

1. Give one important reason why large-sample tests may be unacceptable for small samples. What does 'small' mean in this context? Why is a small-sample test so called?

2. How does the definition of *t* differ from the definition of *Z*?

3. When applying tests for the mean, what is assumed about the distribution of *x* in the population for (a) a large sample and (b) a small sample?

4. Give three characteristics of the *t*-distribution.

5. What are degrees of freedom? What symbol is used to denote their number?

6. If there are 10 values of *x* and one constraint, what is the value of *t* corresponding to a 95% two-tailed confidence interval?

7. What is the formula for a 95% two-tailed confidence interval for the mean?

29.5 **Testing the origin of a small sample**

Frequently we wish to know whether a mean derived from a small sample is compatible with a hypothesised population mean.

⬭ In **Example 29.2** the mean mass of bricks coming from the designated brickworks is $\mu = 4.13$ kg. A random sample of 10 taken from a new delivery has $\bar{x} = 3.98$ kg, with $s = 0.14$ kg. Does this support the hypothesis that the sample is from the right brickworks? In other words, is $\bar{x} = 3.98$ compatible with the null hypothesis that $\mu = 4.13$?

Here we are not really concerned about the mass of the bricks. We are concerned with their origin, of which their weight is some evidence. Is the sample mean weight sufficiently *different from* 4.13 for us to decide that they probably did not come from those works? Only the magnitude of the difference matters. We use a two-tailed test.

In testing the null hypothesis we use the *t*-statistic defined in equation [29.3]. As $n = 10$, $v = 9$. We decide to reject the statement that the bricks have come from

⬭

the prescribed work if $\bar{x}$ differs from the hypothesised μ with a 5% level of significance. As it is a two-tailed test, we find $t_{0.975}$ from the table, using $v = 9$; it is 2.26. Equation [29.3] gives

$$t = \sqrt{10 - 1}(3.98 - 4.13)/0.14 = -3.214$$

We ignore the minus sign. The value of t exceeds 2.26, which corresponds to a 5% significance in the two-tailed test with $v = 9$. We therefore reject the hypothesis that the bricks have come from the stated place.

Note that if we had adopted the 1% significance level and used $t_{0.995}$ in the two-tailed test, we would have just accepted the hypothesis, as the t-table would then have given a critical value of 3.25 instead of 2.26.

Box 29.2

How to ... **use a small-sample mean to test the origin of the sample**

1. Set up the null hypothesis that the mean μ of the population from which the sample comes is equal to the mean μ_0 of the population from which it should come, i.e. $\mu = \mu_0$.

2. State the alternative hypothesis, usually one of these two:
   (a) $\mu \neq \mu_0$ (two-tailed)
   (b) $\mu > \mu_0$ or $\mu < \mu_0$ (both one-tailed)

3. Take the number of degrees of freedom $v = n - 1$, and choose the level of significance α.

4. From the t-table, using the row for the correct value of v, find the critical value for t corresponding to α (if it is a one-tailed test) or $\alpha/2$ (if it is a two-tailed test). Call this value t_c.

5. Calculate the test-statistic

   $$t = \sqrt{n - 1}(\bar{x} - \mu)/s$$

6. Reject the null hypothesis if the numerical value of t exceeds the critical value t_c.

29.6 **Using small-sample means to compare two population means**

A slightly different problem arises when we have two independent samples and wish to see whether the difference in sample means is significant.

Example 29.3 states that a sample of $n_1 = 8$ bags filled on Monday has a mean mass of 19.40 kg and a standard deviation of $s_1 = 0.10$ kg. A sample of $n_2 = 10$ bags filled on Tuesday has a mean mass of 19.70 kg and a standard deviation of $s_2 = 0.08$ kg. Has the setting of the nominal mass been changed?

We assume that, on any day, the masses are normally distributed about the nominal mass as mean. We use the *t*-test to see whether the two population means for Monday and Tuesday are the same. The null hypothesis is that the means of the two populations are equal, i.e. $\mu_1 - \mu_2 = 0$.

The alternative hypothesis is that the setting has been changed, which implies the two population means differ, i.e. $\mu_1 - \mu_2 \neq 0$, which requires a two-tailed test. Application of the test to two samples is strictly valid only if the two populations have a common variance (Notebox 29.3). The wording of the question suggests that this is so, and we estimate it from the two observed sample variances by using the formula

$$\hat{\sigma}^2 = \frac{n_1 s_1^2 + n_2 s_2^2}{n_1 + n_2 - 2} \qquad\qquad [29.5]$$

We have to be careful about the degrees of freedom. We shall be using a value of *t* defined in terms of $\hat{\sigma}$, and equation [29.5] tells us that we use the two sample standard deviations to calculate $\hat{\sigma}$. Each of these standard deviations involves the loss of one degree of freedom, since they are obtained using a calculated mean that becomes fixed, thereby imposing a restriction on our freedom. The number of degrees of freedom is therefore

$$\nu = n_1 + n_2 - 2 = 8 + 10 - 2 = 16$$

We return to this shortly, but let us now take it as agreed. Suppose we choose a 5% level of significance. We now use the *t*-table to find the critical value for *t* corresponding to 5% and 16 degrees of freedom. As it is a two-tailed test, we want the value of *t* that cuts off 2.5% at each tail. The table shows it is 2.120.

Using equation [29.5] we calculate $\hat{\sigma}^2$. The question tells us that $n_1 = 8$, $n_2 = 10$, $s_1^2 = 0.10^2 = 0.0100$ and $s_2^2 = 0.08^2 = 0.0064$. It follows from equation [29.5] that

$$\hat{\sigma}^2 = \frac{8 \times 0.0100 + 10 \times 0.0064}{8 + 10 - 2} = 0.009$$

We therefore assume that the masses of all the bags of food have this variance on both Monday and Tuesday. Its square root, which we need to use in equation [29.6], is 0.094 87.

We now calculate the value of *t* and see how it compares with the critical value of 2.120:

$$t = \frac{\bar{x}_1 - \bar{x}_2}{\hat{\sigma}\sqrt{(1/n_1) + (1/n_2)}} \qquad\qquad [29.6]$$

$$= \frac{19.40 - 19.70}{0.094\,87 \times \sqrt{0.125 + 0.100}} = -0.30/0.0450$$

which is clearly much greater numerically than the critical value of 2.120. We conclude that the observed difference between the two sample means is big enough for us to reject, with 95% confidence, the null hypothesis that there is no difference between the two population means. We cannot accept that the setting has not been changed.

Note that in performing this calculation we have deliberately worked to a higher number of decimal figures than turned out to be necessary. As the calculation proceeds, we have seen that great accuracy is not required and we have been able to work to fewer figures. Eventually we see there is no need to evaluate t. It is clearly larger than the critical value of 2.120.

Box 29.3

How to ... use two small-sample means to compare two population means if there is a common variance

1. Set up the null hypothesis that the means μ_1 and μ_2 of the two populations are equal, i.e. $\mu_1 - \mu_2 = 0$.

2. State the alternative hypothesis, usually one of these two:
   (a) $\mu_1 - \mu_2 \neq 0$ (two-tailed)
   (b) $\mu_1 - \mu_2 > 0$ or $\mu_1 - \mu_2 < 0$ (both one-tailed)

3. Take the number of degrees of freedom $v = n_1 + n_2 - 2$, where n_1 and n_2 are the sample sizes, and state the level of significance α that is to be used.

4. From the t-table, using the row for the correct value of v, find the critical value for t corresponding to α (if it is a one-tailed test) or $\alpha/2$ (if it is a two-tailed test). Call this t_c.

5. Calculate the test statistic

$$t = \frac{\bar{x}_1 - \bar{x}_2}{\hat{\sigma}\sqrt{(1/n_1) + (1/n_2)}} \qquad [29.6]$$

   where

$$\hat{\sigma} = \sqrt{\frac{n_1 s_1^2 + n_2 s_2^2}{n_1 + n_2 - 2}}$$

   $\bar{x}$ denotes the sample mean and s denotes the sample standard deviation.

6. Reject the null hypothesis if the numerical value of t exceeds the critical value t_c.

Before proceeding to the next chapter, we go back to our comment on the number of degrees of freedom. The essential point is that every time sample data have to be used to provide an estimate of a population parameter, it imposes a restriction on the data, which now have to comply with that estimate; the restriction means a loss of one degree of freedom. Thus, if we have 20 independent observations and these data are used to estimate 2 population parameters in order to find the value of some test statistic, this imposes 2 restrictions and $v = 20 - 2 = 18$.

As we shall see later, restrictions may also arise in other ways. Then careful examination will show that the number of *independent* observations is less than n, which should be kept in mind when reading and using Box 29.4.

Box 29.4

How to ...   **find the number of degrees of freedom associated with a test statistic**

Although not always easy to apply, the rule is as follows:

1. Count the number k of population parameters that must be estimated from sample data.

2. Subtract k from the number n of *independent* observations in the sample; this gives $n - k$, which is the number of degrees of freedom v.

It is useful to remember how many degrees of freedom are involved in each of the more common tests. If the sample is large, the number of degrees of freedom is virtually the same as the number of observations.

Consolidation exercise 29B

1. Give one way of testing the origin of a small sample.

2. Two population means are compared by using the *t*-test. What assumptions are made about (i) the distribution of x and (ii) the variances?

3. What is the formula for estimating the variance used in comparing two population means in a small-sample test?

4. If there are n values of the variable and k constraints, what is the number of degrees of freedom? What does 'constraint' mean in this context?

Notebox 29.1

Student

The *t*-test was discovered in 1908 by W.S. Gosset, who worked for Guinness Brewery, where he found it necessary to use small samples. He published it under the pseudonym Student because of company policy.

Notebox 29.2

Degrees of freedom

By stipulating that the values must have a specified mean, we impose a constraint (or restriction) on our freedom to choose these values. Under some circumstances we may impose more than one constraint. The number of degrees of freedom is then reduced by 1 for every constraint. If there are n values of the variable and k constraints, the number of degrees of freedom is $v = n - k$.

Notebox 29.3

The t-test and sample variance

Application of the test to two samples is strictly valid only if the two populations have a common variance. Two samples are unlikely to have identical variances. Whether the two variances differ significantly, implying that the population variances are different, can be formally tested by use of the F-test described in the next chapter. Alternatively we can rely on work which suggests that if the larger sample variance is not more than about three times the smaller, then the use of the t-test remains fairly reliable. If the samples are of very similar sizes, a slightly greater disparity is acceptable.

Summary

1. Large-sample tests break down for small samples, but small-sample tests can be used for samples of all sizes. Unfortunately, they are more complicated.

2. For testing the small-sample mean, we define the test statistic

$$t = \frac{x - \mu}{s/\sqrt{n-1}}$$

   where s is the sample standard deviation (calculated with n in the denominator). Its distribution, known as Student's t-distribution, depends on the number of degrees of freedom v (which is n reduced by the number of restrictions, as explained in Notebox 29.2). For large n it is approximately normal.

3. A 95% confidence interval for the mean is given by equation [29.4]. The procedure is summarised in Box 29.1 A small-sample mean can be used to test the origin of a sample by applying the t-test (Box 29.2). The t-test can also be used to compare two population means (Box 29.3), provided there is a common population variance, but see Notebox 29.3.

4. Noteboxes report on the name 'Student' and comment on degrees of freedom and on the dependence of the t-test on the ratio of sample variances.

30

Comparison of variances and standard deviations

30.1 Introduction

Example 30.1 FLASH decorates 41 identical houses, taking an average of 47.3 half-days per house, with a variance of 2.80. PLOD decorates 31 houses, with a mean time of 57.5 half-days and a variance of 0.95. Assuming the houses inspected can be taken as a random sample of work done by these decorators, does FLASH have a larger variance than PLOD (and show less reliability so far as finishing time goes)?

Example 30.2 A new washing-up liquid (LUSH) is being compared with a well-established product (WOSH). A test trial involves counting the number of cups that can be satisfactorily washed by a small measure of the liquid. A total of 31 trials with LUSH show a mean of 50.3 cups with a variance of 1.50, whereas 41 trials with WOSH show a mean of 50.5 cups and a variance of 0.95. In prolonged and very large-scale use, would these liquids be equally reliable?

Example 30.3 Daphne de Bencha wants to encourage her customers to pay regular bills by direct debit. A random sample of 20 customers shows a mean of 19 monthly direct debit payments, with a variance of 6. Her area manager says that in Upper Sternum a sample of 15 customers showed a mean of 16 with a variance of 22. Hoping that the mean for Upper Sternum is not significantly lower than the mean for Statingham, she decides to use the *t*-test. With luck this will enable her to tell her area manager to go and play with his abacus. Can she validly use it?

In Chapter 29 we emphasised that strictly speaking it is wrong to use the *t*-test for comparing the means of small samples if the population variances are not the same. This is one reason why we may sometimes wish to test whether two populations have the same variances, as in Example 30.3. At other times we may want to know whether one standard deviation is significantly greater than another, as in Examples 30.1 and 30.2, possibly with an eye on the reliability of a product or a performance.

By definition, the standard deviation is the square root of the variance. Any test for one of them implies a test for the other. In what follows we refer only to the variance. Note that, except where we clearly say otherwise, everything in this chapter is applicable to samples of all sizes.

The test we use is the *F*-test and the reasoning behind it is set out in Notebox 30.1.

The *F*-test

The equality of two population variances is tested by using their ratio as a test statistic. To avoid confusion we call the larger sample variance s_1^2 and the smaller s_2^2. We denote their ratio by F, defined to be

$$F = s_1^2/s_2^2 \qquad [30.1]$$

It cannot be less than unity as the top is larger than the bottom. We denote the degrees of freedom associated with s_1^2 and s_2^2 by v_1 and v_2 respectively. For each sample variance the number of degrees of freedom is one fewer than the number of observations, because in calculating each variance it has been necessary first to use the data to calculate one parameter, the mean. Thus

$$v_1 = n_1 - 1 \quad \text{and} \quad v_2 = n_2 - 1 \qquad [30.2]$$

where n_1 is the size of the sample that has the larger variance. Be careful not to suppose that n_1 is the larger n.

By definition, F cannot be less than unity. If F is close to unity, there is little evidence to suggest the population variances are different; but if it is much greater than unity, we may think there are grounds to suspect that a real difference exists. We need to test the hypothesis that the two population variances are the same.

Tables D6a and D6b give the 5% and 1% significant values of F for various pairs of values v_1 and v_2. The column headings show increasing values of v_1 and the rows are in order of increasing values of v_2. Similar tables exist for other levels of significance.

As with other tests of two sample statistics, we may be concerned with one of two questions: Is one population variance significantly *bigger than* the other (implying a one-tailed test)? Is it significantly *different from* the other (implying a two-tailed test)? We explain in Notebox 30.1 that to conduct a two-tailed *F*-test at a specified level of significance (e.g. 2%), we simply conduct a one-tailed test at half that level of significance (e.g. 1%). We illustrate this shortly.

 In **Example 30.1** the larger variance is attached to FLASH, so we give this the subscript 1. To reduce the chance of later error, we clearly summarise our data with the appropriate subscripts

$$n_1 = 41, \ s_1^2 = 2.80 \qquad n_2 = 31, \ s_2^2 = 0.95$$

We are asked if the data suggest that FLASH is less reliable than PLOD. In other words, is s_1^2 significantly greater than s_2^2? This implies a one-tailed test. ⬭

To set up the hypotheses, we think of two populations: one is all the houses painted by FLASH and the other is all the houses painted by PLOD. Our null hypothesis is that these two populations have equal variances, $H_0: \sigma_1^2 - \sigma_2^2 = 0$. The alternative hypothesis is $H_1: \sigma_1^2 - \sigma_2^2 > 0$.

Feeling charitable to FLASH, we decide to adopt a 1% significance level, so we reject the null hypothesis only if the evidence against it is pretty overwhelming.

We use F as a test statistic. We know that the critical value of F printed on the 1% page of the F-table depends on the two degrees of freedom. Having chosen FLASH to have the subscript 1 in the variance, we must now denote FLASH's degrees of freedom by v_1. Thus we have

$$v_1 = (n_1 - 1) = 40 \quad \text{and} \quad v_2 = (n_2 - 1) = 30$$

The 1% F-table shows that for $v_1 = 40$ and $v_2 = 30$ the critical value of F is 2.30. In other words, an F of 2.30 cuts off a tail area of 1%, leaving a 99% probability of F being less than this value. If F exceeds 2.30 we argue that an improbably high value of F has arisen, so the null hypothesis is to be rejected.

We therefore calculate F:

$$F = \frac{s_1^2}{s_2^2} = \frac{2.80}{0.95} = 2.947$$

This falls into the rejection zone. We reject, at the 1% level of significance, the null hypothesis that FLASH and PLOD have equal variances, in favour of the alternative hypothesis that FLASH has a greater variance than PLOD. FLASH's finishing times are almost certainly more variable than PLOD's, so they are likely to be less predictable.

Notice that if we had *wrongly* put $v_1 = 30$ and $v_2 = 40$, we would have found a critical value not of 2.30 but of 2.20. In this particular case, we would have been led to the same conclusion, but we might not have been so lucky. Suppose we have $v_1 = 10$ and $v_2 = 4$; then the entry in the 1% table is 14.55. But if we reverse them, writing $v_1 = 4$ and $v_2 = 10$, we find an entry of 5.99. Any calculated value of F lying between these values would fail the test with one choice of v_1 and pass it with the other. It is essential to attach v_1 to the sample with the larger variance.

Consolidation exercise 30A

1. Give two reasons why we may sometimes wish to compare two variances.

2. What do we call the test statistic used to compare variances? How is it defined?

3. What is assumed about the populations from which the samples are drawn?

4. Sample A is of size 20, mean 30 and variance 40. Sample B is of size 25, mean 32 and variance 50. What is the one-tailed 5% significant value of F?

5. What is the lowest possible value of F?

⬭ **Example 30.3** involves a two-tailed test. We are wondering if there is *any difference between* the variances for the two populations. This implies a two-tailed test. Our null hypothesis is that the two populations have equal variances, $H_0: \sigma_1^2 - \sigma_2^2 = 0$. The alternative hypothesis is the two-tailed hypothesis that LUSH and WOSH have different population variances $H_1: \sigma_1^2 - \sigma_2^2 \neq 0$.

We use as our test statistic

$$F = s_1^2/s_2^2$$

where s_1^2 is the larger of the two sample variances.

We decide to reject the null hypothesis if the value of *F* is so large that it has a probability of only 2% of being reached if the null hypothesis is correct.

We know that the critical value of *F* depends on the two degrees of freedom. Since the larger sample variance must have the subscript 1, this subscript must be attached to the degrees of freedom for LUSH. Thus we have

$$v_1 = (n_1 - 1) = 30 \quad \text{and} \quad v_2 = (n_2 - 1) = 40$$

We now have to find the critical value of *F* such that there is only a 2% chance of it being reached (or exceeded) if the null hypothesis is to be preferred to the two-tailed alternative hypothesis.

As we wish to conduct a two-tailed test at 2% significance we look up the critical value at 1% in a one-tailed test. With $v_1 = 30$ and $v_2 = 40$ the 1% *F*-table gives 2.20 as the critical value. An *F* larger than this falls in the rejection zones for a two-tailed 2% test.

We now calculate *F* and obtain

$$F = s_1^2/s_2^2 = 1.50/0.95 = 1.579$$

This does not fall in the rejection zone. The evidence is therefore not strong enough to warrant rejection at the 2% level of the hypothesis that LUSH and WOSH have equal variances, in favour of the alternative hypothesis that they have different variances.

We now summarise the procedure for applying the *F*-test to compare two variances.

Box 30.1

How to ... **compare two population variances or standard deviations**

1. Set up the null hypothesis that the two populations have the same variance, i.e. $\sigma_1^2 - \sigma_2^2 = 0$, identifying the population whose sample has the larger variance by the subscript 1.

2. State the alternative hypothesis, usually one of these two:
   (a) $\sigma_1^2 - \sigma_2^2 \neq 0$ (two-tailed)
   (b) $\sigma_1^2 - \sigma_2^2 > 0$ (one-tailed) ⬭

3. Take as degrees of freedom $v_1 = n_1 - 1$ and $v_2 = n_2 - 1$, and choose a level of significance α. Be certain that v_1 relates to the sample that has the larger variance.

4. Select the F-table corresponding to the value of α if it is a one-tailed test, or to $\alpha/2$ if it is a two-tailed test.

5. From the F-table, using the top reading for v_1 and the side reading for v_2, find the critical value for F_c.

6. Calculate the test statistic

$$F = s_1^2 / s_2^2 \qquad\qquad [30.1]$$

where the larger variance is in the numerator.

7. Reject the null hypothesis if $F > F_c$.

Example 30.3 involves small samples. The main interest is in the difference between two means, and Daphne is thinking of applying the t-test. However, the sample variances suggest that perhaps the population variances are different, so a basic assumption about the application of the t-test does not hold. We therefore have to test whether the difference between the sample variances is significant.

To test whether the variances are different, she adopts a two-tailed test at 2% significance, which in the F-test is the same as a one-tailed test with 1% significance. Check that $v_1 = (n_1 - 1) = 14$ and $v_2 = (n_2 - 1) = 19$.

The 1% level of significance for F with these degrees of freedom is not given in the tables published in the appendices. The nearest we can get is for $v_1 = 12$ and $v_2 = 19$. The critical value for these degrees of freedom is 3.30. We also have that for $v_1 = 24$ and $v_2 = 19$ the critical value is 2.92. Clearly the critical value we want is a little less than 3.30, perhaps around 3.25.

Check that the calculated value of F is $22/6 = 3.67$. This clearly exceeds the critical value. We therefore reject the hypothesis that the variances are the same. This means that application of the t-test in a comparison of these small sample means is unjustified, so Daphne has to be disappointed.

Consolidation exercise 30B

1. How do you conduct a two-tailed test for the significance of the ratio of two variances?

2. When is it necessary to perform an F-test while comparing the means of two samples?

Notebox 30.1

The *F*-distribution

We earlier defined *F* to be necessarily greater than unity. The reason for this, which is linked to the relationship between one-tailed and two-tailed tests, is now explained.

Take independent random samples *A* and *B* from two populations with equal variances. The sample variances are likely to differ, so their ratio (var *A*/var *B*) is likely to differ from unity. We call this ratio *F*, without stipulating that it must be more than unity.

The probability distribution of the various possible values of *F* will depend on the two values of the degrees of freedom associated with the two sample variances. Figure 30.1 shows this *F*-distribution for selected pairs of values of the degrees of freedom. Figure 30.2 shows it when sample *A* has 100 degrees of freedom and sample *B* has 10. The probability of the ratio being less than say 0.5 is the area to the left of P, whereas the probability of the ratio being more than say 2.5 is the area to the right of Q.

In Figure 30.3 we take a typical *F*-distribution curve and shade two tails, each to indicate values of *F* that cut off 2.5% of the area under the curve. Call these values F_1 and F_2. If the two samples come from populations with equal variances, there is a chance of only 2.5% that the ratio var *A*/var *B* is less than F_1 and a chance of only 2.5% that the ratio var *A*/var *B* is more than F_2.

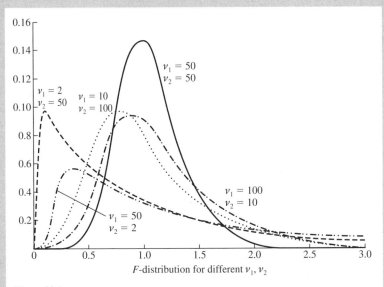

F-distribution for different v_1, v_2

Figure 30.1

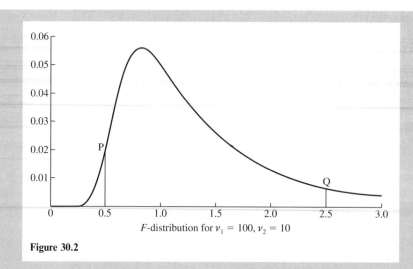

Figure 30.2

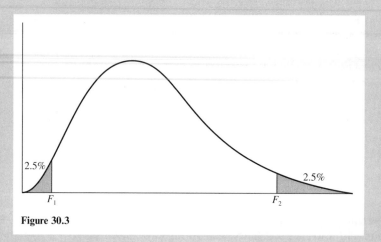

Figure 30.3

On the face of it, we therefore need tables that will give the areas of both tails. However, it can be shown that the chance of getting var A/var B as low as F_1 equals the chance of getting var B/var A as high as F_2 on an F curve redrawn for this new ratio. This means that if we always ensure that the ratio is bigger than 1, by dividing the larger variance by the smaller, we need only look at the area beyond F_2.

For any distribution, if we want to conduct a two-tailed test at the 5% level of significance, we conduct two one-tailed tests each with a 2.5% level. The critical value for a two-tailed 5% test is thus the same as the critical value for a one-tailed 2.5% test.

Summary

1. Two reasons for comparing variances are that (i) strictly speaking the *t*-test should be used only when the two populations have a common variance, and (ii) because we are interested in spread, we may wish to know which of two populations has the greater variance.

2. If s_1^2 and s_2^2 are two sample variances with $s_1^2 > s_2^2$ (take note!), then their ratio is denoted by F, which cannot be less than 1, by definition. The reason for this definition is given in Section 30.2, which may be postponed on first reading.

3. F has a well-studied distribution and is used as a test statistic. The greater its value, the greater the disparity between the variances. To test whether one variance is significantly greater than the other, we look at areas under the appropriate F-curve to assess whether the probability of getting such a high ratio is so low that the difference must be regarded as significant.

4. The F-distribution is shown in Figure 30.1. There is a different curve for every pair of degrees of freedom v_1 and v_2, where v_1 is the number of degrees of freedom for the variable with the greater variance. Significant values of F at commonly used degrees of significance have been tabulated for combinations of v_1 and v_2. We give them only for 1% and 5%.

5. The test just described is for one tail. It is shown in Section 30.2 that to conduct a two-tailed F-test at a specified level of significance, we simply conduct a one-tailed test at half that level of significance; a two-tailed F-test at 2% significance would be conducted as a one-tailed F-test at 1% significance. The general procedure is summarised in Box 30.1.

6. A notebox explains the reasoning behind the F-test.

31

Correlation and significance

Introduction

Example 31.1 A random sample of 18 undergraduates from Statingham shows a product-moment correlation of $r = 0.35$ between the distance from the university to the parental home and the level of expenditure in the students' union bar. The welfare officer says that obviously there is a positive correlation applicable to all the university, and students living far from home are drinking because they miss their parents. The bar manager says there is no relationship at all between distance from home and spending at the bar. Who is right?

Example 31.2 How big must a sample be if we want a correlation of 0.50 to be significant at 5%?

Example 31.3 Lady Agatha is told that a sample of 28 ladies has a correlation coefficient of 0.6 between their ages and their expenditures in beauty parlours. She asserts that obviously they do not come from Upper Sternum, where the correlation coefficient is only 0.3. Is she justified?

Example 31.4 Survey data are used to examine the correlation between houshold expenditure on garlic and the age of the oldest member of the household. A sample of 28 households in Statingham shows a coefficient of 0.6, whereas a sample of 39 households in Upper Sternum shows 0.5. Do these results support Lady Agatha's contention that households of Upper Sternum have more adventurous tastes?

Example 31.5 A random sample of 19 items gives a correlation coefficient of 0.6. Find a 95% confidence interval.

Anybody seeing a sample correlation coefficient needs to know whether it is meaningful – whether it is bigger than might easily arise by chance, even if there is no correlation in the population from which it has been taken. Example 31.1 illustrates this concern.

The questions illustrated by the other examples arise less frequently. Each has a section of the text devoted to it. We end with remarks on spurious and zero correlation.

Is there really any correlation?

If we have sample data, we can expect a small non-zero correlation to arise by chance, without implying a significant relationship between the two variables. But we are unlikely to get a coefficient close to unity by chance, unless the number of ·observations is small. (If there are only two observations the correlation coefficient is bound to be unity, as two points are bound to lie on a straight line.)

The significance of a sample correlation coefficient is tested by using a test statistic that can be shown to have a t-distribution. We can therefore find critical values for various levels of significance and degrees of freedom. Throughout we assume that the sample is random.

The significance probability can be found as described in Notebox 31.1. Here, however, we will be concerned with testing hypotheses about the correlation coefficient. Our null hypothesis is that the sample comes from a population in which the correlation, indicated by ρ, is zero.

If we are asking whether a calculated coefficient is significant, then our interest is in whether it differs from zero by more than is likely to arise through an accident of sampling, so a two-tailed test is implied.

If we are asking whether it shows a significant positive association between the two variables, then we ask whether it is significantly bigger than zero, so a one-tailed test arises. A simlar comment is true if we are testing for a significant negative correlation.

◖ **Example 31.1** illustrates the problem, with $n = 18$ and $r = 0.35$. There are two questions:

(i) Does the sample result imply there is a positive correlation in the population of all students?

(ii) Is the sample result compatible with the statement that there is no relationship at all between these variables at a population level?

In both cases we begin by stating the null hypothesis that the correlation coefficient ρ for the population is zero, $H_0: \rho = 0$.

In case (i) the alternative hypothesis is $H_1: \rho > 0$, implying a one-tailed test. We adopt a 5% significance level and use the test statistic

$$t = r\sqrt{(n-2)/(1-r^2)}$$ [31.1]

which has the t-distribution with $n - 2$ degrees of freedom.

The critical value of t at 5% (one-tailed) for 16 degrees of freedom is 1.746. The calculated value of t is

$$0.35\sqrt{(18-2)/(1-0.35^2)} = 0.35 \times (4/0.9367) = 1.495$$

This is less than the critical value of 1.746, so we do not reject the null hypothesis of no correlation in favour of the alternative hypothesis that the population correlation coefficient is positive. A coefficient of +0.35 has more than a 5%

◖

chance of occurring in a random sample of 18 drawn from a large population in which there is no correlation. The welfare officer is not supported.

In case (ii) the alternative hypothesis is $H_1: \rho \neq 0$. This implies a two-tailed test. The 5% two-tailed significance critical level of t with $18 - 2 = 16$ degrees of freedom is 2.120.

Once again the calculated value is less than the critical value, so we do not reject the null hypothesis (of no correlation) in favour of the alternative hypothesis that there is a non-zero correlation in the population. We have no grounds for rejecting the bar manager's assertion that there is no relation at all.

Box 31.1

How to ... **test the significance of a product-moment sample correlation coefficient**

1. Set up a null hypothesis that a sample of size n comes from a population whose coefficient ρ is zero.

2. State the alternative hypothesis, usually one of these two:
   (a) $\rho \neq 0$ (two-tailed)
   (b) $\rho > 0$ or $\rho < 0$ (both one-tailed)

3. Take the number of degrees of freedom $v = n - 2$, and choose the level of significance α.

4. From the t-table, using the row for the correct value of v, find the critical value for t corresponding to α (if one-tailed) or to $\alpha/2$ (if two-tailed). Call this value t_c.

5. Calculate the test-statistic

   $$t = r\sqrt{(n - 2)/(1 - r^2)}$$

   with $n - 2$ degrees of freedom, where r is the sample coefficient.

6. Reject the null hypothesis if the numerical value of t exceeds the critical value.

Consolidation exercise 31A

1. What is meant by a significant correlation? How does it differ from a significant negative correlation?

2. What test statistic is used to determine whether there is a significant correlation? How many degrees of freedom does the test statistic have?

3. How many tails should be used in testing (a) whether a coefficient of 0.3 is significant and (b) whether a coefficient of -0.4 is significantly negative?

31.3 **How big must the sample be if we want a correlation of 0.50 to be significant at 5%?**

 The answer to **Example 31.2** is obtained most easily by using trial and error. First we must interpret the question carefully. We assume that we want a correlation of 0.5 and a given sign to be significantly greater (if the given sign is $+$) or significantly less (if it is $-$) than zero. This implies a one-tailed test.

With $r = 0.50$ and a sample size n, the test statistic given in Box 31.1 has the value

$$0.50\sqrt{n-2}/\sqrt{1-0.50^2} = 0.577\,35\sqrt{n-2}$$

For the correlation to be significant, this must exceed the 5% (one-tailed) critical value for $n - 2$ degrees of freedom (d.f.). Thus we want the minimum n such that

$$0.577\,35\sqrt{n-2} > t_{0.05} \quad \text{for} \quad n-2 \text{ d.f.}$$

The procedure now takes longer to explain than it takes to perform. First we try an arbitrarily chosen size, say 18; $n = 18$, gives d.f. $= 16$, so $t_{0.05} = 1.75$. Then we have $0.577\,35\sqrt{18-2} = 2.309$, which exceeds the critical value. Thus $n = 18$ is high enough to ensure the coefficient of 0.5 is significant, but there may be a lower size of sample that will also do it.

Try $n = 11$, hence d.f. $= 9$; this gives $t_{0.05} = 1.83$. Now $0.577\,35\sqrt{n-2} = 1.732\,05$, so the calculated value is less than the critical value. We cannot say that the coefficient is significant at the 5% level. A sample of 11 is too small.

Going on in this way we find that a sample of 12 is the smallest sample that will allow a coefficient of 0.5 to be significant at 5%.

31.4 **Is the correlation coefficient more than 0.3?**

Sometimes we want to know whether a sample correlation coefficient differs significantly from a known or hypothesised population value. We may be testing a hypothesis that a sample correlation is so high it is unlikely the sample comes from a population in which the correlation is 0.3, as in Example 31.3. We are not talking about whether two sample values differ; that comes later.

First we resort to a little mathematical sleight of hand. We begin this by using the value of the coefficient to calculate the value of something else. Then we compare this with what it ought to be if the hypothesis is right. If it is significantly different from what it ought to be, then it means that the sample correlation coefficient that gave rise to it is also significantly different from what it ought to be, and so we suspect the hypothesis.

This trick is very easy to perform, but at first sight the formula is a little off-putting. We work through an example which shows how easy it is, and then summarise the procedure.

In **Example 31.3**, $n = 28$ and the calculated correlation coefficient is $r = 0.6$. We want to test the hypothesis that the sample came from a population in which the correlation coefficient is 0.3. Use ρ to denote the correlation coefficient of the population from which the sample comes. Then our null hypothesis is $H_0: \rho - 0.3 = 0$.

We take a one-tailed alternative hypothesis $H_1: \rho - 0.3 > 0$ and choose a 5% level of significance.

We now use the sample result $r = 0.6$ to calculate the value of a statistic usually denoted by Z. Do not be put off by the appearance of the basic formula given below. The notation is described in in Notebox 31.2. The procedure for finding Z is given in Box 31.2, which shows that finding Z is very easy if you have a calculator and it is almost as easy to find it by using logarithm tables.

The basic formula is

$$Z = \tfrac{1}{2} \ln \left(\frac{1 + r}{1 - r} \right)$$

[31.2]

which can also be written as

$$Z = 1.1513 \log \left(\frac{1 + r}{1 - r} \right)$$

where ln and log are explained in Notebox 31.2.

Box 31.2

How to ... find **Z** when testing the significance of correlation coefficients

If your calculator has a [ln] button

1. Divide $(1 + r)$ by $(1 - r)$.
2. Press [ln] with the answer to step 1 still on the screen.
3. Press [=].
4. Multiply by 0.5; the answer is Z.

If your calculator has a [log] button

1. Divide $(1 + r)$ by $(1 - r)$.
2. Press [log] with the answer to step 1 still on the screen.
3. Press [=].
4. Multiply by 1.1513; the answer is Z.

There are other methods but this procedure is probably the easiest.

Continuing **Example 31.3**, with $r = 0.6$ we have

$$(1 + 0.6)/(1.06) = 1.6/0.4 = 4.0$$

so $\quad Z = \frac{1}{2}\ln\left(\dfrac{1 + 0.6}{1 - 0.6}\right) = \frac{1}{2}\ln 4.00 = 0.6931$

or $Z = 1.1513\log 4.0 = 1.1513 \times 0.6021 = 0.6931$.

We now have to compare this value of $Z = 0.6931$, derived from our sample coefficient, with the value Z would have in a population with a correlation coefficient of 0.3. This is

$$\frac{1}{2}\ln\left(\dfrac{1 + 0.3}{1 - 0.3}\right) = \frac{1}{2}\ln 1.8571 = 0.3095$$

Obviously $Z = 0.6931$ (based on $r = 0.6$) is different from $Z = 0.3095$ (based on $r = 0.3$). What we have to find out is whether the difference between them $(0.6931 - 0.3095)$ is significant at the 5% level.

It can be shown that if the difference between the sample Z and the population Z (hypothesised or actual) is multiplied by $\sqrt{n - 3}$, the resulting statistic (denoted by z) has a standard normal distribution with a mean value of 0 and unit variance. This allows us to test significance by using the area under the normal curve.

 Following this procedure in **Example 31.3**, we get

$$z = (0.6981 - 0.3095) \times \sqrt{28 - 3} = 1.943$$

At a 5% level of significance with a one-tailed test, the null hypothesis is to be rejected if $z > 1.64$. Since z is significant, so is r. It is bigger than we would expect by chance if the sample came from a population with a coefficient of 0.3. We reject the null hypothesis at the 5% level. Lady Agatha is right.

The procedure just used is summarised in Box 31.3 below.

Box 31.3

How to ... **compare a product-moment correlation r with a hypothesised non-zero value ρ_0**

1. Formulate the null hypothesis that the sample of size n comes from a population with a coefficient ρ which is equal to the hypothesised coefficient ρ_0, $H_0\colon \rho - \rho_0 = 0$. And formulate a one-sided alternative hypothesis $H_1\colon \rho - \rho_0 > 0$.

2. Adopt a level of significance and find the critical value of z_c that cuts off the corresponding percentage of the area under the normal curve.

3. After referring to Box 31.2 if necessary, calculate

$$Z = \tfrac{1}{2}\ln\left(\frac{1+r}{1-r}\right) \quad \text{or} \quad 1.1513\log\left(\frac{1+r}{1-r}\right)$$

4. Calculate

$$Z_0 = \tfrac{1}{2}\ln\left(\frac{1+\rho_0}{1-\rho_0}\right)$$

5. Multiply $Z - Z_0$ by $\sqrt{n-3}$ to obtain z. Compare z with the critical value z_c found in step 2.

6. Reject the null hypothesis if $z > z_c$.

Consolidation exercise 31B

1. The test statistic for the significance of a correlation coefficient involves the size of the coefficient and the size of the sample. What is it? How is it distributed?

2. How many degrees of freedom are used in testing the significance of a correlation coefficient?

3. To test whether a sample correlation is different from an actual or hypothesised population correlation requires a sleight of hand; what is it? What formula is used?

4. In conducting the test, we form the product of the difference between the values and an expression involving the sample size. What is this expression? How is the product distributed?

31.5 Comparison of two correlation coefficients

Another frequent question can be answered by a similar test. One sample of size n_1 produces a correlation of r_1 while another sample of size n_2 yields a correlation of r_2. Are these coefficients so different that it is unlikely the two samples come from the same population?

It can be shown that if the two samples come from the same parent population, the difference between values of Z calculated for each sample is normally distributed about a zero mean with a standard deviation of

$$\sqrt{\left[\frac{1}{n_1 - 3} + \frac{1}{n_2 - 3}\right]}$$

Thus, if we divide the difference in the Zs by this quantity, the resulting z can be compared with the appropriate area under the standard normal curve.

We illustrate this by looking at Example 31.4. In Notebox 31.3 we make an important comment on this kind of question.

We have two correlation coefficients: $r_1 = 0.6$ from a sample of $n_1 = 28$, and $r_2 = 0.5$ from a sample of $n_2 = 39$. Is the difference significant?

We choose a 5% level of significance. As we are asking about a difference, regardless of sign, we need a two-tailed test of the hypothesis that if many pairs of samples are taken, the mean value of Z_1 equals the mean value of Z_2. The normal table gives the critical value of z for 5% as 1.96, and we reject the null hypothesis involving Z if $z > 1.96$ or $z < -1.96$.

First we use Box 31.2 to calculate

$$Z_1 = \tfrac{1}{2} \ln\left(\frac{1+0.6}{1-0.6}\right) = \tfrac{1}{2} \ln 4.00 = 0.6931$$

$$Z_2 = \tfrac{1}{2} \ln\left(\frac{1+0.5}{1-0.5}\right) = \tfrac{1}{2} \ln 3.00 = 0.5493$$

This gives us $Z_1 - Z_2 = 0.1438$.

The expression for the standard deviation gives

$$\sqrt{\left[\frac{1}{28-3} + \frac{1}{39-3}\right]} = 0.2603$$

Dividing $Z_1 - Z_2$ by this gives

$$z = 0.1438/0.2603 = 0.5525$$

This is much less than 1.96, so the values of Z are not significantly different. It follows that the values of r are not significantly different. Whether even a significant difference could have been used to justify Lady Agatha's claim is debatable. But the debate would be fruitless, as the difference is not significant; see Notebox 31.3.

Box 31.4

How to ... **compare two correlation coefficients r_1 and r_2**

1. Use the expression in Box 31.3, step 1 to calculate Z_1 and Z_2 corresponding to r_1 and r_2.

2. The null hypothesis is that the population coefficients ρ_1 and ρ_2 are equal, i.e. H_0: $\rho_1 - \rho_2 = 0$.

3. The alternative hypothesis H_1 is usually one of these two:
   (a) $\rho_1 - \rho_2 \neq 0$ (two-tailed)
   (b) $\rho_1 - \rho_2 > 0$ or $\rho_1 - \rho_2 < 0$ (both one-tailed)

4. Choose a level of significance and read the critical value from the normal table (e.g. $z_c = 1.96$ for 95% with a two-tailed test). ◯

5. Calculate the test statistic

$$z = \frac{Z_1 - Z_2}{\sigma} \qquad\qquad [31.3]$$

where

$$\sigma = \sqrt{\left[\frac{1}{n_1 - 3} + \frac{1}{n_2 - 3}\right]} \qquad\qquad [31.4]$$

6. Reject the null hypothesis if $|z| > z_c$. The sign makes no difference in the two-tailed test, but remember that the critical value significant at 5% for a one-tailed test is significant at 10% for the two-tailed test. If you are applying a one-tailed test, then for rejection of the null hypothesis using H_1: $\rho_1 - \rho_2 > 0$, the value of z should exceed the positive value of z_c; for rejection of the null hypothesis using H_1: $\rho_1 - \rho_2 < 0$, the value of z should be more negative than the negative value of z_c.

31.6   Confidence interval for *r*

We can also obtain a confidence interval for the correlation coefficient. We use the fact that the statistic Z defined by equation [31.2] is approximately normally distributed with a standard error of $1/\sqrt{n-3}$. The procedure, summarised in Box 31.5, involves finding a confidence interval for Z, then converting it into a confidence interval for *r*.

Box 31.5

How to ...   **find a 95% confidence interval for a correlation coefficient, *r***

1. Referring if necessary to Box 31.3, calculate Z from

$$Z = \tfrac{1}{2}\ln\left(\frac{1+r}{1-r}\right) = 1.1513 \log\left(\frac{1+r}{1-r}\right)$$

2. Calculate σ_Z, the standard error of Z, from $1/\sqrt{n-3}$.

3. Obtain the confidence limits for Z as $Z \pm 1.96\sigma_Z$. Denote the lower limit by z_1 and the upper limit by z_2.

4. Calculate the two values of *r* corresponding to these values of Z. This is done most easily from $r = (e^{2z} - 1)/(e^{2z} + 1)$ by proceeding as below. The constant e is discussed in Notebox 31.2.

  (i) Double z_1 to get $2z_1$.

  (ii) Use your calculator to get e^{2z} (with $z = z_1$). This may be done by using either the $[e^x]$ button with $x = 2z_1$, or the $[y^x]$ button with $y = 2.71828$ and $x = 2z_1$. ⬤

> **5.** The lower limit of the confidence interval for r is given by $(e^{2z} - 1)/(e^{2z} + 1)$.
>
> **6.** Repeat (i) and (ii) with $z = z_2$ to obtain the upper limit of the confidence interval for r.

Example 31.5. illustrates the procedure of Box 31.5. Here $n = 19$ and $r = 0.6$, and we are seeking a 95% confidence interval.

1. $Z = \frac{1}{2}\ln(1.6/0.4) = \frac{1}{2}\ln 4 = 0.6931$

2. The statistic z is normally distributed about Z with a standard error of $\sigma_z = 1/\sqrt{n-3} = 1/4 = 0.25$.

3. The confidence limits for Z are $0.6931 \pm 1.96 \times 0.25$.

4. This gives a lower limit $z_1 = 0.6931 - 0.4900 = 0.2031$, and an upper limit $z_2 = 0.6931 + 0.4900 = 1.1831$.

5. Corresponding to z_1 we have a lower limit on the confidence interval for r of

    $$(e^{0.4062} - 1)/(e^{0.4062} + 1) = 0.501/2.501 = 0.200$$

    and an upper limit corresponding to z_2 of

    $$(e^{1.1831} - 1)/(e^{1.1831} + 1) = 2.2645/4.2645 = 0.531$$

6. The required 95% confidence interval for r is therefore $(0.200, 0.531)$.

31.7 Spurious correlation

The existence of a significant correlation is not proof of a causal relationship. One possibility is that two variables have a high (positive or negative) correlation because each of them is related to a common third variable rather than to the other. For example, suppose a survey showed a high correlation between people's masses and the numbers of rooms in their houses: heavier people have bigger houses. Before searching for a direct causal relationship, we might speculate that there may be an age factor. Do older people tend to be heavier than young people? Do older people tend to have bigger houses? If the answers to both questions were affirmative, we would expect the correlation we have observed. Alternatively there may be some other common factor, such as income. If they are time series they may both be displaying a growth factor due to technical, demographic or other changes.

Many international comparisons are bedevilled by this kind of error. Suppose we have data for many countries about the numbers of cars sold and the consumption of salt. We do a correlation analysis and find a significant positive correlation. A question that arises immediately is how much of this correlation is due to the size of the population. Countries with large populations are likely to have more cars and to consume more salt than smaller countries.

There may also be instances of a sample showing a significant correlation that defies all explanation, despite considerable thought and search. Once in a while we must expect such correlations to arise. A correlation coefficient that is just significant at the 5% level is, by definition, likely to arise 5% of the time by pure accident.

31.8 Zero correlation

A zero value for the product-moment correlation coefficient may indicate a completely random scattering that cannot be said to lie even loosely about any straight line. However, such a scattering is more likely to produce a small non-zero value.

Getting a coefficient of zero would be remarkable, and would be regarded with great suspicion. There could be something wrong with the data or with the sampling process. A zero, or almost zero, value could also arise from the application of the product-moment correlation coefficient to data quite closely grouped around some curve rather than a straight line, such as a circle. This is one reason for drawing a scattergram first.

Consolidation exercise 31C

1. Suppose we are comparing two correlation coefficients. (a) What are we testing? (b) How many values of Z do we calculate? (c) If the samples are of sizes 12 and 28, what is the variance of the difference between the Zs?

2. In finding a confidence interval for a correlation coefficient, it is necessary first to find a confidence interval for something else; what is that something? What is the formula for converting the other confidence interval into a confidence interval for r?

3. Does a significant correlation prove the existence of a link?

Notebox 31.1

Significance probability of a correlation coefficient

In Example 31.1 we can approximate the significance probability of the correlation coefficient by seeking 1.495 in the body of Table D5 for 16 degrees of freedom. For the one-tailed test, the critical value at 10% is 1.337 and at 5% it is 1.746. The value of 1.495 lies between them, so the significance probability of this correlation coefficient is between 5% and 10%. There is a chance of between 5% and 10% of a coefficient as high as this arising when the population shows no correlation.

> **Notebox 31.2**

e, log and ln

The mathematical constant e is a bit like π

Just as $\pi = 3.141\,592\,6536\ldots$, which is just the beginning of an expression that goes on for ever, so $e = 2.718\,281\,8246\ldots$, which also goes on for ever. One way to define it is the sum of the following series:

$$1 + \frac{1}{1!} + \frac{1}{2!} + \frac{1}{3!} + \frac{1}{4!} + \frac{1}{5!} + \frac{1}{6!} + \frac{1}{7!} + \frac{1}{8!} + \ldots$$

This value occurs in the equation for the normal distribution and in many other statistical and mathematical formulae.

Expressions involving log are closely related to expressions involving powers

For example, $3^2 = 9$ relates 9 to 3. The same relationship may be expresed as $\log_3 9 = 2$, which is read as 'log to the base 3 of 9 is 2'.

To a mathematician both expressions mean the same thing: you can get 9 by raising 3 to the power of 2. The two columns below contain further examples.

$3^2 = 9$	$\log_3 9 = 2$
$3^3 = 27$	$\log_3 27 = 3$
$3^4 = 81$	$\log_3 81 = 4$
$3^5 = 243$	$\log_3 243 = 5$
$3^6 = 729$	$\log_3 729 = 6$

It may help to think of them as columns of phrases in English and the equivalent phrase in some other language. You can learn the equivalence of the phrases without learning the whole language. After studying the first few entries, you should be able to complete the second column by looking at the first.

These expressions all involve powers of 3, or $\log_3$. We could have had powers of any other number, which would have led to log with a different subscript. For example, powers of 10 lead to expressions with $\log_{10}$, such as $\log_{10} 100 = 2$.

Before the widespread use of calculators, tables of values of $\log_{10}$ were used to speed up multiplication and division. These tables are still widely used and readily available. Usually the base 10 is omitted. Many calculators have buttons that give the value of $\log_{10}$ for any number. For example, pressing the button [log] followed by 3 will give $\log_{10} 3.000 = 0.4771$. An equivalent way of writing this is $10^{0.4771} = 3.000$.

More advanced mathematics often uses logs to the base e

These relate to powers of e, which are very important in a wide range of problems. Instead of writing $\log_e$ we usually write ln, which is shorthand for log-normal and is pronounced 'lon'.

Calculators also have buttons marked [ln]. If you press [ln] then 3.000, you will get $\ln 3 = 1.0986$, which you can think of as $\log_e 3 = 1.0986$. This is equivalent to $e^{1.0986} = 3$.

Notebox 31.3

What inference can be drawn?

The question in Example 31.4 is whether the difference between the two correlation coefficents supports an assertion about adventurousness in tastes.

Statisticians examine whether the difference between the coefficients is significant. If it is significant, the interpretation of this finding should be performed by behavioural experts (who, in the manner of experts, are likely to differ). All Lady Agatha's statistician can say is that, with a specified level of probability, the difference is greater than can be explained by chance.

However, if the difference is non-significant, the statistician has to say that it is easily explained as a chance effect, and so does not support any statement based on the existence of a difference. Thus it is possible that Lady Agatha will be contradicted by the statistician, but she cannot hope for support for that particular assertion. Even if the difference is significant, it could be for reasons that have nothing to do with adventurousness of taste.

Summary

1. Even if there is no correlation in the population, a sample may suggest that a correlation exists. Whether it is so big that it is unlikely to have arisen by chance is answered by considering its significance. It depends on the size of the sample.

2. Whether the observed sample coefficient differs significantly from zero is tested by calculating

   $$t = r\sqrt{(n-2)/(1-r^2)}$$

   which has a t-distribution with $n-2$ degrees of freedom. The test can be one-tailed or two-tailed and is summarised in Box 31.1.

3. The same test statistic allows us to find the sample size that will allow a coefficient of stated size to be significant at a stated level, as described in Section 31.3.

4. We can also use it to test whether the coefficient is significantly different from a hypothesised value. The procedure is summarised in Box 31.3, and uses a normally distributed test statistic Z. This looks a little off-putting but is easily calculated with the help of Box 31.2.

5. The same test statistic is used to compare two correlation coefficients as in Box 31.4 and to find a confidence interval for the coefficient as in Box 31.5.

6. Even when a coefficient is significant, it can still be spurious, as explained in Section 31.7. A zero or very small coefficient has to be suspected.

7. Noteboxes deal with the significance probability of a correlation coefficient, e and the log and ln notation, and making inferences from a correlation coefficient.

32

A significant regression

32.1 Introduction

Example 32.1 A random sample of 17 car owners from Statingham leads to Table 32.1 showing average annual expenditure on routine car servicing (x) over the last few years and the amount spent so far this year on car repairs (y).

(i) Calculate the regression equation of y on x, first confirming from a scattergram that it is appropriate to do so.
(ii) Is the regression coefficient significantly greater than zero? What is a 95% confidence interval for it?
(iii) Use the regression equation to estimate the repair expenditure so far this year for a Statingham-owned car on which servicing costs have averaged £220. What is a 95% confidence interval for this?

Table 32.1

x (£)	80	250	40	210	270	140	150	60	300	330	100	290	230	210	180	120	130
y (£)	360	150	400	235	200	230	250	350	100	85	310	90	190	175	180	270	310

Example 32.2 With the help of data obtained from the larger motoring organisations, the National Office for Statistics, and *Moore's Almanac*, the Association of Disgruntled Drivers has formed the opinion that, for the nation as a whole, the regression coefficient of current repair costs on recent average maintenance costs is -0.866. Is the coefficient from the Statingham survey significantly different? What does this mean?

These questions ask about estimates made on the basis of a sample regression equation. We consider why there is so much emphasis on the regression coefficient b; how to test whether it is significantly different from zero, or from some other hypothesised quantity; how to find a confidence interval for the regression coefficient; and how to find confidence intervals for estimates based on regression equations. The chapter ends with a few cautionary reminders.

32.2   Is the regression coefficient significant?

The data in **Example 32.1** are scaled-up versions of the data in Example 14.1. Thus there are parallels between the calculations, which provide an excuse for not redrawing the scattergram] and can be used to check the arithmetic. Calculations of the kind described in Chapter 14 produce the following sums (quoted to several significant figures as some of the answers depend on differences, squares, and such like that may give inaccurate final answers if there is too much rounding at this stage): $\sum x = 3090$, $\sum y = 3885$, $\sum x^2 = 686\,900$, $\sum xy = 577\,050$ and $\sum y^2 = 1\,034\,775$. These sums lead to

$$\bar{x} = 181.765 \qquad \bar{y} = 228.529 \qquad a = 415.894 \qquad b = -1.0308$$

Thus we have a sample regression equation of

$$\hat{y} = a + bx = 415.894 - 1.0308x \qquad [32.1]$$

To anticipate part (iii) of the question, this estimates the current repair expenditure on a car that has incurred servicing costs averaging £220 annually as

$$415.894 - 1.0308 \times 220 = £189.12$$

We consider the reliability of this estimate later, but first we need to look more carefully at the value of b. Our estimate depends on both a and b, and any inaccuracy in either will lead to inaccuracy in the estimate. However, the coefficient b also has another use. It is an estimate of how much y will change if x changes by a unit amount. Equation [32.1] tells us that if we increase the average annual amount spent on servicing by £100, we can expect the repair bill to be reduced by $1.0308 \times £100 = £103.08$. Often it is this change in y associated with a change in x that is of interest, rather than the actual value of y. This is one reason why statisticians are more interested in the accuracy of b than in the accuracy of a.

Another reason is that the value of a is defined to be the value that ensures a straight line with a slope given by the calculated value of b goes through the point $(\bar{x}, \bar{y})$. Thus, if the sample means are exactly correct estimates of the population means, any discrepancy between the sample value of a and the population value will arise only from errors in the value of b. In fact, the sample means are unlikely to be exactly correct estimates, so it is sometimes necessary to test the value of a, as well as of b; but we shall not do so in this book.

We turn to the significance of b. Is there a significant chance of a sample of 17 producing a regression coefficient of -1.0308 when there is no regression in the population from which the sample comes? If there is, then the value of b is not signicantly different from zero, and the regression equation is of no use in estimating y from x, since it would then imply a constant value for y.

To examine the significance of the estimated regression slope b, we use the test statistic introduced below, which has a t-distribution with $n - 2$ degrees of freedom. It can be written in several ways, all giving the same answer.

Version 1

Written in its simplest terms, this involves the sums of (a) the differences between every value of x and the mean $\bar{x}$, and (b) the differences between every value of y and the corresponding estimated value $\hat{y}$. These sums can be derived quickly from results already used in the calculation of the regression coefficient, as in equation [32.1]. The basic formula is

$$t = \left(b\sqrt{n-2}\right)\sqrt{\frac{\sum(x-\bar{x})^2}{\sum(y-\hat{y})^2}}$$ [32.2a]

In this equation the two sums can be evaluated from

$$\sum(x-\bar{x})^2 = \sum x^2 - n\bar{x}^2$$
$$\sum(y-\hat{y})^2 = \sum y^2 - a\sum y - b\sum xy$$

Example 32.1 provides an illustration. We set up the null hypothesis that the regression coefficient for the population is zero. Thus we ask whether b is significantly *different from* 0, implying a two-tailed test.

As $n = 17$ the test statistic has a t-distribution with 15 degrees of freedom. The 5% significance level for a two-tailed test is therefore 2.13, which becomes our critical value of t in all three versions of the test.

We use the sums already listed in the derivation of equation [32.1]. Inserting them in the above formulae, we have

$$\sum(x-\bar{x})^2 = \sum x^2 - n\bar{x}^2 = 686\,900 - 17(181.76)^2 = 125\,276.1$$

and, using $a = 415.894$, $b = -1.0308$, we obtain

$$\sum(y-\hat{y})^2 = \sum y^2 - a\sum y - b\sum xy = 13\,849.95$$

The formula for t therefore gives

$$t = \left(-1.0308 \times \sqrt{15}\right)\sqrt{\frac{125\,276.1}{13\,849.95}} = -12.0 \text{ (to 1 d.p.)}$$

This is much larger than the 5% critical value of 2.13, so we reject the null hypothesis. In other words, there is less than a 5% chance of getting such a (numerically) high value of the regression coefficient in a sample of 17 if there is no regression in the population.

Version 2

Version 2 of the formula for t is

$$\left(b\sqrt{n-2}\right)\sqrt{\frac{S_{xx}}{S_{yy}-bS_{xy}}}$$ [32.2b]

where S_{xx}, S_{yy} and S_{xy} are as defined in Notebox 14.1; in this example they have values of 7367.4, 8643.61 and -7594.46 respectively. Check that the same value of t is obtained.

Version 3

Version 3 looks very similar to version 2. It replaces bS_{xy} in the denominator by $b^2 S_{xx}$, giving

$$\left(b\sqrt{n-2}\right)\sqrt{\frac{S_{xx}}{S_{yy}-b^2 S_{xx}}} \qquad \text{[32.2c]}$$

The procedure is summarised in Box 32.1

Box 32.1

How to ... **test the significance of a regression coefficient b**

1. Set up a null hypothesis that the population has a value of $b = 0$.

2. State the alternative hypothesis, usually one of these two:
   (a) $b \neq 0$ (two-tailed)
   (b) $b > 0$ or $b < 0$ (both one-tailed)

3. Take the number of degrees of freedom $v = n - 2$ and state the level of significance α.

4. From the t-table, using the row for the correct value of v, find the critical value for t corresponding to α (if it is a one-tailed test) or $\alpha/2$ (if it is a two-tailed test). Call this critical value t_c.

5. Calculate the test statistic by using one of the following equations:

   (i) $t = \left(b\sqrt{n-2}\right)\sqrt{\dfrac{\sum(x-\bar{x})^2}{\sum(y-\hat{y})^2}}$

   where $\sum(x-\bar{x})^2 = \sum x^2 - n\bar{x}^2$

   $\sum(y-\hat{y})^2 = \sum y^2 - a\sum y - b\sum xy$

   (ii) $t = \left(b\sqrt{n-2}\right)\sqrt{\dfrac{S_{xx}}{S_{yy}-bS_{xy}}}$

   (iii) $t = \left(b\sqrt{n-2}\right)\sqrt{\dfrac{S_{xx}}{S_{yy}-b^2 S_{xx}}}$

   where S_{xx}, S_{yy} and S_{xy} are as defined in Notebox 14.1.

6. Reject the null hypothesis if the calculated value is numerically bigger than the critical value.

32.3 Is the regression coefficient significantly greater than β?

In Example 32.2 we ask whether the sample from Statingham has produced a result that suggests things are different locally from what they are nationally. The test of whether a sample result differs significantly from a given (actual or hypothesised) population value β is very similar to the test used in Section 32.2 to determine whether the sample value differed significantly from zero. It differs only in the statement of the hypotheses and the replacement of b by $b - \beta$ in the numerator of the expression for the test statistic, but *not* in the denominator.

⬭ **Example 32.2** illustrates this. We test whether the sample result of -1.0308 is significantly different from the population value of -0.866. We set up the null hypothesis $H_0: b - \beta = 0$ and the two-tailed alternative hypothesis $H_1: b - \beta \neq 0$.

We use as our test statistic

$$t = \sqrt{n-2}(b - \beta)\sqrt{\frac{\sum (x - \bar{x})^2}{\sum (y - \hat{y})^2}} \qquad [32.3]$$

or one of the other versions. This has a *t*-distribution with $n - 2$ degrees of freedom.

We decide to adopt a strict level of significance, rejecting the null hypothesis only if there is a chance of less than 1% of the population value being exceeded by so much if the null hypothesis is genuine. The 1% two-tailed critical value of t for 15 degrees of freedom is 2.947. This leads to the calculation

$$t = \sqrt{15}[-1.0308 - (-0.866)]\sqrt{\frac{125\,276.1}{13\,849.95}} = -1.9196$$

This is numerically less than the 1% critical level of 2.947, so we do not reject the null hypothesis. The evidence does not warrant rejection of the null hypothesis that the relationship between servicing expenditure and repair costs is the same in Statingham as nationally.

Box 32.2

How to ... **test whether a regression coefficient differs from a hypothesised value**

Proceed as in the test for significance of a regression coefficient in Box 32.1, but now use the test statistic

$$t = \sqrt{n-2}(b - \beta)\sqrt{\frac{\sum (x - \bar{x})^2}{\sum (y - \hat{y})^2}}$$

where β is the hypothesised value, or one of the other versions given in Box 32.1, similarly modified by replacing the first b by $(b - \beta)$.

> **Consolidation exercise 32A**
>
> 1. Give two reasons for being more interested in the significance of *b* than *a*.
> 2. How many degrees of freedom are there in the significance test for *b*?
> 3. The test statistic for *b* involves the ratio of two sums. What are they? Which one goes at the top of the ratio?
> 4. What is the null hypothesis in the test for the significance of *b*? Is it a two-tailed test or a one-tailed test?
> 5. How does the test for whether *b* differs significantly from a stated value β differ from the test for whether *b* is significant?

32.4 A confidence interval for the regression coefficient

Although a significance test enables us to assess whether there is really a statistical relationship between the two variables, it does not enable us to answer the question: In what range does the population regression coefficient probably lie? This is answered by a statement of the confidence interval for *b*.

The confidence interval has to have an associated probability, or level of confidence. Taking a two-sided 95% confidence interval as an example, it can be shown that the confidence interval for the regression coefficient of the population is

$$b \pm \left(t_c / \sqrt{n-2}\right) \sqrt{\frac{\sum (y - \hat{y})^2}{\sum (x - \bar{x})^2}} \qquad [32.4]$$

where *t* has the value appropriate to a two-sided 95% confidence interval with the appropriate number of degrees of freedom – 2.13 for 15 degrees of freedom.

In this expression *t* is multiplied by the inverse of the coefficient of *b* in the definition of the test statistic, and the two summation expressions are again given by

$$\sum (x - \bar{x})^2 = \sum x^2 - n\bar{x}^2$$
$$\sum (y - \hat{y})^2 = \sum y^2 - a\sum y - b\sum xy$$

⬭ In **Example 32.1** the calculated regression coefficient was -1.0308. A 95% confidence interval for this is obtained by making use of our earlier calculations, which contain the values of $\sum (x - \bar{x})^2$ and $\sum (y - \hat{y})^2$. We thus get a 95% confidence interval of

$$-1.0308 \pm \left(2.13 / \sqrt{15}\right) \sqrt{\frac{13\,849.95}{125\,276.1}} = -1.0308 \pm 0.1829$$

A 99% confidence interval can be similarly calculated with 2.95 in place of 2.13. This leads to a two-tailed confidence interval of -1.0308 ± 0.2533, which embraces the national value of 0.866.

How to ... **find a confidence interval for the estimated coefficient b**

1. Choose the level of significance you wish to use
2. Find the critical value of t_c for this level of significance, using $n-2$ degrees of freedom.
3. Calculate the interval

$$b \pm \left(t_c/\sqrt{n-2}\right)\sqrt{\frac{\sum(y-\hat{y})^2}{\sum(x-\bar{x})^2}}$$

32.5   Confidence intervals for regression estimates

In Section 32.2 we anticipated part (iii) of Example 32.1 and estimated the repair expenditure as £189.16 for a Statingham-owned car on which servicing costs have averaged £220. Now we consider how to find a 95% confidence interval for this estimate.

The estimate of £189.16 is the 'best' estimate of the *mean* repair expenditures on all Statingham-owned cars for which servicing costs have averaged £220 per annum.

We can calculate a confidence interval for this mean in basically the same way as for other means, but the formula is a little more complicated. It is not as bad as it looks, because most of the calculation of its various bits will already have been done in making the estimate.

The formula for a confidence interval for the best estimate of y corresponding to a particular value x_0 of x is

$$(a + bx_0) \pm t_c s \left[\frac{(x_0 - \bar{x})^2}{\sum(x - \bar{x})^2} + \frac{1}{n}\right]^{\frac{1}{2}} \qquad [32.5]$$

where $(a + bx_0)$ is the estimated value (189.12 in our example), t_c is the value of t for $(n-2)$ degrees of freedom and the selected level of confidence (2.13 for 15 d.f. and 5% in our example) and $s^2 = \sum(x - \bar{x})^2/(n-2)$

As we now see, performing this calculation is quite easy:

- Both the definition of s^2 and the term in the square bracket involve $\sum(x - \bar{x})^2$. We have already calculated the value of this (on page 363) as 125 276.1.
- To find s we divide 125 276.1 by 15 then take the square root, getting 91.388.
- We know the values of x_0 and $\bar{x}$, so we can calculate the term $(x_0 - \bar{x})^2$ as $(220 - 181.765)^2 = 1461.92$.

We therefore obtain the confidence interval

$$189.12 \pm (2.13)(91.388)\sqrt{\left[\frac{1461.92}{125\,276.1} + \frac{1}{17}\right]} = 189.12 \pm 51.68$$

Thus there is a 95% chance that the *mean* of the many values of repair costs associated with a servicing expenditure of £220 lies between 137.44 and 240.80.

This is a fairly wide interval. If there had been more terms in the formula, the interval would have been narrower for two reasons: (i) the term $\sum(x - \bar{x})^2$ would have been larger, thereby reducing the value of the square bracket, and (ii) the term $1/n$ would have been smaller, having the same effect. The effect on the value of *s* is uncertain.

Box 32.4

How to ... **find a confidence interval for a mean regression estimate**

1. We assume that $\sum x^2$ and $\bar{x}$ are known, and the regression equation has been calculated. If not, proceed as in Box 14.1. Thus the best estimate of *y* corresponding to a specified value x_0 is $a + bx_0$. There are *n* observations.

2. Specify the level of significance and use Table D5 to find the critical value t_c for this level with $n - 2$ degrees of freedom.

3. Calculate $\sum(x - \bar{x})^2$ given by $\sum x^2 - n\bar{x}^2$.

4. Use this to calculate $s = \sqrt{\sum(x - \bar{x})^2/(n - 2)}$

5. Calculate $(x_0 - \bar{x})^2$. Notice that no summation is involved here.

6. Insert the values of $(a + bx_0)$ (from step 1), t_c (from step 2), the results of steps 3, 4 and 5, and the value of *n* in the equation

$$(a + bx_0) \pm t_c s \sqrt{\left[\frac{(x_0 - \bar{x})^2}{\sum(x - \bar{x})^2} + \frac{1}{n}\right]} \qquad [32.5]$$

which gives the confidence interval for the mean of many estimates of *y* corresponding to a given x_0.

In seeking a 'confidence interval' for the repair costs for Lady Agatha's car, which is *one of the many* on which servicing had averaged £220, we have a slightly different problem.

We have been able to find a confidence interval for the *mean* cost of repairing cars with service costs of £220, because the mean is a parameter, and confidence intervals can always be defined and calculated for parameters. However, the cost of repairing Lady Agatha's car *is not a mean*; it is a *single value* that we are *predicting*. Here we use **predicting** in the special sense of statistical jargon to mean making a statement about a single value that either has not yet arisen or has not yet been notified to us. In such a case we call the equivalent to a confidence interval a **prediction interval**. It is an interval within which *a single value* will lie with a stated probability.

The formula for the prediction interval is exactly the same as for the confidence interval, except the square bracket is increased by the addition of 1, which can make a big difference. The formula is

$$(a + bx_0) \pm t_c s \sqrt{\left[\frac{(x_0 - \bar{x})^2}{\sum (x - \bar{x})^2} + \frac{1}{n} + 1 \right]}$$

which is just as before, except for the square bracket, which now comes to 1.07049 instead of 0.07049.

Recalling that $t_c = 2.13$ for the chosen level of significance and the appropriate degrees of freedom, and that $s = 91.388$ (from p. 367), this leads to 189.12 ± 201.40 which is so wide that it may seem to be useless. On the other hand, as this is a 5% two-tailed interval, we can interpret the upper bound of 390.52 as defining a single 2.5% tail, and argue there is a chance of only 2.5% that Lady Agatha's repair costs will exceed £390.52, which she may be glad to know.

Box 32.5

How to ... **find a prediction interval for a single regression prediction**

1. We assume that $\sum x^2$ and $\bar{x}$ are known, and that the regression equation has been calculated. If not, proceed as in Box 14.1. The single prediction of y corresponding to a specified value x_0 is $a + bx_0$. There are n observations.

2. to 5. Proceed as in steps 2 to 5 of Box 32.4.

6. Proceed as in step 6 of Box 32.4, but use the formula

$$(a + bx_0) \pm t_c s \sqrt{\left[\frac{(x_0 - \bar{x})^2}{\sum (x - \bar{x})^2} + \frac{1}{n} + 1 \right]}$$

which differs from the formula in Box 32.4 only by the inclusion of 1 in the square bracket. This gives the prediction interval for a single prediction of the y corresponding to x_0.

Consolidation exercise 32C

1. How many degrees of freedom are there in determining the value of t_c for use in the confidence interval for b?

2. The confidence interval for b involves the ratio of two sums; what are they? Which goes at the top of the ratio?

3. What are (a) a confidence interval for a mean regression estimate and (b) a prediction interval for a regression prediction? How do the formulae differ?

Assumptions behind the *t*-test of *b*

Take any value of x. According to regression analysis, the best estimate of the associated y is given by the point on the regression line corresponding to that value of x. It is $a + bx$.

But that is only the best estimate. In fact there are many values that the y associated with that x could have, some of them more than $a + bx$ and some less. We assume these values are normally distributed around the point $a + bx$ on the regression line, so that for every value of x, the 'errors' between the actual values of y and the value given by the regression equation are normally distributed. We also assume these distributions, one for every x, all have the same variance.

These assumptions allow us to use a t-test in various calculations involving the regression coefficient b.

Two cautions

Going too far

A regression equation should not be assumed to be valid for values of x lying outside the range of the observed values. There is no evidence that the linear relationship indicated by the scattergram and calculated by use of the least squares equations holds outside this range. In some cases it obviously cannot. For instance, in Example 32.1 equation [32.1] cannot hold for $x = 500$, which would imply negative spending on repairs. Careful consideration of the real relationship (rather than the observed statistical relationship) often indicates that the straight line is bound to bend for larger (or smaller) values of x.

Trouble at the origin

Particular care is necessary when values of x are near zero. If there are very low observed values, it may be found that if observations near the origin are omitted, the scattergam and calculations suggest a different linear relationship. It is not uncommon for there to be a structural break near the origin, and the procedures described in Chapter 13 should be observed.

Summary

1. The correctness of an estimate based on the regression equation depends on both a and b. If the means are correctly determined then any error in a will arise from errors in b, and this is what receives most attention. The regression coefficient b also measures the linear relationship between changes in the two variables.

2. Whether b differs significantly from zero is tested by using the test statistic of equation [32.2], which has the t-distribution with $n - 2$ degrees of freedom. Three versions are given and the procedure is summarised in Box 32.1. The test may be one-tailed or two-tailed.

3. A similar test of whether b is significantly different from a hypothesised β uses equation [32.3] and is summarised in Box 32.2.

4. The expression for the test statistic appears in the formula for a confidence interval for b, given by equation [32.4] and Box 32.3.

5. If we use the regression equation to 'predict' the value of y corresponding to a specific value x_0 of x, then the confidence interval for a single prediction, more properly called a prediction interval, is given by equation [32.6]. However, if we view the prediction based on the regression equation as 'best' in the sense that if x_0 occurs many times then the equation gives the best estimate of the mean of the corresponding ys, the confidence interval is narrower and is given by equation [32.5].

6. Noteboxes deal with the assumptions underlying the use of t-test on b and two cautions about using a regression equation.

Introducing non-parametric tests

Introduction

Example 33.1 Thirty-one cases of alleged driving offences were tried by the Statingham magistrates. The 11 acquittals (*A*) and 20 convictions (*C*) occurred in the following order:

AAAA CC A C A CCCCCCCCCC AAAA CCC A CCCC

Lady Agatha said the bunching of convictions and acquittals suggested that things were not as they should be. Perhaps the 31 people had not been put up in random order; perhaps the decisions were influenced by outside events or a wish to maintain a balance between verdicts. Do the data support this view?

Example 33.2 Pupils of two teachers, A and B, sit a public examination; their marks are given in Table 33.1. Do the marks support the hypothesis that one teacher is better than the other?

Table 33.1

A's pupils	35	62	73	47	58			
B's pupils	28	46	77	63	48	61	67	59

Example 33.3 In Statingham General Hospital 300 patients selected at random were monitored for the onset of minor illnesses during the third week of February. The numbers of visitors to each patient in the two preceding weeks was also noted. Does the summary of results in Table 33.2 suggest there was some relationship between number of visitors and onset of illness?

In most of the tests considered so far, we have used our knowledge of (or calculated) the mean or variance of a probability distribution. These tests depend on knowledge of a parameter, hence they are called parametric tests. They cannot be used unless we can calculate the relevant parameters. In the above examples (which we consider in later chapters) this is not possible, so we have to use non-parametric tests.

First we consider when to use non-parametric tests. This depends partly on the the level of measurement of the data, as described in Section 33.2 and especially in

Table 33.2

	Number of visitors			
	$\leqslant 8$	9–12	$\geqslant 13$	Total
Minor illness	42	30	38	110
No minor illness	108	50	32	190
Total	150	80	70	300

Notebox 33.1. Then we summarise the advantages and disadvantages of non-parametric tests. How to choose between tests involves looking at their power and efficiency, as explained in Notebox 33.4. Finally we indicate how we are going to deal with non-parametric tests in this book and how to integrate them with other aspects of statistics.

Most of the more technical material has been put into noteboxes, which should not be regarded as compulsory reading, but it will be useful to get some feeling for their content.

33.2 Levels of measurement and choice of test

All statistics provide some kind of measurement. In Chapters 5 and 15 we mentioned levels of measurement (or types of numerical information). We distinguish between four levels (or scales) summarised below and treated in more detail in Notebox 33.1.

- *Nominal or classificatory*. Symbols or numbers may be used simply to classify.
- *Ordinal or ranking*. Items can be placed in order according to magnitude, preference or another criterion.
- *Interval*. Items can be placed in order and the magnitudes of their differences measured (such as 29 °C, 26 °C, 19 °C).
- *Ratio*. There is an interval scale plus a true (non-arbitrary) zero (such as temperatures measured relative to absolute zero, or masses measured in grams).

Conventionally this list of levels of measurement is considered to be in ascending order, so that nominal is the lowest level and ratio the highest. High levels are described as **better** than low levels.

Whether a parametric test can be used depends on the level of measurement. The main points are made in Notebox 33.2. Briefly, non-parametric tests are particularly useful when we have nominal or ordinal data. However, almost all non-parametric tests can also be used with data on the interval scale. All statistical tests are valid on the ratio scale, provided other considerations (usually to do with the probability distribution) do not invalidate them.

Before applying a test, consider the level of measurement; then, if necessary, use Notebox 33.2 to help you to decide whether a parametric test can be used.

33.3 Advantages and disadvantages of non-parametric tests

Most non-parametric tests use very simple calculations. Sometimes the volume of simple calculation becomes off-putting with large samples; but then the assumptions for a parametric test are probably valid, and it is quicker to use that.

All non-parametric tests lead to probability statements that are exact, regardless of the population distribution. This means they are also useful for samples that are too small for the distributional assumptions of a parametric test to be valid. For example, strictly speaking, the t-test is valid for small samples only if the data come from a normal population. But the non-parametric tests are not restricted in this way. Moreover, some are valid for samples made up of observations from several different populations. Thus the non-parametric tests are useful over a much wider variety of data than the parametric tests.

Non-parametric tests have two main disadvantages. One is merely a matter of convenience. For small samples it is usually necessary to use special tables or formulae, which are sometimes less widely available or less easily summarised than parametric tables, such as tables of the normal distribution.

However, if the sample is large enough (and what this means varies from test to test and is mentioned in the appropriate place), it is often possible to express the result in terms of a binomial variable. Then we can test its significance by using an important result that links the non-parametric discrete binomial distribution to the parametric continuous normal distribution, as described in Chapter 28.

The other disadvantage is that non-parametric tests do not make full use of the available data and this usually makes them in one sense inferior to parametric tests. More specifically they are **less powerful** in the sense that they are less likely than some other test involving the same sample size to reject the null hypothesis when it is false. We consider this further in Notebox 33.4. Although it is a matter to be kept in mind when choosing between tests, it does not in any way complicate the understanding and use of any of the tests we are about to describe.

Consolidation exercise 33A

1. Name four levels of measurement. Explain how (a) the interval scale differs from the ordinal scale, and (b) the ratio scale differs from the interval scale.

2. Why are parametric tests so called? How do non-parametric tests differ from parametric tests?

3. With what levels of measurement can you usually use (a) all parametric tests and (b) most non-parametric tests? Why do we say 'usually'?

4. With what scale(s) may the geometric mean be used?

33.4 Non-parametric tests in the rest of this book

We consider only a few of the ever growing battery of non-parametric tests. We do so in a way that shows how they integrate with other parts of the subject.

Both parametric tests and non-parametric tests usually require data obtained from random samples, so in Chapter 34 we give two non-parametric tests of whether a sample is random. One is a simple extension of what we already know about binomial events, but it does not use knowledge of the parameters of the distribution, so it is a non-parametric test. The other is a runs test.

Whether a sample is likely to come from a stated population is often decided by using a parametric test of the sample mean. The question can also be answered by using a non-parametric test of the sample median; often simpler, sometimes this is all that can be done. We describe several tests of the median in Chapter 35. Some are designed for use with data for matched pairs, a term we explain later on.

A frequent question is whether there is any association between the values of two variables. The data may not be amenable to the product-moment correlation technique described in Chapter 14. It may involve non-numerical classifications, such as colour of hair and nationality. In Chapter 36 we consider how to handle such problems when the data are presented in a contingency table and can be analysed by use of the χ^2 (chi-squared) test. This is a highly important test to which we make several references in later chapters. It is easy to calculate. One use of the χ^2 test is in *before and after* tests, as described in Chapter 37. If the two variables can be ranked, their association can be examined by using non-parametric correlation techniques, as in Chapter 38.

In Chapter 39 we temporarily break away from non-parametric tests in order to describe some important probability distributions useful in the analysis of chance events and risk. We begin with the discrete uniform and geometric distributions. The continuous rectangular and triangular distributions are considered in Chapter 40.

Then we look in Chapter 41 at the Poisson distribution, which is particularly important in the analysis of random events and can be used (as explained in Chapter 42) to obtain good approximations to the binomial distribution. Under certain circumstances the normal distribution can be used to approximate the Poisson distribution. We note in Chapter 43 that intervals between random events have an exponential distribution, which we also discuss.

An important question in parametric statistics, especially when small samples are involved, is how well the data fit the distribution that underlies the theory of the test. However, there are no parametric tests of **goodness of fit**. Fortunately, there are several non-parametric tests, and we describe some of them, including tests of whether data have a random distribution. It is an area where non-parametric tests come to the rescue of the parametric. Chapter 44 describes how the χ^2 test is used to measure goodness of fit and gives a briefer account of another test of whether data fit a normal distribution.

Notebox 33.1

Levels of measurement

Classification and the nominal scale

The simplest, or weakest, level (or scale) of measurement is classification. Examples are credit card and telephone numbers, and the use of 1, 2, 3 to indicate adult male, adult female and child.

Numbers used in these ways are simply convenient labels that could equally well be letters or words. They are in the nominal or classificatory scale. They cannot meaningfully be added, multiplied or put in order of magnitude. Although they can be put in ascending numerical order, as a device for listing or reference, the order has no real meaning.

Ranking and the ordinal scales

A stronger, higher or better level of measurement exists when the classes can be meaningfully placed in order, as when students are awarded different classes of degree, or runners are placed first, second, third, etc. The classes may be ordered through the use of such words as 'bigger', 'better', 'richer' or any other comparative. Each class may contain several members, and there may be variety within the class. We do not say that all members of class 4 are as rich as each other. But all members of class 4 are richer than all members of class 3, and poorer than all members of class 5.

As with nominal numbers, ordinal numbers cannot meaningfully be added or multiplied, therefore many statistical techniques cannot be used with such data.

Interval scales

This is like the ordinal scale, with the additional property that the distances between any two numbers are of a known size. The unit of measurement and the zero point are arbitrary.

We cannot meaningfully add or multiply the measurements. A temperature of 32 °F added to itself will not produce a temperature of 64 °F. Nor is 64 °F twice as hot as 32 °F. However, a temperature *difference* of 12 °F (between say 40 °F and 52 °F) can be added to a temperature *difference* of 8 °F (between 52 °F and 60 °F) to yield a *difference* of 20 °F (between 40 °F and 60 °F). In an interval scale, the measurements cannot be subjected to the laws of ordinary arithmetic, but differences between any two measurements can be. This allows us to find the mean of two temperatures. A temperature of 46 °C is different from 0 °C by 46 °C. When we average 46 °C and 50 °C, we are averaging two differences.

Note that the information yielded by data on an interval scale will be unchanged if the arbitrary zero and unit of measurement are changed, as when we convert temperatures from Celsius to Fahrenheit. The statement that the temperature at noon yesterday was 10 °C tells us exactly the same as the statement that it was 50 °F. ⬭

Ratio scales

The highest level of measurement is the ratio scale. This is an interval scale with a true, non-arbitrary zero. Mass and absolute temperature in kelvins are examples: 20 g is twice as heavy as 10 g, and 10 g plus 20 g equals 30 g; 400 K is twice as hot as 200 K. All the laws of arithmetic are applicable. An object that has twice the mass of another object when measured in grams also has twice the mass when measured in pounds. The unit of measurement does not matter.

The arithmetic mean is valid only if the raw data can properly be added. This is not true of data measured on the nominal or ordinal scales; and with the interval scale it is valid only for differences. But it is always valid on the interval and ratio scales.

Notebox 33.2

Level of measurement and choice of test

If data exist only on the nominal or ordinal scale, parametric tests cannot be used. These data cannot be added or multiplied, and it is impossible to calculate parameters such as the mean and the variance. This rules out the use of parametric tests when the data exist only as ranks.

In principle, parametric tests can be used whenever the data are on the ratio scale. Occasionally a test embodies an assumption that is not valid for the data we have or for the distribution from which it comes. Non-parametric tests are also valid on the ratio scale.

All the common parametric tests can also be used when the data are on the interval scale, although some statistics, such as the geometric mean, cannot be meaningfully defined (Notebox 33.3).

Almost all non-parametric tests can also be used on the interval scale.

To sum up, non-parametric tests are particularly useful when we have data on the nominal scale or the ordinal scale. However, almost all non-parametric tests can also be used with data on the interval scale. All statistical tests are valid on the ratio scale, provided other (usually distributional) considerations do not invalidate them.

Notebox 33.3

The interval scale has no geometric mean

Consider two temperatures measured in degrees Celsius, 4 °C and 100 °C. Their equivalents in degrees Fahrenheit are 39.2 °F and 212 °F. The arithmetic means are 52 °C for the Celsius readings and 125.6 °F for the Fahrenheit readings. These arithmetic means are consistent with each other, as they satisfy the conversion formula $F = 32 + 1.8\,°C$ (where F is the Fahrenheit temperature and C is the Celsius temperature). 〰

However, the geometric mean of the two Celsius temperatures is 20 °C whereas the geometric mean of the two Fahrenheit temperatures is 91.16 °F. These temperatures do not satisfy $F = 32 + 1.8C$.

Both Celsius and Fahrenheit temperature scales have arbitrary origins, which means they are not ratio scales. However, the temperatures measured using these scales can be placed in order and the magnitudes of temperature differences within the scales can be measured. They are therefore interval scales. As we have seen, the geometric mean (and possibly some other statistics) cannot be meaningfully calculated.

Notebox 33.4

Power and efficiency

The power of a test is defined to be the probability of correctly rejecting the null hypothesis, i.e. of rejecting it when it is false. It can be shown that any given test has two properties:

(i) The power increases as the sample n increases.
(ii) A one-tailed test will be more powerful than a two-tailed test, and which of them is used depends on the nature of the alternative hypothesis.

If the data satisfy the assumptions for a parametric test, that test will almost always be more powerful than any non-parametric test.

Sometimes we wish to compare the power of a particular test (test B) with the power of the most powerful test (test A) that can be used on the data being examined. This is done by determining the power of test A when used with a sample of size n_A, and then finding the (larger) size of sample n_B that will enable test B to have the same power. The **power efficiency** of test B is defined to be

$(n_A/n_B) \times 100\%$

Summary

1. Whether the parametric tests described in earlier chapters can properly be used depends on the level of measurement of the data. Very roughly, it depends on whether you can meaningfully add the data. A brief explanation is given early in the chapter and fuller explanations are in Notebox 33.1.

2. The non-parametric tests described in later chapters may be used when parametric tests are inappropriate. Usually these non-parametric tests use very simple calculations, but their volume may become off-putting if the samples are large. If the samples are large, the assumptions underlying a parametric test may be valid, and a parametric test can be used more quickly. The non-parametric tests are useful over a wider variety of data than the parametric tests, especially for small samples.

3. When used with small samples, non-parametric tests usually require access to special tables. If the sample is large enough, it is often possible to express the result in terms of a binomial variable, allowing its significance to be tested.

4. Non-parametric tests do not make full use of the available data, so in this sense they are usually inferior to parametric tests (if parametric tests are valid). Non-parametric tests are less powerful in the sense that they are less likely than some other test involving the same sample size to reject the null hypothesis when it is false (Notebox 33.4).

5. Appropriate selection of non-parametric tests enables us (i) to test the randomness of a sample, (ii) to compare the origins of samples by comparing medians, and to test matched pairs of differences, (iii) to examine associations and rank correlations, (iv) to perform before and after tests and (v) to check on the goodness of fit between observed data and a hypothesised distribution, which is often necessary in the application of parametric tests.

6. Noteboxes give more detail about (1) levels of measurement, (2) the choice of test, (3) the existence of the geometric mean and (4) power and efficiency.

Two tests for random samples

34.1 Introduction

Example 34.1 A sample of 100 residents of Statingham for an opinion poll contained 40 women and 60 men. Given that 55% of the population of Statingham are female, do these figures suggest that it was not a random sample?

Example 34.2 In a second survey 30 people were selected at random by one interviewer. The record showed that the gender of the interviewees according to the order of interview was MMMFF MFFMF FFMMF MMMMF FMMFF FFFFM. Assuming the population contains just as many men as women, does the record support the view that the sample was random?

Example 34.3 In a third survey, by a different interviewer, the genders of the interviewees were listed in order of interview as MFMFM FMFMM FFMFM FMMFF MFMFF MFMFM. Making the same assumption as in Example 34.2, does the record support the view that the sample was random?

Many statistical enquiries and tests are based on the assumption that the sample is random. Too often this is taken for granted. We illustrate two ways of testing for randomness by using the above examples and Example 33.1. As we shall see, a sample may pass one test and fail another. We consider the implications of this. Later we meet other uses of these tests.

34.2 The binomial test

We saw in Chapters 18 and 19 that if a trial has possible outcomes S and F, occurring at random with a constant probability $P(S) = p$ and $P(F) = 1 - p = q$, then we know three things:

(1) The probability of getting r successes in n trials is given by $^nC_r p^r q^{n-r}$ (Box 18.1).
(2) The probability of getting r successes or fewer is the sum of the probabilities of getting $0, 1, 2, \ldots, r$ successes.
(3) If n is not too large, this sum may be obtained from tables of cumulative binomial probabilities (Box 18.2).

Chapter 28 showed that an approximation to the cumulative probability mentioned (item 3) may be obtained by using the normal distribution (Box 28.4).

The sample in **Example 34.1** was not taken in order to estimate the proportion of females in the population; this proportion is already known. We are interested in political opinions and that is the motivation for the sample survey. We know that the sample was taken from the population. Before using it to make statements about political opinions in the whole population, we want to test whether the sample is random.

If the sample is random and every member of it was chosen independently, the proportion of females should not be very different from 0.55, which is known to be the proportion of females in the population. Does the proportion of 0.40 found in the survey differ so much from 0.55 that we have to reject the hypothesis it was a random sample?

The answer depends on the level of significance we adopt. If we adopt 5% then we reject the hypothesis of randomness if a difference between the observed proportion p and 0.55 is so great that there is a chance of less than 5% of it being achieved. The test of randomness therefore involves calculating the probability of getting an observed proportion at least that much different from the observed proportion. It is a two-tailed test.

We expect to get 55 women in a sample of 100. In fact, we have only 40, which differs from 55 by 15. By using the binomial distribution, we can calculate the probability of getting a number so different from the expected number of 55 if the sampling has been random with a constant probability (of 0.55) of success. If we find that this probability is very low, we are faced with the choice of *either* accepting that a highly unlikely event has occurred *or* concluding that the sample has not been random from a population with $p = 0.55$.

We formulate the null hypothesis that the sample has been chosen in this way with $p = 0.55$, giving H_0: random with $p = 0.55$ and the two-tailed alternative hypothesis H_1: not random with $p = 0.55$.

One way to obtain the probability of being out by 15 or more is to calculate lots of separate probabilities and then total them (using (1) and (2) on p. 380). If we choose a 5% level of significance, we need to see whether the total probability of there being 40 women or fewer, or 70 (from $55 + 15$) or more is less than 5%. If so, H_0 is to be rejected.

To avoid such tedious calculation, we use Box 28.4 to approximate the binomial distribution B(100, 0.55) by using the normal distribution N(55, $100 \times 0.55 \times 0.45$), which is N(55, 24.75). We use a two-tailed test of the difference between the actual number of women and the expected number.

We need a continuity correction. Since 40 has to be read as 39.5–40.5, the difference from the mean is $55 - 40.5 = 14.5$. We are therefore interested in the tail areas under the normal curve N(55, 24.75) to the left of 40.5 and to the right of 69.5.

Denoting the number of women by the random variable X, we define a standardised variable

$$Z = \frac{X - 55}{\sqrt{24.75}} = \frac{X - 55}{4.9749}$$

We use Z as a test statistic, with a two-tailed critical level of 1.96. The probability that X has a value of 40.5 or less is

$$P(x \leqslant 40.5) = P\left(z \leqslant \frac{40.5 - 55}{4.9749}\right) = P(z \leqslant -2.915)$$

Thus the calculated value of Z far exceeds (numerically) the critical value of 1.96, so it is highly unlikely that the sample consisted of 100 people chosen at random, each with the same probability of being chosen, from a population in which the proportion of women was 0.55.

Notice that this does not necessarily suggest there was a deliberate attempt to underrepresent women in the sample. It could be simply that the sample was conducted by stopping people in places where men were more likely to be. What we are relying on is that if everybody had the same chance of being sampled, then as 55% of the population are female, we would expect the proportion of women in the sample not to be significantly different from 55%.

Notice also that if the proportion in the sample had not been significantly different from 0.55, this would not prove that the sample was random. It would simply fail to justify rejection of the hypothesis that it was. The surveyor could have deliberately selected 55 females and 45 males without any regard to randomness.

Box 34.1

How to ... use the binomial test of whether a sample consists of n randomly selected items all with the same known probability of being chosen

1. Specify the null hypothesis that the items have been randomly chosen all with a probability p_0 of success so that H_0: random with $p = p_0$.

2. The alternative hypothesis is H_1: not random with $p = p_0$.

3. Choose a level of significance.

4. Multiply p_0 by the sample size n to provide the expected number of successes np_0. The actual number of successes r differs from this.

5. If n is not too big for convenience of calculation, use Box 18.1 and aggregation, or Box 18.2, to find the total cumulative probability of r being so much less than or so much more than np_0. If this is less than the level of significance, reject the null hypothesis.

6. If n is too big for convenient use of Boxes 18.1 or 18.2, proceed as in Box 28.5, testing the significance of the difference between r and np_0 by using the normal approximation with mean np_0 and variance $np_0(1 - p_0)$. Use a continuity correction.

34.3   A one-sample runs test

We now consider Example 34.2. Here we have a sample of 30 people that contained 16 women. If the binomial test is applied by use of Box 34.1, we find that being 1 different from the expected number of 15 is not unlikely, so this piece of information does not warrant rejection of the hypothesis that the sampling was random.

However, we also have other information – the individuals were chosen for the sample in the listed order. Might this suggest it was not taken at random, with the choice of one person being influenced by what had already happened? Possibly there was a conscious or unconscious attempt to get the balance of the genders 'about right', so even though the data will pass a binomial test, there could still be bias in the sample.

To check on this we use the one-sample runs test. Suppose we record the successive outcomes of a multi-trial experiment as successes or failures. If the successes cluster more than we would expect if each outcome were truly independent of those that precede it, then we become suspicious.

To take an extreme case, if we had *SSSSSSSSSSSFFFFFFFFFFF* we would suspect the existence of some kind of bias in the experiment, a bias that suddenly changes halfway through. Equally, however, we would be suspicious of organised results if we had *SFSFSFSFSFSFSFSFSFSFSF*, where there is far less clustering than we might expect.

To put these hunches on a firmer basis, we apply a test that is valid for ranked or ordinal data. It assumes the result of each trial is independent of the results of all other trials, and that in each trial the probability of a success is the same as in repeated tossing of a (possibly biased) coin. We have to state a level of significance; here we take it as 5%.

The test is based on the concept of a **run**, which is defined as any sequence of one or more occurrences of the same event. For example, using the data of Example 33.1, in the sequence

AAAA CC A C A CCCCCCCCC AAAA CCC A CCCC

breaks have been introduced to indicate the end of one run and the beginning of another. There are 11 *A*s, 20 *C*s and 10 runs.

Suppose we arrange 11 *A*s, 20 *C*s in order at random many times. The number of runs will vary from one arrangement to the next. And suppose we can can find a number N_{min} such that we would expect there to be fewer than N_{min} runs only 2.5% of the time. Let there also be a number N_{max} that we would expect the number of runs to exceed only 2.5% of the time.

In other words, we would expect the number of runs to lie within the range $N_{min} - N_{max}$ 95% of the time. If it lies outside this range, either an unlikely event has occurred or at least one of the assumptions is wrong: perhaps the occurrences are not independent of each other or perhaps there is not a constant probability of event *A*.

The sizes of N_{min} and N_{max} will depend on the chosen level of significance and on the numbers of *A*s and *C*s. We shall suppose that there are n_1 occurrences of one kind (such as *A*) and n_2 of the other kind (such as *C*). Two cases arise.

Both n_1 and $n_2 \leqslant 20$

This means that neither A nor C occurs more than 20 times. In this case we can use tables to find N_{min} and N_{max}.

⟹ In **Example 34.4**, with the ordering *AAAA CC A C A CCCCCCCCCC AAAA CCC A CCCC*, we have $n_1 = 11$, $n_2 = 20$ and $r = 10$. The null hypothesis is that the number of runs arises out of a random process in which each event is independent of all others and there is a constant probability of success.

The tabulated 5% level of significance (Table D8) shows that the null hypothesis should be rejected if r lies outside the range 10–20. In our case the number of runs $n = 10$ is within this range, so it is neither small enough nor big enough to warrant rejection of the null hypothesis.

At least one of n_1 and n_2 exceeds 20

In this case we can use formulae given below to estimate the mean and standard error of the number of runs, then apply a test based on the normal distribution.

⟹ As an example, we suppose that an experiment produces a sequence with A occurring $n_1 = 16$ times and B occurring $n_2 = 24$ times, with these occurrences in such an order that there are $r = 9$ runs. Check that with these values the formulae in step 4 of Box 34.2 show that, if the assumptions are valid, the number of runs should be approximately normally distributed with mean $\mu_r = 20.2$ and a variance $\sigma_r^2 = 8.96$, giving $\sigma_r = 2.99$.

We now test whether the observed number of runs $r = 9$ is further from the mean of this normal distribution N(20.2, 8.96) than we would expect on the basis of our hypothesis. The test statistic is

$$z = \frac{r - \mu_r}{\sigma_r}$$

and the 5% (two-tailed) critical value is 1.96.

The calculated value of z has to allow for a continuity correction. The discrete value of 9 runs has to be interpreted as between 8.5 and 9.5. The value nearer to the expected value of 20.2 is 9.5, giving

$$z = \frac{9.5 - 20.2}{2.99}$$

which exceeds (numerically) the critical value of 1.96. We therefore reject the hypothesis. Either the results are not independent of each other, or there is not a constant probability of success for some other reason.

This test does not involve any knowledge of the population. It is simply a test of whether the observed numbers of occurrences of two kinds are likely to have arisen in the observed order if they occurred at random. If the test of randomness is passed, the proportion of occurrences of one kind could still be very different from the proportion in any population from which the sample was believed to have been taken.

Box 34.2

How to ... test the randomness of a sequence of events using the one-sample runs test

1. Specify the null hypothesis H_0: the events of the two different kinds occur in random order; and the alternative hypothesis H_1: the events do not occur in random order.

2. Choose a level of significance α (which will be 5% if Table D7 is to be used).

3. Denote the number of events of one kind by n_1 and the number of events of the other kind by n_2. Count the number of runs r.

4. If both n_1 and n_2 are under 21, consult Table D7. This shows the critical numbers of runs for various values of n_1 and n_2. A value of r outside the stated range is significant at 5%

5. If either n_1 or n_2 is greater than 20, a good approximation is given by a normal distribution with

$$\text{Mean } r = \mu_r = \frac{2n_1 n_2}{n_1 + n_2} + 1$$

$$\text{Variance} = \sigma_r^2 = \frac{2n_1 n_2(2n_1 n_2 - n_1 - n_2)}{(n_1 + n_2)^2(n_1 + n_2 - 1)}$$

6. Calculate the values in step 5, hence obtain the test statistic

$$z = \frac{r - \mu_r}{\sigma_r}$$

The significance of z can be assessed from the normal distribution table as in Box 26.1, using a continuity correction which reduces the numerator by 0.5.

The solution of Examples 34.2 and 34.3 is left an an exercise.

34.4 Contradictory results?

A sample may pass one of these tests and fail the other. When test results differ, the usual advice is to accept the verdict of the more powerful test. Here this advice is invalid because the two tests are not of the same thing, even though they both have implications for randomness.

If the runs test is failed, the sample is unlikely to be random; but passing the runs test says nothing about the relationship of the sample to any population.

If the binomial test is failed, the sample is unlikely to be random with a constant proportion of sucesses equal to the hypothesised proportion. The failure may be due to a lack of randomness, due to a varying proportion or due to a constant proportion that is different from the hypothesised portion. If the test is passed, the proportion of successes is acceptable, but a runs test is still desirable if we are to check on randomness.

For example, suppose we believe that we have a random sample from a population in which $p = 0.5$. A sequence such as $MFMFMFMFMFMFMF$ is clearly not random, even though it satisfies the binomial test for $p = 0.5$. Similarly, a random assortment of M and F with 10 Ms and 40 Fs would pass the runs test, but it would not be a random sample from a population with $p = 0.5$.

Consolidation exercise 34A

1. In what sense does can the binomial test be viewed as a test of randomness?
2. What is a run? What is its smallest size? What assumptions are made for the runs test, and what hypothesis is tested by it?
3. How should you choose between use of tables or a formula in the runs test?
4. A pack of 52 cards is shuffled and the cards put one at a time on the table, face upwards. The colour of the card is recorded as black (event B) or red (event R); this produces a sequence such as $BBB\ RR\ B\ RRRRR \ldots$. Why cannot the runs test be properly applied?

Summary

1. Two complementary tests for randomness of a sample are the binomial test and the one-sample runs test.

2. The binomial test asks whether the observed data are compatible (at a stated level of significance) with the hypothesis that the sample is a random sample in which every selection has the same stated probability of being a success. It may be failed if the probability differs sufficiently from the stated probability, or if the sampling is not random. Passing the test is not a guarantee of randomness. The procedure is summarised in Box 34.1.

3. The one-sample runs test concentrates on the order in which successes and failures occur. A run is defined as any sequence (even as small as 1) of occurrences of one kind. For a given number of successes and failures, the number of runs may be improbably small or improbably large for a random selection.

4. If no occurrence arises more than 20 times, tables can be used to determine whether the number of runs indicates rejection of the hypothesis of randomness at a stated level of significance. Otherwise formulae given in Box 34.2 (which summarises the general procedure) enable the normal distribution to be used as the basis of a test.

5. Both tests need to be interpreted carefully.

Non-parametric median tests

Introduction

▶ **Example 35.1** It is thought that Statingham magistrates allow their trials to be too long. Ten cases sampled from several hundred over the last few years took the following times (in minutes):

  64 78 116 83 87 89 98 93 123 44

Does this information support the hypothesis that the median duration of the trial was 89 minutes?

▶ **Example 35.2** Ten owners of two dogs were asked to take part in a test of the effect of a food additive. (It was considered that dogs of the same breed and sex and roughly the same age belonging to one owner were likely to be treated similarly and form acceptable matched pairs.)

One dog in each pair was randomly chosen for the additive (A). The other had an additive known to be a placebo (B). Later the dogs were examined and graded on a five-point scale according to their amount of tooth decay.

The results are tabulated in Table 35.1, where a low rating means little decay. Has the additive made a significant difference?

These questions, along with Example 33.2, illustrate the use of tests involving the median, which has a role in non-parametric tests similar to that of the mean in parametric tests.

The question asked in Example 35.1, concerning only one sample, can be answered by using a very simple **Signs Test**. A good approximation can be quickly obtained for larger samples. Another test based on a single sample is **Wilcoxon's Signed Rank Test**.

Table 35.1

Owner	a	b	c	d	e	f	g	h	j	k
Given A	1	4	3	4	4	3	5	2	4	1
Given B	3	2	5	5	3	3	2	5	4	3

Examples 35.2 and 33.2 use the median to compare two samples. Example 35.2 uses the **Sign Test for the Medians of Matched Pairs**. Another test applied to matched pairs is the **Wilcoxon Signed Rank (Matched Pairs) Test**, which we describe with the aid of a further example.

Example 33.2 demonstrates the **Mann–Whitney U-Test** to determine whether two independent samples have been drawn from the same population or from populations that have different medians.

35.2   Two sign tests for the median

The sign test

Example 35.1 involves testing whether the median of some sample data is significantly different from a hypothesised population median. The quickly conducted sign test requires measurements of a continuous variable, such as time or mass, but unlike the t-test it does not require a normal distribution.

We count how many sample items are higher than the hypothesised population median and how many are lower. The probability of getting these numbers can then be found from cumulative binomial probability tables or, for larger samples, by using a normal approximation. We consider both cases.

 In **Example 35.1** we are asked if data for the 10 trial durations support the hypothesis that over recent years the median duration of all trials was 89 minutes. As time is a continuous variable, we can use the test.

If we use $+$ and $-$ respectively to indicate values greater than and less than the hypothesised median of 89, and ignore any equalities, we can convert the information as follows:

64	78	116	83	87	89	98	93	123	44
$-$	$-$	$+$	$-$	$-$	(=)	$+$	$-$	$+$	$-$

There are three $+$ signs and six $-$ signs. If our hypothesis is correct, we would expect as many $+$ signs as $-$ signs, so the probability of a $+$ sign, $p(+)$, should equal the probability of a $-$ sign, $p(-)$, i.e. $p(+) = p(-) = 0.50$. We therefore test the null hypothesis H_0: $p(+) = p(-)$ against the two-tailed alternative hypothesis H_1: $p(+) \neq p(-)$. We choose a 5% level of significance.

The occurrence of a $+$ sign or a $-$ sign is a binomial event. The critical level for testing H_0 against a two-tailed alternative is simply half the significance level, so it is 0.025. We will reject the null hypothesis if the probability of getting as many as six $-$ signs (or as few as three $+$ signs) out of a total of nine signs is less than 0.025.

This probability can be found by using equation [18.1] or from cumulative binomial probability tables (Box 18.2) to be 0.2539. It is well in excess of 0.025, so the probability of getting as many as six $-$ signs is not so low that the null hypothesis should be rejected. The test does not cast doubt on the hypothesis that the population median is 89.

The procedure for use with small samples is summarised in Box 35.1. Note the phrase 'adjusted sample size' and its definition in step 2.

Box 35.1

How to ... **use the sign test for the median of a population (adjusted sample size ⩽ 10)**

1. We wish to test whether observed sample data are compatible with a hypothetical population median value of m.

2. Give a $+$ sign to all sample values greater than m. Count their number n_+. Give a $-$ sign to all sample values less than m. Count their number n_-. The sum of n_+ and n_- is the **adjusted sample size**. It is the original sample size reduced by the number of items equal to the hypothetical median.

3. Set up the null hypothesis that the probability of a $+$ sign, $p(+)$, equals the probability of a $-$ sign, $p(-)$, i.e. $H_0: p(+) = p(-) = 0.5$, and the alternative hypothesis that they are not equal, i.e. $H_1: p(+) \neq p(-) \neq 0.5$.

4. Choose the significance level, usually 5%.

5. Choose the lower of n_+ and n_-. Call this number r, and call the corresponding sign a success. Using the binomial test (Box 18.1 or 18.2) find the probability of r successes or fewer in $n_+ + n_-$ trials with $p = 0.5$.

6. Reject the null hypothesis at 5% if this probability is less than half of 5%, i.e. less than 0.025. Rejection means that the median is not significantly different from m.

If the total frequency is too large for use of published tables, then it is large enough for us to approximate to the binomial solution by using the normal distribution (Chapter 28). We now illustrate this.

⊖ **Example 35.3** Statingham smallholders have been wrongly told that EU regulations require carrots to have a median length of 20 cm at time of sale. A sample of 42 carrots shows that 32 are longer than this $(+)$, 8 are shorter $(-)$ and 2 are equal to it. Does this suggest that the median for all carrots grown by members is not 20 cm?

⊖ We ignore the carrots that have lengths equal to the value under test. Our interest is in the balance of lengths on either side of this. Again we write $H_0: p(+) = p(-) = \frac{1}{2}$. We take a two-tailed alternative hypothesis $H_1: p(+) \neq p(-)$.

We choose a 5% level of significance, and agree to reject the null hypothesis if the probability of getting as few as eight $-$ signs out of forty $+$ or $-$ signs is less than 0.025.

Calculating binomial probabilities with $n = 40$ is an unattractive idea. Instead we approximate to the binomial distribution by using a normal distribution with mean $\mu = np = 40 \times \frac{1}{2} = 20$ and variance $npq = 40(\frac{1}{2})(\frac{1}{2}) = 10$. ⊖

Thus the number of negative signs (or the number of positive signs) is normally distributed about a mean of 20 with a standard deviation of $\sqrt{10}$.

Since the number of negative signs is a discrete variable, use of the continuous normal distribution requires a continuity correction. 'As few as 8' means 'as few as 8.5. An ordinate with this value will cut off a tail with an area proportional to the probability of getting 8 or fewer − signs. We reject the null hypothesis if this is less than 0.025, since this is only one of the two tails. From tables of the standard normal curve, this corresponds to a critical value of $z = 1.96$. Standardising the variable, we get

$$z = (x - \mu)/\sigma = (8.5 - 20)/\sqrt{10} = -3.639$$

This is numerically far greater than the critical value of 1.96, so the null hypothesis should be rejected. We cannot deny that the median length of the population of all carrots fails to comply with the regulations.

The procedure when the adjusted sample size exceeds 10 is summarised in Box 35.2

Box 35.2

How to ... **use the sign test for the median of a population (adjusted sample size > 10)**

1. to **4.** As in Box 35.1

5. Use the normal approximation to the binomial test (Box 28.5) to find whether the probability of r successes or fewer in $n_+ + n_-$ trials with $p = 0.5$ produces a test statistic less than the critical value. Use $\mu = \frac{1}{2}(n_+ + n_-)$ and $\sigma^2 = (n_+ + n_-)(\frac{1}{2})(\frac{1}{2})$ and test statistic $z = (x - \mu)/\sigma$, where x is the lower of n_+ and n_- adjusted by a continuity correction of 0.5 that reduces the numerical value of z.

Wilcoxon's signed rank test

The test just described takes account of magnitudes only in determining whether a score is more or less than a stated level. The test we now describe ranks the differences from this stated level, so it makes better use of the data.

⬭ **Example 35.4** We have the same sample of trial lengths as in Example 35.1 and we are wondering whether the trial of Anna Liszt, for speeding, which lasted for 92 minutes was no longer than the population median.

⬭ We note that the length of a trial recorded in Table 35.1 is a continuous variable, and we assume the population of trial lengths from which the sample is drawn is symmetric. ⬭

Table 35.2

Trial length, x	64	78	116	83	87	89	98	93	123	44
$x - 92$	−28	−14	+24	−9	−5	−3	+6	+1	+31	−48
Rank of $\|x - 92\|$	8	6	7*	5	3	2	4*	1*	9*	10

*Asterisks indicate positive values of $x - 92$.

We set up the null hypothesis that the median is at least as great as 92, with the alternative hypothesis that it is less than 92. Then we subtract 92 from every observation, and rank the differences, ignoring the sign, as in Table 35.2. Then we add the ranks of the positively signed terms, starred in the table, getting a total of 21, which we denote by T. Some writers use w.

We now argue as follows. If the median for the sample is very much lower than the hypothesised population median of 92 then the alternative hypothesis is probably true, and in this case relatively few of the trial lengths will be as long as 92 minutes. Consequently, there will be few starred terms in Table 35.2, so *their total T will tend to be small*. Thus small values of T will point to rejection of the null hypothesis.

The critical values of T for samples of various sizes are given in Table D7. We see that in this example the 5% critical value for the lower one-tailed test, with $n = 10$, is 10. Our calculated value of 21 is greater than 10, so we do not reject the null hypothesis that the population median is at least 92.

Box 35.3

How to ... use Wilcoxon's signed rank test for the median

1. Check that the data are measured on a continuous variable and the population distribution is symmetric.

2. Set up the null hypothesis that the population median is at least k, and the null hypothesis that it is less than k.

3. Choose a level of significance.

4. Subtract k from each observation to get n values of $x - k$.

5. Rank the absolute values $|x - k|$.

6. Add the ranks of the observations greater than k to get T.

7. Reject the null hypothesis if T is equal to or less than the value corresponding to n and the level of significance in Table D7.

Consolidation exercise 35A

1. What hypothesis is tested by the sign test for the median?
   What is counted in order to apply the test?

2. To apply the test at 5% significance (two-tailed) with a small sample, how is the
   critical value found? If the sample is large, what alternative procedure may be used?

3. How does the the Signed Rank Test for the median differ from the sign test?
   What is ranked in it?

4. How is *T* obtained in Wilcoxon's Signed Rank Test?

35.3 Two tests for matched pairs

Example 35.2 introduces the idea of **matched pairs**, which are often used in testing
the difference between the effects of two **treatments.** We use this term in a general
sense, embracing treatment for illness, treatment of plants with fertilisers, seeing or
not seeing a specified film, and any other instance of something or somebody
undergoing one of two different experiences (of which one may be simply the
absence of the other).

In examining the effects of a new drug, we may take two groups of people. We
give the new drug (treatment A) to one group and we give placebos (treatment B)
to the other group, called the **control group**. By definition placebos are expected to
have no medical effect. We then count the number of people in each group who
have shown a defined response, or obtain some other comparative measurement
that can be tested statistically.

However, even if the persons are divided into two groups at random, certain
factors may be more prevalent in one group than in another, which may invalidate
the comparison.

To overcome this we can divide the patients into pairs, ensuring that, as far as
possible, each member of a pair has the same characteristics – they are matched.
Group A for treatment A is then formed by selecting one member *at random* from
each pair. This automatically determines group B, consisting of all other persons.
We compare the measurement scored by each A person with that scored by their B
match, as we illustrate shortly.

In some instances, such as in **before and after** tests, it is possible to improve upon
this by matching all persons with themselves. Every matched pair consists of one
individual assessed (for Group A) before treatment again (for Group B) after the
treatment. Each person is his or her **own** control.

If we have data on an interval or ratio scale (Notebox 33.1) and the right conditions
prevail (especially that the differences between the scores are normally distributed) we
will be able to use the parametric *t*-test; otherwise one of the tests we are about to
describe may be used. An important feature of these tests is that they do not require
all pairs to be drawn from the same population. We describe two tests.

The sign test for matched pairs

The sign test compares members of matched pairs, or individuals with themselves when acting as their own controls. It assumes that all individuals are rated, or put into ordered classes, on the basis of a continuous variable. Each pair is then given a + or − sign according to whether the A member or the B member of that pair has the greater score. The balance (or imbalance) of signs is then considered. Note that the magnitudes of the differences are not taken into account.

On the basis of the null hypothesis that the treatment makes no difference, we would expect as many + signs as − signs. Put another way, the two groups have the same median score. This is unlikely to arise exactly. We therefore examine the significance of any imbalance between signs.

We set up the null hypothesis $H_0: p(+) = p(-) = \frac{1}{2}$, and the alternative hypothesis $H_1: p(+) \neq p(-) \neq \frac{1}{2}$. We also specify a level of significance, say 5%, in a two-tailed test.

Eliminating pairs where there are equal ratings, let n be the number of remaining matched pairs, with a + sign or a − sign in the last row. In Example 35.2 we have $n = 8$.

We can test the imbalance by using the binomial distribution, taking $p = q = \frac{1}{2}$ as the probability of a stated sign. We then test the significance of x, the observed number of plus signs or minus signs (whichever is the smaller); in Examples 35.2 we have $x = 3$. In our example we use a two-tailed test, but a one-tailed test can be performed if the alternative hypothesis requires it.

⊃ In **Example 35.2** the data are as in Table 35.3, where we have added + and − signs as just explained. We have $n = 8$ and $x = 3$. The probability of getting an x of 3 or less is read off from tables of the cumulative binomial probability as 0.3633. Notice that, by symmetry, the probability of getting an x of 5 or more is also 0.3633, so the probability of getting 5 signs of one kind and 3 of another (for a two-tailed test) is $2 \times 0.3633 = 0.7266$. We therefore have a result that is very likely to arise by chance, and certainly is much more probable than 0.05 which we have chosen as our level of significance. We have to conclude that the treatment makes no significant difference one way or the other.

The procedure when the adjusted sample size is not more than 10 is summarised in Box 35.4.

Table 35.3

Owner	a	b	c	d	e	f	g	h	j	k
Given A	1	4	3	4	4	3	5	2	4	1
Given B	3	2	5	5	3	3	2	5	4	3
	−	+	−	−	+	=	+	−	=	−

> ### Box 35.4
>
> *How to ...* **use the sign test for the median of a population of differences with matched pairs (adjusted sample size ⩽ 10)**
>
> 1. List all the scores in two rows (as in Table 35.3) or two columns, with the scores for the individuals in a pair being one above the other or side by side.
>
> 2. Calling the scores in the two rows (or columns) A and B, give all pairs for which $A > B$ a + sign, and all pairs for which $A < B$ a − sign. Ignore pairs where $A = B$. Call the total number of + and − signs the adjusted sample size n.
>
> 3. Set up the null hypothesis that the probability of a + sign, $p(+)$, equals the probability of a − sign (which means that the population medians are equal). Thus $H_0: p(+) = p(-) = 0.5$. The alternative hypothesis may be that the probabilities are not equal, resulting in a two-tailed test, $H_1: p(+) \neq p(-) \neq 0.5$; or it may be that they are different in a stated direction, resulting in a one-tailed test, $H_1: p(+) > p(-)$ or $H_1: p(+) < p(-)$.
>
> 4. Choose the significance level, usually 5%.
>
> 5. Choose the lower of n_+ and n_-. Call this number r, and call the corresponding sign a success.
>
> 6. Use the binomial test (Box 18.1 or 18.2) to find the probability of r successes or fewer in n_+ or n_- trials with $p = 0.5$.
>
> 7. (a) If a two-tailed test has been chosen, reject the null hypothesis (of equal medians) at 5% in favour of the two-tailed H_1 if this probability is less than 2.5%.
>
>   (b) If a one-tailed test has been chosen, reject the null hypothesis at 5% in favour of the one-tailed H_1 if this is less than 5%.

If n is larger than about 10 and suitable tables of cumulative binomial probabilities are not at hand, we use the normal approximation to the binomial distribution (Box 28.4).

It can be shown that if there are n pairs with different signs, and n is larger than 10, then the number of positive signs will be approximately normally distributed with mean $= \mu_x = \frac{1}{2}n$ and standard deviation $= \sigma_x = \frac{1}{2}\sqrt{n}$. This allows us to use the test statistic

$$z = \frac{x - \frac{1}{2}n}{\frac{1}{2}\sqrt{n}}$$

which is normally distributed with zero mean. Because we have discrete data, we have to apply a continuity correction which reduces by 0.5 the difference between the observed number of + signs (or − signs) and the expected number (which is the mean).

Suppose we have a longer version of Table 35.3, with 36 paired samples, and suppose the differences lead to 25 + signs and 11 − signs. The null hypothesis is $H_0: p(+) = p(-) = \frac{1}{2}$ and the alternative two-tailed hypothesis is $H_1: p(+) \neq p(-)$. We choose a 5% level of significance, which gives 1.96 as the critical value of z.

Now, instead of calculating the binomial probabilities, we use the approximation given by the normal distribution with $\mu = 36/2 = 18$ and $\sigma = \sqrt{36}/2 = 3$.

Remembering the continuity correction, we narrow the difference by subtracting 0.5 in the numerator of the test statistic, which becomes

$$z = \frac{x - 0.5 - \frac{1}{2}n}{\frac{1}{2}\sqrt{n}} = \frac{25 - 0.5 - 18}{3} = 2.17$$

This exceeds the critical level, so the null hypothesis is rejected. We conclude that it is unlikely at the 5% level that the additive makes no difference.

The procedure for applying this test with larger adjusted sample sizes is summarised in Box 35.5.

Box 35.5

How to ... **use the sign test for the median of a population of differences with matched pairs (adjusted sample size > 10)**

1. to 5. As in Box 35.4.

6. Use the normal approximation to the binomial test (Box 28.5) to find whether the probability of r successes or fewer in $n_+ + n_-$ trials with $p = 0.5$ produces a test statistic of less than the critical value. For details see step 8 of Box 35.2.

The Wilcoxon matched pairs signed rank test

Like the sign test, the Wilcoxon matched pairs signed rank test requires ordinal or better data. However, it makes greater use of the data than the sign test since it also requires ordinal measurement of the differences between pairs. These differences, whose distribution should be symmetrical and continuous, have to be ranked rather than simply marked + or −. We describe the test by referring to Example 35.4. Essentially it is an application of the signed rank test described above. The differences within the matched pairs form a single sample, and using the same table, we test whether this has a median of zero.

⬭ **Example 35.4** A sociologist tests the effectiveness of a new rehabilitation course for young offenders. Offenders are required to react to a variety of stimuli before being given a score. A high score indicates appreciable rehabilitation.

The sociologist is confident that a person who scores 80 has done better than one who scores 40, but not convinced that 80 means twice as well as 40.

⬭

Table 35.4

Pair	a	b	c	d	e	f	g	h	j	k
Group A	79	83	56	65	72	59	79	69	65	56
Group B	58	67	65	60	59	46	85	45	68	52

However, the sociologist is confident that the difference between 60 and 80 is greater than the difference between 50 and 60.

Ten pairs of young offenders are selected with great care, so that as far as possible, the members of a pair have identical, or very close, backgrounds and character.

One member of each pair is sent on the new rehabilitation course, forming group A. The other is given the traditional treatment, forming group B. After three months all 20 offenders are tested. Do the scores awarded (Table 35.4) show a significant difference in the degree of rehabilitation between the two groups?

The test we are about to illustrate requires that the differences come from a continuous symmetrical distribution. The approximate symmetry of the differences should be checked with a dot plot.

⟹ To answer the question, we write down the differences between the pair members, subtracting one row from the other, as in Table 35.5.

The ranks of the *absolute* values of these differences are written down, beginning with 1 for the lowest numerical difference. Tied ranks are given the average of the ranks they would get if they were very slightly different.

Thus the pairs are ranked according to the magnitude of the difference between the two members. These ranks are then prefaced with a + or − sign according to whether the difference $A - B$ is positive or negative. This is usually done in the same row, but in this explanation it is shown separately in the last row.

This enables us to pick out the pairs in which B does better than A, and also to take account of the relative magnitudes of the differences. Three out of the 10 pairs have − signs, but all of these negative ranks are relatively low (being 1, 4 and 5 out of 10). What does this mean?

Table 35.5

Pair	a	b	c	d	e	f	g	h	j	k
Group A	79	83	56	65	72	59	85	69	65	56
Group B	58	67	65	60	59	46	79	45	68	52
A − B	21	16	−9	5	13	13	−6	24	−3	4
Rank of |A − B|	9	8	5	3	$6\frac{1}{2}$	$6\frac{1}{2}$	4	10	1	2
Signed rank	+9	+8	−5	+3	$+6\frac{1}{2}$	$+6\frac{1}{2}$	−4	+10	−1	+2

In essence our method now is to compare (i) the total of the ranks of differences with positive signs with (ii) the total of the ranks of the differences with negative signs. If the null hypothesis is true, we would expect these totals to be equal. Thus we have

H_0: total of positive ranks = total of negative ranks
H_1: total of positive ranks $\neq$ total of negative ranks

Another way of wording these statements is

H_0: median of the population differences $A - B$ is zero
H_1: median of the population differences $A - B$ is not zero.

Instead of comparing the two totals (which must add to the sum of the ranks $1, \ldots, n$, where n is the number of signed differences), we consider the probability of the smaller of the two totals (denoted by T, although some writers use w) being as small as it is or smaller, given the value of n.

> In **Example 35.4** the sum of the negative ranks is smaller and is $T = 1 + 4 + 5 = 10$ and $n = 10$. Critical values of T are given in Table D8, which shows that for $n = 10$ there is a .05 (or 5%) chance of getting $T = 8$ or less if the null hypothesis is true. In fact, we have $T = 10$, which means that if H_0 is true our value of T has more than a 5% chance of arising. We cannot reject the null hypothesis..

The table is given for values of n up to 26 and selected high values.

___

Box 35.6

How to ... **test the median of a population of differences for matched pairs by using the Wilcoxon signed rank test (adjusted sample size $\leqslant 25$)**

1. This test for matched pairs requires ordinal (or better) data. It also requires ordinal data about the differences between pairs, to permit ranking, and that these differences come from a continuous symmetric distribution.

2. List the A and B scores in two rows or columns.

3. Form a third row (or column) of the signed differences between the scores. Ignore all ties (for which the difference will be zero).

4. Disregarding the signs, rank in row 4 the numerical values of the differences, with the smallest (non-zero) numerical difference given rank 1.

5. Preface each rank in row 4 with the sign of the corresponding difference in row 3.

6. Sum all the positive ranks in row 4 to obtain T_+ and all the negative ranks in row 4 to obtain T_-. Check that $T_+ + T_- = \frac{1}{2}n(n+1)$ where n is the number of non-zero differences. ⬭

7. Select the smaller of T_+ and T_- and call it T_{calc}.

8. The null hypothesis is that the median of the population of differences is zero, which means that the median of the A scores equals the median of the B scores – H_0: med $A =$ med B.

9. The alternative hypothesis may be either that the two medians are not equal, H_1: med $A \neq$ med B (two-tailed); or it may be that the medians are different in a stated direction, H_1: med $A >$ med B or H_1: med $A <$ med B (both one-tailed).

10. Choose the significance level, usually 5%.

11. For the value of n (defined in step 6) and the chosen level of significance and number of tails, use Table D8 to find the value of T_{crit}.

12. Compare T_{calc} found in step 7 with T_{crit} found in step 11. Reject the null hypothesis if $T_{calc} \leqslant T_{crit}$.

If $n > 25$ the values of T are approximately normally distributed, as described in Box 35.7.

Box 35.7

How to ... test the median of a population of differences for matched pairs by using the Wilcoxon signed rank test (adjusted sample size > 25)

1. to 10. As in Box 35.6.

11. Test whether the calculated value of T differs significantly from its expected value (given below) by using the fact that if n is large enough, T is approximately normally distributed with mean and standard deviation

$$\mu_T = n(n+1)/4$$

$$\sigma_T = \sqrt{n(n+1)(2n+1)/24}$$

Use the test statistic

$$z = \frac{T - \mu_T}{\sigma_T}$$

and proceed as in Box 25.1.

Consolidation exercise 35B

1. When are matched pairs used? How are the two groups chosen?

2. What are (a) the assumptions and (b) the null hypothesis for the sign test for matched pairs? When does this test require special tables? ◯

3. When the normal distribution is used in the sign test, the mean is $n/2$. What is n? How is the continuity correction applied in this test?

4. What assumptions underlie the Wilcoxon matched pairs signed rank test? How are the observations ranked in this test? What happens to the signs?

35.4 The Mann–Whitney *U*-test

This test is for data from for *two independent samples*, in which the concept of matched pairs does not arise. It uses the ranks of the individual items when combined into one group to determine, on the basis of the median, whether two independent samples have been drawn from identical populations, or from populations that differ because their medians differ.

It assumes that the variable is continuous. We develop the reasoning behind this test while considering Example 35.4. One might be tempted to apply the *t*-test to the information in this example; but if a non-parametric test is preferred, we can proceed as below. Especially for small samples, use of the *t*-test requires that in each of the two populations the measured variable is approximately normally distributed, and the two population variances are the same.

Basic idea and statement of hypotheses

We form the null hypothesis H_0 that there is no difference between the populations from which the two samples come. This implies a two-tailed test. The alternative hypothesis is that the two populations have different medians.

We use a test statistic denoted by U, which we define and calculate below. We decide to use a 5% significance. (We could have formulated a one-tailed test, in which case we would have used the fact that a result significant at 5% for a two-tailed test is significant at 2.5% with a one-tailed test.)

The first step in the calculation of the test statistic is to arrange the pupils in order in one supergroup. If we denote the highest score by rank 1, we have that the group *A* and group *B* pupils have the ranks shown in Table 35.6.

The test statistic U requires the calculation of two values, one defined to be the total number of times that the *B*s are preceded by *A*s; and a second the other way around. There is a laborious procedure for finding this, but it is easier to use the formulae

$$U_1 = n_1 n_2 + n_1(n_1 + 1)/2 - R_1$$
$$U_2 = n_1 n_2 + n_2(n_2 + 1)/2 - R_2$$

[35.3]

Table 35.6

Rank	1	2	3	4	5	6	7	8	9	10	11	12	13
Mark	77	73	67	63	62	61	59	58	48	47	46	35	28
Group	B	A	B	B	A	B	B	A	B	A	B	A	B

where n_1 is the size of the smaller sample (in this case sample A) and R_1 is the total of the ranks of the members of this sample A, with similar definitions for n_2 and R_2 arising from the larger group. We should check that $U_1 + U_2 = n_1 n_2$.

Finding the critical values and applying the test

Since $U_1 + U_2 = n_1 n_2$, if there were two groups of equal size and with no appreciable difference between the populations then each U would be close to $n_1 n_2 / 2$. If, in the case of two groups of equal size, one U were much larger than the other, we would suspect the existence of a difference between the populations. When do our suspicions becoming overwhelming?

If the groups are not of equal size, we can expect some difference between the Us even if the populations are identical, but once again there will come a degree of inequality between the Us when H_0 has to be rejected.

However, as the sum of the Us is always equal to $n_1 n_2$, we can study this inequality by concentrating on the value of one of the Us. We choose, for this purpose, the smaller of U_1 and U_2, which we call U.

Tables D9a and D9b show the critical values of U at stated levels of significance. For any given level, the critical value depends on the sizes n_1 and n_2 of the two groups.

 In **Example 35.4** $R_1 = 2 + 5 + 8 + 10 + 12 = 37$ leading by use of equations [35.3] to $U_1 = 18$; we also find that $U_2 = 22$. To check, note that $18 + 22 = 40 = 5 \times 8$. We choose U as the smaller of 18 and 22.

Using $n_1 = 5$ and $n_2 = 8$ we find from Table D9a that for a two-tailed test at 5% significance the critical value of U is 6. If $U \leqslant 6$ the test implies rejection. In our example $U > 6$, so there are no grounds for rejecting the null hypothesis.

Use with larger samples

If either sample is larger than 20, the table is replaced by using the fact that if the null hypothesis is true then the smaller of the two values of U will be approximately normally distributed with a mean and variance given in Box 35.8. If some ranks are tied, the variance is given by the formula in step 9 of Box 35.8.

Box 35.8

How to ...   **conduct the Mann–Whitney U-test of the null hypothesis that two samples are drawn from the same population**

1. This test is for two independent samples. It can be used when the t-test is inapplicable. It is used to determine whether two independent samples have been drawn from the same population. The alternative is that the medians are different. It uses the ranks of the individual items when combined into one group, and therefore requires at least ordinal data. ◯

2. The hypotheses are H_0: the two populations are identical, so they have identical medians; and H_1: either the two populations have different medians (two-tailed) or the population medians differ in a stated direction (one-tailed).

3. Letting the sample sizes be n_1 and n_2, combine all the observations into one supergroup and arrange them in descending order, so that the highest has rank 1.

4. Calculate the values of U_1 and U_2 as follows:

(i) Sum the ranks, letting the total of ranks for the sample of size n_1 be R_1 and the total of ranks for the sample of size n_2 be R_2.

(ii) Then calculate

$$U_1 = n_1 n_2 + n_1(n_1 + 1)/2 - R_1$$
$$U_2 = n_1 n_2 + n_2(n_2 + 1)/2 - R_2$$

[35.3]

5. Check that $U_1 + U_2 = n_1 n_2$ and use the symbol U to denote the smaller of U_1 and U_2.

6. Specify a level of significance. In doing so, note that if neither n_1 nor n_2 is greater than 20, the test involves a table that we print only for 5% and 1% for a two-tailed test, and at half these levels for a one-tailed test.

7. If neither n_1 nor n_2 exceeds 20, use Tables D9a and D9b to find the critical level of U appropriate to the chosen level of significance, and reject the null hypothesis if U is less than or equal to the critical value.

8. If either (or both) of n_1 and n_2 exceeds 20, use the fact that U is approximately normally distributed with

Mean $\mu_u = \frac{1}{2} n_1 n_2$

Variance $\sigma_u^2 = n_1 n_2(n_1 + n_2 + 1)/12$

In this case use the normally distributed test statistic

$$z = (U - \mu_u)/\sigma_u$$

and proceed as in Box 25.1.

9. If there are tied ranks, the variance is given by

$$\frac{n_1 n_2}{N(N-1)} \left(\frac{N^3 - N}{12} - \sum T_i \right)$$

where $N = n_1 + n_2$, $T_i = (t_i^3 - t_i)/12$ and t_i is the number of observations that share the rank i.

Consolidation exercise 35C

1. What is the Mann–Whitney test? When is it used? What is assumed about the data?

2. There are two values of U. How are they defined? What does $U_1 + U_2$ equal? Which U is used in the test?

Summary

1. In parametric statistics we usually compare two samples by seeing if their means are significantly different. When non-parametric tests are used, we are more likely to ask about the medians.

2. The sign test for the median asks whether a sample median is significantly different from a hypothesised population median. It requires measurements of a continuous variable, but unlike the t-test, it does not require a normal distribution. We count the number of sample items higher than the hypothesised population median and the number lower than it. The probability of getting these numbers can be found from cumulative binomial probability tables (Box 35.1), or if $n > 10$ by using a normal approximation (Box 35.2).

3. Wilcoxon's signed rank test for the median (Box 35.3) makes better use of the data. It ranks the absolute differences between the observations and a hypothesised median, then adds the ranks of the positive differences to obtain a total whose magnitude is tested against a critical value obtained from a special table.

4. Matched pairs are often used in testing the difference between the effects of two treatments (or experiences), where one treatment may be simply the absence of the other. As far as possible, each member of a pair has the same characteristics. Group A for treatment A is then constructed by selecting one member, at random, from each pair. This automatically determines group B, consisting of all other persons. We then compare the measurement scored by each A person with that scored by their B match. If we have data on an interval or ratio scale and the right conditions prevail (especially that the differences between the scores are normally distributed), we will be able to use the parametric t-test; otherwise one of the tests described in this chapter may be used. These tests do not require all pairs to be drawn from the same population.

5. The sign test for matched pairs (or individuals acting as their own controls) tests whether the median scores are the same for both sets. It involves ranking the members of both sets then comparing the ranks of the two members of each pair. If $n \leqslant 25$ we use the binomial distribution with either a one-tailed or a two-tailed test (Box 35.6). If $n > 25$ we approximate to the binomial distribution by the normal distribution, with a continuity correction as described in Box 35.5.

6. The Wilcoxon signed ranks test (of the median of a population of differences for matched pairs) makes better use of the data. It requires ordinal (or better) data, and ordinal measurement of the differences between pairs, which must have a continuous symmetrical distribution.

    Box 35.6 describes how to test the significance of the differences with special tables, if *n* is between 6 and 25. If $n > 25$ a normal approximation can be used as described in Box 25.6.

7. The Mann–Whitney *U*-test does not involve matched pairs. It is used to determine whether two independent samples have been drawn from the same population or whether they come from populations that have different medians.

    The members of the two groups are merged into one ordered set and we examine whether members of one group precede members of the other group significantly more than we would expect on the basis of a null hypothesis. The calculations are simple. If neither sample exceeds 20 we use special tables to obtain critical values. Otherwise we use the fact that the test statistic *U* (calculated as in the main text or as in Notebox 35.1) is approximately normally distributed as described in Box 35.7.

One-sample contingency and chi-squared

36.1 **Introduction**

Example 36.1 Table 36.1 shows, by calendar quarter, the numbers of privately built houses completed in Scotland and Wales during 1967–1969. Does the seasonal pattern differ between countries?

Example 36.2 A headteacher believes there is a tendency for pupils with certain colours of hair to specialise in certain subjects. Does Table 36.2, which summarises the headteacher's observations over the last two years, support this belief?

Table 36.1

	Jan–March	April–June	July–Sept	Oct–Dec	Total
Scotland	5682	5125	6098	7639	24 544
Wales	6341	7360	7549	7227	28 477

Source: *Housing Statistics, Great Britain*, Nos 14 (1969) and 18 (1970), HMSO, London.

Table 36.2

	Subject of specialism		
Hair colour	Academic arts	Practical arts	Sciences
Black	24	41	25
Brown	32	13	15
Auburn	8	7	15
Blond	16	19	15

In one sense the subject of this chapter is similar to correlation in Chapter 14. In both chapters we examine the possibility of a relationship between two variables. But here we have data of a very different kind.

The tables in the above examples, and in Example 33.3, are **contingency tables**. Ignoring rows and columns marked 'total', each table presents information in a two-way classification of m rows and n columns, with $m \times n$ cells. Thus Table 33.3 has 2 rows and 3 columns, giving 6 cells. The rows and columns may be classified according to the values of some numerical variable (such as number of visitors) or by some **attribute** (such as falling ill, country, specialism and hair colour).

Note carefully that every cell entry is a frequency – the number of times the combination indicated by the appropriate row and column headings has occurred or been observed. In Table 36.1 a cell shows the number of times the completion of a house has occurred in that country in that time interval. Thus every entry must be zero or a positive integer.

The questions amount to whether the mn cells have the frequencies that one would expect simply on the basis on the row and column totals. Answering such questions is easiest if there are more than two rows, or more than two columns, or both. A 2×2 classification is a little trickier to analyse. We begin by looking at the 2×3 classification in the question about hospital visitors, introducing the essentials of the method as we proceed. Later we consider a few refinements before going on to consider the special procedures for 2×2 classifications, including the use of Fisher's exact probability when one or more of the expected frequencies is less than 5 (Notebox 36.4).

 Degrees of freedom in a contingency table

In Table 33.1 we have six cell frequencies arranged in a way that produces row and column totals. As explained in Notebox 36.1, we could alter any two of these six frequencies with complete freedom, provided we then adjusted the other four in order to preserve the row and column totals. Because of this, we say there are two **degrees of freedom**. We demonstrate in Notebox 36.1 that for an $m \times n$ contingency table the number of degrees of freedom is $(m-1)(n-1)$ provided both m and n are greater than 1.

36.3 **Testing for an association: calculating χ^2_{calc}**

Null hypothesis and level of significance

The basic method is to set up a null hypothesis H_0 that there is **no association** between the two factors forming the basis of classification. The alternative hypothesis is that there is an association. We then calculate the frequencies we would expect if H_0 is true, and if the row and column totals are preserved, as explained in Box 36.1. Then we compare these expected frequencies with the actual (or observed) frequencies by using a test statistic whose value depends on the differences between them and whose critical value depends on the number of degrees of freedom and the chosen level of significance. The procedure is given in Boxes 36.1 and 36.2.

Finding the expected frequencies

We go through the very simple steps that produce the value of the test statistic. The reasons for these steps will become clear later; concentrate on the process for now.

⬤ We reprint Table 33.1 as the upper panel of Table 36.3. It shows the **observed cell frequencies**. In the lower panel of the table we show **expected frequencies** derived from the null hypothesis H_0 that there is no association between number of visitors and experience of minor illness. We now see how to obtain them.

We have to calculate how many people we would expect to find in each cell if H_0 is true. Looking at the observed frequencies, we can see from the total row that half the patients (150 out of 300) had 8 visitors or fewer. If there is no association between visitors and minor illness we would expect half the 110 people who became ill to have few visitors, and the same to be true of the 190 people who did not become ill. Thus we would expect the first column of this table to contain entries of $110 \times \frac{1}{2} = 55$, and $190 \times \frac{1}{2} = 95$, instead of 42 and 108. These values have been entered as expected frequencies in the first column of the lower part of the table.

The other expected frequencies have been obtained in the same way. A practical rule is given in Box 36.1. Use it to check Table 36.3. Note that we have not rounded our expected frequencies to make them whole numbers. Our reason is given in Notebox 36.2.

Table 36.3

		Number of visitors		
	$\leqslant 8$	9–12	$\geqslant 13$	Total
Observed				
Minor illness	42	30	38	110
No minor illness	108	50	32	190
Total	150	80	70	300
Expected				
Minor illness	55	29.33	25.67	110
No minor illness	95	50.67	44.33	190
Total	150	80	70	300

> **Box 36.1**
>
> *How to ...* **calculate expected frequencies for a contingency table**
>
> To find the expected frequency for any cell, multiply the total frequency of the row containing it by the total frequency of the column containing it; then divide by the grand total.

Comparing the observed and expected frequencies

The null hypothesis is tested by using the observed and the expected frequencies to calculate a test statistic. We will use o to denote an observed frequency, which will vary from cell to cell, and e to denote an expected frequency. In most books o and e appear as o_i and e_i to remind the reader that each cell has its own value. We omit the subscripts to avoid cluttering up the page with symbols, but remember that each cell has its own value of o and e.

To calculate the test statistic, we perform the following steps on each cell:

(i) Find the numerical difference between o and e.
(ii) Square this difference, getting $(o - e)^2$.
(iii) Divide this square by the expected frequency e, getting a value of $(o - e)^2/e$

As there are six cells there will be six values of the expression in step (iii); they are shown in Table 36.4 (in their cell positions). Notice that we have not worked out the actual values. As we see later, this is not always necessary and there is no point in doing unnecessary work.

Table 36.4

$\dfrac{(42 - 55)^2}{55}$	$\dfrac{(30 - 29.33)^2}{29.33}$	$\dfrac{(38 - 25.67)^2}{25.67}$
$\dfrac{(108 - 95)^2}{95}$	$\dfrac{(50 - 50.67)^2}{50.67}$	$\dfrac{32 - 44.33)^2}{44.33}$

Completing the calculation of the test statistic

We can see from this procedure that if every observed value o were equal to the expected value e, then all the terms in Table 36.4 would be zero. In all other cases, some or all of the terms will be greater than zero; they must be positive because of the squaring. As the difference increases between the observed frequencies o and the expected frequencies e, the value of $(o - e)^2/e$ also increases. We define our test statistic to be the sum of the terms $(o - e)^2/e$. The calculated value of the test statistic is denoted by χ^2_{calc}, pronounced 'ki square calc'; it may be defined as the sum of the values of $(o - e)^2/e$, which can be written symbolically as

$$\chi^2_{calc} = \sum (o - e)^2/e \qquad\qquad [36.1]$$

There is one value of $(o - e)^2/e$ for each cell.

The reason behind it

We need to say why we use this strange-looking notation and procedure for calculating the test statistic. It can be shown that (under certain conditions mentioned later) the value calculated from equation [36.1] has a discrete distribution that approximates closely to a well-studied theoretical continuous distribution known to mathematicians as the **chi-squared** (or χ^2) distribution (without the subscript calc). Tables have been produced that show the probabilities of chi-squared having various values. We use them to see whether the calculated value of the test statistic χ^2_{calc} is unacceptably improbable, as it will be if it exceeds a critical value that we obtain from the tables.

Chi-squared and χ^2 will be used to refer to the theoretical continuous distribution, whereas χ^2_{calc} refers to the calculated value derived from discrete data using equation [36.1] or [36.4]. Our calculated values of the test statistic χ^2_{calc} are compared with the tabulated values of the theoretical χ^2 or chi-squared distribution. Some writers use different notation.

36.4 Comparing χ^2_{calc} with the critical value of χ^2

The shape of this continuous χ^2 distribution depends on the number of degrees of freedom (Figure 36.1). Thus the critical value of our test statistic depends not only on the chosen level of significance but also on the number of degrees of freedom.

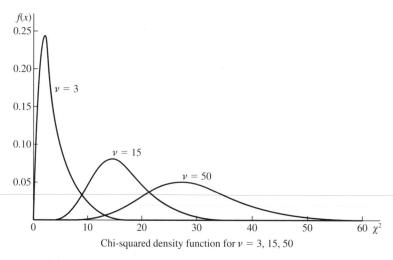

Chi-squared density function for $\nu = 3, 15, 50$

Figure 36.1

The basic idea now is to use our knowledge of the number of degrees of freedom to choose the right curve (or the table that it is based on), from which we find the critical value appropriate to our chosen level of significance. If the calculated value exceeds this, the null hypothesis is rejected.

The χ^2 distribution tables are usually printed as in Table D10 with one row for each selected number of degrees of freedom. The entries in a row show the value of chi-squared that will cut off a tail-end area corresponding to a stated probability.

⬭ In answering **Example 33.3** we have chosen a 5% (or 0.05) significance level and there are $(2-1)(3-1) = 2$ degrees of freedom. The column headed 0.05 in Table D10 shows that for $v = 2$ the critical value of χ^2 is 5.99. If our calculated value exceeds this, we reject the null hypothesis.

Before resorting to accurate calculations, look again at Table 36.4. Notice the top right-hand cell; it contains one of the six values that have to be added to get χ^2_{calc}. Roughly speaking, the value in this cell is $12^2/25 = 144/25$, which is well over 5. The cell beneath it has an approximate value of $12^2/44 = 144/44$, which is over 3.

These two cells alone give a total of well over 8, which exceeds the critical value of 5.99. H_0 has to be rejected.

Use of this technique has saved a great deal of work. But if the answer is not so obvious, the contributions from the cells have to be worked out more carefully. By keeping a running total of what they add up to so far, it may still be possible to avoid working out all of them.

Before summarising the procedure, we need to note the conditions that have to hold for the test to be valid. A more careful statement appears later, but roughly speaking the conditions are as follows:

(i) The total number of observations N is large.
(ii) All the expected values are at least 5.

If some of the expected frequencies are less than 5, the safest thing to do is to amalgamate cells, as in the examples below. Conditions (i) and (ii) clearly hold in Example 33.3.

Box 36.2

How to ... compare observed data with expected data in a table of r rows and c columns, where no expected frequency is less than 5 and there are at least 2 degrees of freedom

1. State the null hypothesis H_0 of no association on which the expectations are based, and state a level of significance.

2. Calculate the number of degrees of freedom v as in Notebox 36.1 (being careful to reduce the number of cells if there have been amalgamations). ⬭

3. Use Table D10 to find the critical value of χ^2 for the chosen level of significance and the calculated value of v.

4. For each cell, perform the following calculations:

(i) Calculate the expected frequency e on the basis of this hypothesis, as indicated in Box 36.1.

(ii) Find the numerical difference between o (the observed frequency) and e.

(iii) Square this difference, getting $(o - e)^2$.

(iv) Divide this square by the expected frequency e, getting a value of $(o - e)^2/e$ for each cell.

5. Sum the values obtained in step 4(iv); call this sum χ^2_{calc}.

6. If χ^2_{calc} exceeds the critical value of χ^2 from step 3, reject the null hypothesis.

Before leaving this account of the basic procedure we consider briefly Examples 36.1 and 36.2.

In **Example 36.1** there are big numbers, but otherwise the problem is very like Example 33.3. The number of degrees of freedom is $(2 - 1)(4 - 1) = 3$. With 1% significance, the critical value of χ^2 is 11.34. The null hypothesis is that the seasonal pattern is the same in both countries.

Despite the big numbers, it is very important not to approximate, perhaps by writing the frequencies correct to the nearest hundred or thousand. Apart from the fact this may lead to big errors in the values of $(o - e)^2/e$, working in anything other than the original units will reduce the value of χ^2_{calc} by a scaling factor. For example, working in units of 10 will mean that the calculated value will need multiplication by 10 before being compared with the critical value.

Table 36.1 shows that more houses were completed in Wales than in Scotland in each of the first three quarters. But in the fourth quarter this was reversed. In finding χ^2_{calc} it may be useful to begin by calculating the part of it that arises from the fourth quarter.

Applying the procedure of Box 36.1 to the fourth-quarter column of Table 36.1, we get the expected frequencies for the two cells in this column. We can check by calculation that each will differ from the observed frequency by about 758.

If we now calculate the values of $(o - e)^2/e$ for either of these two cells, we will get something very much higher than the critical value of 11.34. Thus there is no need to work out the contributions to χ^2_{calc} from the other cells.

Inspecting the data first may pay off.

In **Example 36.2** there are $(4 - 1)(3 - 1) = 6$ degrees of freedom. All expected frequencies are well over 5 and there is no complication.

Consolidation exercise 36A

1. What is a contingency table? All the observed entries in the cells of a contingency table have one thing in common. What is it?

2. Which of the following could appear in a contingency table of observed entries: (a) 1745, (b) 14, (c) 8.5, (d) 0, (e) -3, (f) -3.5 (g) $\frac{1}{2}$?

3. What is the null hypothesis in a χ^2 test?

4. How do we calculate the expected cell frequencies? Which values in Question 2 cannot be cell frequencies?

5. How do we calculate the number of degrees of freedom for an $m \times n$ table?

6. What is the formula for calculating χ^2_{calc}?

7. How do we find the value of χ^2 with which χ^2_{calc} has to be compared?

8. What conditions are necessary for the χ^2 test to be applied?

9. Table 36.5 is for 200 Statingham undergraduates. (a) Frame an appropriate null hypothesis. (b) Calculate the number of degrees of freedom. (c) Find the 5% critical value of χ^2, (d) Calculate the expected frequencies. (e) Calculate $(o - e)^2/e$ for each cell and then find χ^2_{calc}. (f) Decide whether to reject your null hypothesis.

Table 36.5

| | Time of rising | | | |
	Before 8.00	8.00–9.00	After 9.00	Total
Red eyes at midday	10	30	80	120
No red eyes	30	20	30	80

36.5   Combining cells

We have mentioned that the χ^2 test requires each expected cell frequency to be at least 5. If an expected cell frequency is less than 5, the usual practice is to combine cells. We illustrate this with an example, discussing as we go along a few points that arise. This section can be omitted at a first reading. Although the explanation may seem lengthy, the process can be performed quite quickly.

Example 36.3 Table 36.6 shows the results of an enquiry in a medium-sized organisation about the employees' opinions of the new boss. Each employee gave the boss a mark between 1 and 5, with 1 indicating highly favourable, 2 favourable, 3 neutral, 4 unfavourable and 5 highly unfavourable. Is there some association between opinion and pay?

Table 36.6

Employee pay level	Opinion of a new boss					
	1	2	3	4	5	Total
High	0	2	4	0	2	8
Medium	4	6	5	6	1	22
Low	6	4	16	7	7	40
All levels	10	12	25	13	10	70

In principle the first step is to formulate a hypothesis of no association, from which to calculate the expected frequencies; but if we go about it correctly, we can save work. The expected frequency is the product of the row and column totals divided by the grand total, so the highest value in the top row will correspond to the column with the highest total. This is the third cell, for which the expected frequency will be $(8 \times 25)/70$, a little less than 3. Clearly no cell in the top row will have a frequency as high as 5, yet we need this for the χ^2 test.

We must amalgamate the top row with another. It seems to make sense to amalgamate highly paid employees with moderately paid employees, rather than highly paid with lowly paid. If two rows suffer from low cell frequencies, there may be no alternative to combining them if there is to be any test at all. Always consider whether the amalgamation is sensible.

The new table with the top two rows amalgamated is shown in Table 36.7. We decide to test the null hypothesis that there is no association between pay level and opinion of boss at 5% significance. We cannot determine the number of degrees of freedom without knowing how many cells there will be in the table we finally use, which depends on whether we may yet have to amalgamate some cells. We therefore begin to work out the expected frequencies, following our rule in Box 36.1 These are shown in Table 36.8.

The two expected frequencies marked with an asterisk are below 5. If we try to get over this by amalgamating rows, we destroy the problem. Instead we have to amalgamate columns. We can do this in two ways.

Amalgamate each end column with the one next to it

We can amalgamate each end column with the one next to it, so that we merge very favourable with favourable, and very unfavourable with unfavourable. If we do that, and work out the fractions, our new table of expected frequencies is Table 36.9. This has two degrees of freedom, so the critical value for χ^2 at 5% significance is 5.99. The initial calculations give the six terms in Table 36.10.

None of these terms looks as big as 1.00, so it seems unlikely that their total will be nearly as big as 5.99. If we want to play safe, we can calculate them, and find that the total is well below the 5% critical value. According to this test, there is no evidence to reject the null hypothesis that there is no association between pay levels and opinion of the boss. ⬤

Table 36.7

| Employee pay level | Opinion of a new boss | | | | | |
	1	2	3	4	5	Total
High or medium	4	8	9	6	3	30
Low	6	4	16	7	7	40
All levels	10	12	25	13	10	70

Table 36.8

Expected frequencies

| Employee pay level | Opinion of a new boss | | | | | |
	1	2	3	4	5	Total
High or medium	300/70*	360/70	750/70	390/70	300/70*	30
Low	400/70	480/70	1000/70	520/70	400/70	40
All levels	10	12	25	13	10	70

*The two expected frequencies marked with an asterisk are less than 5.

Table 36.9

| Employee pay level | Opinion of a new boss | | | |
	1 or 2	3	4 or 5	Total
High or medium	9.43	10.71	9.86	30.00
Low	12.57	14.29	13.14	40.00
All levels	22.00	25.00	23.00	70.00

Table 36.10

$$\frac{(12-9.43)^2}{9.43} \qquad \frac{(9-10.71)^2}{10.71} \qquad \frac{(9-9.86)^2}{9.86}$$

$$\frac{(10-12.57)^2}{12.57} \qquad \frac{(16-14.29)^2}{14.29} \qquad \frac{(14-13.14)^2}{13.14}$$

Combine the two end columns to get a single column

Alternatively, we can combine the two end columns (each with a low-frequency cell) to get a single column containing people who have extreme opinions of the boss. We would then have have four columns labelled as below, with the expected frequencies shown in Table 36.11. For a test on this table we have $(4-1)(2-1) = 3$ degrees of freedom. The 5% critical value with $v = 3$ is 7.82. Complete the calculation, but first consider carefully whether there is any way of rewording the hypothesis. Or is the existence or non-existence of 'an association' all that can be achieved?

This question becomes particularly pertinent if we recognise that by merging the last three columns of this new table, we could test the hypothesis that there is no association between pay level and whether or not an extreme opinion is held. Table 36.12 shows the observed values (calculated from the original table) and the expected values. There is only 1 degree of freedom. And now we run into a problem. A slightly different procedure has to be used for 2×2 tables. Before completing the discussion of this example, we have to look at this.

Table 36.11

Expected frequencies

Extremely favourable or extremely unfavourable	Favourable	Neutral	Unfavourable
8.57	5.14	10.71	5.57
11.43	6.86	14.29	7.43

Table 36.12

	Holding an extreme opinion		Not holding an extreme opinion	
	observed	expected	observed	expected
Well paid	7	8.57	23	21.42
Less well paid	13	11.43	27	28.58

36.6 ⬤ **2×2 contingency tables**

It is particularly easy to use the χ^2 test wrongly with 2×2 tables. First note that the χ^2 test should be applied to a 2×2 table only if $N > 20$ and no expected cell frequency is less than 5. If either of these conditions does not hold, use Fisher's exact probability test described in Notebox 36.4.

If the conditions do hold, there is still a problem; we are using discrete data, but the chi-square distribution is continuous. This introduces a possibility of error. Fortunately, it can usually be ignored, even with as few as 2 degrees of freedom. But it becomes important when there is only 1 degree of freedom, as is always the case with a 2 × 2 table.

The recommended procedure is to use Yates' continuity correction. This means that every value of $o - e$ is reduced numerically by 0.5, so equation [36.1] becomes

$$\chi^2_{\text{calc}} = \sum (|o - e| - 0.5)^2 / e \qquad [36.2]$$

where $|o - e|$ indicates the (positive) numerical value of the difference. On average it tends to lead to overcorrection. But it is especially important to make this correction if the expected frequency size is between 5 and 10.

Another feature of a 2 × 2 table simplifies calculations by making it unnecessary to calculate the expected frequencies. If we denote the four observed frequencies in a 2 × 2 table by A, B, C, D and the overall total by N (Table 36.13), then some algebra shows that the formula for χ^2_{calc} can be rewritten as

$$\chi^2_{\text{calc}} = \frac{N(|AD - BC| - N/2)^2}{(A + B)(C + D)(A + C)(B + D)} \qquad (v = 1) \qquad [36.3]$$

In this form the continuity correction appears as $N/2$. It is normally easier to use equation [36.3] than to go back to equation [36.1] modified by the continuity correction, especially as equation [36.3] removes the need to calculate expected frequencies. As an aid to memory, the top part of this expression contains the difference between the two diagonal products AD and BC, whereas the bottom is the product of all four row and column totals. If $N > 20$ then equation [36.3] may be used with a 2 × 2 contingency table only if all the expected frequencies are at least 5. With 1 degree of freedom the more commonly used critical values of χ^2 are

$\alpha = 1\%$ $\qquad\qquad \chi^2_{\text{crit}} = 6.635$

$\alpha = 5\%$ $\qquad\qquad \chi^2_{\text{crit}} = 3.841$

$\alpha = 10\%$ $\qquad\qquad \chi^2_{\text{crit}} = 2.706$

Table 36.13

		Total
A	B	$A + B$
C	D	$C + D$
Total $\quad A + C$	$B + D$	N

Box 36.3

How to ... conduct a 2 × 2 contingency test

1. A 2 × 2 contingency test should be performed only if N is greater than 20 and all expected frequencies are at least 5.

2. Check that the condition in step 1 holds by checking N and working out the smallest expected frequency from

$$\text{Smallest expected frequency} = \frac{\text{row total} \times \text{column total}}{N}$$

3. Specify the null hypothesis, choose a level of significance and find the critical value of χ^2 for that level with 1 degree of freedom.

4. Use the following formula to calculate χ^2_{calc}:

$$\chi^2_{calc} = \frac{N(|AD - BC| - N/2)^2}{(A + B)(C + D)(A + C)(B + D)} \qquad (v = 1)$$

where A, B, C, D are the observed frequencies in the four cells arranged as

$$
\begin{array}{cc}
A & B \\
C & D
\end{array}
$$

5. Reject the null hypothesis if χ^2_{calc} exceeds the critical value of χ^2.

When we apply this test to the data in the 2 × 2 table (Table 36.13) we obtain $\chi^2_{calc} = 0.328$, which is not significant at the chosen level of 5%, so it does not warrant rejection of the null hypothesis.

We can now review our findings about the data presented in Table 36.6:

- We have to combine two salary levels before we can begin the analysis, getting Tables 36.7 and 36.8.
- We then have to combine some rows. If we do this as in Table 36.9, χ^2_{calc} is below the 5% critical level and does not warrant rejection of H_0. This is also true if we combine as in Table 36.11, or by forming a 2 × 2 table.
- In all three cases we reach non-significant values. Rejection of the null hypothesis is not warranted.

Sometimes we find that one or more ways of combining the data do not warrant rejection of the null hypothesis but another way does. In that case the null hypothesis should be rejected.

Note that when we apply a chi-squared test to a contingency table, we are testing whether there is some association between the two attributes, in the sense that the frequencies are distributed between the cells in a way that would be unlikely to occur by chance. Except in the 2×2 case, the test does not take account of any ordering of the variables. Only by considering the sizes of the terms arising from each cell can we determine whether there is any pattern in the association.

36.7 Suspiciously small

Sometimes a calculated value of χ^2 may be so close to zero that one becomes suspicious. Observed values will rarely provide an almost perfect fit to a theoretical distribution, except perhaps when very large numbers are involved. Table D10 gives values of χ^2 which are almost bound to be exceeded. For example, it shows that if there are 6 degrees of freedom then there is a 95% chance of getting a calculated value of at least 1.64; and with 10 degrees of freedom there is a 99% chance of getting a value of at least 2.56. If the calculated value falls below these critical values, there are grounds to suspect an error of calculation, fiddling of the data, or some other force that removes the randomness of the event.

36.8 The additive property of χ^2

Sometimes an experiment is performed more than once. It is perfectly in order to sum the values of χ^2 obtained from each performance of the experiment, to sum the degrees of freedom, and then to test the significance. If this is done, it may be found that although each separate performance of the experiment yields a χ^2 which does not warrant rejection at say the 5% level of significance, the new value arising from summation does.

If each performance involves only 1 degree of freedom, the fact that application of Yates' correction tends to lead to overcorrection means it is usually omitted when the additive property is used.

Consolidation exercise 36B

1. When is it necessary to combine cells in performing a χ^2 test?

2. How do you locate the cell that is going to have the smallest expected frequency? How do you combine cells?

3. In examining a contingency table, when is it necessary to consider use of a continuity correction? How is it applied in the version of the formula for χ^2_{calc} that (a) involves and (b) does not involve the expected frequencies?

4. What is the main term on the top of the formula that does not involve expected frequencies? The bottom of the formula is the product of four brackets. What do they contain?

5. What may a very small value of χ^2_{calc} suggest? How can you check whether a value of χ^2_{calc} is too small?

6. If an experiment is repeated several times and each time there is a non-significant χ^2_{calc}, how do you combine the results to make better use of the total information? Should Yates' correction be used in these circumstances?

7. When should Fisher's exact probability test be used?

Notebox 36.1

Degrees of freedom for a χ^2 test

The number of degrees of freedom is the number of cell frequencies that can be chosen with complete freedom without violating the row and column totals. Suppose we have a table of 2 rows and 3 columns. The cells are empty but we have the row and column totals. If we now put any figure in any cell, the remaining figure in that column has to make the column total right. In completing that column, we have only 1 degree of freedom.

If we now choose a second column, we can again write any number in one cell, but the remaining cell has to make the column total right. Thus, for the two columns, we have a total of 2 degrees of freedom.

When we turn to the third (and last) column, we find we have no freedom at all. Both entries in this column must make the row totals right. In all we have only 2 degrees of freedom.

Repeating this exercise with a 3×4 table should convince you of the truth of the rule that if there are m rows and n columns, the number of degrees of freedom is $(m-1)(n-1)$ provided both m and n are greater than 1.

Notebox 36.2

To round or not to round?

People have different opinions about whether these expected frequencies should be rounded to whole numbers. It can be argued that nobody would expect 25.67 patients to have many visitors and to suffer minor illness. If this argument is accepted then 25.67 needs to be rounded up to 26, with similar adjustments of the other expected frequencies.

On the other hand, the expected frequencies are the averages of what we would expect to happen if we were able to repeat this experiment many times. In this sense, when you toss 5 coins the expected number of heads is $2\frac{1}{2}$; but nobody would expect to see $2\frac{1}{2}$ heads in a single trial.

We will accept this argument. However, in most cases calculation is a bit easier if the frequencies are rounded, and no harm is done if the calculated value is well clear of the chosen critical value. It may be convenient and acceptable practice to round the expected frequencies for the purpose of a first calculation, but to be ready to perform a more accurate calculation unless this is clearly unnecessary.

Notebox 36.3

An alternative formula for χ^2_{calc}

Another way of stating equation [36.1] is

$$\chi^2_{calc} = \sum \frac{\sigma^2}{e} - N \qquad [36.4]$$

where N is the total frequency. Using this sometimes facilitates calculations, although it involves squaring larger numbers. However, equation [36.1] is more useful in that it allows quick identification of cells that make large (or small) contributions to the value of χ^2_{calc}.

Notebox 36.4

Fisher's exact probability test

If the total frequency N is not greater than 20, or if any expected cell frequency is less than 5, the χ^2 test should not be used on a 2×2 contingency table. Instead we use a test due to Fisher, in which we calculate (i) the probability of getting the observed cell frequencies, given the row and column totals; (ii) the probabilities of all equally or less likely sets of expected frequencies compatible with the row and column total; and (iii) the totals of these probabilities, which gives the probability of getting the observed results or 'even worse'. This is just what we do when calculating the significance probability (Chapter 28), but we can also adapt the procedure to hypothesis testing. We now illustrate the method, which can be speeded up by using a few tricks, but it takes a little space to explain them.

As an example, suppose we are asked whether the observed frequencies in Table 36.14 suggest the existence of some relationship or association between X and Y. We set up H_0 that there is no association, and H_1 that there is an association. This is a two-tailed hypothesis. We could equally well set up a one-tailed alternative hypothesis that (in this case) there is a positive association, with Y being high or low according to whether X is high or low.

A typical 2×2 table, with four cells, two row totals, two column totals and a grand total is Table 36.15, where $N = a + b + c + d$. Note carefully the positions of the four letters a, b, c and d. The marginal (or row and column) totals are $(a + b)$, $(c + d)$, $(a + c)$ and $(b + d)$. We may know all of these or be able to infer them from a knowledge of some of them and of N.

Table 36.14

	X_1	X_2	Total
Y_1	8	4	12
Y_2	3	10	13
Total	11	14	25

Table 36.15

a	b	a + b
c	d	c + d
a + c	b + d	N

Several possible values of a, b, c and d are compatible with given row and column totals. It can be shown that if we are given (or have inferred) the marginal totals, then the probability of the cells having the four values stated above is

$$P(a, b, c, d) = \frac{(a+b)!\,(c+d)!\,(a+c)!\,(b+d)!}{N!\,a!\,b!\,c!\,d!}$$

Notice that the top consists of the factorials of the four marginal totals, whereas the bottom is the product of the factorials of the grand total and the four cell frequencies. Thus, in our example of Table 36.14, we have

$$(a+b) = 12 \qquad (c+d) = 13 \qquad (a+c) = 11 \qquad (b+d) = 14$$

and $N = 25$. Therefore the probability of $a = 8$, $b = 4$, $c = 3$ and $d = 10$ is

$$P(8, 4, 3, 10) = \frac{12!\,13!\,11!\,14!}{25!\,8!\,4!\,3!\,10!}$$

With the help of a calculator this is easily shown to be 0.031 76. This is the probability of the marginal totals producing exactly the observed cell frequencies if the null hypothesis is right. However, to test the hypothesis, we need the probability of getting the observed result 'or worse'. This is equivalent to saying that when we apply a normal test (or t-test), we are not content with knowing the probability of getting exactly the critical value. We want the probability of getting the critical value or something 'worse' – something even more different from the expected value.

We need therefore to look at the other sets of values of a, b, c and d that are compatible with our mariginal totals. A couple of tricks help us here, as we now demonstrate.

We go back to Table 36.14. We can increase 8 to 9, 10 or 11 without violating any total. Whichever we do, the other cell frequencies are determined by the totals, so we get the following three arrangements:

A	9	3	B	10	2	C	11	1
	2	11		1	12		0	13

We can also decrease 8 to 7, 6, 5, 4, 3, 2, 1 and 0, getting the following eight arrangements:

D	7	5	E	6	6	F	5	7	G	4	8
	4	9		5	8		6	7		7	6
H	3	9	J	2	10	K	1	11	L	0	12
	8	5		9	4		10	3		11	2

There are no other arrangements compatible with the marginal totals. We need the probabilities of all these arrangements. They can be derived from the basic formula, but if we are systematic we can save a lot of work. For example, the probability of the observed set of numbers, given in the first table, was written as

$$P(8, 4, 3, 10) = \frac{12!13!11!14!}{25!8!4!3!10!} = 0.031\,76$$

Now we seek the probability of getting arrangement A; it is

$$P(9, 3, 2, 11) = \frac{12!13!11!14!}{25!9!3!2!11!}$$

If we compare this with the previous expression, we see that they differ only in that the bottom has been altered from 25!8!4!3!10! to 25!9!3!2!11!.

Using the fact that $9! = 9 \times 8!$ and similar expressions, we see that for A

$$(9, 3, 2, 11) = P(8, 4, 3, 10) \times (4 \times 3)/(9 \times 11) = 0.031\,76 \times 12/99 = 0.003\,85$$

Similarly we can get $P(10, 2, 1, 12) = 0.000\,19$ (to 5 d.p.) for arrangement B, and $P(11, 1, 0, 13) = 0.000\,00$ (to 5 d.p.) for arrangement C.

Notice that we have obtained A, B and C from the original set of frequencies by successively increasing the top left value in steps of 1. For A we have obtained $P(9, 3, 2, 11)$ from $P(8, 4, 3, 10)$ by multiplying $P(8, 4, 3, 10)$ by the middle two terms of the old probability (4, 3) and dividing it by the two end terms of the new probability (9, 11). We have obtained the probabilities for B and C by similar adjustments of the previous value. Table 36.16 summarises the probabilities so far calculated, correct to four decimal places.

We now consider the probabilities for arrangements D to L. These sets have been obtained from the original set by decreasing the top left value, the reverse of the procedure that produced A, B and C.

Table 36.16

Arrangement	Original	A	B	C
Probability	0.0318	0.0038	0.0002	0.0000

Table 36.17

Arrangement	E	F	G	H	J	K	L
Probability	0.2668	0.3049	0.1906	0.063	0.0159	0.0008	0.000

Continuing in this way through sets E to L, we obtain the probabilities (correct to 4 d.p.) in Table 36.17. As a check we note that, except for a small rounding error, the probabilities add to 1. We are interested only in those probabilities that are equal to or less than the probability of the observed set of frequencies; these are

0.0318, 0.0038, 0.0002, 0.0000, 0.0159, 0.0008, 0.0000

They total 0.0525. Thus there is a chance of 5.25% of getting the observed set of frequencies or worse if the null hypothesis of no association between the variables is correct.

Alternatively we can argue in terms of the alternative hypotheses. In the two-sided case, the probability of getting the observed result or worse exceeds the chosen 5% critical level, so it does not warrant the rejection of H_0.

But if we take the one-sided alternative hypothesis of a positive association, we count only the the observed probability and the three other probabilities in Table 36.14, all of them having been obtained for arrangements showing an even more extreme positive association, and all of them less than the observed probability. The probability of the data showing such a strong association as they do, or stronger, is the sum of these probabilities, i.e. 0.0358. Thus there is less than a 5% probability of getting such a large positive association as that indicated, so the null hypothesis is rejected in a one-sided test.

There are two points to note. The evidence does not warrant the rejection of the null hypothesis unless one looks not at the difference between the observed and expected frequencies, but at the probability of the difference in the positive direction being so marked.

Sometimes the probability of the original frequencies is so great that the significant level is clearly exceeded, and there is no need to look at the probabilities of what we have called the other arrangements.

Summary

1. A contingency table classifies a total frequency according to two variables, which may be numerical scores or non-numerical attributes. An $m \times n$ table will have m rows and n columns, forming mn cells, each showing the frequency with which those values of the two variables occur jointly. Note that the entries in the cells are always frequencies. They should not be rounded in larger units, such as hundreds or millions.

2. The number of degrees of freedom associated with an $m \times n$ table is $(m-1)(n-1)$ provided neither m nor n is unity.

3. To test for an association between the two variables, we compare the actual frequencies with those expected on the basis of the null hypothesis of no association. These are easily calculated as in Box 36.1. We use the observed and expected frequencies to calculate the value of a test statistic χ^2_{calc}, which is then compared with a critical value derived from tables of a theoretical χ^2 distribution. This depends on the number of degrees of freedom and chosen level of significance. The test is described in Box 36.2.

4. To apply the test there should be at least two degrees of freedom and no expected frequency should be less than 5. If any frequency is less than 5, cells should be combined.

5. The test should be used with a 2×2 table only if the total frequency is at least 20 and no cell frequency is less than 5. In this case the calculation can be simplified by proceeding as in Box 36.3, which also incorporates the continuity correction that becomes necessary with a table of this size.

6. A value of χ^2_{calc} may be suspiciously small, i.e. the observed and expected results are too close to be credible (Section 36.8).

7. If an experiment is repeated, the values of χ^2_{calc} can be added (Section 36.8). It may be that no single experiment produces a significant result, but their aggregation does.

8. If the frequency conditions do not hold, consider using Fisher's exact probability test. This calculates the exact probability of getting the obtained frequencies and all equally or less likely frequencies, given the need to preserve the row and column totals. The procedure is explained in Notebox 36.4.

9. Other noteboxes deal with degrees of freedom, rounding and an alternative formula.

Before and after

37.1 Introduction

Example 37.1 A studio audience in the TV programme *A Mind to Change* was asked to vote for or against the penalisation of pedagogical obscurity by a tax on long words. The recorded votes were 73 for and 27 against. After debate several people changed their minds and a vote recorded 70 for and 30 against. Ten of those who first voted in favour ended up voting against, and 7 of those who first voted against ended up voting in favour. Did the debate have an overall effect on people's opinions?

Example 37.2 The employees of a medium-sized office vote for working (1) more or (2) fewer hours, with pay adjustments: 45 vote for (1) and 75 for (2). The implications for travel time are then explored and a second vote is taken. Now 83 people vote for (1) and 37 for (2), as in Table 37.1. Did the study of travel time have a significant effect on opinion?

These examples ask questions about the effect of an experience or treatment, in this case on opinions. They are solved by testing the significance of a change. In this chapter we describe two commonly used tests; one is a version of the χ^2 test due to McNemar and the other is a sign test.

37.2 The McNemar test

The McNemar test for significance of changes is especially useful in testing before and after differences. It may be used with nominal or higher-level data. Each

Table 37.1

		After study of travel time		
		(1)	(2)	
Before study of	(2)	63	12	75
travel time	(1)	20	25	45
		83	37	

424

person is his or her own control. The test is a very simple χ^2 test with a continuity correction.

Esentially this is a test of whether, in response to some experience or treatment, the number of items changing in a specified way differs significantly from the number changing in the opposite way. We illustrate this in answering Example 37.1, but before becoming involved with the numerical work, we need to look carefully at the procedure.

The numbers of people voting for and against the motion before and after the debate are recorded in a 2×2 table (Table 37.2) where the letters A, B, C and D represent the cell frequencies. Note carefully that people who changed their minds (shown by A and D) are recorded in the diagonal from top left to bottom right. In using this test and the associated formulae, we must always follow this order of letters, row titles and column headings.

We frame a null hypothesis that the debate makes no difference to opinions. In this case, since $A + D$ people changed their views, we would expect half of them to change from for to against, and half from against to for; hence the expected frequencies in the cells that have actual frequencies of A and D would each be $\frac{1}{2}(A + D)$. The null hypothesis can therefore be stated as $A = D = \frac{1}{2}(A + D)$ or as $A - D = 0$. We compare these two expected frequencies, each of $\frac{1}{2}(A + D)$, with the actual frequencies A and D by using the χ^2 test.

The statistic we have to calculate emerges after a little algebra as

$$\chi^2_{\text{calc}} = \frac{(A - D)^2}{A + D} \qquad (v = 1)$$

However, this has no continuity correction. Yates' correction, which is especially important if all the expected cell frequencies are small, and works on average with a tendency to overcorrect, is to subtract 1 from the numerical value of $A - D$, so the revised formula is

$$\chi^2_{\text{calc}} = \frac{(|A - D| - 1)^2}{A + D} \qquad (v = 1) \qquad [37.1]$$

Notice how this test statistic depends only on A and D. To perform the test, we need to know only the numbers of changes in both directions. We do not need to know the row and column totals or the total frequency.

Table 37.2

		AFTER DEBATE	
		For	Against
BEFORE DEBATE	Against	A	B
	For	C	D

We state our hypotheses, beginning with the null hypothesis H_0: $A - D = 0$, and the alternative hypothesis H_1: $A - D \neq 0$. We decide to use a 5% level of significance. This leads to the critical value (from tables) of $\chi^2 = 3.84$. A value of χ^2_{calc} exceeding this should lead to rejection of the null hypothesis.

⬭ In **Example 37.1** $A = 7$ and $D = 10$. The expected cell frequency of $\frac{1}{2}(A + D)$ is 8.5, which suggests it may be safer to use the correction. Using equation [37.1] we calculate $\chi^2_{calc} = 4/17 = 0.2353$, which is much less than the critical value. There are no grounds for rejecting the null hypothesis. If we had not used the correction, we would have obtained $9/17 = 0.5294$, which leads to the same conclusion.

Anybody using this test at all frequently could usefully memorise some or all of the following critical values:

$\alpha = 10\%$ $\chi^2_{crit} = 2.71$
$\alpha = 5\%$ $\chi^2_{crit} = 3.84$
$\alpha = 2\%$ $\chi^2_{crit} = 5.41$
$\alpha = 1\%$ $\chi^2_{crit} = 6.64$
$\alpha = 0.1\%$ $\chi^2_{crit} = 10.83$

As an aid to memory, notice that the 0.1% value is double the 2% value, which in turn is double the 10% value.

Box 37.1

How to ...   **use the McNemar test for the significance of changes**

1. The test is especially useful in testing 'before and after' differences. It may be used with nominal or higher-level data. If the expected frequency $\frac{1}{2}(A + D)$ is less than 5, the McNemar test should not be used.

2. Call the number of items changing in a specified way A and the number changing in the reverse way D. These numbers may be given data. If they are not, it will be possible to infer them if you know the row and column totals and the entry in any one cell of a table such as Table 37.2, with the cells corresponding to changes appearing in the diagonal from top left to bottom right.

3. Specify the acceptable level of significance for testing the null hypothesis that treatment has no effect.

4. Calculate

$$\chi^2 = \frac{(|A - D| - 1)^2}{A + D} \qquad (\nu = 1)$$

which includes Yates' correction.

5. Consult tables of the significant values for χ^2, and reject the null hypothesis if the estimated value equals or exceeds the critical value at the chosen level of significance.

If the expected frequency $\frac{1}{2}(A+D)$ is less than 5, the McNemar test should not be used. Instead the binomial test can be used, with $n = A + D$ and the smaller of the two frequencies A or D taken as the value of x.

For example, if $A = 3$ and $D = 5$ we can use a one-tailed binomial test on the probability of getting $x = 3$ or fewer with $n = 8$.

Consolidation exercise 37A

1. What is tested by the McNemar test? It uses a table of 4 cells. In which cells do the numbers show changes? What is the null hypothesis for this test?

2. What formula gives the calculated value of the test statistic? What tables are consulted for the critical value? How is the continuity correction applied?

3. When should the McNemar test be replaced by a binomial test?

37.3 The sign test

The sign test can also be used in 'before and after' studies. It cannot be used with nominal data. The following solution of Example 37.2 further illustrates the normal approximation to the binomial distribution.

The data are summarised in Table 37.1. This shows that 63 changed their opinions from fewer to more, and 25 changed from more to fewer. This information could also be analysed using the McNemar test. But here we use the data to illustrate the sign test with a large sample.

 With numbers of this size we can use the normal distribution, testing the chance of getting as many as 63 changes from fewer to more out of a total of 88 changes, with a mean expectation of $np = 88 \times 0.5 = 44$, and a standard deviation of $\sqrt{npq} = \sqrt{88 \times 0.5 \times 0.5} = 4.69$. With a continuity correction the calculated test statistic becomes

$$z = \frac{(63 - 0.5) - 44}{4.69} = 3.945$$

indicating a deviation from the mean of almost four standard errors, which is extremely unlikely if the null hypothesis is true. The evidence suggests that knowledge of the travel time implications shifted opinion more than is likely to have arisen by chance. (If we take a one-tailed test, we can say that it did so in favour of more hours.)

Box 37.2

How to ... **use the sign test for significance of a change**

1. The sign test compares members of matched pairs, or individuals with themselves. It requires ordinal (or higher-level) data. It assumes all individuals are rated, or put into ordered classes, on the basis of a continuous variable.

2. For each pair (or for each person in each circumstance) determine the sign of the difference.

3. Formulate the null hypothesis that the number of changes in one direction is no different from the number in the other (or $A = D$). Decide on whether to have a one-tailed or two-tailed alternative hypothesis and state a level of significance.

4. Count the number of pairs n showing a sign (i.e. with a non-zero difference).

5. If n is not more than 10, or if suitable tables are not at hand, use the binomial test to assess the probability of getting a number of sign changes as small as or smaller than x, where x is the lower of the number of $+$ signs and the number of $-$ signs, referring to Box 18.2 with $p = 0.5$. Otherwise calculate z from

$$z = \frac{2(A - 0.5) - (A + D)}{\sqrt{A + D}}$$

   where A is the number changing one way and D is the number changing the other way. Use the normal distribution table to obtain the critical value appropriate to the chosen level of significance, with mean $\frac{1}{2}(A + D)$ and variance $\frac{1}{4}(A + D)$.

6. Reject the null hypothesis if the probability is equal to or less than the chosen level of significance.

Consolidation exercise 37B

1. How does the table used in the sign test for before and after differ from the table for the McNemar test? What is the null hypothesis for this test? What effect do the frequencies in the upward sloping diagonal (C and B) have?

2. The binomial test is used if n is not more than 10. What is n?

Summary

1. The McNemar test and the sign test indicate whether some experience or treatment has produced more changes in one direction than in the opposite direction. This is a common question in 'before and after' studies. The tests use only data about changes; numbers recording no change are irrelevant.

2. The McNemar test may be used with nominal or higher-level data, provided the total number of changes is at least 10. It is a very simple χ^2 test with a continuity correction that is embodied in equation [37.1]. Note especially the arrangement of the data in Table 37.2, with the numbers of changes recorded in the diagonal from top left to bottom right.

3. If the total number of changes is less than 10, a binomial test should be used.

4. The sign test cannot be used with nominal data. It uses the same tabular arrangement as the McNemar test, and tests the imbalance of change by using either a binomial test or the normal approximation to it (Box 37.2).

38

Non-parametric correlation

Introduction

▷ **Example 38.1** As part of an American-aided programme to ensure the moral well-being of students, Statingham University has asked the Bishop to scrutinise all videos sold on the campus to check their suitability for sophomores. The editor said that inspecting them for the length of time that Lovely Lily is on screen would produce the same results. To test this, Anna Liszt took a random sample of seven videos from a large catalogue and ranked them according to the Lily factor, giving 1 to the video in which Lily appeared most. Then, hiding this from the Bishop, who was unaware of the editor's assertion, she asked him to rank the same videos in order of suitability, giving 1 to the most suitable. Do the results in Table 38.1 support the editor?

▷ **Example 38.2** For 13 traffic wardens selected at random from the whole country, Table 38.2 shows the recorded numbers of hours during which they were unable to perform their duties last month, and their stated average daily consumption of banana sandwiches. A scattergram does not have a more or less elliptical shape, but we wish to examine their correlation. How do we proceed?

We are again faced with questions about the extent of association between two variables. We have already seen that one measure is provided by the χ^2 test. This chapter begins by describing an important if not entirely successful attempt to convert the value of χ^2 into a correlation coefficient, called the **contingency coefficient**.

Table 38.1

Video	K	L	M	N	P	Q	R
Suitability rank	1	7	5	2	4	6	3
Lily factor rank	2	6	7	1	3	5	4

Table 38.2

Warden	A	B	C	D	E	F	G	H	J	K	L	M	N
Hours	18	13	25	4	19	21	160	20	158	22	58	33	65
Sandwiches	4	5	6	3	4	10	11	12	13	6	7	8	9

Then we look at two other non-parametric correlation coefficients that are particularly useful with ranked data but may profitably be used in certain other cases. One is due to Spearman and the other to Kendall. Each test has its own significance table for small numbers of observations. For larger numbers the significance of Spearman's coefficient can be found by using the *t*-test, and the significance of Kendall's coefficient can be found by using the normal distribution.

These rank correlation coefficients should also be used when the scattergram is of unacceptable shape, as in Example 38.2.

38.2   The contingency coefficient

A two-variable frequency table of *r* rows and *c* columns is called an **r × c contingency table**. The variables may be measured on any scale from nominal to ratio, but usually ratio and interval measurements are grouped in order to have a manageable number of cells with frequencies that are not unacceptably low.

Whatever the size of the table, the value of χ^2 (calculated as in Chapter 37) measures the degree of association. It tends to increase as the number of observations increases, and can be quite large without becoming significant. With 20 degrees of freedom, χ^2 has to be as high as 43.77 before becoming significant at 5%. In an attempt to obtain a measure of correlation that will produce values lying between 0 and 1, the **contingency coefficient** has been defined as

$$C = \sqrt{\chi^2/(N + \chi^2)} \qquad\qquad [38.1]$$

where *N* is the total number of observations (which is the total of the cell frequencies).

If there is absolutely no association and $\chi^2 = 0$ (which would be suspicious, see page 417), *C* will be zero. For any given *N*, the value of *C* increases as χ^2 increases, but *C* can never reach unity. Unfortunately, the upper limit may be much lower than this and is not easy to find. For example, to take an easy case, if there are as many columns as rows, so that $r = c$, and if the two variables are perfectly correlated, then the maximum value that *C* can have is $\sqrt{(r-1)/r}$. Thus with a 2 × 2 contingency table its maximum value is $1/\sqrt{2} = 0.707$. With a 5 × 5 table it is 0.89, and with a 10 × 10 table it is 0.949.

Despite these and other limitations, *C* is widely used. However, *C* is significant if and only if χ^2 is significant. Instead of testing the significance of *C*, we test the significance of χ^2, as described in Chapter 36.

Consolidation exercise 38A

1. When should correlation be examined by use of a non-parametric measure, rather than a parametric measure?

2. What kind of data (or level of measurement) can be tested by the contingency coefficient? How is it defined? What is its maximum value?

3. How do you test the significance of the contingency coefficient?

38.3 Spearman's rank correlation

If both variables can be measured ordinally, so that they can be ranked, their correlation can be measured by Spearman's coefficent. We describe this with the aid of a simple example.

Suppose a class of n students sits examinations in subjects X and Y. They are ranked in order of performance, with ranks going from 1 to n. If there is a perfect correlation between performances in the two subjects, then for every student the X rank (X) should equal the Y rank (Y). If, as is likely, the marks differ then the greater the difference, the lower the correlation; but we can expect some difference by chance. We have to measure the difference in a way that enables us to assess its significance.

If we denote the difference between X and Y by d, so that every student has a value of d (which may be zero) then **Spearman's rank correlation coefficient** is defined by

$$r_s = 1 - \frac{6 \sum d^2}{(n^3 - n)}$$ [38.2]

where $\sum d^2$ is the sum of the squares of all the ds (one for each student), and there are n students. This formula is derived from the product moment correlation coefficient with the variables replaced by ranks (Notebox 38.2).

If the students are a random sample from a large population we may test whether the calculated value of r_S indicates that there is probably an association between X and Y in the population.

If there are *tied ranks* (such as when two or more persons have the same rank in one or more of the tests), the tying members should all be given the average of the ranks they would have had if they had been very slightly different from each other. (Thus if we have eight people ranked 1, 2, 2, 4, 4, 4, 7, 8 then the two people ranked 2 should both be ranked $(2 + 3)/2 = 2\frac{1}{2}$, and the three people ranked 4 should all have the rank $(4 + 5 + 6)/3 = 5$, so we get 1, $2\frac{1}{2}$, $2\frac{1}{2}$, 5, 5, 5, 7, 8.)

It can be shown that if the number of ties is only a small proportion of the number of persons n, the effect on r_S will be trivial. The general procedure in this case is given by Box 38.1. But if the number of ties forms a large proportion of the total, use the correction in Box 38.2.

Box 38.1

How to ...   calculate and test Spearman's rank correlation coefficient if there are not many ties

1. Specify the null and alternative hypotheses and significance level in the usual way.

2. Rank the n members according to their X scores (X) and their Y scores (Y).

3. For each member calculate the difference between ranks d and square it.

4. Sum the values of d^2 to get $\sum d^2$

5. If the number of ties is not large compared with n, calculate r_S from

$$r_S = 1 - \frac{6 \sum d^2}{(n^3 - n)}$$ [38.2]

6. If n is between 4 and 30, consult Table D11 for the significant value of r_S. If the calculated value is less than the tabulated value, do not reject the null hypothesis of no correlation.

7. If $n \geqslant 10$ we can use the alternative test statistic

$$t = r_S \sqrt{(n-2)/(1-r_S^2)}$$

and tables for the t-test, with $n - 2$ degrees of freedom.

We illustrate the procedure of Box 38.1 by tackling Example 38.1.

1. We specify the null hypothesis H_0 that there is no correlation between the data in the two rows. It has been said that the Lily test is a good substitute for the Bishop's assessment. This could be true if there is a sufficiently high positive or negative correlation. It could mean either that if the film has lots of Lily, the Bishop will declare it to be unsuitable, or that lots of Lily will guarantee episcopal praise. Thus the alternative hypothesis has to be two-tailed. We adopt a 5% level of significance.

2. and 3. The ranking has already been done. The given ranks, together with the values of d and d^2 are shown in Table 38.3.

4. From the last row we get $\sum d^2 = 10$.

5. There are no ties, so we put $n = 7$ and $\sum d^2 = 10$ in equation [38.2] to calculate

$$r_S = 1 - (6 \times 10)/(343 - 7) = 1 - 0.1786 = 0.8214$$

6. Table D11 shows that for $n = 7$ (or $\nu = 5$) the two-sided critical value at the 5% level of significance is 0.786. This is the value we would expect 5% of the time even if there were no correlation in the population. But we would not expect our calculated value so often. We therefore reject the null hypothesis in favour of the alternative that there is a correlation, which supports the editor's assertion.

Table 38.3

. . . .

Video		K	L	M	N	P	Q	R
Suitability rank (X)		1	7	5	2	4	6	3
Lily factor rank (Y)		2	6	7	1	3	5	4
$d = X - Y$		−1	1	−2	1	1	1	−1
d^2		1	1	4	1	1	1	1

. . . .

With larger numbers we use the t-test to assess the significance of the Spearman rank correlation coefficient, as we now illustrate. Example 38.2 uses the data shown in Table 38.2. On the face of it we might try a parametric correlation, with a few reservations about the reliability of the data, but the scattergam is not right. We therefore use a rank correlation analysis.

➦ 1. We specify the null hypothesis H_0 that there is no correlation between hours away from workstation and consumption of banana sandwiches; and a one-sided alternative hypothesis $H-1$ that there is a positive correlation. We adopt a 5% level of significance.

2. and **3.** The scores in the original table are now replaced by ranks. It is convenient to give the lowest score in each row the rank 1. These ranks are shown in Table 38.4, along with values of their difference d and then d^2.

Table 38.4

Warden	A	B	C	D	E	F	G	H	J	K	L	M	N
Hours (X)	3	2	8	1	4	6	13	5	12	7	10	9	11
Sandwiches (Y)	$2\frac{1}{2}$	4	$5\frac{1}{2}$	1	$2\frac{1}{2}$	10	11	12	13	$5\frac{1}{2}$	7	8	9
$d = X - Y$	$\frac{1}{2}$	2	$2\frac{1}{2}$	0	$1\frac{1}{2}$	4	2	7	1	$1\frac{1}{2}$	3	1	2
d^2	$\frac{1}{4}$	4	$6\frac{1}{4}$	0	$2\frac{1}{4}$	16	4	49	1	$2\frac{1}{4}$	9	1	4

4. From the last row we get $\sum d^2 = 99$.

5. We will argue (for the moment) that 2 ties in 13 observations is not a high proportion. Shortly we will rework this example without saying this. Using $n = 13$ and $\sum d^2 = 99$ in equation [38.2] we calculate

$r_s = 1 - (6 \times 99)/(2197 - 13) = 1 - 0.2720 = 0.7280$

There are $13 - 2 = 11$ degrees of freedom. Table D11 does not give critical values for $v = 11$, but for $v = 10$ it gives a 5% two-tailed value of 0.591, and for $v = 12$ it gives a value of 0.544. The value we seek is between these two, so the calculated value of 0.7280 is well above the critical value. We reject the null hypothesis.

An alternative significance test, useful when n is large enough, is given in Notebox 38.1

Taking account of ties

We now solve the example without ignoring the existence of 2 ties, both in the bottom row (the Y row). It can be shown that the Spearman rank correlation coefficient can also be derived from the formula

$$r_S = \frac{\sum x^2 + \sum y^2 - \sum d^2}{2\sqrt{(\sum x^2)(\sum y^2)}} \quad [38.3]$$

where $\sum x^2$ and $\sum y^2$ are obtained as we now describe with the help of our example.

In **Example 38.2** each tie has 2 observations. There could have been 3 or more. To be general for a moment, denote the number of items tying in the first tie by t. Then, for this first tie, we calculate $T_Y = (t^3 - t)/12$. In the example $t = 2$, so $T_Y = 6/12 = 0.5$. The second tie also contains 2 observations, so when we repeat the calculation we get the same answer. Thus, in our example, we can write $\sum T_Y = 0.5 + 0.5 = 1$. We apply the same procedure to the top row (the X row). This has no ties, so $t = 0$, giving $\sum T_X = 0$.

We now use the following formulae, which can be looked upon as defining the values of $\sum x^2$ and $\sum y^2$ for use in equation [38.3]:

$$\sum x^2 = (n^3 - n)/12 = \sum T_X$$
$$\sum y^2 = (n^3 - n)/12 - \sum T_Y$$

Inserting $n = 13$, $\sum T_X = 0$ and $\sum T_Y = 1$, we get

$$\sum x^2 = [(13^3 - 13)/12] - 0 = 182$$
$$\sum y^2 = [(13^3 - 13)/12] - 1 = 181$$

We can now calculate the Spearman rank correlation coefficient from

$$r_S = \frac{\sum x^2 + \sum y^2 - \sum d^2}{2\sqrt{(\sum x^2)(\sum y^2)}} \qquad [38.3]$$

$$= \frac{182 + 181 - 99}{2\sqrt{182 \times 181}} = 264/363 = 0.7273$$

Comparison with the previous result (0.7280) shows that the existence of two ties, both in one row, in 13 pairs of observations affects the value of r_S only very slightly. If there are many ties, or multiple ties (with several items all having the same value) the effect is greater.

Box 38.2

How to ... **calculate and test Spearman's rank correlation coefficient if there are many ties**

1 to 4. Proceed as in Box 38.1.

5. If there are several ties, or multiple ties, proceed as follows:

    (i) For each rank with an X-tie let t be the number of items tying (usually 2 or 3 but it can be higher). Calculate $T_X = (t^3 - t)/12$

    (ii) Total the values of T_x to get $\sum T_X$

    (iii) Calculate $\sum x^2 = (n^3 - n)/12 - \sum T_X$.

(iv) Repeat steps (i) to (iii) for the Y-ties, getting $\sum T_Y$ and $\sum y^2 = (n^3 - n)/12 - \sum T_Y$.

(vi) Calculate the correlation coefficient from

$$r_S = \frac{\sum x^2 + \sum y^2 - \sum d^2}{2\sqrt{(\sum x^2)(\sum y^2)}}$$ [38.3]

6 and **7.** As in Box 38.1.

Consolidation exercise 38B

1. In the formula for Spearman's rank correlation coefficient, what does d represent?

2. How are tied ranks written? How does the procedure depend on their number?

3. How would you test the significance of a rank correlation coefficient if there are (a) 8 and (b) 18 items? What is the minimum number of items you must have in order to use your chosen test for case (b)?

38.4 **Kendall's rank correlation**

Kendall's rank correlation coefficient τ takes a different approach. When r_S can be used, so can τ, and vice versa. We use it to examine the same data. It is quite different in concept from Spearman's coefficient, and a different value must be expected. What matters is the significance of the value, when appropriately tested, rather than the value itself. It is quite possible for one test to show a correlation significant at say the 5% level, whereas the other does not show it.

The first step in the calculation of **Kendall's rank correlation coefficient** is to rank the scores, which has already been done in the calculation of the Spearman coefficient.

The next step is to shuffle the columns so that the wardens appear in increasing order of X, as shown in Table 38.5. If X and Y were perfectly correlated, the third row would be in increasing order of Y, but it is not. We measure the extent of departure.

Compare the first Y entry with each of the entries to its right in turn. If the entry to the right is greater, give it a score of $+1$; if it is less, give it a score of -1; if it is equal, give it a score of 0, as in the first row of Table 38.6.

Table 38.5

Warden	D	B	A	E	H	F	K	C	M	L	N	J	G
Hours (X)	1	2	3	4	5	6	7	8	9	10	11	12	13
Sandwiches (Y)	1	4	$2\frac{1}{2}$	$2\frac{1}{2}$	12	10	$5\frac{1}{2}$	$5\frac{1}{2}$	8	7	9	13	11

Table 38.6

for D	$+1+1+1+1+1+1+1+1+1+1+1+1$	$=12$
for B	$-1-1+1+1+1+1+1+1+1+1+1$	$=7$
for A	$0+1+1+1+1+1+1+1+1+1+1$	$=9$
for E	$+1+1+1+1+1+1+1+1+1+1$	$=9$
for H	$-1-1-1-1-1-1+1-1$	$=-6$
for F	$-1-1-1-1-1+1+1$	$=-3$
for K	$0+1+1+1+1+1$	$=5$
for C	$+1+1+1+1+1$	$=5$
for M	$-1+1+1+1$	$=2$
for L	$+1+1+1$	$=3$
for N	$+1+1$	$=2$
for J	-1	$=-1$

Then add these scores. Since the first entry is 1, all twelve entries to its right give $+1$, so the total score for D is 12. We now do the same for the second Y entry, looking only at entries to its right. Going on in this way, we complete the table. The grand total, or **score**, is 44; in general, we denote this by S.

It can be shown that with n cells the maximum possible score obtainable in the case of perfect agreement is $\frac{1}{2}n(n-1)$. Kendall's coefficient compares the actual score with the maximum possible score. If there are no ties, it is defined to be

$$\tau = \frac{S}{\frac{1}{2}n(n-1)} \tag{38.4}$$

In our case the denominator is $\frac{1}{2}(13)(12) = 78$, so τ would be $44/78 = 0.564$ except for the fact there are two Y-ties. We therefore have to modify the formula, as described in step 10 of Box 38.3, which gives $\tau = 0.571$.

The significance of the coefficient can be found from tables, as described in Box 38.3. If $n > 10$ we can use the normal distribution with the test statistic

$$z = \frac{\tau}{\sqrt{2(2n+5)/9n(n-1)}} \tag{38.5}$$

In our case, using the value that takes account of ties, this gives us

$$z = \frac{0.571}{\sqrt{(2 \times 31)/(9 \times 13 \times 12)}} = 2.727$$

indicating a highly significant correlation.

Box 38.3

How to ... calculate and test Kendall's rank correlation coefficient

1. Let there be two lists of ranks, list A and list B, with an A value and a B value for each of n items or members.

2. Write list A as row A with the members arranged in ascending order of the A values.

3. Write list B beneath it, in row B, ordered so that every member's B score is beneath their A score.

4. Take the leftmost entry in row B; call this the first entry.

5. Count the number of entries to its right that are higher than it. Subtract the number of entries to its right that are lower than it. Call the difference the first entry score.

6. Move one place to the right. Repeat the procedure in step 5 to obtain the next entry score.

7. Repeat step 6 until you have a score for each entry in row B. The last entry score will necessarily be 0.

8. Add all the entry scores to get a total entry score. Call this S.

9. If there are no tied ranks, define the rank correlation coefficient to be

$$\tau = \frac{S}{\frac{1}{2}n(n-1)} \qquad\qquad [38.4]$$

10. If there are ties, the denominator of equation [38.4] becomes

$$\tau = \frac{S}{\sqrt{[\frac{1}{2}n(n-1) - T_X][\frac{1}{2}n(n-1) - T_Y]}}$$

In this expression T_X and T_Y are obtained as follows. Suppose for the A values that t_a observations are tied at rank a, t_b observations are tied at rank b, and so on. Then we define $T_X = \frac{1}{2}[t_a(t_a - 1) + t_b(t_b - 1) + \ldots]$, which can be summarised as $T_X = \frac{1}{2}\sum t(t - 1)$ where there is a value of t for every rank that has ties. T_Y is similarly defined for the B values.

11. Use Table D11 to test the significance of τ. If $n > 10$ we can use the normal distribution with the test statistic

$$z = \frac{\tau}{\sqrt{2(2n + 5)/9n(n - 1)}}$$

following the procedure described in Box 25.1.

Consolidation exercise 38C

1. Suppose that when used on the same ranked data, Kendall's method and Spearman's method give different values for the rank correlation coefficent. Which is correct?

2. After ranking the data, what is the next step in calculating Kendall's coefficient? Do you need to take account of ties?

3. How do you test the significance of Kendall's coefficient?

Notebox 38.1

An alternative significance test for r_S

There is an alternative way of testing the significance of the coefficient. If $n \geqslant 10$, instead of using Table D11 we can use the t-test, with t defined by

$$r_S \sqrt{(n-2)/(1-r_S^2)} \qquad (v = n-2)$$

Notebox 38.2

Reconciling the formulae

Although equations [38.2] and [38.3] look very different, they have both been derived from the basic formula for the product-moment correlation. If there are no ties, then $\sum T_X$ and $\sum T_Y$ may both be put equal to zero. If we then substitute the modified versions of $\sum x^2$ and $\sum y^2$ (which will be equal) into equation [38.3], we quickly get equation [38.2].

Summary

1. The product-moment correlation coefficient is inapplicable to nominal and ordinal data. We consider three non-parametric tests of wider application.

2. The first, applicable to nominal data, is the contingency coefficient [38.1]. It has important limitations. It is closely related to the χ^2 test and uses the same table to test significance.

3. The second and third tests are rank correlation tests, so they require at least ordinal data. One is due to Spearman. An example illustrates the procedure, using equation [38.2]. If $n \geqslant 10$ we can test the significance of the coefficient by using the t-test, but consider using Table D11 if $4 \leqslant n \leqslant 30$. The procedure is summarised in Box 38.1. A slight modification arises if there are tied ranks, as summarised in Box 38.2.

4. The other test is due to Kendall. His rank correlation coefficient compares the actual score with the maximum possible score. It is defined by equation [38.4]. The significance of Kendall's coefficent can be found from Table D11. For $n > 10$ we can use the normal distribution with the test statistic given by equation 38.5. If there are tied ranks, the formula has to be slightly modified. Box 38.3 summarises the whole procedure.

5. Note that the rank correlation coefficients should also be used with interval and ratio data if the basic assumption underlying the product-moment correlation coefficient does not hold. The assumption is a joint normal distribution of the values x and y, giving a roughly elliptical scattergram.

The uniform and geometric distributions

39.1 Introduction

Example 39.1 By rolling a well-balanced ten-sided die, an examiner gives marks out of 10 to a large number of candidates for the Certificate in Skiving. What are the expected arithmetic mean and standard deviation of the marks?

Example 39.2 I go on throwing a well-balanced six-sided die until I get a 4. What is the average number of throws I have to make?

Example 39.3 A purchaser of coffee beans uses a scoop to take samples from sacks of various crops. The purchaser wants only one sack and will accept the first to meet these requirements: the sample has to contain between 90 and 120 beans, and the aroma has to be right. Past experience indicates a 10% chance of there being fewer than 90 beans in a sample, and a 15% chance of there being more than 120. It also shows that 60% of samples have failed the aroma test. What is the probability of (i) the first five sacks being rejected, (ii) the sixth sack being chosen?

Example 39.4 Amanda finds from long experience that her chance of winning a prize by scoring a bull's-eye with a dart is 0.2, and this is true even after many throws.

(a) What is the probability that Amanda will miss the bull's-eye for the first three throws but hit it on the fourth?

(b) A game ends when Amanda hits the bull's-eye. She then starts another game, which ends with the second hit, and so on. What are (i) the arithmetic mean and (ii) the standard deviation of the number of throws?

(c) If every prize is worth £1, how much should Amanda be willing to pay for a game? If the charge for a game were to double, what would her new probability of hitting the bull's-eye have to be for her to break even? What would be the new standard deviation of the number of throws per game?

We have already distinguished between discrete and continuous variables. The number of people waiting in a bus queue is a discrete variable; the length of time one of them has been waiting is a continuous variable. If these values of the variable arise at random, the variables are random variables, discrete in the first case, continuous in the second.

As we saw in Chapter 18, a statement of the probabilities of a discrete random variable having its various possible values is called the variable's probability distribution. It is found that most discrete random variables occurring in practical and theoretical problems have one of a very small number of different probability distributions. One already considered is the binomial distribution. In this chapter we pay particular attention to two more (very simple) ones, using the above examples to illustrate them. In later chapters we apply very similar techniques to study other important discrete distributions, and a few continuous probability distributions (including the normal distribution)

We also look at the **expected values** and **variance** of these discrete random variables. We end with some important results about new discrete random variables composed of the sum or difference of other discrete random variables.

39.2 The discrete uniform distribution

In Example 39.1 there are 10 possible outcomes, all arising at random with equal probabilities. If X denotes the variable (in this case the mark) then the probability distribution for X is shown in Table 39.1. This is an example of a **discrete uniform distribution** (Notebox 39.1). All the probabilities are the same. Before looking at them more generally, we continue with the solution of Example 39.1, introducing a few points as we go along.

 The probability distribution in Table 39.1 can be written as $P(X = x) = 0.1$ for $1 \leqslant x$ (an integer) $\leqslant 10$, zero elsewhere.

This statement, which defines the probability distribution, says that the probability P of the variable X having a value x, where x can be any integer between 1 and 10 (inclusive of 1 and 10) is 0.1; whereas the probability of X having any other value is zero.

The graph of this probability distribution is a series of vertical lines located at $x = 1$, $x = 2, \ldots, x = 10$, each of height 0.1. (If it is presented as a column graph, the width of the column is simply a presentational convenience. It

Table 39.1

x	1	2	3	4	5	6	7	8	9	10
$P(X = x)$	0.1	0.1	0.1	0.1	0.1	0.1	0.1	0.1	0.1	0.1

should not be interpreted as implying that any value other than an integer is possible.)

⮕ In **Example 39.1** the examiner throws the ten-sided die many times. Each throw produces a mark. We are interested in what the arithmetic mean of these marks is likely to be if there are very many throws. We saw in Chapter 18 that this is called the **expected mark**. We are also interested in the spread of the marks, which is measured by the **expected variance** of very many marks obtained in this way.

There are simple formulae for the expected mark and the expected variance, but first we will calculate the answers by using the basic results that

$$\bar{x} = E(x) = \sum xP(x) \quad \text{and} \quad \text{var}(X) = \sum x^2 P(x) - \bar{x}^2$$

We begin by constructing Table 39.2. It follows that

$$\bar{x} = \sum xP(x) = 5.5$$

$$\text{var}(X) = \sum x^2 P(x) - \bar{x}^2 = 38.5 - 5.5^2 = 8.25$$

Thus the **expected mark**, is = 5.5. Obviously this mark never occurs, but it is the result we expect to get when we calculate the arithmetic mean of very many marks awarded by the examiner. The variance of the marks is 8.25, indicating a standard deviation of $\sqrt{8.25} = 2.87$.

Table 39.2

x	1	2	3	4	5	6	7	8	9	10	$\sum$
$P(X = x)$	0.1	0.1	0.1	0.1	0.1	0.1	0.1	0.1	0.1	0.1	1.0
$xP(x)$	0.1	0.2	0.3	0.4	0.5	0.6	0.7	0.8	0.9	1.0	5.5
$x^2 P(x)$	0.1	0.4	0.9	1.6	2.5	3.6	4.9	6.4	8.1	10.0	38.5

These results can be obtained more quickly by using simple formulae that are valid for all discrete uniform distributions. If a random variable has a discrete uniform distribution and the integers a and b are its lowest and highest possible values, it can be shown that the following results are true. It may help if you keep thinking of $b = 6$ and $a = 2$ as you read them.

(i) $1 + b - a$ integers are *possible*.
(ii) The *uniform probability* of any possible integer x arising is

$$P(X = x) = \frac{1}{1 + b - a} \tag{39.1}$$

(iii) The *expected value* of X (or the average of the values of X we would expect to get if we repeated the experiment very many times) is

$$\mu = E(X) = \frac{a + b}{2} \tag{39.2}$$

(iv) The *variance* of the values of X is given by

$$\text{var}(X) = \frac{(1+b-a)^2 - 1}{12}$$ [39.3]

⬭ In **Example 39.1** $a = 1$ and $b = 10$. Equations [39.1] to [39.3] lead to $P(X = x) = 0.1$, $\mu = E(X) = 5.5$ and $\text{var}(X) = 8.25$, agreeing with the results obtained by longer calculation.

It can also be shown, and is easily checked, that the **cumulative distribution function** giving the total probability of x having a value up to c is

$$F(c) = \frac{1+c-a}{1+b-a} \qquad (a \leqslant c \leqslant b)$$ [39.4]

We can do a quick check by noting that this has the value given by equation [39.1] when $c = a$ and the value of unity when $c = b$.

⬭ Applied to **Example 39.1**, equation [39.4] means that the probability of getting a mark of up to 5 is

$$F(5) = \frac{1+5-1}{1+10-1} = 0.5$$

In the case of simple uniform distributions with few possible values, equations [39.2] and [39.4] may sometimes seem to be complicating the obvious, but we shall see that they are important later on. Equation [39.3] is far from obvious.

It is important to note that these formulae relate to the *discrete* uniform distribution.

Consolidation exercise 39A

1. What is a discrete uniform distribution? What is meant by the expected mean of such a distribution?

2. A discrete uniform distribution can take all integral values from 3 to 8 inclusive. What are (a) its expected value, (b) its expected variance and (c) the probability of getting a value not more than 6?

39.3 **The geometric distribution**

Another discrete probability distribution is illustrated by a simple gambling example. Although seemingly trivial, it embodies the essential features of many important practical problems.

In **Example 39.2** the variable is not the score (as it was in Example 39.1) but the number of throws I have to make in order to get 4. We now consider its probability distribution.

This is a binomial experiment. I get a 4 or I do not; and in every trial the chance of getting a 4 is the same.

Suppose that in my first attempt I go on throwing for x_1 throws before getting a 4. Then I start again, and this time make x_2 throws before getting a 4. I do this many times, noting the number of throws I have to make each time, thus generating a long series of numbers $x_1, x_2, x_3, \ldots, x_n$.

If n is very great, what is the average value of all the xs? In other words, what is the average number of throws I have to make in order to get a 4?

First we write down the probability of getting a 4 in the first throw, when $x = 1$. If we denote getting a 4 on the first throw by A, we know that for a well-balanced die the probability of success on the first throw is 1/6, so this is also the probability that $x = 1$. We can write

$$P(A) = P(X = 1) = 1/6$$

Now consider the probability of failing to get 4 on the first throw but getting it on the second, when $x = 2$. I make a second throw only if I fail on the first, so the probability of making a second throw is $P(\overline{A}) = 5/6$.

The probability of the second throw *succeeding* (event B) is 1/6. Therefore the probability of making a second throw and of this second throw succeeding (which means that $x = 2$) is

$$P(x = 2) = P(\overline{A} + B) = P(\overline{A})P(B) = (5/6)(1/6)$$

We now consider the probability of x being 3. I have to fail to get 4 on both the first and the second throw, and then get it on the third. The probability of failing on both the first and the second throw is

$$P(\overline{A} + \overline{B}) = P(\overline{A})P(\overline{B}) = (5/6)(5/6) = (5/6)^2$$

This is therefore the probability of my making a third throw. The probability of the third throw succeeding (event C) is 1/6. The probability of making a third throw and it succeeding, which means $x = 3$, is therefore

$$P(x = 3) = P(\overline{A} + \overline{B} + C) = P(\overline{A})P(\overline{B})P(C) = (5/6)^2(1/6)$$

It is easy to see that $P(X = 4) = (5/6)^3(1/6)$ and more generally the probability of needing x throws and succeeding only on throw x is

$$P(X = x) = (5/6)^{x-1}(1/6)$$

We have worked out this example in terms of 1/6 and 5/6. More generally, if we repeat an experiment many times, and each time the chance of success is the constant p, and the chance of failure $q = 1 - p$, then the probability of needing exactly n trials of the experiment to achieve a success is

$$P(X = n) = q^{n-1}p \qquad\qquad [39.5]$$

Box 39.1

How to ... **find the probability of needing exactly *n* trials of a binomial experiment in order to achieve a single success**

1. Denote the chance of success in a single performance of the experiment by p, and of failure by $q = 1 - p$. Determine the value of p and then of q.

2. Calculate q^{n-1}; then multiply this by p to obtain the chance of needing exactly n trials of the experiment.

The first few results of the experiment described in Example 39.2 can be summarised as in Table 39.3, in which the last column is explained later. Thus every probability is a constant multiple of the preceding probability, so the probabilities form a geometric series (defined to be a series such that every term is a constant multiple of the preceding term). The probabilities of needing different numbers of trials before achieving success are therefore said to have a **geometric distribution** and the probability of needing x trials, where x can be any positive integer, is $P(X = x) = q^{x-1}p$. If a random variable X is distributed in this way, we sometimes write $X \sim \text{Geo}(p)$.

We now state two important results for the geometric distribution.

Expected value

Table 39.3 above gives us the probabilities of x having various values. If, for every possible value of x, we multiply each value by its probability and add the answers, getting

$$\sum xP(X = x) = \sum xq^{x-1}p$$

then this expression is the **expected value of x**, denoted by $E(x)$.

Table 39.3

x	$P(x = x)$		$xP(X = x)$
1		$(1/6) = 0.166\,6667$	0.166\,667
2	$(5/6)\,(1/6) = (5/6)\,P(x = 1) = 0.138\,8889$		0.277\,778
3	$(5/6)^2(1/6) = (5/6)\,P(x = 2) = 0.115\,7407$		0.347\,222
4	$(5/6)^3(1/6) = (5/6)\,P(x = 3) = 0.096\,4506$		0.385\,802
5	$(5/6)^4(1/6) = (5/6)\,P(x = 4) = 0.080\,3755$		0.401\,878
6	$(5/6)^5(1/6) = (5/6)\,P(x = 5) = 0.066\,9796$		0.401\,878
7	$(5/6)^6(1/6) = (5/6)\,P(x = 6) = 0.055\,8163$		0.390\,714
8	$(5/6)^7(1/6) = (5/6)\,P(x = 7) = 0.046\,5136$		0.372\,109
⋮	⋮	⋮	⋮
10	$(5/6)^{19}(1/6) = (5/6)\,P(x = 19) = 0.005\,2168$		0.104\,336
⋮	⋮	⋮	
		$\sum xP(X = x)$	6.000\,00

The first few values, corresponding to $p = 1/6$, are shown in the last column of this table. There is no *guarantee* that we will ever get a 4, so this series goes on indefinitely. It can be shown that, as we go on taking more and more terms, they become progressively smaller and their sum (which gives the value of $E(x)$) will get closer and closer to $1/p$, where p is the probability of a success in a single trial. For a very large number of terms, the difference between the actual value and $1/p$ will be negligible. Symbolically we write

$$\text{if} \quad X \sim \text{Geo}(p) \quad \text{then} \quad E(X) = 1/p \qquad\qquad [39.6]$$

Thus in **Example 39.2** the average number of throws needed for a 4 (which is another way of describing the expected number) is $1/(1/6) = 6$.

Variance

We can also calculate the standard deviation of the values of x. As with the uniform distribution, this can be derived by using a table to evaluate $\text{var}(X) = \sum x^2 P(x) - \bar{x}^2$, but it is quicker to use the result (which we do not prove) that for the geometric distribution

$$\text{var}(X) = q/p^2 \qquad\qquad [39.7]$$

Thus in **Example 39.2** the variance of the number of throws I must make before getting a 4 is

$$\frac{1 - (1/6)}{(1/6)^2} = 30$$

and so the standard deviation of the number of throws is $\sqrt{30} = 5.48$.

To illustrate further the use of these formulae, we tackle Example 39.3.

The beans have to pass two tests. To pass the first test the number of beans in the sample has to be between 90 and 120. We are told there is a 10% chance of there being fewer than 90 beans in a sample, and a 15% chance of there being more than 120. It follows that the probability of the number of beans N being between 90 and 120 is given by

$$P(90 \leqslant N \leqslant 120) = 1 - P(N < 90) - P(N > 120)$$
$$= 1 - 0.1 - 0.15 = 0.75$$

The beans also have to have the right aroma. The probability of this is $P(\text{smell good}) = 1 - 0.60 = 0.40$.

The probability of the beans passing both tests is the product of these probabilities, which is $0.75 \times 0.40 = 0.30$.

We have been asked about the probability of the first five sacks being rejected. Before using the formula to obtain a quick answer, we summarise the long way of doing it. We have P(first sack being rejected) $= 1 - 0.30 = 0.70$. If (and only if) the first sack fails the second is tested, and so P(second sack being tested) $= 0.70$. Also P(second sack failing if it is tested) $= 0.70$, so P(2nd sack being tested and failing) $= 0.70 \times 0.70 = 0.70^2$.

The third sack is tested if (and only if) a second sack is tested and fails, so P(third sack being tested and failing) $= 0.70^2 \times 0.70 = 0.70^3$. Similarly P(fourth sack being tested and failing) $= 0.70^3 \times 0.70 = 0.70^4$ and P(5th sack being tested and failing) $= 0.70^4 \times 0.70 = 0.70^5$.

However, the fifth sack is tested only if the four previous sacks have failed, so $0.70^5 = 0.168$ is also the probability of the first five sacks failing, answering the first part of the question.

We are also asked about the probability of the sixth sack being chosen. This is given by

P(sixth sack being tested and passing)
$= P$(sixth sack being tested) $\times P$(sixth sack passing if it is tested)
$= P$(first five sacks failing) $\times 0.30$
$= 0.70^5 \times 0.30 = 0.050$

which answers the second part of the question.

If, instead, we use the formula, we have that the probability of failing the first five tests is $0.70^5 = 0.168$, as before. From Box 39.1 the probability of needing exactly 6 samples is $q^5p = (0.70)^5(0.30) = 0.050$.

Finally we use these ideas to indicate the solution to Example 39.4. This problem involves repeated attempts to hit the bull's-eye with a constant probability of success in each try. The variable is the number of tries needed to hit the bull's-eye. This is discrete, random and with probabilities summarised in the geometric distribution with $p = 0.2$. Here are the answers to the various parts.

⬭ (a) The probability of missing 3 times then succeeding on the fourth try is $(0.8)^3(0.2) = 0.1024$.
(b) The arithmetic mean number of throws needed for success is given by equation [39.6] and is $1/0.2 = 5$; the standard deviation is the square root of the variance given by equation [39.7], so it is $\sqrt{0.8/0.2^2} = \sqrt{20} = 4.47$.
(c) Since Amanda expects to win on average once in 5 games, she should be prepared to pay at most $100p/5 = 20p$. If the price of a game is 40p, then to break even she has to win on average 1 game in $1/(0.40) = 2.5$ games. Thus her probability of success in each game must be $1/2.5 = 0.40$.
(d) In this case, with $p = 0.40$, the variance will be $0.6/0.4^2 = 3.75$, leading to a standard deviation of 1.94.

39.4 Combining random variables

Sometimes two or more random variables are added or subtracted to form another variable. For example, suppose that on Monday morning phone calls (X) and visitors (Y) are nuisances, both occurring at random. It can be shown that

(i) $Z = X + Y$ is also a random variable.
(ii) If $E(X)$, $E(Y)$ and $E(Z)$ denote the expected values of X, Y and Z then
      $E(Z) = E(X + Y) = E(X) + E(Y)$.
(iii) If X and Y are independent then $\text{var}(Z) = \text{var}(X + Y) = \text{var}(X) + \text{var}(Y)$.

Notice carefully that if we define a new variable W as the *difference* between two *independent* random variables, the result for the expected value is unremarkable:

(iia) $E(W) = E(X - Y) = E(X) - E(Y)$

But the result for the variance may seem surprising, as it involves a plus sign:

(iiia) $\text{var}(W) = \text{var}(X - Y) = \text{var}(X) + \text{var}(Y)$

   thus $\text{var}(X - Y) = \text{var}(X + Y)$

These and similar formulae are discussed further in Chapter 45. The simpler results are mentioned now since we shall occasionally use them.

Consolidation exercise 39B

1. What is a geometric distribution? In an experiment whose outcome has a geometric distribution, if the probability of a success is p, what is the probability of the first success arising on the fourth attempt?

2. What is (a) the expected value of the outcome and (b) the expected variance in Question 1?

3. The expected value of a raffle is £5 and the variance is £1.50^2. The expected value of a lucky dip is £2 and the variance is 60p^2. (a) If I take part in both the raffle and the lucky dip, what is the total expected value, and what is the expected variance? (b) I take part in the raffle and my partner plays the lucky dip. I agree to repay to the lucky dip organisers anything won by my partner. What is my expected net gain? What is the expected variance?

Notebox 39.1

Uniform

The word 'uniform' puzzles some students. Statisticians use it to mean three things: all the same, unvarying or of the same size. Some statisticians describe only a continuous random variable as uniform and have no special name for the distribution we describe here. We prefer to call the comparable continuous distribution 'rectangular', as we do in Chapter 40. Since some writers consider 'uniform' and 'rectangular' to be interchangeable, we use 'discrete uniform' to avoid confusion.

Summary

1. We define a discrete random variable X to be a discrete variable capable of having only specified values (typified by x_i) with associated probabilities p_i adding to unity. A statement (verbal or mathematical) that allocates probabilities to all possible values of the variable is called its probability distribution.

2. The discrete uniform distribution has a finite number of equally spaced possible values. If these are consecutive integers ranging from a to b, the number of possible values, the mean (or expected) value, the variance and the cumulative distribution function are given by equations [39.1] to [39.4].

3. If a binomial event has probability p of happening in a single trial, and the trials are repeated until success is achieved, then the probability of needing exactly x trials to achieve success is $P(X = x) = (1 - p)^{x-1} p$.

4. A random variable X with the distribution $P(X = x) = (1 - p)^{x-1} p$ is said to have a geometric distribution and we write $X \sim \text{Geo}(p)$. The mean and variance of a geometric distribution are given by equations [39.6] and [39.7].

5. If X and Y are two independent discrete random variables and new variables Z and W are formed by addition or subtraction, then these new variables are also random and some important results hold. The results for the variance are valid only if X and Y are independent. More results of this kind are given in Chapter 45.

Continuous distributions: rectangular and triangular

40.1 Introduction

Example 40.1 Cub membership of Clever Clown Clubs is restricted to persons aged 17 and over but under 21. Within this range the ages of the national membership are evenly spread. What is the probability that a member chosen at random will be older than 18.36 years? What are the expected arithmetic mean and standard deviation of the ages of cub members?

Example 40.2 The extrusion plant of ScrawlBall Pens produces lengths of plastic tubing. The minimum length is precisely 10 cm and the maximum precisely 10.5 cm. The probability of a tube having any specified length between these limits is independent of the length. What are (i) the expected length of a tube selected at random? (ii) the probability that a tube selected at random will have a length of more than 10.4 cm? (iii) the expected variance?

Example 40.3 The vice-chancellor of Statingham University asserts that he is never too busy to see an undergraduate, and any student calling to see him on Saturday afternoons in term will be guaranteed an audience within 15 minutes of arriving.

As a matter of principle, he keeps everybody waiting at least 3 minutes and never spends more than 12 minutes with anybody. Knowing that obscurity is a virtue, he makes the following claims:

- The most probable waiting time is 3 minutes, and nobody has to wait more than 15 minutes. Between those durations the probability distribution function falls smoothly.
- The chance of having to wait between 6 and 7 minutes is less than the chance of having to wait between 5 and 6 minutes by just as much as the chance of having to wait between 5 and 6 minutes is less than the chance of having to wait between 4 and 5 minutes.

What are (i) the probability of having to wait for 7 minutes at most, (ii) the expected waiting time and (iii) the expected variance of the waiting times?

These examples involve *continuous* random variables, which can take any value between specified lower and upper limits. Between any two possible values, no matter how close, there are infinitely many other possible values. Keep this in mind whenever continuous variables are being studied. We have already met this idea when discussing the normal distribution, but now we look at it more formally. We concentrate on two simple continuous distributions, but we begin by making a few points that are true for all continuous random variables.

40.2 Continuous probability density functions

A continuous random variable is studied with the aid of its **probability density function** (p.d.f.). This is a statement, often mathematical, that

- Specifies the lower and upper extremes of the possible values that the random variable X may take. The lower extreme is often denoted by a and the upper extreme by b.
- Enables the probability to be found for X having a value of less than or equal to any value c.

Continuous probability density functions can be represented diagrammatically by curves (a term used by mathematicians to include straight lines) such as those shown in Figure 19.4. They are drawn only between the lower and upper limits. We choose the vertical scale so that the total area under the curve is unity (representing a total probability of 1). It follows that the area beneath the curve and to the left of c gives the probability of X having a value of c or less.

The equation of the curve is the mathematical statement of the probability density function. It is usually written as $f(x)$ where $f(x)$ is used to denote the height of the curve at the point corresponding to the chosen value of x. In this chapter we shall use only distributions that have very simple equations. For example, the distribution shown in Figure 40.1(a) has the equation

$$f(x) = \begin{cases} \frac{1}{3} & \text{for } 2 \leqslant x \leqslant 5 \\ 0 & \text{elsewhere} \end{cases}$$

We can always use the equation or the diagram to find the probability of X having a value not more than c, as specifed above. It is the area to the left of c. In the case of Figure 40.1(a) the probability of x having a value not more than 4.1 is shown by the shaded area, which is $(4.1 - 2.0) \times 1/3 = 0.7$.

The equation showing this area up to the point c is called the **cumulative probability density function** and is usually denoted by $F(c)$. In the case of Figure 40.1(a) we have

$$F(c) = (c - 2) \times 1/3 = \frac{c - 2}{5 - 2}$$

as illustrated in Figure 40.1(b), with $F(c) = 0.7$ for $c = 4.1$.

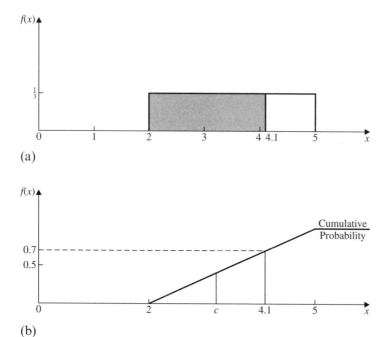

(a)

(b)

Figure 40.1

Notice that although we have spoken of the variable X having a value of *not more than* 4.1, we have not mentioned the probability of it being *precisely* 4.1. The reason is illustrated in Figure 40.2. The area up to the heavy line at $x = 4.20$ shows the probability of x being not more than 4.20. Similarly the area up to the heavy line at 4.00 shows the probability of x being not more than 4.00. Thus the area between the heavy lines represents the probability of x being between 4.00 and 4.20. It can be calculated as $(4.20 - 4.00) \times 1/3 = 0.066\,666\ldots$.

The area between the lighter lines at 4.05 and 4.15 shows the probability of x being between 4.05 and 4.15 as $(4.15 - 4.05) \times 1/3 = 0.033\,333\ldots$. Similarly the probability of x being between 4.09 and 4.11 is $0.02 \times 1/3 = 0.006\,666\ldots$.

In this way it is possible to find the probability of x lying between any two values very close to 4.1 and on either side of it. As those values get closer and

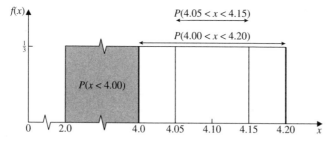

Figure 40.2

closer, the probability falls. But the probability of x being *precisely* 4.1 will be less than the area of the narrowest of these ever narrowing rectangles.

In studying the probabilities of continuous variables, if we become interested in a precise value, we must replace it by a very narrow range of values.

At the end of Chapter 39 we gave results for the expected value and variance of a new discrete random variable formed by the addition or subtraction of two or more other discrete random variables. The same results hold for continuous random variables.

40.3 Rectangular distributions

The distribution drawn in Figure 40.1(a) is an example of a **rectangular distribution** also called a **(continuous) uniform distribution**. It shows a probability density function such that there is a constant probability of X having a value between any two values a fixed distance apart, wherever this pair of values may be, provided they lie between the lower and upper limits.

Symbolically $P(a \leqslant x \leqslant b) = k$ (where a, b and k are constants). The only way of satisfying this is to have a probability density function whose diagram is a straight horizontal line between the lower and upper limits (Figure 40.3). Along with its limits, this defines a rectangle of length $b - a$. Since its area must be unity (to represent the total probability), its height must be $1/(b - a)$ for all values of x.

However, we saw that the height of the curve at the point x is denoted by $f(x)$. Thus, for the rectangular distribution, the probability density function is

$$f(x) = \begin{cases} 1/(b - a) & \text{for } a \leqslant x \leqslant b \\ 0 & \text{elsewhere} \end{cases} \tag{40.1}$$

If a variable X has this probability density function, we sometimes write $X \sim R(a, b)$. It can be shown (and is perhaps fairly easy to see) that the associated cumulative probability density function, giving the area of the rectangle to the left of c is

$$F(c) = \frac{c - a}{b - a} \tag{40.2}$$

This is also the probability $P(x \leqslant c)$ of x being equal to or less than c.

It can also be shown that, for the rectangular distribution, the **expected value** of the variable (or the arithmetic mean we would expect to find for very many values drawn at random from this distribution) is

$$E(X) = \frac{b + a}{2} \tag{40.3}$$

and the expected variance (or the variance we would expect to find for very many values drawn at random from this distribution) is

$$\text{var}(X) = \frac{(b - a)^2}{12} \tag{40.4}$$

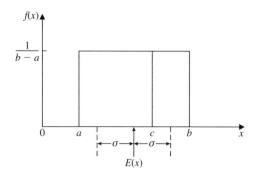

Figure 40.3

so the expected standard deviation, which is the square root of this, is $0.289(b - a)$. We now apply these ideas to two examples.

⬭ **Example 40.1** describes a distribution with lower and upper limits of precisely 17 and precisely 21. The constant probability implies a rectangular probability distribution. Using our notation this is indicated by $X \sim R(17, 21)$.

From equation [40.1] the probability density function is therefore

$$f(x) = 1/(21 - 17) = \tfrac{1}{4} \quad \text{for} \quad 17 \leqslant x \leqslant 21 \qquad f(x) = 0 \text{ elsewhere}$$

The cumulative probability density function [40.2] gives us the probability of an age being less than or equal to 18.36 years as

$$P(x \leqslant 18.36) = F(18.36) = \frac{18.36 - 17}{21 - 17} = 0.34$$

The expected age of a member, obtained from equation [40.3], is

$$E(X) = (21 + 17)/2 = 19 \text{ (precisely)}$$

and the expected variance of the ages is

$$\text{var}(X) = \frac{(21 - 17)^2}{12} = 1.333$$

giving an expected standard deviation of $\sqrt{1.333} = 1.155$ years.

⬭ **Example 40.2** can be tackled very similarly. The distribution defined in the question is $X \sim R(10, 10.5)$ and use of $a = 10$, $b = 10.5$ in our equations will provide answers to the questions.

The probability of a tube having a length greater than 10.4 cm is found by first finding the probability of the length being less than or equal to 10.4 cm, and then taking this away from unity. This uses the result

$$P(x > c) = 1 - P(x \leqslant c) = 1 - F(c) \qquad [40.5]$$

Consolidation exercise 40A

1. What is (a) a continuous random variable and (b) a probability density function?

2. If a random variable with values between 1 and 4 has a p.d.f. given by $f(x)$, what does $f(3)$ represent?

3. What is (a) the cumulative probability density function, (b) a rectangular distribution?

4. If the extreme values of a continuous random variable having a rectangular distribution are a and b, what is the formula for the probability density function? What are the expected value and variance of the variable?

5. What is meant by $F(c)$? What is its value in Question 2?

40.4 Triangular distributions

The distribution shown in Figure 40.4(a) is often called **the triangular distribution**. We prefer to say that it is an example of a triangular distribution. Other examples are shown in Figure 40.4.

In any triangular probability distribution the area of the triangle has to be unity. Since the area is given by $\frac{1}{2}$(base × height) and this equals 1, it follows that for a triangular probability distribution of base $(b - a)$ the height must be $2/(b - a)$, as shown in the diagrams.

We now indicate how to derive a few important results. It is the results that matter. Taking Figure 40.4(b) as an example, it can be shown by geometry that the equation of the probability density function (graphed by the line LM) is

$$f(x) = \begin{cases} \dfrac{2(b - x)}{(b - a)^2} & \text{for } a \leqslant x \leqslant b \\ 0 & \text{elsewhere} \end{cases} \tag{40.6}$$

It can also be shown that, for this triangular distribution, the cumulative probability density function is

$$F(c) = 1 - \frac{(b - c)^2}{(b - a)^2} \qquad (a \leqslant x \leqslant b) \tag{40.7}$$

This gives the probability $P(x \leqslant c)$ of X having a value equal to or less than c. Using these equations (or geometry) we can show that the expected value of X is

$$a + \tfrac{1}{3}(b - a) = \tfrac{2}{3}a + \tfrac{1}{3}b$$

and the expected variance is

$$(b - a)^2/18 \tag{40.9}$$

This gives an expected standard deviation of $0.236(b - a)$.

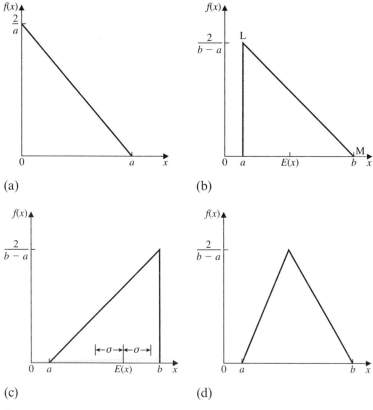

Figure 40.4

If the triangle is reversed, so that it has its lowest point at a and its highest at b, the expected value is slightly different:

$$a + \tfrac{2}{3}(b - a) = \tfrac{1}{3}a + \tfrac{2}{3}b$$

but the variance is unchanged. We now demonstrate the use of these results by considering Example 40.3.

⬭ The vice-chancellor's statement suggests there is a triangular distribution of the type shown in Figure 40.4(b), with the highest point occurring when the waiting time X has the value 3 (so that $a = 3$) and the lowest point when X has the value 15, so that $b = 15$. Such a distribution clearly fits (i) and (ii) of the vice-chancellor's statement, and we shall see shortly that it also fits (iii). Taking this on trust for the moment, we can quickly answer the three questions.

(i) The probability of having to wait 7 minutes at most is given by the appropriate value of the cumulative function [40.7]:

$$F(c) = 1 - \frac{(b - c)^2}{(b - a)^2} \qquad (a \leqslant c \leqslant b)$$

From this we get $F(7)$ by putting $a = 3$, $b = 15$ and $c = 7$. This gives the probability as $1 - 8^2/12^2 = 5/9 = 0.555\,556$. The probability of having to wait 6 minutes at most is $F(6) = 1 - 9^2/12^2 = 0.4375$. Similarly $F(5) = 0.305\,556$ and $F(4) = 0.159\,722$. These results can be used to support part (ii) of the vice-chancellor's statement.

(ii) The expected waiting time is given by equation [40.8]:

$$E(X) = \tfrac{2}{3}a + \tfrac{1}{3}b$$

$$= 6/3 + 15/3$$

$$= 21/3 = 7\,\text{min}$$

(iii) The expected variance of the waiting times is given by equation [40.9]

$$(b - a)^2/18 = 12^2/18 = 8\,\text{min}^2$$

Consolidation exercise 40B

1. Draw four examples of a triangular distribution. In each case give your chosen numerical extreme values and the height.

2. If a triangular distribution is valid for $a \leqslant x \leqslant b$ and has a maximum at $x = a$, what are (a) its height, (b) the cumulative p.d.f., (c) the expected value and (d) the expected variance? Which of the answers will change if the distribution is reversed, so that it has a maximum at $x = b$?

Summary

1. The chapter begins with a restatement of the salient points about continuous random variables made earlier. Between any two values there can lie an infinite number of other values. The probability density function (p.d.f.) specifies the lower and upper limits between which X can lie, and enables the probability of x having a value less than or equal to c to be found from the area beneath its graph to the left of c. The function giving this area is the cumulative probability density function. The total area under the graph is unity.

2. The rectangular distribution is defined by $P(a \leqslant x \leqslant b) = k$. It is easily shown that $k = 1/(b - a)$. The definition can be restated as

$$f(x) = \begin{cases} 1/(b - a) & \text{for } a \leqslant x \leqslant b \\ 0 & \text{elsewhere} \end{cases}$$

3. The expected value of a random variable X with this distribution is $E(X) = \tfrac{1}{2}(b - a)$ and the expected variance is $\text{var}(X) = (b - a)^2/12$.

4. The triangular distribution is defined by

$$f(x) = \begin{cases} \dfrac{2(b-x)}{(b-a)^2} & \text{for } a \leqslant x \leqslant b \\ 0 & \text{elsewhere} \end{cases}$$

in the case of Figure 40.4(b), or in the same way with a and b interchanged in the case of Figure 40.4(c).

5. The expected value of a random variable X with the distribution of Figure 40.4(b) is $E(X) = (2a+b)/3$ and the expected variance is $\text{var}(X) = (b-a)^2/18$. If the triangle is reversed, as in Figure 40.4(c), then $E(x) = (a+2b)/3$ but $\text{var}(x)$ is unchanged.

The Poisson distribution and random events

41.1 **Introduction**

Example 41.1   The daily number of reported cases of violence in Statingham is random. On average there are 3.5 reported cases per day. What is the chance of having 5 outbreaks tomorrow?

Example 41.2   The roads leading into Statingham have potholes occurring at random. On a total of 8 miles of road there are 272 potholes. What is the chance of coming across fewer than 3 potholes on a given stretch of road measuring one-quarter of a mile?

Example 41.3   In their away matches, Statingham Strikers score an average of 3.2 goals. Tomorrow they are playing away against the Gentlemen of Upper Sternum. Assuming the numbers of goals are random, what is the chance of Statingham Strikers scoring (i) at most 4 goals? (ii) more than 4 goals?

Example 41.4   The fire station gets on average 4 calls per hour. What are the chances of (a) 6 calls between 2.00 pm and 3.00 pm, (b) 3 calls between 3.00 pm and 4.00 pm, (c) a total of 9 calls between 2.00 pm and 4.00 pm.?

Example 41.5   *Scrawlball Scribbles* is a weekly newspaper for the employees of Scrawlball Pens Inc. It has three pages written by management and one page written by captains of works sports teams. Spelling mistakes by both writing teams occur at random. Management spelling mistakes have averaged 1.3 per page, but on the sports pages mistakes have averaged 1.1 per page. What is the probability that next week's paper will have more than 6 mistakes?

These examples involve events scattered at random over time or space. Although no unit of time or space contains more than a few events, in principle there is no upper limit to the number there could be – or at least the upper limit is very large. An essential tool in the analysis of such events is the Poisson distribution. In this chapter we define it, show how to use it in the analysis of random events, and introduce a few of its properties. In the next chapter we show how to approximate it using the normal distribution, and how to use it to obtain approximate solutions to binomial problems.

41.2 Random events

If an event occurs at random intervals (such as plane crashes), or is randomly scattered in space (such as the outbreaks of fire in a large suburban area), then a knowledge of the mean number of occurrences per period (or per unit area) will enable us to calculate the probabilities of there being any stated number of occurrences in a given period (or area). If there are on average 3 incidents per week (or per grid square) we can state the chances of next week (or a selected grid square) containing 8 incidents.

When used for this purpose, the Poisson distribution gives exact results. Under certain circumstances approximate probabilities for Poisson events can be obtained from the normal distribution, as described in Chapter 42.

Suppose we know the mean number of times that a randomly occurring event is likely to occur in a specified period or space. How can we calculate the probability of this event not occurring at all in a given period, or of it occurring more than say 8 times?

It can be shown that if an event occurs at random intervals or is randomly scattered in space, in such a way that

(i) the arithmetic mean number of occurrences per unit of time or space is λ,
(ii) there is no upper limit or a very high upper limit to the possible number of occurrences per unit of time or space,

then the probability of there being x occurrences in a unit period or a unit space is given by

$$P(X = x) = \frac{\lambda^x e^{-\lambda}}{x!} \qquad (x = 1, 2, 3, \dots) \qquad [41.1]$$

where the exponential constant e is defined in Notebox 31.2 and $e^{-\lambda}$ is explained in Notebox 41.1. Future references to units of space apply equally to units of time; and future references to units of time apply equally to units of space.

Equation [41.1] defines the Poisson distribution. Sometimes it is defined or used with r instead of x. The only conditions necessary for occurrences to have a Poisson distribution are that they are randomly scattered and have no upper limit or a very high upper limit to their possible frequency.

 Example 41.1 illustrates the use of this result. Since there are on average 3.5 cases per day and we are interested in the probability of 5 cases, we put $\lambda = 3.5$ and $x = 5$. Using equation [41.1] we get that the probability of 5 reported cases on a stated day is

$$P(X = 5) = \frac{3.5^5 \times e^{-3.5}}{5!}$$

Tables and pocket calculators give values of $e^{-\lambda}$ for various values of λ, which is the parameter of the distribution. We find that $e^{-3.5} = 0.030\,20$, so

$$P(X = 5) = \frac{525.22 \times 0.030\,20}{120} = 0.1322$$

Thus there is a chance of just over 13% that 5 cases will be reported tomorrow.

Shortly we shall see that tables enable us to reach this result without performing these calculations. But first we do another example, then look at a few important properties of the **Poisson distribution** defined by equation [41.1], where λ has a value that is defined by the problem being considered.

○ In **Example 41.2** the potholes are distributed at random. As there are 272 in 8 miles of road, there are on average 34 per mile or 8.5 per quarter-mile. We can therefore use the Poisson distribution with $\lambda = 8.5$ to find the probability that a stretch of road measuring one-quarter of a mile contains fewer than 3 potholes. Later we see how to use tables to do this very easily, but it will help us if we get used to the formula.

The probability of getting fewer than 3 potholes is the probability of getting 0, 1 or 2 potholes. This is the sum of the separate probabilities of getting 0 potholes, of getting 1 pothole and of getting 2 potholes. Symbolically

$$P(X \leqslant 3) = P(X = 0, 1 \text{ or } 2) = P(X = 0) + P(X = 1) + P(X = 2)$$

We can easily find these three probabilities from the formula. Note the trick of postponing the awkward part of the calculation (involving $e^{-\lambda}$) until the last moment, which cuts down the work. We have

$$P(X = 0) = e^{-\lambda}\left(\frac{\lambda^x}{x!}\right) = e^{-8.5}\left(\frac{8.5^0}{0!}\right)$$

$$P(X = 1) \qquad\qquad = e^{-8.5}\left(\frac{8.5^1}{1!}\right)$$

$$P(X = 2) \qquad\qquad = e^{-8.5}\left(\frac{8.5^2}{2!}\right)$$

Thus, by addition, we have

$$P(X = 0) + P(X = 1) + P(X = 2) = e^{-8.5}\left(\frac{8.5^0}{0!} + \frac{8.5^1}{1!} + \frac{8.5^2}{2!}\right)$$

Recall that anything raised to the power 0 has a value 1 (Notebox 41.1) and that $0! = 1$ (Notebox 17.2). Thus we can write this expression as

$$e^{-8.5}\left(\frac{1}{1} + \frac{8.5}{1} + \frac{8.5^2}{2}\right) = e^{-8.5}(1 + 8.5 + 36.125)$$

$$= e^{-8.5}(45.625) = 0.000\,2035 \times 45.625 = 0.009\,28$$

There is a chance of less than 1% (0.01) of getting fewer than 3 potholes in one-quarter of a mile.

Table 41.1

Number of occurrences	0	1	2	3	4	5	6
Probability		0.030	0.106	0.185	0.216	0.189	0.132	0.077

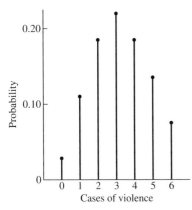

Figure 41.1

As the number of occurrences r increases from 0 to infinitely many, we obtain a whole succession of such terms, each giving a probability. For example, by using the basic formula, we can find in Example 41.1 the probabilities of getting various numbers of reported cases of violence (Table 41.1). These probabilities can be represented by vertical bars of heights proportional to the probabilities, as in Figure 41.1.

As is clear from both the table and the diagram, after initially increasing, the probabilities tail off rapidly, but they never quite reach zero since $P(X = r) = \lambda^r e^{-\lambda}/r!$ must always be positive.

Whatever the number of reported cases of violence may be on any given day, it will have a bar in this diagram, which represents all possibilities. The total of all bar heights represents the sum of the probabilities of all possible numbers of occurrences, and must therefore be unity.

If a discrete random variable X has a Poisson distribution with parameter λ we write $X \sim \text{Po}(\lambda)$. In Example 41.2 if X is the number of potholes in a quarter-mile stretch of road then $X \sim \text{Po}(8.5)$.

We may note that λ, the parameter of the distribution, is the **arithmetic mean** of the distribution. Thus, for a Poisson distribution, we have $\mu = E(X) = \lambda$. It can be shown that λ is also the variance, so

$$\sigma^2 = \text{var}(X) = \lambda = E(X) = \mu \qquad [41.2]$$

We shall often use this important result.

Figure 41.2 illustrates the behaviour of the distribution for various values of λ, all drawn to the same scale.

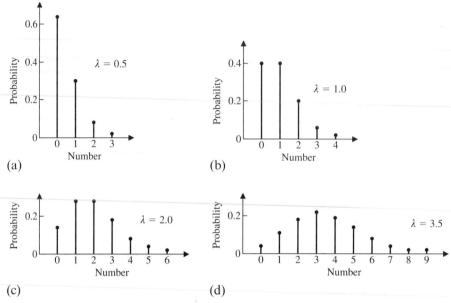

Figure 41.2

Sometimes we have a frequency distribution showing the numbers of units of time or space that have 0, 1, 2, ... occurrences, and we wonder whether these occurrences are randomly distributed. We know that if they are, then the given frequency distribution should not be significantly different from a Poisson distribution. In Chapter 43 we describe how to test this. In this chapter we concern ourselves with occurrences that are known to be random (or can safely be assumed random).

Consolidation exercise 41A

1. What probabilities are shown by the Poisson distribution? What information provides its parameter?

2. If a random event occurs on average 9 times per week, what is the standard deviation of the weekly number of occurrences? What is the formula giving the probability of it occurring 4 times next week?

(**41.3**)   **Cumulative Poisson probabilities**

In many cases we can obtain Poisson probabilities more or less directly from tables such as D3. These show cumulative Poisson probabilities for selected values of λ, making use of the fact that r, the number of occurrences, must always be an integer.

Table 41.2

Cumulative Poisson probabilities for $\lambda = 3.5$

$r = 0$	$P = 0.0302$
1	0.1359
2	0.3208
3	0.5366
4	0.7254
5	0.8576
6	0.9347
7	0.9733
8	0.9901
9	0.9967
10	0.9990
11	0.9997
12	0.9999

Table 41.2 is an extract from the table of cumulative Poisson probabilities. For $\lambda = 3.5$ it shows the cumulative probabilities of there being *r or fewer* occurrences per unit of space or time, where *r* takes all possible values from 0 to 12. (The reason for stopping at 12 becomes clear below.) Thus the probability of there being 5 or fewer occurrences of an event that follows the Poisson distribution with a mean of 3.5 is $P(X) \leqslant 5) = 0.8576$. This is the sum of the lengths of the first six vertical lines in Figure 41.2(d).

Table 41.2 also shows that the probability of there being 4 or fewer occurrences is $P(X) \leqslant 4) = 0.7254$. It follows that the probability of there being exactly 5 occurrences is

$$P(X = 5) = P(X \leqslant 5) - P(X \leqslant 4) = 0.8576 - 0.7254 = 0.1322$$

which agrees with the result we obtained at the beginning of Section 41.2 by using equation [41.1].

Since $P(X > r) = 1 - P(X \leqslant r)$ the tables can also be used to provide quick answers to questions about the probability of the number of occurrences exceeding a stated number. Thus, in the example just considered, the probability of there being more than 5 occurrences is

$$P(X > 5) = 1 - P(X \leqslant 5) = 1 - 0.8576 = 0.1424$$

⬭ **Example 41.3** provides a further illustration of the use of cumulative tables. In this the mean number of goals per away match is 3.2. To obtain the probability of 4 goals or fewer, we use the tables of cumulative Poisson probabilities with $\lambda = 3.2$ and $r = 4$, getting $P(x \leqslant 4) = 0.7806$. The probability of more than 4 goals is $1 - 0.7806 = 0.2194$.

The unit of the interval of space or time

In using these tables, or answering Poisson questions by other methods, we have to be particularly careful about one aspect. The numerical value of the parameter λ of a Poisson distribution is related to a unit of time. It tells us the mean number of occurrences per unit. If we wish to find the probability of a specified number of occurrences in a specified place or time, we need first to consider how many units it contains. For example, we may be told that $\lambda = 3$ accidents per day. If we wish to know the chances of there being more than 10 accidents next week, we have to allow for the fact that next week consists of 7 units of one day, so we use $\lambda = 21$ accidents per week.

⬭ **Example 41.4** illustrates this. The fire station gets on average 4 calls per hour. We are interested in the chances of it receiving (a) 6 calls between 2.00 pm and 3.00 pm, (b) 3 calls between 3.00 pm and 4.00 pm and (c) a total of 9 calls between 2.00 pm and 4.00 pm. These can be found from tables, but we do it the other way to show the method.

Assuming the arrival times of calls are randomly distributed, without any tendency to bunch, we use equation [41.1] with $\lambda = 4$.

(a) We have

$$P(r \text{ calls}) = \frac{\lambda^r e^{-\lambda}}{r!} \quad \text{with} \quad \lambda = 4$$

Therefore the probability of 6 calls in a given hour is

$$p_6 = P(X = 6) = \frac{4^6 e^{-4}}{6!}$$

Here $4^6 = 4096$, $6! = 720$ and $e^{-4} = 1/e^4 = 1/54.598$. Inserting these values gives

$$p_6 = \frac{4096}{720 \times 54.598} = 0.104$$

(b) A similar calculation with the same value of λ but $r = 3$ shows that the probability of there being three calls in an hour is $p_3 = 0.195$.

(c) When we consider a period of two hours, which is twice as long as the unit period to which $\lambda = 4$ relates, we begin by putting $\lambda = 8$ for this part of the question. We then obtain the chance of getting 9 calls in a two-hour period from

$$P(r \text{ visitors}) = \frac{\lambda^r e^{-\lambda}}{r!}$$

with $\lambda = 8$ and $r = 9$. This can be evaluated as above (or by using tables) to obtain $p_9 = 0.1144$.

The use of these tables is summarised in Box 41.1.

Box 41.1

How to ... **use cumulative Poisson probability tables to find the probability of a random event occurring (a) *r* times or fewer, (b) more than *r* times and (c) exactly *r* times in a given period or space, if the mean number of occurrences per unit of time or space is λ and the maximum possible number is very high**

1. Decide how many units of time or space there are in the given period or space. Multiply this by the given (or derived) value of λ to obtain a new value of λ for this question.

2. Check whether the table has a column for this new value of λ. If so, proceed as in step 3; if not, either use the formula in Box 41.2 or use an approximation based on interpolation of probabilities obtained using step 3 for slightly higher and slightly lower values of λ. Interpolation is not always possible.

3. (a) To obtain $P(X \leqslant r)$, the probability of the event happening r times or fewer, select the column for the new value of λ and read off the entry corresponding to $P(X \leqslant r)$.
   (b) To obtain $P(X > r)$, the probability of it happening more than r times, perform (a) and subtract the answer from 1.
   (c) To obtain $P(X = r)$, the probability of it happening exactly r times, select the column for the new value of λ and read off the values of $P(X \leqslant r)$ and $P(X \leqslant r - 1)$. The required probability is the difference between these two values.

The procedure for using equation [41.1] rather than tables is summarised in Box 41.2.

Box 41.2

How to ... **find, without using tables, the probability of a random event occurring (a) exactly *r* times, (b) *r* times or fewer and (c) more than *r* times in a given period or space, if the mean number of occurrences per unit of time or space is λ and the maximum possible number is very high**

1. Decide how many units of time or space there are in the given period or space. Multiply this by the given (or derived) value of λ to obtain a new value of λ for this question.

2. (a) To obtain $P(X = r)$ evaluate $\lambda^r e^{-\lambda}/r!$ in which r has the value specified by the question and λ has the new value found in step 1.
   (b) To obtain $P(X \leqslant r)$ evaluate $\lambda^x e^{-\lambda}/x!$ for x taking all values $0, 1, 2, \ldots, r$ and add them. Use step 3 to reduce work.
   (c) To obtain $P(X > r)$ first obtain $P(X \leqslant r)$ as in (b) then subtract the answer from 1. ⟳

3. Calculations in (b) and (c) can be simplified by using the recurrence formula

$$P(X = n) = \left(\frac{\lambda}{n}\right) P(X = n - 1) \tag{41.3}$$

so $P(X = 1) = \left(\frac{\lambda}{1}\right) P(X = 0)$

You can substitute the numerical value of λ at each recurrence. The next recurrence is

$$P(X = 2) = \left(\frac{\lambda}{2}\right) P(X = 1) = \left(\frac{\lambda}{2}\right)\left(\frac{\lambda}{1}\right) P(X = 0)$$

and continue as in Notebox 41.4 to write every probability in terms of $P(X = 0)$. Then add them, as in the notebox. Replace $P(X = 0)$ by its numerical value, which is obtained from $e^{-\lambda}$, and multiply as in the notebox.

41.4 The sum of two (or more) independent Poisson variables

It can be shown that if two *independent* variables have Poisson distributions with parameters λ_1 and λ_2, their sum also has a Poisson distribution and its parameter is $\lambda_1 + \lambda_2$. This can be written as follows:

if $X \sim Po(\lambda_1)$ and $Y \sim Po(\lambda_2)$ then $X + Y \sim Po(\lambda_1 + \lambda_2)$ [41.4]

The result can be generalised to the sum of any number of independent Poisson distributions. A corollary is that the difference between two independent Poisson distributed variables also has a Poisson distribution, with a new parameter equal to the difference between the component parameters.

Example 41.5 illustrates this. The paper has three pages written by the management and one page written by captains of works sports teams. Spelling mistakes occur at random. Management spelling mistakes have averaged 1.3 per page, but mistakes have averaged 1.1 per page on the sports pages. We are interested in the probability that next week's paper will have more than 6 mistakes

First we must note that the given averages are for mistakes per page, but the question is about mistakes per paper. Since management mistakes average 1.3 per page and there are 3 management pages, we have that $\lambda_{man} = 3.9$ per paper. As there is only one sports page, $\lambda_{sport} = 1.1$ per paper. The total number of mistakes per issue will therefore follow a Poisson distribution with parameter

$$\lambda = \lambda_{man} + \lambda_{sport} = 3.9 + 1.1 = 5.0$$

We can use either equation [41.1] or tables to show that the probability of more than 6 mistakes is $1 - 0.7622 = 0.2378$.

41.5 The mode of a Poisson distribution

By using the recurrence formula it can be shown that the mode of a Poisson distribution can be found from the following rule

if λ is an integer, the Poisson distribution has a double mode for $X = \lambda$ and $\lambda - 1$

if λ is not an integer, then the mode is the single integer lying between the values λ and $\lambda - 1$

For example, if $\lambda = 3$ step 3 of Box 41.2 gives the first few probabilities as

$P(X = 0)$
$P(X = 1) = 3.0 \times P(X = 0)$
$P(X = 2) = 4.5 \times P(X = 0)$
$P(X = 3) = 4.5 \times P(X = 0)$
$P(X = 4) = 3.375 \times P(X = 0)$

Thus the probability reaches its highest value for $X = 2$, stays at this value for $X = 3$ and then falls; there is a double mode for $X = 3$ and $X = 2$.

A similar table with $\lambda = 3.2$ instead of 3.0 will show a single mode for the integer lying between 3.2 and 2.2, i.e. 3. Note that there are double modes in Figure 41.2(b) and (c).

Consolidation exercise 41B

1. In a table of cumulative Poisson probabilities, what probability is shown by the entry opposite $r = 5$ in the column headed $\lambda = 3$? Answer in words.

2. How can the table be used to find the probability of an event occurring (a) more than 5 times? (b) exactly 5 times?

3. A house is divided into two flats. If the number of letters delivered in a day to flat 1 has a Poisson distribution and averages 2.3 letters per day, and the number delivered to flat 2 has a Poisson distribution with an average of 3.7 letters per day, what can be said of the total number of letters delivered to the house? Include in your answer a statement of any assumption you make. What is the most frequent number of letters delivered in a day to (a) flat 1, (b) the house?

4. Some of the letters delivered to flat 2 (of Question 3) have to be readdressed unopened to a former occupier. If the number being readdressed on any day is random and averages 1.6, what can be said of the distribution of the number of letters remaining for the present occupier to open?

Notebox 41.1

y^0 and $e^{-\lambda}$

Whatever the value of y may be, $y^0 = 1$. Also $y^{-k} = 1/y^k$ for all values of k. Note that $y^2 = 1 \times y \times y$, $y^1 = 1 \times y$ and $y^0 = 1$. Also $y^{-1} = 1/y$ and $y^{-2} = 1/y^2$. Similarly $e^{-\lambda} = 1/e^\lambda$ for all values of λ.

Notebox 41.2

Decimal places in the Poisson table

The cumulative probability of there being 12 occurrences or fewer is 0.9999 correct to four decimal places. The cumulative probability of 13 or fewer is quoted in the full table as 1.0000 correct to four decimal placess. Obviously it is possible for there to be more than 13 occurrences. Therefore the exact probability of there being 13 or fewer must be less than unity, but to record it we would need more than four decimal places. This is because the probability of there being more than 13 occurrences is less than 0.000 05. Thus, if we are working to four decimal places, then with $\lambda = 3.5$ we need not consider values of r greater than 13.

Notebox 41.3

Interpolation

If the value of λ lies between the values at the heads of adjacent columns in the table, interpolation will give an approximate answer. For example, suppose we have columns for $\lambda = 2.6$ and for $\lambda = 3.0$, but not for $\lambda = 2.8$. For $\lambda = 2.6$ we can read from the table that $P(X \leqslant 4) = 0.8774$, and for $\lambda = 3.0$ we can find $P(X \leqslant 4) = 0.8153$. We may infer that as 2.8 is midway between 2.6 and 3.0, the entry in a column for $\lambda = 2.8$ would be midway between 0.8774 and 0.8153, which is 0.8464. In fact, it should be 0.8477, as use of equation [41.1] confirms.

Notebox 41.4

A recurrence formula

If tables of cumulative Poisson probabilities are not readily available, it saves a lot of work if you use the following formula, which enables you to use the calculated probability of X having one value to find the probability of it having another stated value. The formula is

$$P(X = n) = \left(\frac{\lambda}{n}\right) P(X = n - 1) \qquad [41.4]$$

⮑

Thus if $\lambda = 3$ and $r = 4$ we first use our tables (or calculators) to find $P(X = 0)$:

$P(X = 0) = e^{-3} = 0.049\,787$

Then we use equation [41.4] to obtain

$$P(X = 1) = \left(\frac{3}{1}\right) P(X = 0) = 3P(X = 0)$$

$$P(X = 2) = \left(\frac{3}{2}\right) P(X = 1) = \left(\frac{9}{2}\right) P(X = 0)$$

$$P(X = 3) = \left(\frac{3}{3}\right) P(X = 2) = \left(\frac{9}{2}\right) P(X = 0)$$

$$P(X = 4) = \left(\frac{3}{4}\right) P(X = 3) = \left(\frac{27}{8}\right) P(X = 0)$$

The total probability of there being 4 or fewer occurrences is the sum of these four values:

$$(1 + 3 + 9/2 + 9/2 + 27/8)P(X = 0) = 16.375 \times 0.049787 = 0.8153$$

which can be verified from tables.

Summary

1. The Poisson distribution is an important discrete probability distribution. It arises in two different ways, discussed in this and the next chapter.

2. If an event occurs at random intervals or is randomly scattered in space, with no upper limit to the possible number of occurrences, in such a way that the arithmetic mean number of occurrences per unit of time or space is λ, then the probability of there being r occurrences in a unit period or space is given by

$$P(X = r) = \frac{\lambda^r e^{-\lambda}}{r!}$$

3. A random variable with this distribution is said to have the Poisson distribution, and we write $X \sim \text{Po}(\lambda)$. The equation involves λ, which is the parameter of the distribution with a value defined by the problem being considered. It is both the arithmetic mean of the distribution and the variance.

4. Provided that λ is a convenient number, it is possible to read off from tables the cumulative probability of there being r or fewer occurrences, hence the probability of there being more than r occurrences. The tables can also be used to estimate the probability of there being exactly r occurrences (Box 41.1).

5. Other cases require equation [41.1] but often the work can be reduced by using the recurrence formula (Notebox 41.5) as illustrated in examples and summarised in Box 41.2.

6. If λ is an integer, the Poisson distribution has a double mode for $X = \lambda$ and $X = \lambda - 1$. If λ is not an integer, the mode will be the single integer lying between λ and $\lambda - 1$.

7. If two independent variables have Poisson distributions, their sum also has a Poisson distribution; so has their difference. This result can be generalised to the sum of any number of Poisson distributions.

8. In using the Poisson distribution it is especially important to keep in mind the nature and magnitude of the units of time and space. For example, if λ is given as 3 occurrences per hour and we are interested in the probabilities of the numbers of occurrences per working day of 8 hours, we need to use $\lambda = 24$.

9. Noteboxes recall some mathematical results and deal with the consequences of working to four decimal figures in the cumulative table, with interpolation in the tables, and with the recurrence formula mentioned in item 5 above.

Poisson and approximation

42.1 Introduction

Example 42.1 The number of undergraduates visiting the medical officer with symptoms of nervous exhaustion is random but averages 28 per five-day week. Find the probability that next week the number will be (a) exactly 30 and (b) between 25 and 30 inclusive. (c) Assuming that one day is like another, what is the chance of there being no more than 3 visits tomorrow?

Example 42.2 A large estate agent has found that 3% ($p = 0.03$) of clients have to receive 'final reminders' before they pay their bills. Find the probability that if the estate agent takes a random sample of $n = 100$ clients, precisely 2 will need reminders. Do this (a) by using an exact method and (b) by using a good approximation.

Example 42.3 The probability of a certain kind of structure revealing weakness within five years is 0.001. What are the probabilities that if 4000 structures of this kind are erected, the number revealing weakness within five years will be (a) exactly three and (b) more than three?

Example 42.1 is just like those of the previous chapter, except that λ is higher than is likely to be found in readily available tables. This means having to use the formulae, which can become troublesome when large numbers are involved. In this chapter we show how a good approximation to a Poisson distribution can be obtained by using the normal distribution, provided certain conditions hold.

Then we turn the tables, and show how the Poisson distribution can itself be used to approximate the binomial distribution. This allows us to tackle the large-number binomial problems of Examples 42.2 and 42.3 more easily.

We refer to an approximation we discussed in Chapter 28, where the normal distribution is used to approximate the binomial distribution, and we include a table that summarises the use of one distribution as an approximation to another.

42.2 Using the normal distribution to approximate the Poisson distribution

In all the problems tackled in Chapter 41, the Poisson distribution gives exact probabilities. Deriving them involves finding the values of $\lambda^r e^{-\lambda}/r!$, which can present computational problems if λ is large. Many tables stop at $\lambda = 20$, and for λ as large as this, pocket calculators can begin to run into problems if the calculations are not performed in the recommended order. Fortunately if $\lambda > 20$ a good approximation to the Poisson distribution is given by the normal distribution with a mean and variance equal to those of the Poisson distribution it is approximating. We know from equation [41.2] that both of them are λ. In using this approximation we need to apply a *continuity correction* for the usual reason. The procedure is illustrated by the solution to Example 42.1.

⬭ **Example 42.1** has as its random variable 'the number of visits per week', which has the Poisson distribution Po(28).

(a) We are asked to calculate the probability $P(X = 30)$. We do this in two ways: exactly, using the Poisson distribution, and approximately, using the normal distribution. From equation [40.1]

$$P(X = r) = \frac{\lambda^r e^{-\lambda}}{r!}$$

Since $\lambda = 28$ and $r = 30$ we have the exact result

$$P(X = 30) = \frac{28^{30} e^{-28}}{30!} = 0.068 \text{ (to 3 d.p.)}$$

If we use the *normal approximation* with the same mean and variance as the Poisson distribution (so that both equal λ, which is 28) the required probability is given by the area under the normal curve N(28,28) between $X = 29.5$ and $X = 30.5$. Taking account of the continuity correction, we have

$$P(29.5 \leq X \leq 30.5) = P\left(\frac{29.5 - 28}{\sqrt{28}} < \frac{X - 28}{\sqrt{28}} < \frac{30.5 - 28}{\sqrt{28}}\right)$$

$$= P(0.2835 < Z < 0.4725)$$
$$= 0.6817 - 0.6116 \text{ (from Table D4)}$$
$$= 0.070 \text{ (to 3 d.p.)}$$

which should be compared with the previous result.

(b) We are asked for the probability that the number of visits per week will be between 25 and 30 inclusive. We again have $\lambda = 28$, too high for use of Table D3. For an *exact solution* we have to make repeated use of the formula. We want to calculate the probability $P(25 \leq X \leq 30)$. From equation [40.1]

$$P(X = r) = \frac{\lambda^r e^{-\lambda}}{r!}$$

$$P(X = 25) = \frac{28^{25} e^{-28}}{25!} = 0.067\,306$$

From the recurrence formula [41.3]

$$P(X = n) = \left(\frac{\lambda}{n}\right) P(X = n - 1) \tag{42.1}$$

⬭

we have that

$$P(X = 26) = \left(\frac{28}{26}\right)(0.067\,306) = 0.072\,484$$

Further use of the formula gives $P(X = 27) = 0.075\,169$, $P(X = 28) = 0.075\,169$, $P(X = 29) = 0.072\,577$ and $P(X = 30) = 0.067\,739$, then by addition $P(25 \le X \le 30) = P(X = 25) + P(X = 26) + \ldots + P(X = 30) = 0.430$ (to 3 d.p.).

If we use the normal approximation with mean and variance equal to λ, which is 28, we have that the probability is given by the area under the normal curve N(28,28) between $X = 24.5$ and $X = 30.5$:

$$P(24.5 \le X \le 30.5) = P\left(\frac{24.5 - 28}{\sqrt{28}} < \frac{X - 28}{\sqrt{28}} < \frac{30.5 - 28}{\sqrt{28}}\right)$$

$$= P(-0.6614 < Z < 0.4725)$$

$$= 0.6817 - 0.2541$$

$$= 0.428 \text{ (to 3 d.p.)}$$

(c) To solve this we find a new value of λ for the mean number of visits per day, and then proceed as before.

Box 42.1

How to ... **approximate the Poisson distribution by using the normal distribution**

1. If a random variable X has a Poisson distribution with parameter λ then, provided $\lambda \ge 20$, the probability of X (a) lying between specified values or (b) having a specified value can be estimated approximately by using the normal distribution with both mean and variance equal to λ.

2. The probability $P(A \le X \le B)$ of X lying between A and B (with $B > A$) is given by the area under the normal curve between

$$\frac{A - 0.5 - \lambda}{\sqrt{\lambda}} \quad \text{and} \quad \frac{B + 0.5 - \lambda}{\sqrt{\lambda}}$$

3. The probability $P(X = A)$ of X having the value A is given by the area under the normal curve between

$$\frac{A - 0.5 - \lambda}{\sqrt{\lambda}} \quad \text{and} \quad \frac{A + 0.5 - \lambda}{\sqrt{\lambda}}$$

4. The areas specified in steps 2 and 3 may be found by following the procedure of Box 28.1.

Consolidation exercise 42A

1. When is it appropriate to approximate the Poisson distribution by using the normal distribution? What are (a) the mean and (b) the variance of the normal distribution if the Poisson parameter is λ? Is it necessary to use a correction of any kind? If so, what is it and how is it appplied?

 42.3 ## Using the Poisson distribution to approximate the binomial distribution

Although the Poisson distribution gives exact results for random events, and can be approximated by a normal distribution, the Poisson distribution can also be used to give approximate solutions to binomial probability problems.

We know from the binomial probability distribution that if there are n identical trials, in each of which the probability of a success is p, then the probability of getting r successes in the n trials is

$$P_r = {}^nC_r p^r q^{n-r} = \frac{n!}{r!(n-r)!} p^r q^{n-r}$$

If n is large this can lead to tedious calculations. However, it can be shown that a good approximation to the (exact) binomial result is given by use of the Poisson distribution, provided that n is big enough and a success is very rare (low p) or highly likely (low q).

In practice, the Poisson distribution is safely used as an approximation to the binomial distribution if $n \geq 50$ and either np or nq is less than 5. A case that just fails to satisfy these conditions is $n = 50$ and $p = 0.10$. If $n \gg 50$ it is considered safe to use the Poisson distribution as an approximation if there is a probability of 10%.

Notice that since both the binomial and the Poisson distributions are discrete, no continuity correction is necessary when one is used to approximate to the other.

The Poisson distribution is sometimes used as an approximation to the binomial distribution, with less accuracy and justification, even when n is considerably less than 50. We illustrate the dangers by considering four versions of Example 42.2, differing only in the values of n and p.

Version 1. The estate agent has found that 3% ($p = 0.03$) of clients have to receive 'final reminders' before they pay their bills. The estate agent takes a random sample of $n = 100$ clients and we are asked to calculate the probability that precisely 2 will need reminders by (a) using the binomial distribution and (b) using the Poisson distribution as an approximation.

We begin by deriving an *accurate result* through using the binomial distribution:

$$P(2 \text{ slow clients in } 100) = {}^{100}C_2(0.03)^2(0.97)^{98}$$
$$= 4950 \times 0.0009 \times 0.050\,54$$
$$= 0.225\,15$$

As $n = 100 \gg 50$ and $np = 100 \times 0.03 = 3.0 \ll 5$, the recommended conditions for use of the Poisson approximation are well observed. We get

$$P(2 \text{ slow clients in } 100) = p_2 = \frac{(3.0)^2 e^{-3}}{2!} = 4.5/e^3$$

$$= 4.5/20.0855 = 0.224\,04$$

Here the percentage error (between 0.225 and 0.224) is only 0.5%.

⊙ **Version 2** We alter the example by putting $p = 0.08$ and $n = 60$. Here we have $n = 60 > 50$ and $np = 60 \times 0.08 = 4.8 < 5$. The recommended conditions for using the Poisson distribution as an approximation are therefore satisfied, although we are rather close to both limits (of $n \geq 50$ and $np < 5$). You can check that in this case the approximation is less good, with a percentage error of about 5%.

⊙ **Version 3** We put $p = 0.10$ and $n = 10$. With such a small n the approximation should not be used, but we may think that as $\lambda = np = 10 \times 0.1 = 1.0$ (well under 5), there may be some compensation for the low n. In fact, the error from using the Poisson distribution under these not recommended conditions is just over 5%, which is much the same as for Version 2. But it would be unwise to think that we can always violate one condition provided we amply satisfy the other.

⊙ **Version 4** We put $p = 0.25$ and $n = 32$, which badly violates both conditions. You can check that the 'approximation' is nowhere near the correct result, demonstrating the dangers of using the Poisson distribution when the conditions are not observed.

Box 42.2

How to ... **approximate the binomial distribution by using the Poisson distribution**

1. Let the chance of success in a single trial be p and the number of trials be n. We are interested in the probability of getting (a) exactly x successes, (b) x or fewer successes, and (c) between x_1 and x_2 successes. Use the Poisson distribution as an approximation to the binomial distribution only if $n \geq 50$ and $np < 5$. Breaching either of these conditions may lead to unacceptable inaccuracy, but for $n \gg 50$ the approximation is acceptable provided $p \leq 0.1$.

2. Use the Poisson distribution with $\lambda = np$. To find the probability of getting (a) exactly x successes proceed as in Box 41.1. For (b) x or fewer successes, proceed as in Box 41.2. For (c) between x_1 and x_2 successes, proceed as in Box 41.2 to find the probabilities first of getting x_2 or fewer successes and then of getting x_1 or fewer successes. Then take the difference between these probabilities.

Example 42.3 is left as an exercise. The answers are (a) 0.1954 and (b) 0.5665.

42.4 **Binomial, normal and Poisson approximations**

Table 42.1 summarises the conditions under which one of these three distributions of a random variable can be used as an approximation to another. The second entry does not involve a Poisson distribution. Compare it carefully with the third entry. In using Table 42.1 replace p by $q = 1 - p$ if $p > 0.5$. In general, the larger the value of n the bettter.

Table 42.1

Distribution of X	Conditions	Approximate distribution
Poisson X Po(λ)	Large λ (at least 20) See Section 42.2	Normal $x \sim N(\lambda, \lambda)$ Use Box 42.1
Binomial X B(n, p)	$n > 10$, and p close to 0.5; as a lower limit take 0.4 $n > 30$ and $0.1 < p \leq 0.4$ See Chapter 28	Normal $X \sim N(np, npq)$ Use Box 28.5
Binomial X B(n, p)	Large n (at least 50) and small p (at most 0.1); preferably $np \leq 5$ See Section 42.3	Poisson $X \sim$ Po(np) Use Box 42.2

Summary

1. If $\lambda > 20$ it is possible to approximate a Poisson solution by using the normal distribution with both mean and variance equal to λ. (Box 42.1).

2. A good approximation to the binomial distribution is obtained by using the Poisson distribution, provided the chance of a success is very small (or very large) and the sample is large enough. This can be done if $n \geq 50$ and either np or nq is less than 5. Then the event of a success (or failure, if nq is small) is called a rare event. The procedure is summarised in Box 42.2.

3. Table 42.1 summarises the conditions to be observed in approximating one of these distributions by using another.

The exponential probability distribution

43.1 **Introduction**

⟹ **Example 43.1** Anna Liszt suffers from random painful attacks of *digitus significantus*, also known as statistician's thumb. Over the last few years, the mean length of time between them has been 25 days. She had one this morning and hopes there is not another in five days time, when she has a wedding. What is the probability that (i) the next attack will be within the first four days and (ii) the next attack will be on the fifth day?

⟹ **Example 43.2** A company hires out video recorders which require occasional repair. The interval X between major repairs is random and has a mean of 25 months.

(a) It wants to know the probability that a given recorder will not require a major repair for two years.
(b) It intends to advertise that any set requring two major repairs within M months will be replaced. It wants to choose M so that it does not have to replace more than 1 set in 10. What should M be?

⟹ **Example 43.3** Telephone calls arrive at an office switchboard at random intervals. The number of internal calls averages 2 per five-minute period. External calls average 1 per five-minute period.

(a) What is the probability that there will be more than 2 calls in any period of 3 minutes?
(b) What is the average period between calls?
(c) What is the probability that an interval between calls will be as long as $2\frac{1}{2}$ minutes?
(d) If there has already been an interval of $2\frac{1}{2}$ minutes since the last call, what is the probability of there being at least another half-minute before the next call?
(e) If there has already been an interval of 3 minutes since the last call, what is the probability of there being at least yet another half-minute before the next call?

▶ **Example 43.4**   The lifetime of an electric light bulb is a random variable with a mean of 1000 hours. Making a suitable assumption, find the probability that a bulb (i) is still working after 1250 hours, and (ii) will then last for at least another 250 hours.

In discussing the Poisson distribution, we were interested in the number of times a random event occurred in a given period or space. This was necessarily a positive integer (including 0), so we had a discrete variable. On the other hand, the interval between these events is measured in continuous time, or continuous distance, so its probability distribution is continuous. In this chapter we study the probability distribution of the intervals between random events, which also occurs in many other circumstances and is closely related to decay. We shall see that it has some remarkable properties. Unless you are already familiar with exponential functions, you should read Noteboxes 43.1 and 43.2 before proceeding.

43.2   The exponential distribution

We know from Chapter 42 that if an event occurs at random, with no limit to the possible number of occurrences, the frequencies of occurrences per unit time or space fit a Poisson distribution. Suppose this has a parameter λ, so the probability of there being x events per unit is $e^{-\lambda}(\lambda x/x!)$.

It can be shown that the intervals between these random occurrences will have random lengths with a probability density function given by $f(x) = \lambda e^{-\lambda x}$. An important point is that the value of λ is the same in both expressions and measures the average rate of occurrence.

The function $f(x)$ is called an **exponential probability function**, for reasons explained in Notebox 43.1. It has a graph like one of those in Figure 43.1. Its highest value is λ and it slopes downwards, becoming ever smaller but always remaining positive. It can be shown that the area under the curve is unity, as is necessary for a probability distribution.

In what follows keep in mind that the variable in the Poisson distribution is the number of events in a given unit of space or time, whereas the variable in the exponential distribution is the interval of space or time between consecutive events.

As with all probability distributions, the area under the exponential curve between $x = 0$ and $x = c$ shows the probability of x having a value of c or less. This is measured by the cumulative probability distribution $F(c)$. It can be shown that $F(c) = 1 - e^{-\lambda c}$ (Figure 43.2). Thus

$$F(c) = P(x \leqslant c) = 1 - e^{-\lambda c} \qquad [43.1]$$

It is very easy to evaluate $F(c)$ using a pocket calculator or tables, as we now illustrate.

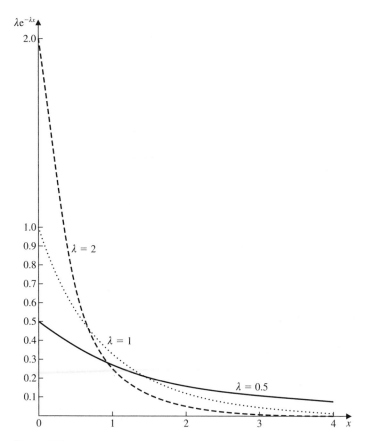

Figure 43.1

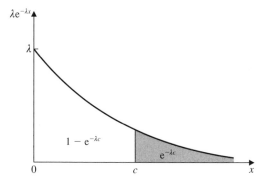

Figure 43.2

Example 43.1 states that attacks occur at random and the mean length of time between them is 25 days. Thus the average rate of occurrence is 1/25 per day, or 0.04 per day. This is our value of λ.

Moreover, when we speak of an attack within the first 4 days we have to think of the first day beginning at 0 and ending at midnight when $X = 1$, so the first four days imply $X \leqslant 4$.

(a) To calculate the probability of an attack within 4 days, we use equation [43.1]. Remembering that $\lambda = 0.04$, the probability of getting an attack within the first 4 days is therefore

$$P(x \leqslant 4) = 1 - e^{-0.04(4)}$$

The easy way to get the value of this with the help of a calculator is described in Box 43.1. We get

$$P(x \leqslant 4) = 0.1479$$

Box 43.1

How to ... **find the value of $e^{-\lambda c}$ (where λ and c are positive)**

1. In this expression e is the usual constant defined in Notebox 31.2.

2. Multiply λ by c and call the product x.

3. If your calculator has a button for e^x, which we denote $[e^x]$:
   (i) Either press $[e^x]$ (it may involve getting into a second function) and follow by entering $-x$ then press $[=]$.
   (ii) Or press $[e^x]$ (it may involve getting into a second function) and follow by entering x. Press $[=]$ then find the reciprocal of the answer by using $[x^{-1}]$.

4. If your calculator does not have a button for e^x, use the button for y^x. Enter 2.7183. Press the button. Then enter the negative value of $-x$ and press $[=]$. Or enter the positive value of x, press $[=]$ then press $[x^{-1}]$.

(b) The probability of an attack of *digitus significantus* occurring on the fifth day means the probability of an attack in the time interval between 4.0 and 5.0 whole days. We already have $P(x \leqslant 4) = 0.1479$. Similarly

$$P(x \leqslant 5) = 1 - e^{-0.04(5)} = 0.1813$$

Thus the probability of an attack on day 5 is

$$0.1813 - 0.1479 = 0.0334$$

Alternatively we could argue

$$P(4 \leqslant x \leqslant 5) = e^{-0.04(4)} - e^{-0.04(5)} = 0.0334$$

Note that the exponential probability distribution defined by $f(x) = \lambda e^{-\lambda x}$ has

mean $= 1/\lambda$ [43.2]

var $= 1/\lambda^2$ [43.3]

which is the square of the mean.

Example 43.2 illustrates the use of these ideas and introduces a few points of more general application.

The company accepts that the interval X (measured in months) between major repairs is random with a negative exponential distribution, parameter $\lambda = 0.04$. We consider in turn the questions that it raises.

(a) It wants to know the probability that a given recorder will not require a major repair for two years. The probability distribution for the interval between repairs is given by

$f(x) = 0.04e^{-0.04x}$

where x is measured in months.

Note that 0.04 relates to months, so 2 years must also be expressed in months. The cumulative frequency up to $x = 24$ shows the total probability of X – the interval between major repairs – being less than (or equal to) 24. This is the value of $F(24)$ where F has the value defined by equation [43.1]. It measures the probability of a set requiring (another) major repair within 24 months. Therefore the probability of it not requiring a major repair within 24 months is $1 - F(24)$. Thus

$P(X > 24) = 1 - F(24) = 1 - (1 - e^{-0.04(24)}) = 0.383$

(b) It intends to advertise that any set requiring two major repairs within M months will be replaced. It wants to choose M so that it does not have to replace more than 1 set in 10. What should M be?

We have to find an M such that the probability of X, the interval between major repairs, being less than M is not more than 0.1. (The reason for this wording will become apparent shortly.)

Thus we want to find M such that $P(X < M) \leqslant 0.1$. We know that $\lambda = 0.04$. Therefore this means that we want M such that $1 - e^{-0.04M} \leqslant 0.1$, which gives $0.9 \leqslant e^{-0.04M}$.

This kind of expression can arise quite often in the solution of problems to do with the exponential probability distribution. Box 43.2 explains it and gives a rule for using it to find the value of M. (In the case of this example, the far from obvious answer is that M must be not more than 2.635. Take that on trust for the moment and finish reading the next few lines before turning to the box.)

Box 43.2

How to ... **find the value of M if $A = e^{-BM}$ (where A and B are positive)**

1. In this expression e is the usual constant defined in Notebox 31.2.

2. If we are told that $A = e^{-BM}$ then the value of M is given by a formula that may look a bit frightening but leads to an easy way of finding it. The formula is

$$M = \frac{\ln A}{B} = 2.3026 \frac{\log A}{B}$$

3. (a) If you have a calculator with a button marked ln, you can get M by pressing [ln], entering the value of A, pressing [/], entering the value of B, then pressing [=].
   (b) If your calculator has no [ln] button but has a [log] button, enter 2.3026 then press [×][log], enter the value of A, press [/], enter the value of B then press [=].
   (c) If you have no [log] button, either use log tables or get a better calculator.

⬭ Thus, if the company says it will replace sets that require repair within 2.635 months of an earlier repair, it will probably have to replace 1 set in 10. If the company insists on a whole number of months, it will have to use $M = 2$ (or 1). Using $M = 3$ would lead to too many replacements. If it thinks that 2 months does not sound very generous, it may choose 10 weeks, which sounds better but is comfortably less than 2.635 months; so is 80 days.

Box 43.3

How to ... **find the probability that an exponentially distributed random variable will have a value of (a) not more than c and (b) more than c**

1. The probability density function is

   $$f(x) = \lambda e^{-\lambda x}$$

   where x is the value of the random variable X, and λ is the parameter of the distribution. From the data given in the problem, decide the value of λ.

2. $P(X \leqslant c)$ is given by the cumulative probability function with $x = c$, i.e. by $F(c) = 1 - e^{-\lambda c}$.

3. $P(X > c)$ is $1 - F(c) = e^{-\lambda c}$.

4. Find these values as explained in Box 43.1.

Consolidation exercise 43A

1. What is the variable that has an exponential probability density function? What, in words, is its parameter λ?

2. If a random variable x has an exponential distribution, what (a) in words and (b) in symbols is the probability of it having a value not more than c?

3. If $0.8 \leqslant e^{-x}$ what is the easy way of finding x?

43.3 Further problems

Now we look at some problems that illustrate this connection between the Poisson and exponential distributions, beginning with Example 43.3. Its solution illustrates the procedures for tackling a wide range of problems and highlights an important feature of the exponential distribution. After tackling each part of the problem, we summarise the procedure used to answer that kind of question and comment on any important general points which arise.

Example 43.3 has telephone calls arriving at an office switchboard at random. Internal calls average 2 per five-minute period. External calls average 1 per five-minute period.

The probability that there will be more than 2 calls in any period of 3 minutes
This involves nothing new. Define a random variable I to be the number of internal calls per three-minute period, and another random variable E to be the number of external calls per three-minute period. Each variable has a Poisson distribution.

Note that the given data are about five-minute periods but the question is about three-minute periods. We choose to work in three-minute periods, and to adapt the data accordingly. Since we expect 2 internal calls in 5 minutes we expect 1.2 in 3 minutes. Thus I has the Poisson distribution Po(1.2). Similarly E has the distribution Po(0.6).

Remember that the sum of two Poisson distributions with parameters λ_1 and λ_2 is another Poisson distribution with parameter $\lambda_1 + \lambda_2$. Thus the total number of calls T has a distribution which is the sum of I and E, so is Po(1.8) with a time unit of 3 minutes.

To find the probability that there are more than 2 calls in a three-minute period, we first find the probability that there are 2 calls or fewer. This can be calculated or obtained from Table D3 as 0.7306

It follows that the probability of there being more than 2 calls in 3 minutes is $1 - 0.7306 = 0.2694$.

The average period between calls
We have just seen that calls of one kind or another arrive according to a Poisson distribution with parameter $\lambda = 1.8$ for a time unit of 3 minutes. (We could equally well work with $\lambda = 3.0$ and a time unit of 5 minutes.)

Therefore the three-minute intervals between calls have an exponential p.d.f. with the same parametric value, 1.8.

We know from earlier in the chapter that if calls arrive at intervals of length x with the exponential p.d.f. $f(x) = \lambda e^{-\lambda x}$ then the mean time between calls is $1/\lambda$ units of 3 minutes. This is therefore $(1/1.8) \times 3$ minutes $= 1.667$ minutes. The same answer is given by $\frac{1}{3} \times 5$ minutes.

This result should be noted. It can also be obtained deductively. In the Poisson distribution λ is the mean number of calls per three-minute period, so the average interval between calls must be the reciprocal. The procedure used in this part of the solution is summarised in Box 43.4.

Box 43.4

How to ... **find the mean interval between Poisson-distributed events**

1. If the Poisson distribution is $\text{Po}(\lambda)$ such that there are on average λ events per unit of space or time, then the mean interval between events is given by $1/\lambda$ units.

⬭ **The probability that an interval between calls will be as long as $2\frac{1}{2}$ minutes**

We know that if x represents the length of an interval between calls then x has a p.d.f. of $f(x) = \lambda e^{-\lambda x}$ where $\lambda = 1.8$ (if we work in periods of 3 minutes).

For this distribution the cumulative distribution function is $F(x) = 1 - e^{-\lambda x}$ from equation [43.1]. This gives the probability of x having a value less than c.

It follows that the probability of x having a value greater than c is $1 - F(c) = 1 - (1 - e^{-\lambda c})$ which is $e^{-\lambda c}$, as illustrated in Figure 43.2.

We have to be careful about the units of time. The question asks about calls lasting at least $2\frac{1}{2}$ minutes, but λ relates to units of 3 minutes. This means that we want to know the probability of an interval lasting $2\frac{1}{2}/3$ time units (of 3 minutes). Thus we have to put $c = 0.8333$.

It follows that the probability of x having a value greater than $2\frac{1}{2}$ minutes (or 0.8333 time units) is given by $e^{-\lambda c}$ with $\lambda = 1.8$ and $x = 0.8333$. This gives $P = 0.223$.

The probability of there being at least another half-minute before the next call when there has already been an interval of $2\frac{1}{2}$ minutes since the last call

Remember we are working in units of 3 minutes. In Figure 43.3 the total area under the curve is unity and the area to the left of A represents $F(0.8333)$, which is the probability of an interval of less than $2\frac{1}{2}$ minutes (or 0.8333 units of 3 minutes)

In this part of the example we know that $2\frac{1}{2}$ minutes have already passed, so the exact duration of the interval is bound to be represented by a point to the right of A.

Let B represent an interval of exactly 3 minutes, so that points between A and B represent extensions of up to, but not more than, half a minute. Points to the right of B represents extensions of more than half a minute. ⬭

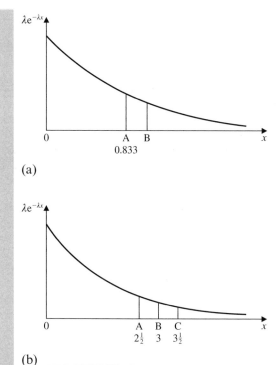

(a)

(b)

Figure 43.3

Since the next phone call is bound to come at a time to the right of A, the chances of its also coming after B are given by the area to the right of B expressed as a fraction of the area to the right of A. We have already seen that the area to the right of A is $e^{-(1.8)(0.8333)} = 0.22313$.

Since we are working in units of 3 minutes, and B corresponds to a total interval of 3 minutes, the value of x for B is 1 and the area to its right is $e^{-(1.8)} = 0.16530$. The chance of the interval lasting at least 3 minutes given that it has already lasted $2\frac{1}{2}$ minutes is therefore

$$\frac{\text{Area to the right of B}}{\text{Area to the right of A}} = \frac{0.16530}{0.22313} = 0.7408$$

Now we may begin to see a highly important result that the figures mask. Note first that we have two time periods – $2\frac{1}{2}$ minutes at A and $2\frac{1}{2} + \frac{1}{2} = 3$ minutes at B. We have obtained the answer 0.74 by dividing two areas.

The derivation can be slightly rewritten. The chance of the interval lasting at least 1 time unit (3 minutes) given that it has already lasted 0.83333 time units ($2\frac{1}{2}$ minutes) are

$$\frac{\text{Area to right of B}}{\text{Area to right of A}} = \frac{e^{-\lambda(1.0)}}{e^{-\lambda(0.83333)}} = e^{-\lambda(0.16667)}$$

⬭

and tables or calculators will show that when $\lambda = 1.8$ this has the value 0.7408, as before.

The remarkable thing about this result will become clear when we answer the next part of this problem.

The probability of there being at least another half-minute before the next call when there has already been an interval of 3 minutes since the last call

This is just like the previous question except that instead of there having already been an interval of $2\frac{1}{2}$ minutes, there has now already been an interval of 3 minutes.

A similar argument will show that the probability of there being another half-minute of silence when there has already been a period of 3 minutes is given by the ratio of two areas, which may now be denoted by areas to the right of B (at 3 minutes) and C (at $3\frac{1}{2}$ minutes) in Figure 43.3(b). Show that this too is 0.7408. Note that your solutions contain $e^{-(1.8)(0.16667)}$, as in the previous question. This 0.16667 is the additional one-sixth of a time unit that is specified in both parts.

We have just seen that the probability of the silence lasting for another half-minute does not depend on how long has already passed since the last call. All that matters is the length of the future interval in which we are interested – in this case 0.16667 time units.

Box 43.5

How to ... **find the probability of an interval continuing for another b units of time or space, if it has an exponential probability distribution**

1. Ignore the length of interval that has already passed.

2. Determine λ either as the mean rate of occurrence of the event that separates the intervals or as the inverse of the mean length of intervals. Use the time (or space) units that have been used to define b.

3. The probability of the interval continuing for b units is given by $e^{-\lambda b}$, which can be evaluated by using Box 43.1.

We now use the result in Box 43.5 to answer another example.

▶ **Example 43.4** is answered by assuming that lifetimes of bulbs are random and follow an exponential distribution with parameter $1/\text{mean} = 0.0001$ if time is measured in hours, or $1/4 = 0.25$ if time is measured in units of 250 hours, which the numbers in the problem suggest may be convenient. This is equivalent to assuming that bulb failures are random events with a mean rate of occurrence of 1 in 1000 hours, as given.

Working in units of 250 hours, we are interested in the probability of a life continuing for 1 time unit (beyond the 4 time units), when the parameter $\lambda = 0.25$. It is $e^{-(0.25)(1)} = 0.779$. That answers part (i); the answer to part (ii) is just the same.

Consolidation exercise 43B

1. If a random event happens on average three times a day, what are (a) the parameter of the distribution of the intervals (measured in hours) between events, (b) the average time between events and (c) the probability that an interval between two consecutive events will be at least 6 hours?

2. In Question 1, given that there has been no event in the last four hours, what is the probability that (a) there will be no event in the next three hours and (b) there will be at least one event in the next three hours?

Notebox 43.1

Exponential functions

In the expression x^2, x is multiplied by itself, squared or raised to the power 2. Another way of saying this is that x has an exponent of 2. Whatever the value of the variable x, it is raised to the power 2 (or squared). Similarly in x^3 there is an exponent of 3, and whatever may be the value of the variable x, it is raised to the power 3. These are examples of powers of the variable x, and the exponent is the number that corresponds to the power.

However, we could get it the other way round, and think of powers of a constant number, such as 2. In this case the expression 2^x would mean 2 raised to the power x. As x is given different values, we get different values for 2^x. Similarly for 3^x as shown in Table 43.1. Here the constant is being raised to ever higher powers that are denoted by the variable x. It is the variable x that is the exponent.

We know that any expression involving x may be called a function of x. Its value depends on x, and x is allowed to vary. Thus x^2, $4x + 3$, $x!$, 2^x, 3^{2x}, 3^{-x}, 3^{-4x} are all functions of x. The last four, with x in the power (or the exponent) are called exponential functions of x.

Table 43.1

x	0	1	2	3	4	5	...
2^x	1	2	4	8	16	32	...
3^x	1	3	9	27	81	243	...

Notebox 43.2

The behaviour of exponential functions

Provided x is positive, the values of 2^x and 3^x rise as x increases. This is also true if we use any other number bigger than 1. We may also have something like 3^{2x}, with very rapidly rising values; for $x = 0, 1, 2, 3$ and 4 they are 1, 9, 81, 729 and 6561. ⟱

However, 3^{-x} behaves very differently. Remember the meaning of the minus sign when it is in a power; x^{-2} means $1/x^2$ and 2^{-x} means $1/2^x$. Keeping this in mind, we can work out some values of 3^{-x} and see that they decline very rapidly: 1, 1/3, 1/9, 1/27, 1/81 . . . .

Exponential functions such as 2^x and 3^{2x} are called **positive exponential functions** because the exponent is positive. Exponential functions such as 3^{-x}, 3^{-4x} have negative exponents, so they are called **negative exponential functions**.

If the constant (such as 3 in 3^{4x}) is greater than 1, a positive exponential function will grow very rapidly (or explode); but a negative exponential function will rapidly decline towards zero. Figure 43.1 shows the behaviour of a negative exponential function that has the constant e in place of 3, namely $\lambda e^{-\lambda x}$ for selected values of λ.

Notebox 43.3

It is only how much longer that counts

This important notebox is included here for ease of reference, but the reference to conditional probability may not be understood until pages 516–17 have been read. The probability density function for a negative exponential distribution is $f(t) = \lambda e^{-\lambda t}$. The probability that $t > a$ is given by

$$P(T > a) = e^{-\lambda a} \qquad [43.4]$$

From our knowledge of conditional probability we have

$$P(T > a + b \,|\, T > a) = \frac{P(T > a + b \cap T > a)}{P(T > a)} \qquad [43.5]$$

$$= \frac{e^{-\lambda(a+b)}}{e^{-\lambda a}} = e^{-\lambda b}$$

which we can see from equation [43.4] is $P(T > b)$.

In Example 43.2 $b = \frac{1}{2}$ minute, so it equals 0.1667 unit where 1 unit represents 3 minutes. Inserting this in the result we have just derived gives $e^{-(1.8)(0.1667)} = 0.74$, which is the same as the answer obtained in the text. Note that it depends only on b. This leads to a highly important observation:

If a random variable X has an exponential probability distribution, the probability of X being greater than $a + b$ when it is already greater than a is given by the right-hand side of equation [43.5] and depends only on the value b. The value of a does not come into it.

Summary

1. If events occur according to a Poisson distribution with parameter λ, the intervals between them have a continuous probability distribution given by $f(x) = \lambda e^{-\lambda x}$. This equation defines the exponential probability distribution, which has the same parameter as the Poisson distribution to which it is related.

2. The curve of the distribution cuts the vertical axis at λ and slopes downwards at a declining but always positive rate that ensures it becomes closer and closer to the x-axis; eventually the curve and the axis become indistinguishable.

3. The mean (or expected) value of a random variable that has an exponential probability distribution is $1/\lambda$ and the variance is the square of the mean.

4. The cumulative probability distribution function is $F(c) = 1 - e^{-\lambda c}$, which can be evaluated with the help of Box 43.1.

5. Questions involving the solution of expressions like $A = e^{-BM}$ for M in terms of the values of A and B can be solved with the help of Box 43.2.

6. The probabilities $P(x \leqslant c)$ and $P(X > c)$ can be found with the help of Box 43.3.

7. Box 43.5 explains how to find the probability that an interval which has already lasted a units of time or space will continue for another b units, if it has an exponential probability distribution. The answer does not depend on the value of a.

8. Noteboxes contain an explanation of exponential functions, an account of the behaviour of exponential functions, a discussion of the point made in item 7 above, and a summary table of the means and variances of distributions studied in this and the last few chapters.

Goodness of fit

44.1 Introduction

Example 44.1 Last week the numbers of customers present in the Calculators' Arms at midday on five consecutive days were as shown in Table 44.1. Do these figures fit the hypothesis that there is no difference between one day and another? Or more formally, that they are generated by a uniform distribution?

Example 44.2 We have the data in Table 44.2 about the observed numbers of people of different nationalities on a plane. Do they support the hypothesis that the four nationalities are represented on flights in the ratio 4:3:1:2?

Example 44.3 ScrawlBall Pens Inc. have a unit making balls for their pens. Over the last month 400 samples of five balls have shown the frequencies of faulty balls given in Table 44.3. We wish to test the two-part hypothesis that the chance of any one ball being faulty is independent of whether any other ball is faulty; and that the proportion of defective balls is 0.2.

Table 44.1

Monday	Tuesday	Wednesday	Thursday	Friday
37	47	43	35	38

Table 44.2

	American	British	Chinese	Dutch	Total
Observed	43	32	7	18	100

Table 44.3

Number of faulty components, x	0	1	2	3	4	5
Observed frequency, f	140	150	100	7	2	1

Example 44.4 Every day the manager of a large office chooses five employees at random and notes the time they spend away from their desks, deciding at the end of the day whether they pass or fail a test for diligence. At the end of 300 days the manager produces Table 44.4 showing the frequencies with which different numbers of people passed the test. Does this pattern conform to a binomial distribution?

Example 44.5 The manager of ScrawlBall Pens Inc. notes that, over the last 300 workdays, the daily number of absentees has been between 0 and 8, with the distribution in Table 44.5. The general manager believes there are dark forces but the works manager says the absences are purely random. Which view, if either, is supported by the data?

Example 44.6 One hundred undernourished undergraduates were weighed, fed daily dollops of Statingham AddMass for 60 days, and then weighed again. The changes in mass are shown in Table 44.6. Test the hypothesis that these changes are normally distributed with a mean of $4\,\text{kg}$ and a variance of $4\,\text{kg}^2$.

Example 44.7 A random sample of 12 examination papers from a large population has marks 34, 48, 52, 55, 55, 58, 61, 62, 65, 66, 68, 83. Does this sample support the hypothesis that the marks in the population are normally distributed?

Table 44.4

Number of people passing, x	0	1	2	3	4	5
Observed number of days, o	20	50	80	100	30	20

Table 44.5

Absentees, x	0	1	2	3	4	5	6	7	8
Frequency, o	20	45	85	65	40	25	10	6	4

Table 44.6

Change in mass (kg)	$x \leq 0$	$0 < x \leq 1$	$1 < x \leq 2$	$2 < x \leq 3$	$3 < x \leq 4$
Frequency	3	6	10	15	22

Change in mass (kg)	$4 < x \leq 5$	$5 < x \leq 6$	$6 < x \leq 7$	$7 < x \leq 8$	$x > 8$
Frequency	17	12	9	4	2

In all these examples we ask whether some observed data about one variable are compatible with (or 'fit') what we would expect under some hypothesis. We want to know how well an observed frequency distribution fits (or is fitted by) a hypothesised distribution.

We answer most of these questions by using the χ^2 test. To do this we need to be thoroughly at home with degrees of freedom, on which we remark in Section 44.2, and concentrate in the first part of each solution. Then we go on to apply the χ^2 test for goodness of fit, including fit to a normal distribution. We answer the last example by applying another test (due to Shapiro and Wilk) for goodness of fit to a normal distribution.

44.2 Degrees of freedom for a single row or column

The basic method for χ^2 goodness-of-fit tests is to use a hypothesis to generate expected values. The observed values are then compared with them by use of the χ^2 test described in Chapter 36. To determine the critical value, we have to know both the selected level of significance and the number of degrees of freedom.

In conducting t-tests we introduced the idea of degrees of freedom and in Box 29.4 defined their number v to be $n - k$, where n is the number of observations and k is the number of restrictions imposed on the data. In applying this to single-row χ^2 tests, we have to ensure that every cell in the row has a frequency of at least 5. Doing this may mean amalgamating cells, so that when we come to apply the test, the number of cells is not n but the **adjusted number** n^*. We therefore write $v = n^* - k$.

Our main examples fall into two categories. In all cases we find our expected frequencies on the basis of a hypothesised distribution. In some cases (such as Example 44.3) we know the parameter of the distribution. In other cases (such as Example 44.4) we do not, so we have to calculate them from the data. Each parameter that we have to calculate imposes a restriction and uses up one degree of freedom. Another is used up by making the totals agree. The procedure and some frequently used results are summarised in Box 44.1.

Box 44.1

How to ... find the number of degrees of freedom for a single row

1. If there are n cells all in one row (or one column), the number of degrees of freedom is $v = n - k$ where k is the number of restrictions that may be necessary to calculate the expected frequencies and will have a minimum value of 1. It will depend on the hypothesis employed.

2. If cells have to be amalgamated, we have to take n as the number of cells in the final table. We denote this by n^*. ⬭

3. Frequently used cases are:

(i) Uniform distribution	$v = n^* - 1$
(ii) Specified ratio	$v = n^* - 1$
(iii) Binomial distribution, p known	$v = n^* - 1$
(iv) Binomial distribution, p unknown	$v = n^* - 2$
(v) Poisson distribution, λ unknown	$v = n^* - 2$
(vi) Normal distribution, μ, σ known	$v = n^* - 1$
(vii) Normal distribution, μ, σ unknown	$v = n^* - 3$

4. More generally, if the expected frequencies can be calculated without having to estimate any population parameters (such as p, μ, σ) from sample statistics then $v = n^* - 1$. But if the expected frequencies can be calculated only after estimating m population parameters then $v = n^* - m - 1 = n^* - k$ where k is the total number of restrictions.

Note that, even though we will sometimes be calculating parameters, we still have a non-parametric test since we do not need to calculate parameters of the probability distribution used in the test. The parameters we calculate are part of the statement of the expected frequencies rather than the procedure for testing whether the observed frequencies differ significantly from them.

44.3 ⬭ Using χ^2 to test goodness of fit

We illustrate the procedures in several different kinds of problem.

A single row (or column) of cells with theoretical frequencies based on the hypothesis of a uniform distribution.

In **Example 44.1** we adopt the null hypothesis that there is a uniform distribution and obtain the expected frequencies by dividing the total (of 200) by the number of cells ($n = 5$). There is no need to calculate any parameter. The only restriction on our numbers is that the observed frequencies should have the same total as the theoretical frequencies, so $k = 1$. As all frequencies are 40, no amalgamation is necessary, so $n^* = n$ and $v = n^* - k = 4$.

We choose a level of significance, say 5%. For this level and $v = 4$ the critical value of χ^2 is 9.49.

The numerical differences between the observed and expected values are as follows:

$$|37 - 40| = 3 \quad |47 - 40| = 7 \quad |43 - 40| = 3 \quad |35 - 40| = 5 \quad |38 - 40| = 2$$

And as all the expected values are 40, these give rise to the following values of $(o - e)^2/e$:

9/40, 49/40, 3/40   25/40, 4/40

Their sum is obviously less than the critical value of 9.49, so the evidence does not refute the null hypothesis.

A single row (or column) of cells with theoretical frequencies based on the hypothesis that they should be in a specified ratio.

In **Example 44.2** to calculate the expected values, we need to impose only one restriction – the total of the expected values is 100. Thus $k = 1$. Making use of the specified ratios, we can then write the expected values as

40, 30, 10, 20

All the frequencies are at least 5, so no amalgamation is needed. The number of degrees of freedom is therefore

$$v = n^* - 1 = 4 - 1 = 3$$

At the 5% level of significance and with $v = 3$, the critical value of χ^2 is 7.82. The observed and expected frequencies can be used as before to find that χ^2_{calc} is clearly below the critical value, so it does not lead to rejection of the null hypothesis.

Check that if the specified ratios had been 12, 9, 1, 3 then with a total of 100, one of the expected frequencies would have been less than 5. Amalgamating the last two entries would lead to only three cells, hence to 2 degrees of freedom and a 5% critical level of 5.99. Complete the calculation.

A single row (or column) of cells with theoretical frequencies based on the hypothesis that they should fit a binomial distribution B(n,p) for which p is known.

In **Example 44.3** we can calculate the theoretical frequencies by hypothesising a binomial distribution $B(n,p)$ with $n = 5$ and $p = 0.2$ This does not involve calculating any parameters, but we do have the restriction that the total of the expected frequencies must be 400. Thus there is one restriction, so $v = n^* - 1$.

We do not know the value of n^* until we have seen whether there have to be amalgamations, and that means working out the expected frequencies by applying the binomial distribution. The frequency of x faulty balls will be

$$f = 400P(X = x) = 400 \times {}^5C_x(0.2)^x(0.8)^{5-x}$$

Application of this formula gives the values in Table 44.7. To ensure that no theoretical frequency is too low, we have to combine the last three cells, giving Table 44.8. It is to this table that we apply the test, and here the number of cells is $n^* = 4$. Thus the number of degrees of freedom is

$$v = n^* - 1 = 4 - 1 = 3$$

If we choose a 5% level of significance the critical value of χ^2 for $v = 3$ is 7.82. We leave it as an exercise to show that the test produces a value of χ^2_{calc} greater than this, whether or not we use rounding.

Table 44.7

Number of faulty balls, x	0	1	2	3	4	5
Calculated frequency	131.07	163.84	81.92	20.48	2.56	0.12
Rounded frequency	131	164	82	20	3	0

Table 44.8

Number of faulty balls, x	0	1	2	3+
Observed frequency	140	150	100	10
Calculated frequency (rounded)	131	164	82	23

We discuss the matter of **rounding** in Notebox 36.2. The procedure used in the last few examples is summarised more formally in Box 44.2.

Box 44.2

How to ... **compare observed data with expected data where there is a single row of cells, no expected result is less than 5 and there are at least 2 degrees of freedom***

1. State the hypothesis H on which the expectations are based and state a level of significance.

2. For each cell calculate the expected frequency e on the basis of this hypothesis.

3. If an expected frequency is less than 5, combine two or more observations (and thereby reduce the value of v used in step 4) so that the new expected frequency is big enough. This should also be done if the number of observations is less than 5 and one or more of the expected frequencies is not much above 5. A modified version of this instruction appears as a footnote.

4. Calculate the number of degrees of freedom v as in Box 44.1; be careful to reduce n if there have been amalgamations.

5. Consult tables to determine the critical value of χ^2 for the stated level of significance and the calculated number of degrees of freedom.

6. For each cell calculate the difference between the observed and expected frequency, $o - e$, where e is now based on step 3 instead of step 2. Use rounded expected frequencies for a first quick appraisal. ⬭

---

* The requirement that no expected result is less than 5 and there are at least 2 degrees of freedom is a simplification. If $v \geq 2$ it is safe to use the test provided that fewer than 20% of the cells have an expected frequency of less than 5, and that no cell has an expected frequency of less than 1. If this requirement is not met cells have to be combined until it is met. Many statisticians adopt a more cautious approach and insist on combining cells until all have expected frequencies of at least 5. Remember that a fundamental assumption of the χ^2 test is that the total number of observations is large. If it is less than about 50, the test is not all that reliable.

7. Square each difference and divide it by the expected frequency, getting for each *cell* a value of $(o - e)^2/e$.

8. Sum these values to get χ^2_{calc}.

9. Compare this with the critical value found in step 5. Check whether the direction of the difference may be due to rounding, and if necessary repeat steps 6 to 8 with the unrounded expected frequencies. Reject the hypothesis if the calculated value exceeds the critical value found in step 5. Note that all χ^2 tests are one-tailed.

A single row (or column) of cells with theoretical frequencies based on the hypothesis that they should fit a binomial distribution for which p is not known

In **Example 44.4** we follow the numbered steps of Box 44.2.

1. We hypothesise that these data fit a binomial distribution and we specify a level of significance, say 5%. The hypothesis is appropriate because for each person observed there can be only two outcomes, pass or fail. If the results are purely random, a binomial distribution should be revealed.

2. To calculate the expected frequencies, we must first estimate the proportion of successes p. The need to do this reduces the number of degrees of freedom. We know that $np = \bar{x}$, but

$$\bar{x} = (\sum fx)/\sum f = 730/300 = 2.433 \quad \text{and} \quad n = 5$$
$$\text{so} \quad p = 2.433/5 = 0.4866$$

This is the probability of any one person passing. It follows that the probability of any one person failing is $q = 1 - 0.4866 = 0.5134$.

Remembering that the experiment was repeated 300 times, we now find the expected frequencies using these values of p and q in

$$f(x) = 300P(X = x) = 300 \times {}^5C_x p^x (1 - p)^{5 - x}$$

getting the values in Table 44.9, which we have also rounded for a quick estimate of χ^2.

3. No expected frequency is less than 5, so we can use Table 44.9 without amalgamation.

4. The number of degrees of freedom obtained from Box 44.1 for the binomial distribution with an unknown p is $v = n - 2$ where n is the number of cells We have $n = 6$ so $v = 4$.

5. The 5% critical value of χ^2 with $v = 4$ is 9.49.

Table 44.9

Number of people passing, x	0	1	2	3	4	5
Number of days (expected), e	10.70	50.71	96.12	91.11	43.17	8.18
Number of days (rounded)	11	51	96	91	43	8

6. Using rounded values, the approximate values of $|o - e|$ are as follows:

$$|20 - 11| = 9 \qquad |50 - 51| = 1 \qquad |80 - 96| = 16$$
$$|100 - 91| = 9 \qquad |30 - 43| = 13 \qquad |20 - 8| = 12$$

7. Each of them has to be squared and divided by the appropriate e, getting the following approximate contributions to χ^2_{calc}:

$$81/11, \ 1/51, \ 256/96, \ 81/91, \ 169/43, \ 144/8$$

8. These approximate contributions have to be added to give χ^2_{calc}, but there is no need to do the whole of this because we can see straight away that the last value taken alone is well above the 5% critical value of 9.49, and would be so even without rounding.

9. We therefore reject at the 5% level the hypothesis that the data fit a binomial distribution. In particular, the number of times the test is passsed by nobody or by everybody (0 or 5 persons in the sample) deviates markedly from what is suggested by a binomial distribution. It would be worth looking into the reasons for this.

Consolidation exercise 44A

1. What is meant by goodness of fit?

2. Why is it necessary to know the number of degrees of freedom?

3. In testing goodness of fit, we may be calculating some parameters. Why can we nevertheless say that we are using a non-parametric test?

4. How do we define the number of degrees of freedom?

5. What is the number of degrees of freedom in testing the goodness of fit to a binomial distribution if (a) p is known and (b) p is not known?

6. When carrying out the test in Question 5, three cells had to be amalgamated into one. What was the effect on the number of degrees of freedom?

A single row (or column) of cells with theoretical frequencies based on the hypothesis that they should fit a Poisson distribution for which λ is not known

Example 44.5 illustrates this. If absence is a random rare event it should follow a Poisson distribution. The derivation of the number of degrees of freedom is very similar to that just presented for the binomial distribution with unknown p. The total of the theoretical frequencies must match the total of the observed frequencies, which gives us one constraint; and to estimate λ the mean of the expected frequencies has to be the same as the mean of the actual frequencies, which gives us a second constraint. Thus $v = n - 2$, where n has to be determined *after any amalgamation of cells*. The procedure of Box 44.2 ensures that we do this, as we now see.

1. The null hypothesis is that the numbers of absences are random, and in that case they should fit a Poisson distribution. Suppose that we wish to reject this hypothesis only if the chance of it being wrong is quite low, so we adopt 1% as the significance level.

2. To calculate the number of degrees of freedom, we have to estimate the value of the parameter λ by finding the mean daily number of frequencies $\sum fx/\sum f$. This gives $\lambda = 2.7633$ and the expected frequencies can be derived from

$$P(X = x) = 300\left(\frac{e^{-\lambda}\lambda^x}{x!}\right)$$

with this value of λ, and x taking the values 0,1,2, ..., 8. If they total less than 300, we add a final (hypothetical) frequency for '9 or more'. Pursuing this procedure gives us Table 44.10. The entry for 9+ is to provide the correct total of 300. This means that there are 10 cells. Rounding (Notebox 36.2) has produced a total of 301, which is corrected in the next step.

3. The entries in the last three cells are all under 5, so we should combine these three cells into one, labelled 7+ and containing an expected frequency of 6.8, which we round to 7; this provides an automatic correction of the rounded total back to 300. (The corresponding observed frequency is $6 + 4 = 10$.) Amalgamation means that we now have only 8 cells.

4. As we have had to estimate one population parameter (λ) the number of degrees of freedom will be 2 fewer than the number of cells, i.e. $v = 6$, as in Box 33.3.

5. We use $v = 6$ and the chosen significance level (1%) to find the critical value of χ^2. Tables give $\chi^2_{0.01} = 16.81$.

6. The rounded numerical values of $o - e$ are as follows:

x	0	1	2	3	4	5	6	7+
$o - e$	1	7	13	2	6	0	2	3

7. Squaring them and dividing by e gives

1/19, 49/52, 169/72, 4/67, 36/46, 0, 4/12, 9/7

8. To obtain χ^2_{calc} we need to add them. But we do so with the critical value of 16.81 in mind. Clearly the total is a great deal less than this. And it is obvious that it would have been less even if we had not rounded.

9. The hypothesis that the data fit a Poisson distribution cannot be rejected at the 1% level of significance, i.e. the differences between observed and expected results are not so great that they have a chance of 1% or less of occurring if the hypothesis (which implies randomness) is right.

Table 44.10

Absentees, x	0	1	2	3	4	5	6	7	8	9+
Frequency, e_i	18.9	52.3	72.3	66.6	46.0	25.4	11.7	4.6	1.6	0.6
Rounded e_i	19	52	72	67	46	25	12	5	2	1

A single row (or column) of cells with theoretical frequencies based on the hypothesis that they should fit a normal distribution for which the mean and variance are known
Example 44.6 provides an illustration. The initial argument is essentially the same as for Example 44.3 and leads to the same conclusion: $v = n^* - 1$ where n^* is determined after any amalgamations. The procedure of Box 44.2 gives the following solution.

1. The hypothesis H is that the observed data come from a population that is normally distributed about a mean of 4 kg with a variance of 4 kg^2. We wish to test this at the 5% level of significance.
2. We now have to calculate the expected frequencies. In all the examples so far we have had a discrete variable, but here the variable (mass) is continuous. This introduces a complication, which makes the solution a little longer but it does not introduce any new ideas. The first point is that we have to calculate the probability of a mass lying between lower and upper bounds that define the cells. Note that the upper bound can be equalled, but the lower bound must be exceeded.

    If we denote the upper boundary of each group by b we can write down the entries in column 2 of Table 44.11. The frequencies now give us the numbers of observations in the groups with these values of b as their upper boundaries.

    If they are normally distributed with mean 4 and variance 4, we can standardise the variable by writing

$$z = \frac{b - 4}{\sqrt{4}}$$

as in column 3. A normal distribution table gives us the probabilities of a random variable Z being less than the values in column 3. These probabilities are shown in column 4.

Table 44.11

(1) Change in mass $(a < x \leq b)$	(2) Upper bound, b	(3) $z = (b - 4)/2$	(4) $P(Z \leq z)$	(5) $P(a < X \leq b)$	(6) Expected frequency, $e = Np$
$x \leq 0$	0	-2	0.0228	0.0228	2.28
$0 < x \leq 1$	1	-1.5	0.0668	0.0440	4.40
$1 < x \leq 2$	2	-1	0.1587	0.0919	9.19
$2 < x \leq 3$	3	-0.5	0.3085	0.1498	14.98
$3 < x \leq 4$	4	0	0.5000	0.1915	19.15
$4 < x \leq 5$	5	0.5	0.6915	0.1915	19.15
$5 < x \leq 6$	6	1	0.8413	0.1498	14.98
$6 < x \leq 7$	7	1.5	0.9333	0.0919	9.19
$7 < x \leq 8$	8	2	0.9772	0.0440	4.40
$8 < x$	∞	∞	1.0000	0.0228	2.28

The entries in column 5 are the successive increments in column 4 (e.g. $0.0440 = 0.0668 - 0.0228$) and they give us the probabilities of the random variable X (change of mass) lying within the ranges indicated in column 1.

Multiplying them by the total number of students (100) will give us the expected numbers of students with mass gain within each range, enabling us to calculate the expected frequencies, as we show in column 6.

3. We see that the first two expected frequencies are lower than 5, as are the last two. We therefore form one new cell labelled $x \leq 1$, with an observed frequency of 9 and an expected frequency of 6.68 (rounded to 7); and another new cell labelled $x > 7$ with an observed frequency of 6 and an expected frequency of 6.68 (rounded to 7). Our observed and rounded expected results can therefore be put in Table 44.12.

4. Having obtained our expected frequencies, we can begin to apply the test. First we need to decide on the value of ν. After amalgamation we have 8 cells. The mean and variance of the hypothesised distribution were given, so we have not had to calculate any parameters of the normal distribution in order to find the estimated frequencies. Therefore the number of degrees of freedom is $\nu = 8 - 1 = 7$.

5. Table D.10 shows that the critical value of χ^2 for 5% significance and $\nu = 7$ is 14.07.

6 to 8. We compare the expected frequencies e_i with the observed frequencies o_i. The comparison and calculation of χ^2 are shown in Table 44.13. A glance at

Table 44.12

Change in mass (kg)	$x \leq 1$	$1 < x \leq 2$	$2 < x \leq 3$	$3 < x \leq 4$
Observed frequency	9	10	15	22
Expected frequency	7	9	15	19

Change in mass (kg)	$4 < x \leq 5$	$5 < x \leq 6$	$6 < x \leq 7$	$x > 7$
Observed frequency	17	12	9	6
Expected frequency	19	15	9	7

Table 44.13

o	e	$\lvert o - e \rvert$	$(o - e)^2$	$(o - e)^2/e$
9	7	2	4	4/7
10	9	1	1	1/9
15	15	0	0	0
22	19	3	9	9/19
17	19	2	4	4/19
12	15	3	9	9/15
9	9	0	0	0
6	7	1	1	1/7

the last column shows that its sum (which gives the value of χ^2) is well under the critical value of 14.07. There is no need to work it out. We can also check that use of the unrounded expected frequencies does not change this conclusion.

9. The value of χ^2 is not high enough for us to reject the hypothesis that the observed gains in mass follow a normal distribution with mean 4 kg and variance $4\,\mathrm{kg}^2$.

A single row (or column) of cells with theoretical frequencies based on the hypothesis that they should fit a normal distribution for which the mean and variance are not known

We might have been given the same data but asked to test the hypothesis that they are normally distributed with unknown mean and variance.

Now we need to calculate both parameters. Since the total expected frequency must also equal the total observed frequency this means that we have three restrictions, so the number of degrees of freedom is $v = n^* - 3$, where n^* is the number of groups after performing any necessary amalgamations.

We also have to use the estimated mean (instead of the given mean of 4) in standardising our variable. Furthermore, we must use our best estimate of the population variance, which will be the sample variance multiplied by $N/(N-1)$ where in this example $N = 100$.

Once the values of standarised variable have been found, the calculation proceeds as before.

44.4   The Shapiro and Wilk test for normality

Another test is specifically designed for one distribution. It assesses the probability that a sample of observations comes from a normal distribution. It is particularly useful with small samples and is not easily adapted to frequency distributions.

The basis of the method is to calculate the ratio of two quantities that should be estimates of the same thing on the hypothesis that the observations are randomly drawn from a normal population. For any given size of sample, the more the ratio departs from unity the less likely the hypothesis.

The ratio whose value we calculate is

$$W = b^2/S^2$$

where $S^2 = ns^2 = n\sum(x - \bar{x})^2$; s is the sample standard deviation defined (as we have always done) with n in the denominator. b is based on an estimate of the regression of the observed sample values of the associated expected values derived from a normal distribution, but tables allow it to be found very easily. We summarise the procedure in Box 44.3 before using it to solve Example 44.7.

Box 44.3

How to ... **test the normality of a sample by using the Shapiro and Wilk test**

1. Place the n observations in ascending order so that they are $x_1, x_2, \ldots, x_{n-1}, x_n$.

2. Calculate $S^2 = ns^2 = \sum(x - \bar{x})^2$, where s is the sample standard deviation (defined with n in its denominator).

3. Working from both ends, pair off the values and get their differences $(x_n - x_1)$, $(x_{n-1} - x_2)$, and so on. If n is even $(= 2k)$ there is no problem. If n is odd $(= 2k + 1)$ there will remain one term that cannot be paired off, which will be the median. Ignore it. Thus in both cases there will be k bracketed differences.

4. We have to form a weighted sum of these differences, with weights derived from Table D12. Note carefully the labelling of the weights in the sum written as

   $$b = a_n(x_n - x_1) + a_{n-1}(x_{n-1} - x_2) + a_{n-2}(x_{n-2} - x_3) + \ldots$$

5. Obtain the values of the weights a from Table D12 by selecting the column with the sample size n at its head then going down this column, with the first entry being a_n.

6. Use steps 4 and 5 to calculate b then square it to get b^2.

7. Divide b^2 (from step 6) by S^2 (from step 2) to get W.

8. Enter Table D13 at the row corresponding to the sample size n. If the calculated value of W is less than the value in the leftmost column, the sample is very unlikely to come from a normal distribution. If it is higher than the entry in the rightmost column, it is highly likely to do so, perhaps even suspiciously likely. If the calculated value of W lies between the values in two adjacent columns, then the probability of getting that value (or lower) if the sample is from a normal distribution lies between the probabilities given at the heads of the columns.

The tables used in this test are based partly on theoretical work and partly on extensive, careful and well-tested simulations. The original work presents larger tables than we have in Appendix D. Example 44.7 illustrates the procedure just summmarised.

1. The observations are already in order.
2. Calculations give $s^2 = 11.485^2 = 131.91$, so $S^2 = 12 \times 131.91 = 1582.917$.
3 and 4. We obtain $b = a_{12}(83 - 34) + a_{11}b(68 - 48) + a_{10}(66 - 52) + a_9(65 - 55) + a_8(62 - 55) + a_7(61 - 58)$.
5. The values of the weights a are obtained from the column headed $n = 12$ in Table D12 and are, in order, 0.548, 0.333, 0.235, 0.159, 0.092, 0.030.

6. Thus $b = 0.548 \times 49 + 0.333 \times 20 + 0.235 \times 14 + 0.159 \times 10 + 0.092 \times 7$
$+ 0.030 \times 3 = 39.126$, giving $b^2 = 1530.84$.

7. Division gives $W = 1530.84/1582.917 = 0.967$.

8. The row in Table D13 for $n = 12$ gives $W = 0.943$ at $P = 0.50$ and $W = 0.973$ at $P = 0.90$. There is thus a probability of between 50% and 90% of getting a W as low as the calculated value if the sample is from a normal population.

Consolidation exercise 44B

1. You are investigating whether some sample data in a frequency table of eight cells fit a Poisson distribution with an unknown parameter. (a) What further information (if any) is necessary before you can say how many degrees of freedom are involved? (b) What is the maximum number of degrees of freedom? (c) What would be the minimum number of degrees of freedom?

2. You have a frequency table, with a total frequency of 300 spread over seven cells. What test would you use to determine whether the sample comes from a normal distribution?

3. How do you calculate expected cell frequencies for a hypothesised normal distribution?

4. What test would you use to determine whether a sample of 14 observations comes from a normal distribution?

5. In the Shapiro and Wilk test what is S^2? How is b calculated?

Summary

1. We are concerned with how to test whether an observed statistical distribution is adequately represented by (or has a good fit with) a hypothetical distribution that in most cases will be a well-known theoretical distribution.

2. The main test used for this purpose is the chi-square test already described in Chapter 43. If there is a perfect fit between the observed distribution and the hypothesised distribution, the calculated value χ^2 will be zero. In all other cases the value will be positive, increasing as the observed distribution fits less and less well to the hypothesised distribution.

3. If we have a set of n observations in a single row or column for comparison with a hypothesised distribution, there will be $n - 1 - m$ degrees of freedom where m is the number of parameters of the hypothesised distribution that have to be estimated from the data before the test can be applied. Examples are considered carefully and the results summarised in Box 44.1.

5. We illustrate the testing of closeness of fit by considering whether the differences between observed data and hypothetical data are significant when the hypothetical distribution is (i) the binomial distribution with known p; (ii) the binomial distribution with unknown p; (iii) the Poisson distribution with unknown parameter; the normal distribution with known mean and variance; and (v) the normal distribution with unknown mean and variance. The procedure is summarised in Box 44.2.

6. The Shapiro and Wilk test of whether a sample comes from a normal distribution is based on the ratio of a weighted sum of differences between symmetric pairs of observations and a multiple of the standard deviation, as described in Box 44.3. The weights and significant levels are obtained from Tables D12 and D13.

Further probability

Introduction

Example 45.1 Every Sunday lunchtime the landlord of the Calculators' Arms puts a £10 note into a mug. Then he puts 60 discs with the initials of his regular customers into a box, shakes it, and draws out one disc. If the initials belong to a customer present at that moment, the customer gets the contents of the mug. If the initials belong to anybody else, the note is left in the mug. When the next draw takes place a week later, the mug will contain £20, which may be won by a lucky customer or carried forwards. And so on.

Every Sunday exactly 40 customers are present, always including Tippling Tommy and Guzzling Greg. Last Sunday (May 1) Erratic Eric won £10. (a) What is the chance of Tommy winning £30 on May 22? (b) What is the chance that on May 29 either Tommy or Alan will win £40?

Example 45.2 A bag contains 1 red ball, 2 blue balls and 3 white balls. Three balls are withdrawn one at a time, and set aside so that only three balls remain in the bag. What is the probability of (i) withdrawing one ball of each colour? (ii) a red ball being withdrawn before a blue ball?

Example 45.3 The senate of Statingham University has 100 members: 40 are drinkers, 40 neither drink nor think, and 30 drink but do not think. How many thinkers are there? What proportion of thinkers are also drinkers?

Example 45.4 Last Thursday one-fifth of the university staff fasted. Half had breakfast. One-fifth had midday lunch but not breakfast. Two-fifths had only one meal. Four percent had three meals. Ten percent had both lunch and evening dinner. Twelve percent had lunch and at least one other meal. How many had two meals? What proportion of those having dinner had at least one other meal?

Example 45.5 The probability that an undergraduate has a parental home within 20 miles of the university is 0.30, and the probability that he or she has a part-time job is 0.60. If 0.70 is the probability that an undergraduate has either a parental home within 20 miles or a part-time job, but not both, what is the probability that an undergraduate has (i) both, (ii) neither and (iii) a part-time job, given that the parental home is within 20 miles?

Examples 45.1 and 45.2 can be solved in several ways. The safest is to use a **probability tree**, which is valuable if a probability problem involves a sequence of events.

Examples 45.3 and 45.4 do not involve probability. Their solutions involve logical thought aided by a simple diagrammatic technique. We use this technique in solving Example 45.5, which does involve probability. We also use it to establish some important results, including some hitherto taken on trust. In particular, we look at conditional probability. Noteboxes summarise some results about expectations and look at the National Lottery problem.

45.2   **Probability trees**

One way of solving the first example illustrates the use of probability trees.

⬭ In **Example 45.1**, Tommy can win £30 on May 22 only if there is £30 in the mug when the draw is made. Our first task is to establish the probability of this.

The first experiment (denoted by A in Figure 45.1) is drawing one disc from a bag containing 60 when there is a prize of £10. The probability that it belongs to one of the 40 regular customers present is 40/60. In that case the second experiment (denoted by B1) is a repetition of A, still with a prize of £10. The alternative is that the disc does not belong to anybody present. The probability of this is 20/60 and it leads to the alternative experiment B2, with a prize of £20. This is shown by the three boxes A, B1 and B2 with connecting lines, each carrying the appropriate probability fractions.

If B1 is performed, the two possible outcomes lead to C1 and C2; if B2 is performed, the two possible outcomes lead to C3 and C4. The probability of C1 arising is the product of the probability of B1 arising and the probability of C1 arising if B1 arises. Thus it is the product of the probabilities written on the lines that lead from A to C1; it is $(40/60)^2 = 4/9$. Similarly the probability of C2 arising is $(40/60) \times (20/60) = 2/9$; the probability of C3 arising is $(20/60) \times (40/60) = 2/9$; and the probability of C4 arising is $(20/60)^2 = 1/9$. Note as a check that these probabilities add to unity. One of the four C possibilities must happen. We now answer the two questions asked in Example 45.1.

(a) The probability of there being £30 to be won on May 22 is the probability of C4 arising, which is 1/9. We want to know the probability of Tommy winning this. The two possible outcomes, of Tommy winning and Tommy not winning, are shown by the two lines going to T1 and T2, with probabilities of 1/60 and 59/60. The probability of there being £30 in the mug on May 22 and Tommy winning it is the product of the probabilities written on the lines leading from A to T1, which is 1/540.

(b) For anybody to win £40 on May 29 there must be £40 in the mug, as indicated by D8. This has a probability of $(20/60)^3 = 1/27$. When the next disc is drawn, there is a chance of 2/60 that it belongs to Tommy or Alan, so the probability of one of these two winning £40 on May 29 is $(1/27) \times (2/60) = 1/810$.   ⬭

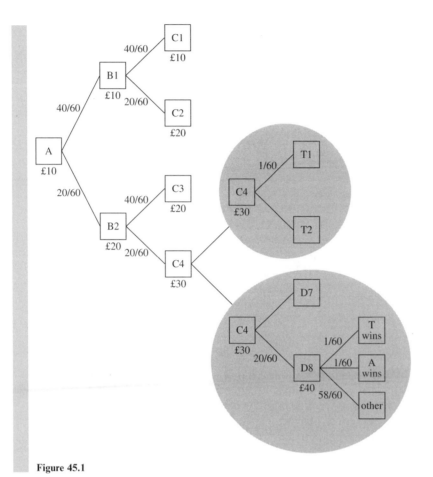

Figure 45.1

These results can be reached quite easily without drawing a tree. An advantage of the tree is that it reduces the risk of forgetting about one or more of the relevant possibilities, especially in more complicated problems such as Example 45.2, for which the probability tree is shown in Figure 45.2 overleaf.

⬭ In **Example 45.2** we start with 1 red, 2 blue and 3 white balls, shown in the box at the left. The three lines from this box, labelled R, B and W, show the chances of drawing a red, blue and white ball. If a red ball is withdrawn, the bag contains only 2 blue balls and 3 white balls, as shown in the box at the end of the R line.

Look carefully at Figure 45.2, tracing the paths from the first box on the left to each of the nineteen boxes on the right. Although they have been reached by different routes, notice that some of them show the same mix of balls.

The probability of withdrawing one ball of each colour is the same as the probability of ending with 0 red, 1 blue and 2 white. There are six boxes marked 1B, 2W, each with a probability of 1/20. Thus the probability of withdrawing one ball of each colour is $6 \times 1/20 = 0.3$. ⬭

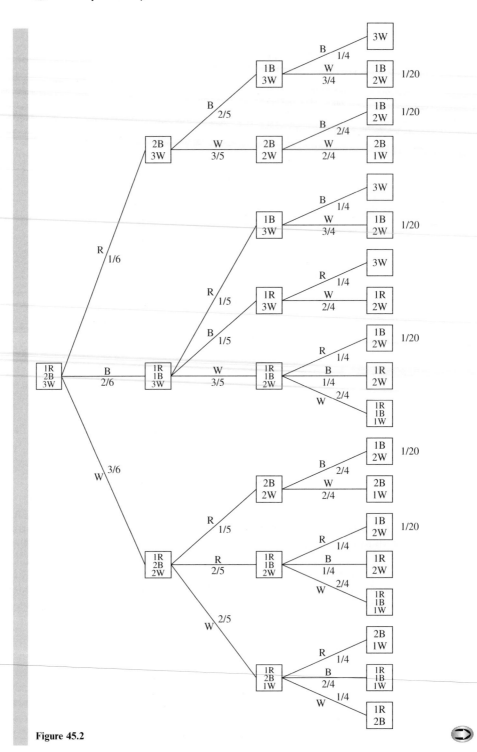

Figure 45.2

The second part of the question asks how often an R precedes a B. It can be answered by looking at the lines first. All sequences starting with the line R have to be taken into account, provided that the second or third line is B. These end in one of the top three boxes on the right, so they have a total probability of $1/60 + 1/20 + 1/20 = 7/60$.

No sequence beginning with the line B can be considered. The only remaining sequences with R before B must begin with W and be WRB, which has a probability of $3/6 \times 1/5 \times 2/4 = 1/20$. Thus the total probability of R preceding B is $7/60 + 1/20 = 1/6$.

Consolidation exercise 45A

1. A six-sided die is rolled. If it scores 3 or 6, Tom has a chance of 0.3 of going home early. If not, he has a chance of 0.2. Calculate the chance of Tom going home early; do this (a) without and (b) with the help of a probability tree.

 45.3 **Venn diagrams**

We solve Example 45.3 with the help of Figure 45.3, a simple Venn diagram – two overlapping circles in a rectangle. This follows the conventional approach, but the shapes are not important. We could have any shapes, regular or irregular.

For **Example 45.3** we imagine that the rectangle contains 100 scattered dots, each representing one member of the senate.

The 40 dots corresponding to the drinkers are within circle D. The unknown number of dots corresponding to the thinkers are in the overlapping circle T.

We identify four regions in this diagram, and use the letters r, s, t, u to indicate the number of dots (or of senators) in each region. For example, there are s dots in the region where the two circles overlap, representing s senators that both drink and think.

We now use the given information to write down as many equations as there are regions in the diagram.

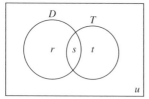

Figure 45.3

The senate has 100 members, so there are 100 dots distributed between the four regions. This gives

$$r + s + t + u = 100 \tag{i}$$

Forty are drinkers, and they are all in circle D, which is divided into two parts. By adding them we get

$$r + s = 40 \tag{ii}$$

Forty are neither drinkers nor thinkers, so they are represented by dots that are in neither circle. They are represented by u dots in the space labelled U, giving

$$u = 40 \tag{iii}$$

The 30 drinkers that are not thinkers are in circle D but not in the part of it that overlaps with T. Thus

$$r = 30 \tag{iv}$$

We now solve these four simultaneous equations to find the four values. We already know r and u. Putting $r = 30$ in (ii) gives us $s = 10$, then we get from (i) that $t = 20$.

Having found the number of people represented by each region (the number of dots in each region), we answer the questions by combining regions.

How many thinkers are there?

All thinkers are represented by circle T, which consists of two parts, s and t. The number of thinkers is therefore $s + t = 10 + 20 = 30$.

What proportion of thinkers are also drinkers?

These people are shown by the overlapping region, which has $s = 10$ people in it. As a proportion of thinkers, they amount to 10/30 or one-third. Notice that it is $s/(s + t)$.

The same approach is used in the solution of Example 45.4, but now the rectangle contains three overlapping circles, representing those who had breakfast, those who had lunch and those who had dinner. This leads to eight regions representing r, s, t, u, v, w, x and y people (or containing these numbers of imaginary dots), as shown in Figure 45.4.

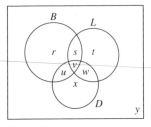

Figure 45.4

Example 45.4 gives eight equations whose correctness should be carefully checked. To encourage this, we give them with less explanation than in the last example. We begin by numbering each piece of information and following it with an equation that restates it in algebraic terms.

(i)   20% of the total staff fasted on Thursdays:

$$y = 20 \tag{i}$$

(ii)   50% had breakfast:

$$r + s + u + v = 50 \tag{ii}$$

(iii)   20% had mid-day lunch but not breakfast:

$$t + w = 20 \tag{iii}$$

(iv)   40% had only one meal:

$$r + t + x = 40 \tag{iv}$$

(v)   4% had three meals:

$$v = 4 \tag{v}$$

(vi)   10% had both lunch and evening dinner:

$$v + w = 10 \tag{vi}$$

(vii)   12% had lunch and at least one other meal:

$$s + v + w = 12 \tag{vii}$$

(viii) Assuming that everybody who did not fast had one, two or three meals, we can say that the percentages add to 100:

$$r + s + t + u + v + w + x + y = 100 \tag{viii}$$

We now solve these equations, beginning with (v), (vi) and (iii) in that order, and getting $v = 4$, $w = 6$ and $t = 14$.

Using these results in the remaining equations, we get $s = 2$, $r = 16$, $u = 28$, $x = 10$ and $y = 20$. It follows that the number who had two meals (but not three) is $s + u + w = 36$. The number who had dinner is $u + v + w + x = 48$. Of these the number who had at least one other meal is $u + v + w = 38$. Therefore the proportion of those having dinner who also had at least one other meal is 38/48.

Notice the importance of the last sentence. It means that if we choose somebody at random, and this person is known (or found) to be one of the 48 persons who had dinner, then the probability that he or she also had at least one other meal is $38/48 = 0.79$.

45.4   Venn and probability problems

As we have begun to see, the data and conclusions in Example 45.4 can be restated in terms of probabilities. If we choose one person at random out of the 100 there is (from (i)) a probability of 20% that he or she fasted. There is also (from $u + v + w + x = 48$) a probability of 48% that he or she had dinner.

We can therefore use Venn diagrams to examine probabilities. The rectangle represents a total probability of 100% (or 1) and the circles represent the probabilities of an item having specified characteristics. It is a little easier to talk about it if we think of the rectangle as having a unit area (representing unit total probability) and the circles as having areas proportional to the probabilities they represent.

We illustrate this by solving the next example, then we go on to show how important theorems (including those in Chapter 17) can be derived very simply by using these diagrams.

⟶ **Example 45.5** uses the same basic diagram as Example 45.3. We let circle D represent the probability of an undergraduate having a nearby parental home, and circle T represent the probability of having a part-time job.

The probability represented by D (having nearby parents) totals $r + s$, which is therefore 0.30. The probability represented by r is that of having nearby parents but not having a job. The probability represented by s is that of having both.

The probability shown by T is $s + t$, which must be 0.60, the probability of having a part-time job. The probability of having a part-time job or a nearby parental home but not both is $r + t$, which must be 0.70.

These three statements lead to three equations, which can easily be solved to give $r = 0.2$, $s = 0.1$ and $t = 0.5$. It follows that the probability an undergraduate has both is given by $s = 0.1$, which answers part (i).

The probability that an undergraduate has neither is represented by u, the area outside both circles. As the probability of being in one circle or the other or both is $r + s + t = 0.8$, the probability of being outside both circles is $1 - (r + s + t) = 1 - 0.8 = 0.2$, so $u = 0.2$; this is the answer to part (ii).

The probability of having a part-time job, given that the parental home is within 20 miles is calculated just as we calculated proportions in Examples 45.3 and 45.4. The probability of having nearby parents is $s + t = 0.6$. The probability of having both nearby parents and a part-time job is given by $s = 0.1$. Therefore the probability of having a part-time job given that there are nearby parents is $s/(s + t) = 0.1/0.6 = 1/6$, which answers part (iii). We return to questions like these towards the end of the next section.

(45.5)   **Venn and probability theorems**

Suppose we are interested in the probabilities of two different events, such as the probability of rain today in Upper Sternum, and the probability of the university refectory serving fish for breakfast tomorrow. We will call these two events A and B, and denote their probabilities by $P(A)$ and $P(B)$.

We draw a rectangle with unit area and inside it draw two circles, one with an area equal to $P(A)$, say 0.6, and the other with an area equal to

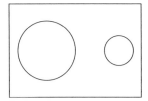

Figure 45.5

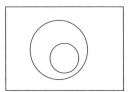

Figure 45.6

$P(B)$, say 0.2. These two circles may overlap (Figure 45.3), they may be quite separate (Figure 45.5) or one of them may be completely inside the other (Figure 45.6).

We now consider the first of these possibilities. We shall soon see that the other possibilities can be looked upon as extreme cases of it. We shall also see that the probabilities of the various possible combinations of these two events depend on whether they are independent events (as described in Chapter 17), but we need not raise that question immediately.

Referring to Figure 45.3, we can see the following results that are true **for any two events**:

$$P(A) = r + s \qquad P(B) = s + t \qquad P(A \text{ and } B) = s$$

$$P(A \text{ or } B \text{ or both}) = r + s + t = (r + s) + (s + t) - s \qquad\qquad [45.1]$$

$$= P(A) + P(B) - P(A \text{ and } B)$$

Usually we omit the words 'or both' on the understanding that they are implied by $P(A \text{ or } B)$. In words equation [45.1] is

The probability of A or B or both is the sum of the probabilities of A and of B reduced by the probability of both (which would otherwise be counted twice, as s).

This proves equations [17.4] and [17.7].

Noting that these results are true for any two events, we now look at some special cases. In the case of **mutually exclusive events** (Figure 45.5) there is no overlap and $s = 0$. Putting $s = 0$ in equation [45.1] gives

$$P(A) = r \qquad P(B) = t \qquad P(A \text{ and } B) = 0$$

$$P(A \text{ or } B) = r + t = P(A) + P(B) \qquad\qquad [45.2]$$

In the case of **contained events**, if B is contained by A then no part of circle B is outside A, so instead of $s = 0$ we have $t = 0$. Putting $t = 0$ in equation [45.1] gives

$$P(A) = r + s \qquad P(B) = s \qquad P(A \text{ and } B) = s = P(B)$$
$$P(A \text{ or } B) = r + s = P(A) \tag{45.3}$$

A new notation

The area where the two circles intersect is called the **intersection of A and B** and denoted by ∩. (You can think of this as a bridge connecting the two circles.) Thus we write $P(A \text{ and } B)$ as $P(A \cap B)$ or $P(AB)$ as in Chapter 17:

$$P(A \text{ and } B) = P(A \cap B) = P(AB) \tag{45.4}$$

Similarly the area within one circle or the other or both $(r + s + t)$ is called the **union of A and B** and denoted by ∪. Thus we write $P(A \text{ or } B) = P(A \cup B)$ or $P(A + B)$, as in Chapter 17. Remember that in the expression $A + B$ the $+$ sign means 'or'.

$$P(A \text{ or } B) = P(A \cup B) = P(A + B) \tag{45.2a}$$

Equation [45.1] can also be written

$$P(A \cup B) = P(A) + P(B) - P(A \cap B) \tag{45.1a}$$

In words this is

The probability of the union of A and B is the sum of the probabilities of A and of B reduced by the probability of the intersection (which would otherwise be counted twice, as s).

For mutually exclusive events $P(A \text{ and } B) = P(A \cap B) = 0$, so

$$P(A \text{ or } B) = P(A \cup B) = P(A) + P(B) \tag{45.1b}$$

Consolidation exercise 45B

1. A Venn diagram has three circles drawn inside a square. What are (a) the maximum, and (b) the minimum number of different areas it can show? Two things can be represented by these areas; what are they?

2. If a Venn diagram shows six different areas, how many equations are needed to find their contents?

3. A Venn diagram has three circles labelled A, B and C. Circles A and B overlap, but circle C does not overlap either of the others. What can you say about the events indicated by the diagram?

Conditional probabilities

Whether one event occurs may depend on whether another occurs. For example, the probability of my getting wet when I go out depends on the probability of it raining. I may get wet by falling in a pool or walking under a ladder. But if it is very likely to rain, I am more likely to get wet than otherwise. This introduces the idea of **conditional probability**: What is the probability of event A given that event B occurs?

The probability of A given that B occurs, or the **conditional probability** of A given B, is derived from an argument that has already been illustrated. We denote this probability by $P(A|B)$.

In terms of the diagram it is the ratio of the areas s and $(s + t)$ as at the end of Section 2 above. Thus we have

$$P(A|B) = \frac{s}{s+t} = \frac{P(A \text{ and } B)}{P(B)} = \frac{P(A \cap B)}{P(B)} \qquad [45.5]$$

It follows from a similar argument that the probability of B given A is

$$P(B|A) = \frac{P(A \cap B)}{P(A)}$$

These two results imply, by cross multiplication, that

$$P(A \cap B) = P(A)P(B|A) = P(B)P(A|B) \qquad [45.6]$$

A similar manipulation leads to the sometimes useful result

$$P(B) = P(B|A)P(A) + P(B|\overline{A})P(\overline{A}) \qquad [45.7]$$

Independent events

Independent events have already been defined as events such that the probability of one happening is unaffected by whether the other happens. Thus for independent events the probability of A happening, given that B happens, is the same as the probability of A happening, given that B does not happen. Using our notation this means that *for independent events* we have

$$P(A|B) = P(A|\bar{B}) \qquad [45.8]$$

If we identify the areas in the Venn diagram corresponding to these probabilities, then for independent events we have that

$$P(A|B) = \frac{s}{s+t} \quad \text{and} \quad P(A|\overline{B}) = \frac{r}{1-(s+t)}$$

and by equation [45.8] these have to be equal. A little algebra reduces this to

$$s = (s+t)(r+s)$$

which translates into a fundamental property of independent events:

$$P(A \text{ and } B) = P(A \cap B) = P(A)P(B) \qquad [45.9]$$

Thus for independent events, equation [45.1b] can be replaced by

$$P(A \text{ or } B) = P(A \cup B) = P(A) + P(B) - P(A)P(B) \qquad [45.1c]$$

More than two events

We can extend our results to three or more events. We illustrate this by presenting the three-event versions of some of the above equations. In place of equation [45.1a] we have

$$P(A \cup B \cup C) = P(A) + P(B) + P(C) - P(A \cap B) - P(B \cap C) - P(C \cap A)$$
$$+ P(A \cap B \cap C)$$

[45.1d]

For three mutually exclusive events, all the intersection terms are zero; then $P(A \cup B \cup C) = P(A) + P(B) + P(C)$.

In place of equation [45.6] we have

$$P(A \cap B \cap C) = P(A)P(B|A)P(C|A \cap B)$$

[45.6a]

For three independent events, we replace equation [45.9] with

$$P(A \cap B \cap C) = P(A)P(B)P(C)$$

[45.9a]

Consolidation exercise 45C

1. What is conditional probability?
2. What does $P(B|A)$ mean? It is equal to the ratio of two other probabilities; what is this ratio?
3. If A and B are two independent events, what is the probability of A or B?
4. What is a contained event?
5. If A contains B, what are (a) $P(A|B)$ and (b) $P(B|A)$?

Notebox 45.1

More expectation theorems

Apart from the results given towards the end of Chapter 39, the following may be useful. In these equations a and b are any constants. Note the signs very carefully.

$$E(b) = b \qquad E(aX + b) = aE(X) + b \qquad E(X \pm Y) = E(X) \pm E(Y)$$

$$E(X^2) = [E(X)]^2 + \operatorname{var} X \qquad E(XY) = E(X)E(Y) + \operatorname{cov}(X, Y)$$

$$\operatorname{var}(X \pm Y) = \operatorname{var} X + \operatorname{var} Y \pm 2\operatorname{cov}(X, Y)$$

In the last two equations, $\operatorname{cov}(X, Y) = 0$ if X and Y are independent.

Notebox 45.2

The National Lottery

The main part of the National Lottery is described in Example 17.2. Chapter 17 calculates the chance of getting six winning numbers as 1 in 13 983 816. Here we

⮑

show how to find the chance of getting (a) no winning numbers and (b) three winning numbers. Then we indicate (c) the chances of the other straightforward prizewinning possibilities, before looking at (d) the 'five plus bonus'.

(a) Imagine 49 balls, of which 6 have been secretly marked. If you select a ball at random, the probability of it not being marked is 43/49. Suppose that happens; there are now 48 left, of which 6 are marked. If you now draw a second ball, the probability of it being unmarked is 42/48. Suppose that happens, and you draw a third ball; the probability of this also being unmarked is 41/47. The probability of the fourth ball also being unmarked is 40/46; for the fifth it is 39/45, and for the sixth it is 38/44. The probability of all six balls being unmarked is the product of these six probabilities, which comes to 0.435 965 (to 6 d.p.) and is the probability that none of the six numbers on your ticket is selected in the official lottery draw.

(b) Now suppose that your first selection gives a marked ball. The chance of this is 6/49. There are 48 balls left, of which 5 are marked. The chance that your second ball is also marked is therefore 5/48. The chance that the third ball is also marked is 4/47. Now there are 46 balls left, and 43 of them are unmarked. The chance that the fourth ball is unmarked is 43/46. The chances that the fifth and sixth balls are also unmarked are 42/45 and 41/44. Thus the chances that the first three balls are marked but the other three are unmarked is the product of these six probabilities, which is 0.000 882 520.

However, this is only one way in which you may choose three marked and three unmarked balls out of six. The number of ways in which you can do this is $^6C_3 = 20$. Therefore the probability of getting three of the six marked balls (or officially chosen numbers) is $20 \times 0.000 882 50 = 0.017 650$, representing a chance of about 1 in 57.

(c) It can be shown similarly that the chance of getting four of the selected numbers (or four marked balls) is

$$15 \times (6/49)(5/48)(4/47)(3/46)(43/45)(42/44) = 0.000 9689$$

or about 1 in 1032. The chance of getting five numbers is

$$6 \times (6/49)(5/48)(4/47)(3/46)(2/45)(43/44) = 0.000 018 45$$

or about 1 in 54 200.

(d) If one of the 43 unmarked balls is secretly marked 'bonus', then if you have chosen only one of the 43 unmarked balls, the chance that this is the bonus ball is 1/43. Therefore the chance of getting five of the six marked balls plus the bonus ball is obtained by dividing 0.000 018 45 by 43, getting 0.000 000 429, which is about 1 in 2 330 000.

Summary

1. The probabilities of the outcomes of a sequence of events can be examined with the help of probability trees.

2. Venn diagrams represent the occurrence of events or attributes by circles or other closed areas, which may or may not overlap depending on whether two (or more) events or attributes can occur. They can be used both to explore the detailed consequences of multiple classifications, such as determining how many people fall into a specified category, and to examine the probabilities associated with these classifications.

3. Venn diagrams can therefore be used to establish probability theorems such as some of those mentioned in Chapter 39 and in Notebox 45.1.

4. If A happens only when B happens then A is contained by B. If A and B can happen together (which means overlapping circles in a Venn diagram) their joint occurrence is denoted by $A \cap B$. It is shown by the intersection of the circles, and called the intersection of A and B. The occurrence of one or other or both events is shown by the total area in one circle or the other and is called the union of A and B denoted by $A \cup B$.

5. If the probability of A depends on whether B has occurred (or occurs) then the probability of A is conditional on B. It is written $P(A|B)$, which is equal to the ratio $P(A \cap B)/P(B)$.

6. The definitions and theorems can be extended to more than two events.

7. Noteboxes contain a summary of expectation theorems and an analysis of the National Lottery problem.

Time series

Introduction

▷ **Example 46.1** Burrowing in the episcopal chest, the Bishop discovers Figure 46.1. He wonders if he can use it to show the ups and downs of marriage. He is told it needs decomposing.

▷ **Example 46.2** The university medical officer finds a rise in the number of students displaying symptoms of clinical apathy. He wonders whether it is part of a long-term increase or just due to the time of year. The Bishop wonders whether there might also be seasonal ups and downs in marriages.

These examples involve time series, typically consisting of statistics for several dates or periods. Usually the dates are more or less regularly spaced, such as the second Tuesday in every month. If the data are for periods, the periods usually have equal lengths (or nominal lengths, as with months); they usually have equal spacings (such as every August, year by year); and they may be contiguous (such as

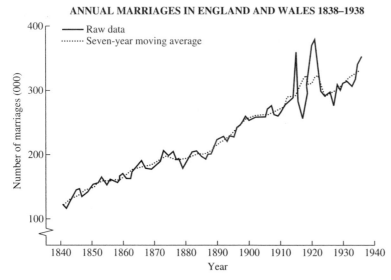

Figure 46.1

521

every August, year by year); and they may be contiguous (such as monthly figures for brick production).

A thorough account of how to handle time series requires at least one book and the use of econometric techniques that involve substantial mathematics. Here we simply indicate briefly a few elementary ideas. The serious user of time series should consult more specialist books.

 Decomposing a time series

The best way of getting to know the data, and of assessing what questions to ask, is usually to plot a graph with time on the horizontal axis, as in Figure 46.1. We can normally expect to see a combination of several different movements:

- *A long-term trend T*, sometimes called a secular trend, shows the underlying direction of movement.
- *Cyclical fluctuations C* occur above and below the long-term trend; the trade cycle is an example. We do not use 'cyclical' to imply a regular periodicity, though it may be present.
- *Seasonal fluctuations S* are usually a special case of cyclical fluctuations and can arise only if the statistics are for time periods shorter than a year. They may be due to natural factors (such as the length of daylight) or to human factors (such as the end of the tax year and the timing of public holidays).
- *Irregular movements R*, possibly random, are due to a large number of factors which may have either direct or indirect effects on the statistics we are examining.

Much of the analysis of time series involves attempts to break down the series into the four components just indicated. Usually it is supposed that every value of the variable can be built up from a knowledge of the components by multiplication, implying $y = T \times C \times S \times R$, or by addition, implying $y = T + C + S + R$.

 The trend

Before summarising procedures for fitting trends, we give an important warning: use of trends in forecasting is dangerous. Trends are summaries of the broad consequences of the interaction of many forces that contribute to economic, social and other change. Unless these forces and their complex interactions are thoroughly understood, one cannot tell when something will happen, gradually or suddenly, to cause the trend to bend or to end.

We summarise three ways of fitting a long-term trend to annual data. The simplest is usually to fit a *freehand trend*, drawing a fairly smooth straight or curved line that seems to summarise the basic underlying movement. There is plenty of room for judgement, as Figure 46.1 indicates. At one level of summary, a trend could be shown by a single straight line for the whole period 1838–1938. At another level, one might feel that although a single straight line can summarise the data from 1838 to about 1877, another would be needed to summarise it from about 1878 to 1914; and in 1915 something quite different appears. The purpose for which the trend is being abstracted and used becomes relevant. This room for judgement is both a merit and a fault, opening freehand trend fitting to the criticisms made about the freehand fitting of regression lines in Chapter 14.

An important objection to fitting trends by **mathematical methods** is that they take no notice of the direction of time. Methods developed for fitting curves to experimental data, in which both axes measure variables in the interval or ratio scales, have been applied to time series in which the *x*-axis represents irreversible time. This failure to take account of an essential property of time often flaws the use of time series.

It is particularly important when we are discussing biological or social variables and using the trend as some kind of norm. If at some date the current value is above the trend, we may expect organisms and people to behave in one way; if it is below the trend, we may expect them to behave in another way.

However, in biological and social terms current events may be influenced by past events themselves, but they can be influenced only by predictions of future events. Yet mathematically determined trends use information for periods both before and after the date for which the trend value is being estimated, giving equal weight to past and future. On that date, people who are deciding about their actions cannot know what the future will bring, even though (on the basis of the past) they may estimate or guess it. Thus a trend fitted to historic data cannot be taken as the trend that was, or could have been, perceived at the time.

Moving averages

The simplest mathematical trend is the **moving average**. This can eliminate a large part of the random fluctuation, and even the cyclical and seasonal fluctuations if we wish. The simplest moving average is of **order 3**, which is defined to be a new series of $n - 2$ terms denoted by m_i such that

$$m_2 = (y_1 + y_2 + y_3)/3$$
$$m_3 = (y_2 + y_3 + y_4)/3 \hspace{2cm} [46.2]$$
$$m_4 = (y_3 + y_4 + y_5)/3$$
$$\vdots$$
$$m_{n-1} = (y_{n-2} + y_{n-1} + y_n)/3$$

The new series is shorter than the original series. A moving average of order 7 fitted over *n* years would consist of $n - 6$ terms, each being the average of 7 consecutive original terms, *centred on* the fourth of those 7 terms. A moving average thus presents at each date the average of a small set of values centred on that date. This is illustrated in Figure 46.1, where a seven-year moving average has been fitted. It reflects the break in trend around 1875 but does not cope well with the violent short-term disturbances in 1915–1921.

A quick method for calculating a moving average is summarised in Box 46.1, which takes a moving average of order 5 to illustrate the procedure.

Box 46.1

How to ... **calculate a moving average of order 5**

1. Arrange the values of *y* in a column, as in Table 46.1

Table 46.1

t	*y*	5-year moving total	5-year moving average
1	43		
2	47		
3	49	$m_3 = 43 + 47 + 49 + 46 + 44 = 229$	$229/5 = 45.8$
4	46	$m_4 = 229 - 43 + 49 = 235$	$235/5 = 47.0$
5	44	$m_5 = 235 - 47 + 52 = 240$	$240/5 = 48.0$
6	49	$m_6 = 240 - 49 + 51 = 242$	$242/5 = 48.4$
7	52		
8	51		

2. Add the first five values of *y*, and write the total in a new column in line 3. Call this total m_3.

3. Add y_6 to this; subtract y_1. Get $m_4 = m_3 + y_6 - y_1$.

4. Add y_1 to this; subtract y_2. Get $m_5 = m_4 + y_7 - y_2$.

5. Continue to the end of the column.

6. Divide each *m* by 5 and write these values of the moving average in a final column.

This method is easily adapted to any other odd order, but some situations require an even-order moving average, such as a four-quarter moving average. The problem is that, although we centre a five-year moving average on year 3 there is no obvious way of centring a four-year (or four-quarter) moving average. The usual way around this is to calculate a **centred fourth-order moving average**. This embraces data from five periods but gives added weight to the middle three. For example, we calculate

$$m_{4a} = (y_2 + y_3 + y_4 + y_5)/4$$
$$m_{4b} = (y_3 + y_4 + y_5 + y_6)/4$$

which we then average to get

$$m_4^* = (y_2 + 2y_3 + 2y_4 + 2y_5 + y_6)/8$$

and locate it in the fourth space.

Other weighted moving averages are sometimes used. One of the disadvantages of moving averages is that they 'lose' data at each end. The degree of 'smoothing' increases with their order, but so does the terminal loss. They are therefore of limited use if we are interested in the early or latest values of a trend.

Least squares trends

Frequently trends are fitted, like regression lines, by the method of least squares (described in Chapter 14) or some other mathematical method. If a straight line does not appear to be appropriate, some other curve can be fitted by essentially the same methods.

The least squares equations for a straight line are simplified if the dates or time periods are evenly spaced. If the first is called 1, the xs of the regression equations become the positive integers $1, \ldots, n$, where there are n observations. Even greater saving arises if we have an odd number of observations and call the middle date 0, counting backwards and forwards from it. This greatly simplifies the equations. A modification of this technique is available when there is an even number of observations, as described in Box 46.2.

Box 46.2

How to ... **fit a straight-line trend to a time series**

For an ODD number n of evenly spaced observations

1. Call the middle date 0, and label all other dates with negative or positive consecutive integers from $-(n-1)/2$ to $+(n-1)/2$, which become the values of t.

2. Calculate

$$a = \frac{\sum y}{n} \quad \text{and} \quad b = \frac{12 \sum yt}{n(n^2 - 1)}$$

3. The equation is $y = a + bt$.

For an EVEN number n of evenly spaced observations

1. Call the middle two dates -1 and $+1$, and label all other dates with negative or positive odd integers from $-(n-1)$ to $+(n-1)$, which become the values of t.

2. Calculate

$$a = \frac{\sum y}{n} \quad \text{and} \quad b = \frac{3 \sum yt}{n(n^2 - 1)}$$

3. The equation is $y = a + bt$

Frequently, and especially if the variable is rapidly growing or declining, or if the time span is great, a better fit is obtained by taking logs of the data, so that in fitting a straight line, the same equations are used but every y is replaced by $\log y$. The easiest way of seeing whether this should be done is to plot the values of y on a log scale, using paper which has the horizontal grid lines drawn at distances representing the logs of the numbers. If the plotted data appear to fit a straight line (or any other curve approximated by a known type of equation), then every y should be transformed into $\log y$, which becomes the new variable. Computer programs calculate the equations of log curves directly from the raw data. If a logarithmic graph shows a straight-line trend, the data are growing at a constant rate.

There are several sources of trouble; we mention only two of them in our condensed treatment of the subject. Do refer to other textbooks for more information. The regression formula gives an estimated trend value $y = a + bt$. The actual value will differ from this by an 'error' term u. Among other assumptions, the mathematical theory assumes two things:

- There is no systematic relationship between the value of u at one date and the value at another – there is no **autocorrelation**.
- The variance of u is not systematically related to the value of y or to the value of t – there is no **heteroscedasticity**.

If these assumptions do not hold, the fitting of trends by essentially regression methods may lead to serious errors.

46.4   Cycles

The cyclical element in a time series is often separated out by expressing every value as a percentage of, or a difference from, the trend value. This means that the graph displays a series of fluctuations around a horizontal line. These fluctuations are likely to have a systematic component and an irregular and possibly random component. Separating them is a hazardous task beyond the scope of this book. However, so far as trend fitting goes, the 'error' term, mentioned in connection with autocorrelation and heteroscedasticity, contains both the random and the systematic component, so it almost certainly invalidates our two assumptions. There are methods for handling such problems.

46.5   Seasonal correction

Often, as in Example 46.2, we wish to estimate how much of the latest value of some statistic is due to the season of the year rather than to any underlying forces. In briefly describing how this is done, we concentrate on quarterly data. Exactly the same methods can be used for monthly or weekly data, except that for the shorter periods there is a greater chance of some holiday or other event moving from period n in one year to period $n \pm 1$ in the next year. Even quarterly data are not immune from this.

We will assume there is a more or less steady seasonal pattern. This might not be so if school summer holidays slowly shortened and winter holidays slowly lengthened, or if annual tax changes suddenly took effect a few months later than in previous years. Consideration of how to deal with such matters is beyond the scope of this book.

There are several methods for estimating and allowing for the seasonal effect. A simple and crude method is given in Box 46.3. Another method is essentially the same but begins by expressing the raw data as *percentages* of the moving average rather than as differences from it. Somewhat different methods use link relatives, in which each quarterly value is expressed as a percentage of the previous quarter's value.

None of these methods has any claim to great accuracy. Unless we can define precisely what we mean by the seasonal factor, and show how to identify it, we cannot hope to measure it accurately. At best it provides an approximate way of adjusting the raw data for an average amount by which past data seem to have been affected by the season. For example, an excessively cold and long winter may lead to an unusual winter decline in house building. The seasonal adjustment for that winter will not fully allow for it. But the low value will have got into the statistics used to calculate future seasonal adjustments, so the effect of a later milder winter may be overestimated. There are more complicated methods for getting over problems like this, but they are not universally accepted.

Box 46.3

How to ... **deseasonalise quarterly data**

1. Calculate a centred four-quarter moving average.

2. For every quarter, subtract the moving-average value from the actual value. There will be both positive and negative differences.

3. Work out the average of all first-quarter differences. If you use the arithmetic mean, ignore any extreme values. Alternatively, take the median value.

4. Work out the averages for the other three quarters.

5. These average differences are considered to be the normal seasonal factors. If they do not add to zero, allocate the total between them so the corrected normal seasonal factors do add to zero.

6. 'Correct' every actual quarterly value by subtracting from it the (corrected) normal seasonal factor.

46.6 **Irregular movements**

We have already commented on the difficulty of identifying irregular movements. Other time series are often invoked to try to 'explain' the movements in one series by movements in one or more other series. This, too, is full of problems and beyond our scope.

Consolidation exercise 46A

1. Why is it dangerous to use a trend for forecasting?
2. What is a logarithmic trend? What is indicated by a straight-line logarithmic trend?
3. What is a moving average?
4. What original terms affect the tenth term of (a) a centred five-year moving average and (b) a centred four-year moving average?
5. What is seasonal correction?

Summary

1. The treatment of time series is a vast and complicated subject. This chapter simply touches it at a very elementary level. It is based on a decomposition of a time series into a long-term trend, cyclical fluctuations, seasonal fluctuations and irregular movements.

2. Using a trend for forecasting is dangerous.

3. Freehand trend fitting allows room for judgement, which is both an advantage and a disadvantage.

4. Mathematical trend fitting ignores the irreversbility of time. The simplest mathematical method is the moving average, which achieves a mechanical smoothing of the data. Another uses least squares (or some other procedure) to fit a linear or curvilinear trend. Logarithmic procedures may be useful if there is rapid growth or a long time span. A straight-line logarithmic trend indicates steady growth. Mathematical trend fitting relies on important assumptions about the 'error' term. If these assumptions are invalid, the procedures will probably give invalid results.

5. Seasonal correction is at best an approximation. Box 46.3 outlines a simple and crude procedure for estimating and correcting for the seasonal factor.

6. Cyclical and irregular movements can be handled by various techniques, but they are not described.

Appendices

Using pocket calculators

Several pocket calculators are programmed to find the mean, standard deviation and regression and correlation coefficients. The exact procedures to be followed depend on the calculator. Here we describe them in detail for two calculators that are widely used but are very different. We also point out an error in the programming of some calculators.

In all cases two steps have to be followed when beginning a new calculation.

(1) The calculator has to be put in the right mode, so that the keys perform the operations appropriate to the kind of calculation you are performing. This is done by pressing [MODE] followed by one of the codes given below.
(2) The memory has to be cleared, so that data from previous calculations do not interfere with your calculation.

- On the **Sharp EL556G** this is done by pressing [2nd F] [CA].
- On the **Casio fx-6300G** this is done by pressing [SHIFT] [SCI] [EXE]. The [SCI] operation is denoted in orange beneath [→] on row 5. By first pressing [SHIFT] you have commanded the calculator to perform the operations printed in the same colour as [SHIFT].

The arithmetic mean and standard deviation

Four observations: $x = 1, 3, 5, 7$

Sharp EL556G

. . . .

	Screen shows	
. . . .

Get into right mode

   [MODE] 3 0 (3 0 is the code we need)   ST0   0

Clear memory

Input the data:

   1 [DATA] 3 [DATA] 5 [DATA] 7 [DATA] n 4

Call up the results

Begin by pressing the blue [RCL] key. This means that the next key you press will perform the operation shown above or below it in blue. The key labelled 4 has a blue $\bar{x}$, whereas 5 has a blue s_x and 6 has a blue σ_x. Thus we proceed

   [RCL] [$\bar{x}$] $\bar{x}$ 4.000

   [RCL] [s_x] s_x 2.582

   [RCL] [σ_x] σ_x 2.2361

. . . .

This calculator uses σ_x to denote the standard deviation of the population of four observations, calculated with n in the denominator (as explained in Chapter 11). It uses s_x to denote the best estimate of the population standard deviation if these four observations are regarded as a sample from a larger population.

Casio fx-6300G

· ·

Screen shows

· ·

Get into right mode
    [MODE] [×] (the multiplication sign)   SD1

Clear memory   SD1   0

Input the data
    1 [DT] 3 [DT] 5 [DT] 7 [DT]
[DT] appears in blue beneath the [$\sqrt[x]{}$] key

Call up the results
The keys for 1, 2 and 3 have $\bar{x}$, $x\sigma_n$ and $x\sigma_{n-1}$ printed beneath them in orange, indicating use of the [SHIFT] key. Thus we proceed
    [SHIFT] [$\bar{x}$] [EXE] $\bar{x}$ 4.
    [SHIFT] [$x\sigma_n$] [EXE] $x\sigma_n$ 2.2361
    [SHIFT] [$x\sigma_{n-1}$] [EXE] $x\sigma_{n-1}$ 2.5820

· ·

This calculator uses $x\sigma_n$ to denote the standard deviation of the population of four observations, calculated with n in the denominator (as explained in Chapter 11). It uses $x\sigma_{n-1}$ to denote the best estimate of the population standard deviation if these four observations are regarded as a sample from a larger population.

A frequency table

x	1	3	5	7
f	2	6	4	1

Sharp EL556G

· ·

Screen shows

· ·

Get into right mode
    [MODE] 3 0   STO   0

Clear memory
Input the data:
    1[×]2[DATA] 3[×]6[DATA] 5[×]4[DATA] 7[×][1] [DATA] n 13

Call up the results
    [RCL] [$\bar{x}$] $\bar{x}$ 3.615
    [RCL] [s_x] s_x 1.710
    [RCL] [σ_x] σ_x 1.643

· ·

Casio fx-6300G

. . . .

Screen shows

. . . .

Get into right mode
  [MODE] [×]   SD1
Clear memory   SD1   0
Input the data
  1[SHIFT] [;]2[DT] 3[SHIFT] [;]6[DT]
  5[SHIFT] [;]4[DT] 7[SHIFT] [;]1[DT]   SD1   7
  [;] appears in orange above [)]
Call up the results.
  [SHIFT] [$\bar{x}$] [EXE] $\bar{x}$ 3.615
  [SHIFT] [$x\sigma_n$] [EXE] $x\sigma_n$ 1.643
  [SHIFT] [$x\sigma_{n-1}$] [EXE] $x\sigma_{n-1}$ 1.710

. . . .

A.2   Regression and correlation

x	1	2	3	4	4	5	6
y	0	2	2	3	5	4	7

Sharp EL556G

. . . .

Screen shows

. . . .

Get into right mode
  [MODE] 3 1 (note the new code)   ST1   0
Clear memory   0
Input the data
  1[x, y]0[DATA] 2[x, y]2[DATA] 3[x, y]2[DATA]
  4[x, y]3[DATA] 4[x, y]5[DATA] 5[x, y]4[DATA]
  6[x, y]7[DATA] n 7
Call up the results
Use the blue [RCL] key. The regression constant a is the blue
function of the key [(]. The regression coefficient b comes from
[)] and the correlation coefficient r from [÷]. Thus we proceed
  [RCL] [a]   −1.121
  [RCL] [b]   1.2339
  [RCL] [r]   0.9263

. . . .

To obtain the best linear regression estimate $\hat{y}$ of y corresponding to $x = 4.3$, we
perform

    4.3 [2ndF] [y']

which gives 4.185. Note that [y'] is the yellow function of [)].

The calculator also has a key [x']. This is designed to give the best linear estimate of x corresponding to a given value of y – but it does not (see below).

Casio fx-6300G

. . . .

Screen shows

. . . .

Get into right mode

    [MODE] [÷]   LR1

Clear memory   LR1   0

Input the data

    1[SHIFT] [,]0[DT] 2[SHIFT] [,]2[DT]

    3[SHIFT] [,]2[DT] 4[SHIFT] [,]3[DT]

    4[SHIFT] [,]5[DT] 5[SHIFT] [,]4[DT]

    6[SHIFT] [,]7[DT]   LR1   6

[,] appears in orange above [( ]

Call up the results

Use the orange [SHIFT]. The keys for 7, 8 and 9 have orange labels for A, B and r. Thus:

    [SHIFT] [A] [EXE] giving a −1.121

    [SHIFT] [B] [EXE] giving b 1.2339

    [SHIFT] [r] [EXE] giving r 0.9263

. . . .

To obtain the best linear regression estimate $\hat{y}$ of y corresponding to $x = 4.3$, we perform

    4.3 [SHIFT] [$\hat{y}$] [EXE]

which gives 4.185. Note that [$\hat{y}$] is the orange function of [÷].

The calculator also has a key [$\hat{x}$]. This is designed to give the best linear estimate of x corresponding to a given value of y – but it does not.

At the time of writing this book, the programs of both models contain the error we warn against in Chapter 14. See Noteboxes 14.3 and 14.4 for advice on how to calculate $\hat{x}$.

Assumed means and new units

This appendix describes some techniques that can save a lot of effort if programmed calculators are not at hand.

 ## Calculating the arithmetic mean

Assumed mean

Use of an assumed mean saves time because it leads to smaller numbers that are easier to manage. The basic idea is to look at the data and then choose a number that is probably close to the true value of the mean, but which is also easy to add to or subtract from all the other values of the variable. Then you find out how wrong you are.

It is easily shown that if a is taken as the assumed mean then

$$\bar{x} = a + \sum(x - a)/n \qquad \text{[B.1]}$$

For example, Table 3.2 has 60 entries, all showing seat heights. If 70 is subtracted from every entry the new heights, measured from the assumed mean of 70.0, produce a table whose first row is

1.0   2.8   5.7   5.9   1.4   1.6   1.7   3.2   2.1   −5.7

There are five similar rows. The total for all six rows is $\sum(x - 70) = 153.2$. We know that there are 60 entries, one for each person whose requirements were measured. Thus, from equation [B.1], the mean is $70 + 153.2/60 = 70 + 2.57 = 72.57$, as in Chapter 6.

The assumed mean can be particularly useful when we have a frequency table. The data of Table 6.3 are shown once more in columns 1 and 3 of Table B.1. We take an assumed mean of 28 and then perform a calculation to see how wrong we are. In Table B.1 we have a new column (column 2) showing the values of a new variable, $(x - 28)$. Then we replace the fx column of Table 6.3 with column 4, which shows values of the product $f(x - 28)$. Thus, if we assume the mean to be 28, we are undercounting the customers by 15, spread over 31 evenings. The corrected mean is

$$\bar{x} = a + \frac{\sum f(x - a)}{\sum fx} \qquad \text{[B.2]}$$

$$= 28 + (15/31) = 28.48$$

> **Table B.1**

Calculation of arithmetic mean using a frequency table and an assumed mean

(1) Number of customers x	(2) New variable (customers in excess of 30) $x - 28$	(3) Number of evenings with this number of customers f	(4) = (3) × (2) Product $f(x - 28)$
20	−8	1	−8
21	−7	0	0
22	−6	0	0
23	−5	1	−5
24	−4	0	0
25	−3	1	−3
26	−2	2	−4
27	−1	4	−4
28	0	6	0
29	+1	5	+5
30	+2	4	+8
31	+3	3	+9
32	+4	3	+12
33	+5	1	+5
		$n = \sum f = 31$	$\sum f(x - 28) = +39 - 24$ $= 15$

We obtain 28.48 customers per evening, as in Chapter 6. We have commented in the main text on the need to consider the number of significant figures.

New units

New units and coded data provide another trick. Changing the units of the variable is sometimes an obvious thing to do. If we have values that are all divisible by 100, the obvious thing to do is to work in units of 100, replacing 300 by 3, and then to multiply the answer by 100. But it may sometimes make sense to work in units of 3, or 2.5 or anything else if it saves work. More generally, if we adopt new units x' such that $x' = x/c$ then we have to multiply the answer by c to convert it back to the old units.

Thus, if we adopt new units such that $x' = x/c$ then

$$\bar{x} = c\bar{x}' = c \sum fx' / \sum f \qquad \text{[B.3]}$$

This trick can be combined with the use of an assumed mean. It would obviously have made sense for Lady Agatha to have combined working in units of 0.1 cm (or 1 mm) with the use of an assumed mean of 700 (in new units) getting a first row of

   10   28   57   59   14   16   17   32   21   −51

which reduces the number of keystrokes by one-third. She would then have needed to divide her answer of 725.7 by 10, which simply means moving the decimal point. More generally, if there is a frequency table, the formula then becomes

$$\bar{x} = c[a + \sum f(x' - a)/\sum f]$$ [B.4]

We illustrate this with another example.

⬭ **Example B.1**   Twenty-five first-class long-distance airline passengers have paid the fares shown in the first two columns of Table B.2. What is the arithmetic mean fare?

⬭ After looking at column 1, we decide to work in units of 10 (which becomes our value of c). The fares are written in the new units in column 3. We also take an assumed mean of 260 (in the new units) and subtract this from every entry in column 3 to get column 4.

When we multiply the entries in column 4 by those in column 2, we get column 5. Beneath the table we have two totals and the values of c and a. These can be inserted into equation [B.4] to give $\bar{x} = 10(260 + 18/25) = 2607.20$ dollars, which looks sensible.

If all the frequencies are 1s, equation [B.4] becomes simply

$$\bar{x} = c(a + \sum (x' - a)/n)$$

Note that c need not be an integer. For example, if the values of x are 2.5, 5.0, 7.5, etc., then it would be convenient to write x' as $x/2.5$ with values 1, 2, 3, etc.

Table B.2

(1) Fare ($) x	(2) Number of passengers paying it f	(3) In new units $x' = x/10$	(4) Measured from assumed mean $x' - 260$	(5) Product $f(x' - 260)$
2470	2	247	−13	−26
2540	3	254	−6	−18
2610	9	261	1	9
2630	6	263	3	18
2670	5	267	7	35
	$\sum f = 25$	$c = 10$	$a = 260$	$\sum f(x' - a) = 62 - 44$
				$= 18$

When we use an assumed mean, new units or both, the data are sometimes said to be **coded**. Using coded data is particularly useful if we have grouped data with class intervals of equal widths.

B.2 Calculating the standard deviation

Assumed mean

Using an assumed mean with ungrouped data is illustrated by a recalculation of the standard deviation of the data in Table 11.4 and used in Chapter 11. This time we take an assumed mean of $a = 4$. The table, reprinted as part of Table B.3, has an extra column, for $(x - a)$, and the calculation proceeds as before with $(x - a)$ being used instead of x. Writing $x - a = x'$, we use the modified equation

$$\text{s.d.} = \sigma = \sqrt{\frac{\sum fx'^2}{n} - \left(\frac{\sum fx'}{n}\right)^2} \qquad [\text{B.5}]$$

This gives

$$\sqrt{\left(\frac{1060}{291}\right) - \left(\frac{-71}{291}\right)^2}$$

$$= \sqrt{3.583}$$

$$= 1.893$$

Table B.3

(1) x Family size	(2) f Number of families of this size	(3) x − a	(4) = (2) × (3) f(x − a)	(5) = (4) × (3) f(x − a)²
1	30	−3	−90	270
2	65	−2	−130	260
3	40	−1	−40	40
4	55	0	0	0
5	50	1	50	50
6	30	2	60	120
7	12	3	36	108
8	5	4	20	80
9+	4	5+	23*	132*
	$n = \sum f = 291$		$\sum f(x - a) = -71$	$\sum f(x - a)^2 = 1060$

*Assuming that 9+ is to be interpreted as 9.75.

which agrees with Chapter 11. Notice that the standard deviation is not affected by the choice of assumed mean, which reflects location not spread.

New units: grouped data

Using new units with grouped data is especially useful when all or almost all the groups have equal width. We calculate the standard deviation in the new units then convert the answer back to the old units, as we now illustrate. The age distribution of Steve's neighbours is shown once more in Table B.4. Column 3 shows the midpoint of each group. In column 4 we rewrite this in units of 2.5, and take these values as our x. We have taken 82.5 as the average age of the four people in the open-ended group 80+ partly for computational convenience.

Using the totals of Table B.4 and equation [B.6], we find

$$\text{s.d.} = \sigma = \sqrt{\frac{\sum fx^2}{n} - \left(\frac{\sum fx}{n}\right)^2}$$

[B.6]

This gives

$$\sqrt{\left(\frac{92\,956}{301}\right) - \left(\frac{4586}{301}\right)^2}$$

$= 8.757$ units   (1 unit $= 2.5$ years)

$= 21.89$ years

Table B.4

(1) Age	(2) Frequency f	(3) Midpoint	(4) In units of 2.5 x	(5) fx	(6) fx^2
0–4	20	2.5	1	20	20
5–9	22	7.5	3	66	198
10–14	20	12.5	5	100	500
15–19	18	17.5	7	126	882
20–29	35	25.0	10	350	3 500
30–39	38	35.0	14	532	7 448
40–49	44	45.0	18	792	14 256
50–59	48	55.0	22	1 056	23 232
60–64	20	62.5	25	500	12 500
65–69	16	67.5	27	432	11 664
70–79	16	75.0	30	480	14 400
80+	4	82.5	33	132	4 356
$n = \sum f = 301$				$\sum fx = 4\,586$	$\sum fx^2 = 92\,956$

This answer looks reasonable. The mean is $\sum fx/n = 38.09$ years. Twice the standard deviation on either side of this contains virtually all the population.

New units and an assumed mean: grouped data

We save further effort if we combine an assumed mean with new units, as the following reworking shows. We take an assumed mean age of 35, which corresponds to $x = 14$ in new units of 2.5. This results in a new column for $x - 14$, and we replace fx of Table B.4 by $f(x - 14)$ and fx^2 by $f(x - 14)^2$. The result is Table B.5.

Using these totals and equation [B.6], we have

$$\text{s.d.} = \sigma = \sqrt{\frac{\sum fx^2}{n} - \left(\frac{\sum fx}{n}\right)^2}$$

$$= \sqrt{\left(\frac{23\,544}{301}\right) - \left(\frac{372}{301}\right)^2}$$

$$= 8.757 \text{ units} \qquad (1 \text{ unit} = 2.5 \text{ years})$$

$$= 21.89 \text{ years}$$

Table B.5

(1) Age	(2) Frequency f	(3) Midpoint	(4) In units of 2.5 x	(4a) $x - 14$	(5) Product $f(x - 14)$	(6) Product $f(x - 14)^2$
0–4	20	2.5	1	−13	−260	3 380
5–9	22	7.5	3	−11	−242	2 662
10–14	20	12.5	5	−9	−180	1 620
15–19	18	17.5	7	−7	−126	882
20–29	35	25.0	10	−4	−140	560
30–39	38	35.0	14	0	0	0
40–49	44	45.0	18	4	176	704
50–59	48	55.0	22	8	384	3 072
60–64	20	62.5	25	11	220	2 420
65–69	16	67.5	27	13	208	2 704
70–79	16	75.0	30	16	256	4 096
80+	4	82.5	33	19	76	1 444

$n = \sum f = 301$ $\sum f(x - 14) = 372$ $\sum fx^2 = 23\,544$

B.3 Regression and correlation

New variables

If the values of one or both of the variables are large, work can often be saved by noting that the slope of the scattergram is unaltered if we subtract a convenient amount (u) from all the values of x and another convenient amount (v) from all the values of y. Calculation with the new variables $X = x - u$ and $Y = y - v$ converts $Y = a + bX$ into $y - v = a + b(x - u)$, giving

$$y = (a + v - bu) + bx$$

Thus the regression coefficient b is unchanged but the constant term is different.

To take an example, suppose we write $X = x - 25$ and $Y = y - 10$, and suppose we obtain a regression equation of

$$Y = 12 + 0.8X$$

then this can be rewritten in terms of the original variables as

$$y = (12 + 10 - (0.8)(25)) + 0.8x = 2 + 0.8x$$

Note that the correlation coefficient will be unchanged.

We illustrate the procedure by showing how to find the regression equation required in Example 14.2. The data are set out as the first two columns of Table B.6 and the construction of the remaining columns is described below.

A quick glance at columns 1 and 2 suggests that the mean value of x is not far from 10, whereas the mean value of y is not far from 550. Moreover, these numbers are fairly easily subtracted from the original values. This enables us to get columns 3 and 4. Except at the very end, the rest of the calculation is based on these columns, with their entries for $x' = (x - 10)$ and $y' = (y - 550)$. The scattergram has new axes but is otherwise unchanged, and the slope of the regression line is unchanged. But we have smaller numbers to handle.

We compile columns 5, 6 and 7, and we use them to find the sums $\sum x$, $\sum y$, $\sum x^2$, and $\sum xy$, which we use according to the instructions of Box 14.1.

New units

If we work in new units, so that $X = gx$ and $Y = ky$ then the calculated regression equation

$$Y = a + bX$$

will have to be rewritten as

$$ky = a + bgx$$

giving

$$y = (a/k) + (bg/k)x$$

Table B.6

(1) Weeks x	(2) Rebate (£) y	(3) $x - 10$ x'	(4) $y - 550$ y'	(5) x'^2	(6) y'^2	(7) $x'y'$
13	567	3	17	9	289	51
6	515	−4	−35	16	1 225	140
1	489	−9	−61	81	3 721	549
17	571	7	21	49	441	147
21	602	11	52	121	2 704	572
1	498	−9	−52	81	2 704	468
14	563	4	13	16	169	52
15	579	5	29	25	841	145
9	532	−1	−18	1	324	18
8	537	−2	−13	4	169	26
$n = 10$		$\sum 5$	−47	403	12 587	2168

so both the regression coefficient and the constant term have been altered. However, once more the correlation coefficient will be unchanged.

Sampling

This appendix is to be read in conjunction with Chapter 16

C.1 Simple random samples

In a simple random sample every item in the population has to have exactly the same chance of being chosen; and whether one item is included should be quite uninfluenced by whether some other item is included. Most of the formulae used in the analysis of data based on a random sample assume that this chance is constant throughout the sampling process (Notebox C.1). The essential stages in taking a random sample are

1. Choose a sampling frame.
2. Decide on the sample size n.
3. Select n random numbers.
4. Use steps 3 and 1 to select the sample.

A **sampling frame** is a list, numbered or capable of being numbered, that can be used as an acceptable substitute for a numbered list of all the items or people in the population under investigation. If we wish to sample from the population of all adults resident in Manchester, we could take the offical numbered list of electors. It would not be perfect, as it would include some people no longer resident and exclude some residents who have lately immigrated or for some other reason never appeared on it. If we use this as a substitute for a list of adult residents, we have to consider whether the exclusions and wrong inclusions make it biased in any particular way.

Possible sampling frames for households would be lists of domestic telephone subscribers and of electricity meters. Both have imperfections, and again it is important to consider whether they introduce a bias.

The **sample size** affects the precision and the cost. Although the precision of a population estimate based on a sample increases with the square root of the sample size, the cost of taking the sample and analysing the data is likely to increase more rapidly. Determining a sample size involves balancing precision against cost and usually speed. Chapters 21 and 22 describe methods of choosing a sample size in order to obtain confidence intervals with a stated width, but they do not consider cost.

Random numbers are best obtained from **random number tables** such as Table D1. They normally consist of one or more pages from which Table C1 could be a

7461	7402	4566	0204	7696	1119	7533	2854	4530	1944	7798	1574	2077
6765	0966	2180	3071	0341	8536	2742	6801	4442	1801	8418	6253	8383
0095	2036	7961	7196	5097	2875	3323	4339	6315	9666	3305	9775	2305
0194	4321	1871	0501	6886	8976	4605	8751	2791	2006	0592	1193	4513

typical extract. This extract consists of 208 digits printed in groups of four simply for convenience. There is no pattern. No digit appears significantly more or less often than it should. No run of two digits (such as 73) appears more or less often than it should; or of three digits, and so on. Starting with any group of digits (such as 0966), it is not possible by the application of any rule to generate the rest of the table. The digits appear at random.

Suppose we want 120 random numbers lying between 0 and 85235. This means we are seeking five-digit numbers. We begin by finding a starting point. One not very good way of doing this is to stick a pin in the page and select the nearest digit. A better way is to note that the table has (say) 64 rows, each of 42 digits. We therefore modify this method by taking the first two digits after the pin mark as the number of a row. If the first two digits exceed 64, we ignore them and take the next two digits. Thus we obtain a row that will contain our starting point.

We find the column containing the starting point in the same way, but we economise in effort by adopting the following rule:

- If the first two digits ≤ 42, use them to locate the column.
- If they lie between 43 and 49, ignore them and take the next two.
- If they lie between 50 and 92, subtract 50 and use the remainder to locate the row.
- If they lie between 53 and 99, ignore them and take the next two.

We now choose sequences of five digits in any systematic way, beginning with the starting point determined by the chosen row and column. For example, if the starting point is the second 8 in row 2 of Table C1 we could form groups of five digits by going to the right, getting 85362 (which is too big so we ignore it), 74268, 01444, 21801, 84186, 25383, 83009, and so on. Alternatively we could go across one digit and down one digit, beginning (from the same starting point) with 85877. In that case we would need a rule for what to do when we reach the end of a row or the bottom of the page.

If our upper limit had been 40235 instead of 85235, we would ignore numbers between 40236 and 49999, and subtract 50000 from all numbers between 50000 and 90235. In this case the above illustration would have generated $85362 - 50000 = 35362$, $74268 - 50000 = 24268$, 01444, 21801, $84186 - 50000 = 34186$, 25383, $83009 - 50000 = 33009$, and so on. We continue with this procedure until we have 120 five-digit random numbers, which is how many we require.

Alternatively we can use the random number facility on a pocket calculator. This produces pseudorandom numbers, which differ from random numbers mainly in that they are produced by a complicated formula, and somebody knowing this could reproduce any sequence, given the first few terms. For ordinary everyday purposes they can be used as random numbers.

Whether we use tables or a calculator, we may come across the same random number more than once. If we are sampling with replacement (Notebox C.1), a number that appears twice means that we take account of that item twice. If we are sampling without replacement (which requires a population very many times larger than the sample) we ignore second and later appearances of the same number and go on sampling until we have the right number of different random numbers.

Having obtained a list of 120 random numbers, we choose from the numbered list in the sampling frame the individuals with those numbers. This provides our random sample of 120 individuals.

 ## C.2   Controlled sampling

The more common forms are:

Cluster sampling
Simple random samples are taken from one or more groups (or clusters), selected sometimes for convenience but preferably at random, rather than from the whole population. A 'national' public opinion poll involving 1100 people may be organised to select from only 20 specified localities.

Quota sampling
This is often used in market research. Interviewers are told how many people of different characteristics (age, sex, appearance, etc.) they are to interview. Within this quota they usually choose at will, but may attempt some element of randomness.

Stratified sampling
The sample is chosen so that it reflects important characteristics of the population. If 20% of the population consists of females living in rural areas then 20% of the sample will consist of females living in rural areas. If this is achieved by selecting the rural females (and other strata) at random, it is a stratified random sample. Usually this is more reliable than a simple random sample. There are formulae for calculating standard errors and confidence intervals.

Systematic sampling
The individuals in a sample are chosen systematically from what amounts to an ordered list, such as every fortieth entry in a telephone directory. In some cases this quick procedure can give satisfactory results but there are always dangers arising out of possible patterns. As no two adjacent individuals will be chosen, the chance of one individual being chosen is not independent of whether another is chosen.

Notebox C.1

Replacement

Most of the formulae used in the analysis of data derived from a random sample assume that the chance of being chosen is constant throughout the sampling process. Suppose we have a bag containing 26 balls lettered A to Z but otherwise identical. Five of these letters are vowels. We shake the bag and select one ball. There is a chance of 5/26 that it carries a vowel.

If we replace the ball then the chance of a second selection producing a vowel is once again 5/26. Suppose, however, that after withdrawing the first ball we put it aside, so that the bag now contains only 25 balls, of which 4 or 5 will be vowels. If the first ball was a vowel, the chance of the second being a vowel is 4/25. If the first was not, then the chance of the second being a vowel is 5/25. The chance of the second draw producing a vowel depends on what happened in the first draw, and a fundamental requirement of random sampling is absent.

If there are very many balls, the difference between what happens with replacement and without replacement is trivial. Unless the population is very large compared with the sample, sampling should always be with replacement. If it is not, then a correction can sometimes be applied, as in Chapter 21.

Statistical reference tables

Table D1

Random numbers*

6201	2434	3732	0863	5029	3849	6349	3203	1957	9356
4648	0471	5232	3267	6340	8994	3833	5795	2191	1739
4318	7787	5637	8519	8341	1630	2189	3671	9722	8935
7767	9525	5085	3769	9728	5503	3891	4689	0757	9639
4556	6499	0402	7403	3381	8938	7327	4550	7739	9536
9614	0376	6858	5908	2221	6655	3970	2339	8956	1791
1079	4989	4852	2498	8309	8345	6457	8251	9472	0034
8659	1285	9260	2376	3151	6689	5342	7313	7482	0377
7864	8699	7517	9272	9366	5180	5550	5508	6011	8953
0365	7944	7825	8450	7185	2258	3576	7976	1825	5159
7903	7225	9150	4668	6644	7621	3312	5332	8646	7240
6226	3468	2150	3747	9422	2134	0346	8767	8978	3762
9982	2836	6481	9327	1156	2347	1459	5366	8597	1207
4970	6095	6280	6991	7348	3446	2141	8369	1692	5555
9735	3988	1031	5244	8845	4134	3106	9901	4516	8877
0032	4927	9686	8308	8875	5019	9903	0808	9824	3602
8088	8371	9930	6153	9091	1694	3245	2738	8386	4761
9264	8673	4071	4657	9325	3559	1060	4064	4752	7116
9851	5663	3120	1743	8526	7963	7137	5112	8280	3609
3251	8751	6275	2894	8633	7190	4086	3824	9468	6032
3550	0407	8280	7674	0330	8112	8210	9670	5667	9026
7896	0826	4027	7125	6619	0649	7088	3908	2140	9126
4649	4293	4473	7640	0905	8478	3612	9329	3123	9398
4288	3089	5181	0275	4028	5400	1721	3112	1033	9406
2866	6545	3287	0162	3373	2877	2700	7328	5290	5853
3157	9288	5383	9510	8554	8604	2981	4071	3131	4513
7520	4795	0855	5653	1546	1917	9391	4968	7173	2156
9519	9329	9610	4278	1398	6642	0847	8061	7173	3896
9797	5131	0649	6762	7511	0119	0630	8263	7360	1255
6423	4338	4264	0294	5926	0737	4854	1428	5681	8337
3934	0821	6507	1108	3621	5080	4893	1696	5189	1765
1313	8377	0135	1761	8860	1413	3638	1716	2651	5643
3711	2512	6239	8554	2626	4485	7195	2665	7206	1801
0790	9836	7350	0892	4524	4061	8416	3648	5902	1262
2894	8427	7753	3909	3858	0981	1410	5441	4075	5017
5248	5660	4988	8555	3088	8492	2039	2072	3831	5556
0876	9195	6649	2330	0943	3111	3519	8530	5468	2978
7821	9799	7247	5263	3829	9253	6563	0597	2835	8112
4573	1545	3291	5441	8078	3830	7703	5350	2393	0712
0633	5981	1039	1180	6829	4643	5612	4491	6109	4925
7212	5289	7970	3305	6016	3335	9594	8881	5673	3944
4265	9421	1650	3860	0451	4057	4974	0066	8272	2885
3549	3628	4680	8092	1312	4891	3334	1356	6180	8753
8141	0091	3494	1480	4139	3038	3229	1402	1813	6383
8492	6730	9755	4871	9742	0949	9919	2838	9612	4900
5132	8678	5875	3308	7539	1707	2520	6300	7476	9927
5315	4432	2134	0437	5795	3028	7512	5277	9671	6495
8497	0999	8660	8283	6833	5130	8710	4199	4095	3990
7815	7374	8774	5166	6248	8663	3814	5596	1086	2036
3299	6819	3439	1907	2524	2591	1504	6183	3762	3209

*Instructions for the use of this table are given in Appendix C.

Table D2

Cumulative binomial probabilities*

	$p =$	0.05	0.10	0.15	0.20	0.25	0.30	0.35	0.40	0.45	0.50
$n = 2$	$r = 0$	.9025	.8100	.7225	.6400	.5625	.4900	.4225	.3600	.3025	.2500
	1	.9975	.9900	.9775	.9600	.9375	.9100	.8775	.8400	.9795	.7500
$n = 3$	$r = 0$	.8574	.7290	.6141	.5120	.4219	.3430	.2746	.2160	.1664	.1250
	1	.9928	.9720	.9393	.8960	.8438	.7840	.7183	.6480	.5748	.5000
	2	.9999	.9990	.9966	.9920	.9844	.9730	.9571	.9360	.9089	.8750
$n = 4$	$r = 0$	.8145	.6561	.5220	.4096	.3164	.2401	.1785	.1296	.0915	.0625
	1	.9860	.9477	.8905	.8192	.7383	.6517	.5630	.4752	.3910	.3125
	2	.9995	.9963	.9880	.9728	.9492	.9163	.8735	.8208	.7585	.6875
	3		.9999	.9995	.9984	.9961	.9919	.9850	.9744	.9590	.9375
$n = 5$	$r = 0$	.7738	.5905	.4437	.3277	.2373	.1681	.1160	.0778	.0503	.0313
	1	.9774	.9185	.8352	.7373	.6328	.5282	.4284	.3370	.2562	.1875
	2	.9988	.9914	.9734	.9421	.8965	.8369	.7648	.6826	.5931	.5000
	3		.9995	.9978	.9933	.9844	.9692	.9460	.9130	.8688	.8125
	4			.9999	.9997	.9990	.9976	.9947	.9898	.9815	.9688
$n = 6$	$r = 0$	.7351	.5314	.3771	.2621	.1780	.1176	.0754	.0467	.0277	.0156
	1	.9672	.8857	.7765	.6554	.5339	.4202	.3191	.2333	.1636	.1094
	2	.9978	.9842	.9527	.9011	.8306	.7443	.6471	.5443	.4415	.3438
	3	.9999	.9987	.9941	.9830	.9624	.9295	.8826	.8208	.7447	.6563
	4		.9999	.9996	.9984	.9954	.9891	.9777	.9590	.9308	.8906
	5				.9999	.9998	.9993	.9982	.9959	.9917	.9844
$n = 7$	$r = 0$	.6983	.4783	.3206	.2097	.1335	.0824	.0490	.0280	.0152	.0078
	1	.9556	.8503	.7166	.5767	.4449	.3294	.2338	.1586	.1024	.0625
	2	.9962	.9743	.9262	.8520	.7564	.6471	.5323	.4199	.3164	.2266
	3	.9998	.9973	.9879	.9667	.9294	.8740	.8002	.7102	.6083	.5000
	4		.9998	.9988	.9953	.9871	.9712	.9444	.9037	.8471	.7734
	5			.9999	.9996	.9987	.9962	.9910	.9812	.9643	.9375
	6					.9999	.9998	.9994	.9984	.9963	.9922
$n = 8$	$r = 0$	.6634	.4305	.2725	.1678	.1001	.0576	.0319	.0168	.0084	.0039
	1	.9428	.8131	.6572	.5033	.3671	.2553	.1691	.1064	.0632	.0352
	2	.9942	.9619	.8948	.7969	.6785	.5518	.4278	.3154	.2201	.1445
	3	.9996	.9950	.9786	.9437	.8862	.8059	.7064	.5941	.4770	.3633
	4		.9996	.9971	.9896	.9727	.9420	.8939	.8263	.7396	.6367
	5			.9998	.9988	.9958	.9887	.9747	.9502	.9115	.8555
	6				.9999	.9996	.9987	.9964	.9915	.9819	.9648
	7						.9999	.9998	.9993	.9983	.9961
$n = 9$	$r = 0$	.6302	.3874	.2316	.1342	.0751	.0404	.0207	.0101	.0046	.0020
	1	.9288	.7748	.5995	.4362	.3003	.1960	.1211	.0705	.0385	.0195
	2	.9916	.9470	.8591	.7382	.6007	.4628	.3373	.2318	.1495	.0898
	3	.9994	.9917	.9661	.9144	.8343	.7297	.6089	.4826	.3614	.2539
	4		.9991	.9944	.9804	.9511	.9012	.8283	.7334	.6214	.5000
	5		.9999	.9994	.9969	.9900	.9747	.9464	.9006	.8342	.7461
	6				.9997	.9987	.9957	.9888	.9750	.9502	.9102
	7					.9999	.9996	.9986	.9962	.9909	.9805
	8							.9999	.9997	.9992	.9980

continues

Table D2 continued

$p =$		0.05	0.10	0.15	0.20	0.25	0.30	0.35	0.40	0.45	0.50
$n = 10$	$r = 0$	.5987	.3487	.1969	.1074	.0563	.0282	.0135	.0060	.0025	.0010
	1	.9139	.7361	.5443	.3758	.2440	.1493	.0860	.0464	.0233	.0107
	2	.9885	.9298	.8202	.6778	.5256	.3828	.2616	.1673	.0996	.0547
	3	.9990	.9872	.9500	.8791	.7759	.6496	.5138	.3823	.2660	.1719
	4	.9999	.9984	.9901	.9672	.9219	.8497	.7515	.6331	.5044	.3770
	5		.9999	.9986	.9936	.9803	.9527	.9051	.8338	.7384	.6230
	6			.9999	.9991	.9965	.9894	.9740	.9452	.8980	.8281
	7				.9999	.9996	.9984	.9952	.9877	.9726	.9453
	8						.9999	.9995	.9983	.9955	.9893
	9								.9999	.9997	.9990
$n = 15$	$r = 0$	.4633	.2059	.0874	.0352	.0134	.0047	.0016	.0005	.0001	.0000
	1	.8290	.5490	.3186	.1671	.0802	.0353	.0142	.0052	.0017	.0005
	2	.9638	.8159	.6042	.3980	.2361	.1268	.0617	.0271	.0107	.0037
	3	.9945	.9444	.8227	.6482	.4613	.2969	.1727	.0905	.0424	.0176
	4	.9994	.9873	.9383	.8358	.6865	.5155	.3519	.2173	.1204	.0592
	5	.9999	.9978	.9832	.9389	.8516	.7216	.5643	.4032	.2608	.1509
	6		.9997	.9964	.9819	.9434	.8689	.7548	.6098	.4522	.3036
	7			.9994	.9958	.9827	.9500	.8868	.7869	.6535	.5000
	8			.9999	.9992	.9958	.9848	.9578	.9050	.8182	.6964
	9				.9999	.9992	.9963	.9876	.9662	.9231	.8491
	10					.9999	.9993	.9972	.9907	.9745	.9408
	11						.9999	.9995	.9981	.9937	.9824
	12							.9999	.9997	.9989	.9963
	13									.9999	.9995
$n = 20$	$r = 0$	.3585	.1216	.0388	.0115	.0032	.0008	.0002	.0000	.0000	.0000
	1	.7358	.3917	.1756	.0692	.0243	.0076	.0021	.0005	.0001	.0000
	2	.9245	.6769	.4049	.2061	.0913	.0355	.0121	.0036	.0009	.0002
	3	.9841	.8670	.6477	.4114	.2252	.1071	.0444	.0160	.0049	.0013
	4	.9974	.9568	.8298	.6296	.4148	.2375	.1182	.0510	.0189	.0059
	5	.9997	.9887	.9327	.8042	.6172	.4164	.2454	.1256	.0553	.0207
	6		.9976	.9781	.9133	.7858	.6080	.4166	.2500	.1299	.0577
	7		.9996	.9941	.9679	.8982	.7723	.6010	.4159	.2520	.1316
	8		.9999	.9987	.9900	.9591	.8867	.7624	.5956	.4143	.2517
	9			.9998	.9974	.9861	.9520	.8782	.7553	.5914	.4119
	10				.9994	.9961	.9829	.9468	.8725	.7507	.5881
	11				.9999	.9991	.9949	.9804	.9435	.8692	.7483
	12					.9998	.9987	.9940	.9790	.9420	.8684
	13						.9997	.9985	.9935	.9786	.9423
	14							.9997	.9984	.9936	.9793
	15								.9997	.9985	.9941
	16									.9997	.9987
	17										.9998

*Values of $P(X \leqslant r)$ where $X \sim B(n, p)$. Probabilities for $r = n$ are 1.000. All other omitted probabilities are 1.0000 to 4 d.p.

Cumulative Poisson probabilities*

$\lambda =$	0.05	0.1	0.2	0.4	0.5	0.6	0.8	1.0	1.2	1.4	1.5
$r = 0$	.9512	.9048	.8187	.6703	.6065	.5488	.4493	.3679	.3012	.2466	.2231
1	.9988	.9953	.9825	.9384	.9098	.8781	.8088	.7358	.6626	.5918	.5578
2		.9998	.9989	.9921	.9856	.9769	.9526	.9197	.8795	.8335	.8088
3			.9999	.9992	.9982	.9966	.9909	.9810	.9662	.9463	.9344
4				.9999	.9998	.9996	.9986	.9963	.9923	.9857	.9814
5							.9998	.9994	.9985	.9968	.9955
6								.9999	.9997	.9994	.9991
7										.9999	.9998

$\lambda =$	1.6	1.8	2.0	2.2	2.4	2.5	2.6	2.8	3.0
$r = 0$	.2019	.1653	.1353	.1108	.0907	.0821	.0743	.0608	.0498
1	.5249	.4628	.4060	.3546	.3084	.2873	.2674	.2311	.1991
2	.7834	.7306	.6767	.6227	.5697	.5438	.5184	.4695	.4232
3	.9212	.8913	.8571	.8194	.7787	.7576	.7360	.6919	.6472
4	.9763	.9636	.9473	.9275	.9041	.8912	.8774	.8477	.8153
5	.9940	.9896	.9834	.9751	.9643	.9580	.9510	.9349	.9161
6	.9987	.9974	.9955	.9925	.9884	.9858	.9828	.9756	.9665
7	.9997	.9994	.9989	.9980	.9967	.9958	.9947	.9919	.9881
8		.9999	.9998	.9995	.9991	.9989	.9985	.9976	.9962
9				.9999	.9998	.9997	.9996	.9993	.9989
10						.9999	.9999	.9998	.9997
11									.9999

$\lambda =$	3.2	3.4	3.5	3.6	3.8	4.0	4.5	5.0	5.5
$r = 0$	.0408	.0334	.0302	.0273	.0224	.0183	.0111	.0067	.0041
1	.1712	.1468	.1359	.1257	.1074	.0916	.0611	.0404	.0266
2	.3799	.3397	.3208	.3027	.2689	.2381	.1736	.1247	.0884
3	.6025	.5584	.5366	.5152	.4735	.4335	.3423	.2650	.2017
4	.7806	.7442	.7254	.7064	.6678	.6288	.5321	.4405	.3575
5	.8946	.8705	.8576	.8441	.8156	.7851	.7029	.6160	.5289
6	.9554	.9421	.9347	.9267	.9091	.8893	.8311	.7622	.6860
7	.9832	.9769	.9733	.9692	.9599	.9489	.9134	.8666	.8095
8	.9943	.9917	.9901	.9883	.9840	.9786	.9597	.9319	.8944
9	.9982	.9973	.9967	.9960	.9942	.9919	.9829	.9682	.9462
10	.9995	.9992	.9990	.9987	.9981	.9972	.9933	.9863	.9747

continued overleaf

Table D3 continued

$\lambda =$	3.2	3.4	3.5	3.6	3.8	4.0	4.5	5.0	5.5
11	.9999	.9998	.9997	.9996	.9994	.9991	.9976	.9945	.9890
12		.9999	.9999	.9999	.9998	.9997	.9992	.9980	.9955
13						.9999	.9997	.9993	.9983
14							.9999	.9998	.9994
15								.9999	.9998
16									.9999

$\lambda =$	6.0	6.5	7.0	7.5	8.0	8.5	9.0	9.5	10.0
$r = 0$	.0025	.0015	.0009	.0006	.0003	.0002	.0001	.0001	.0000
1	.0174	.0113	.0073	.0047	.0030	.0019	.0012	.0008	.0005
2	.0620	.0430	.0296	.0203	.0138	.0093	.0062	.0042	.0028
3	.1512	.1118	.0818	.0591	.0424	.0301	.0212	.0149	.0103
4	.2851	.2237	.1730	.1321	.0996	.0744	.0550	.0403	.0293
5	.4457	.3690	.3007	.2414	.1912	.1496	.1157	.0885	.0671
6	.6063	.5265	.4497	.3782	.3134	.2562	.2068	.1649	.1301
7	.7440	.6728	.5987	.5246	.4530	.3856	.3239	.2687	.2202
8	.8472	.7916	.7291	.6620	.5925	.5231	.4557	.3918	.3328
9	.9161	.8774	.8305	.7764	.7166	.6530	.5874	.5218	.4579
10	.9574	.9332	.9015	.8622	.8159	.7634	.7060	.6453	.5830
11	.9799	.9661	.9467	.9208	.8881	.8487	.8030	.7520	.6968
12	.9912	.9840	.9730	.9573	.9362	.9091	.8758	.8364	.7916
13	.9964	.9929	.9872	.9784	.9658	.9486	.9261	.8981	.8645
14	.9986	.9970	.9943	.9897	.9827	.9726	.9585	.9400	.9165
15	.9995	.9988	.9976	.9954	.9918	.9862	.9780	.9665	.9513
16	.9998	.9996	.9990	.9980	.9963	.9934	.9889	.9823	.9730
17	.9999	.9998	.9996	.9992	.9984	.9970	.9947	.9911	.9857
18		.9999	.9999	.9997	.9993	.9987	.9976	.9957	.9928
19				.9999	.9997	.9995	.9989	.9980	.9965
20					.9999	.9998	.9996	.9991	.9984
21						.9999	.9998	.9996	.9993
22							.9999	.9999	.9997
23								.9999	.9999

continues

Table D3 continued

$\lambda =$	11.0	12.0	13.0	14.0	15.0
$r = 0$	.0000	.0000	.0000	.0000	.0000
1	.0002	.0001	.0000	.0000	.0000
2	.0012	.0005	.0002	.0001	.0000
3	.0049	.0023	.0011	.0005	.0002
4	.0151	.0076	.0037	.0018	.0009
5	.0375	.0203	.0107	.0055	.0028
6	.0786	.0458	.0259	.0142	.0076
7	.1432	.0895	.0540	.0316	.0180
8	.2320	.1550	.0998	.0621	.0374
9	.3405	.2424	.1658	.1094	.0699
10	.4599	.3472	.2517	.1757	.1185
11	.5793	.4616	.3532	.2600	.1848
12	.6887	.5760	.4631	.3585	.2676
13	.7813	.6815	.5730	.4644	.3632
14	.8540	.7720	.6751	.5704	.4657
15	.9074	.8444	.7636	.6694	.5681
16	.9441	.8987	.8355	.7559	.6641
17	.9678	.9370	.8905	.8272	.7489
18	.9823	.9626	.9302	.8826	.8195
19	.9907	.9787	.9573	.9235	.8752
20	.9953	.9884	.9750	.9521	.9170
21	.9977	.9939	.9859	.9712	.9469
22	.9990	.9970	.9924	.9833	.9673
23	.9995	.9985	.9960	.9907	.9805
24	.9998	.9993	.9980	.9950	.9888
25	.9999	.9997	.9990	.9974	.9938
26		.9999	.9995	.9987	.9967
27		.9999	.9998	.9994	.9983
28			.9999	.9997	.9991
29				.9999	.9996
30				.9999	.9998
31					.9999

*Values of $P(X \leqslant r)$ where $X \sim \text{Po}(\lambda)$. Omitted probabilities are 1.0000 to 4 d.p.

Table D4

Area beneath the standard normal curve N(0, 1) to the left of ordinate z

z	0	1	2	3	4	5	6	7	8	9	1	2	3	4	5	6	7	8	9
														ADD					
0.0	.5000	.5040	.5080	.5120	.5160	.5199	.5239	.5279	.5319	.5359	4	8	12	16	20	24	28	32	36
0.1	.5398	.5438	.5478	.5517	.5557	.5596	.5636	.5675	.5714	.5753	4	8	12	16	20	24	28	32	36
0.2	.5793	.5832	.5871	.5910	.5948	.5987	.6026	.6064	.6103	.6141	4	8	12	15	19	23	27	31	35
0.3	.6179	.6217	.6255	.6293	.6331	.6368	.6406	.6443	.6480	.6517	4	7	11	15	19	22	26	30	34
0.4	.6554	.6591	.6628	.6664	.6700	.6736	.6772	.6808	.6844	.6879	4	7	11	14	18	22	25	29	32
0.5	.6915	.6950	.6985	.7019	.7054	.7088	.7123	.7157	.7190	.7224	3	7	10	14	17	20	24	27	31
0.6	.7257	.7291	.7324	.7357	.7389	.7422	.7454	.7486	.7517	.7549	3	7	10	13	16	19	23	26	29
0.7	.7580	.7611	.7642	.7673	.7704	.7734	.7764	.7794	.7823	.7852	3	6	9	12	15	18	21	24	27
0.8	.7881	.7910	.7939	.7967	.7995	.8023	.8051	.8078	.8106	.8133	3	5	8	11	14	16	19	22	25
0.9	.8159	.8186	.8212	.8238	.8264	.8289	.8315	.8340	.8365	.8389	3	5	8	10	13	15	18	20	23
1.0	.8413	.8438	.8461	.8485	.8508	.8531	.8554	.8577	.8599	.8621	2	5	7	9	12	14	16	19	21
1.1	.8643	.8665	.8686	.8708	.8729	.8749	.8770	.8790	.8810	.8830	2	4	6	8	10	12	14	16	18
1.2	.8849	.8869	.8888	.8907	.8925	.8944	.8962	.8980	.8997	.9015	2	4	6	7	9	11	13	15	17
1.3	.9032	.9049	.9066	.9082	.9099	.9115	.9131	.9147	.9162	.9177	2	3	5	6	8	10	11	13	14
1.4	.9192	.9207	.9222	.9236	.9251	.9265	.9279	.9292	.9306	.9319	1	3	4	6	7	8	10	11	13

z											ADD								
											1	2	3	4	5	6	7	8	9
1.5	.9332	.9345	.9357	.9370	.9382	.9394	.9406	.9418	.9429	.9441	1	2	4	5	6	7	8	10	11
1.6	.9452	.9463	.9474	.9484	.9495	.9505	.9515	.9525	.9535	.9545	1	2	3	4	5	6	7	8	9
1.7	.9554	.9564	.9573	.9582	.9591	.9599	.9608	.9616	.9625	.9633	1	2	3	4	4	5	6	7	8
1.8	.9641	.9649	.9656	.9664	.9671	.9678	.9686	.9693	.9699	.9706	1	1	2	3	4	4	5	6	6
1.9	.9713	.9719	.9726	.9732	.9738	.9744	.9750	.9756	.9761	.9767	1	1	2	2	3	4	4	5	5
2.0	.9772	.9778	.9783	.9788	.9793	.9798	.9803	.9808	.9812	.9817	0	1	1	2	2	3	3	4	4
2.1	.9821	.9826	.9830	.9834	.9838	.9842	.9846	.9850	.9854	.9857	0	1	1	2	2	2	3	3	4
2.2	.9861	.9864	.9868	.9871	.9875	.9878	.9881	.9884	.9887	.9890	0	1	1	1	2	2	2	3	3
2.3	.9893	.9896	.9898	.9901	.99036	.9906	.9909	.9911	.9913	.9916									
2.4	.9918	.9920	.9922	.9925	.9927	.9929	.9931	.9932	.9934	.9936									
2.5	.9938	.9940	.9941	.9943	.9945	.9946	.9948	.9949	.9951	.9952									
2.6	.9953	.99547	.9956	.9957	.9959	.9960	.9961	.9962	.9963	.9964									
2.7	.9965	.9966	.9967	.9968	.9969	.9970	.9971	.9972	.9973	.9974									
2.8	.9974	.9975	.9976	.9977	.9977	.9978	.9979	.9980	.9980	.9981									
2.9	.9981	.9982	.9983	.9983	.9984	.9984	.9985	.9985	.9986	.9986									
3.0	.9987	.9987	.9987	.9988	.9988	.9989	.9989	.9989	.9990	.9990									
3.1	.9990	.9991	.9991	.9991	.9992	.9992	.9992	.9992	.9993	.9993									
3.2	.9993	.9993	.9994	.9994	.9994	.9994	.9994	.9994	.9995	.9995									
3.3	.9995	.9995	.9995	.9996	.9996	.9996	.9996	.9996	.9996	.9997									
3.4	.9997	.9997	.9997	.9997	.9997	.9997	.9997	.9997	.9997	.9998									
3.5	.9998	.9998	.9998	.9998	.9998	.9998	.9998	.9998	.9998	.9998									
3.6	.9998	.9998	.9999	.9999	.9999	.9999	.9999	.9999	.9999	.9999									

For these values of z the differences cannot easily be expressed in this form

Table D5

Critical values of t^*

One-tail area	.10	.05	.025	.01	.005	.001	.0005
Two-tail area	.20	.10	.050	.02	.010	.002	.0010
$v = 1$	3.078	6.314	12.71	31.82	63.66	318.3	636.6
2	1.886	2.920	4.303	6.965	9.925	22.33	31.60
3	1.638	2.353	3.182	4.541	5.841	10.21	12.92
4	1.553	2.132	2.776	3.747	4.604	7.173	8.610
5	1.476	2.015	2.571	3.365	4.032	5.893	6.869
6	1.400	1.943	2.447	3.143	3.707	5.208	5.959
7	1.415	1.895	2.365	2.998	3.499	4.785	5.408
8	1.397	1.860	2.306	2.896	3.355	4.501	5.041
9	1.383	1.833	2.262	2.821	3.250	4.297	4.781
10	1.372	1.812	2.228	2.764	3.169	4.144	4.587
11	1.363	1.796	2.201	2.718	3.106	4.025	4.437
12	1.356	1.782	2.179	2.681	3.055	3.930	4.318
13	1.350	1.771	2.160	2.650	3.012	3.852	4.221
14	1.345	1.761	2.145	2.624	2.977	3.787	4.140
15	1.341	1.753	2.131	2.602	2.947	3.733	4.073
16	1.337	1.746	2.120	2.583	2.921	3.686	4.015
17	1.333	1.740	2.110	2.567	2.898	3.646	3.965
18	1.330	1.734	2.101	2.552	2.878	3.610	3.922
19	1.328	1.729	2.093	2.539	2.861	3.579	3.883
20	1.325	1.725	2.086	2.528	2.845	3.552	3.850
21	1.323	1.721	2.080	2.518	2.831	3.527	3.819
22	1.321	1.717	2.074	2.508	2.819	3.505	3.792
23	1.319	1.714	2.069	2.500	2.807	3.485	3.767
24	1.318	1.711	2.064	2.492	2.797	3.467	3.745
25	1.316	1.708	2.060	2.485	2.787	3.450	3.725
26	1.315	1.706	2.056	2.479	2.779	3.435	3.707
27	1.314	1.703	2.052	2.473	2.771	3.421	3.690
28	1.313	1.701	2.048	2.467	2.763	3.408	3.674
29	1.311	1.699	2.045	2.462	2.756	3.396	3.659
30	1.310	1.697	2.042	2.457	2.750	3.385	3.646
32	1.309	1.694	2.037	2.449	2.738	3.365	3.622
34	1.307	1.691	2.032	2.441	2.728	3.348	3.601
36	1.306	1.688	2.028	2.434	2.719	3.333	3.582
38	1.304	1.686	2.024	2.429	2.712	3.319	3.566
40	1.303	1.684	2.021	2.423	2.704	3.307	3.551
60	1.296	1.671	2.000	2.390	2.660	3.232	3.460
120	1.289	1.658	1.980	2.358	2.617	3.160	3.373
∞	1.282	1.645	1.960	2.326	2.576	3.090	3.291

*The tabulated values are the values of t, corresponding to values of v, that give the tail areas shown in the column heading.

Table D6a

F-distribution: 5% points*

v_2 \ v_1	1	2	3	4	5	6	7	8	9	10	12	18	24	∞
1	161.4	199.5	215.7	224.6	230.2	234.0	236.8	238.9	240.5	241.9	243.9	247.3	249.0	254.3
2	18.5	19.0	19.2	19.2	19.3	19.3	19.4	19.4	19.4	19.4	19.4	19.4	19.5	19.5
3	10.13	9.55	9.28	9.12	9.01	8.94	8.89	8.85	8.81	8.79	8.74	8.67	8.64	8.53
4	7.71	6.94	6.59	6.39	6.26	6.16	6.09	6.04	6.00	5.96	5.91	5.82	5.77	5.63
5	6.61	5.79	5.41	5.19	5.05	4.95	4.88	4.82	4.77	4.74	4.68	4.58	4.53	4.36
6	5.99	5.14	4.76	4.53	4.39	4.28	4.21	4.15	4.10	4.06	4.00	3.90	3.84	3.67
7	5.59	4.74	4.35	4.12	3.97	3.87	3.79	3.73	3.68	3.64	3.57	3.47	3.41	3.23
8	5.32	4.46	4.07	3.84	3.69	3.58	3.50	3.44	3.39	3.35	3.28	3.17	3.12	2.93
9	5.12	4.26	3.86	3.63	3.48	3.37	3.29	3.23	3.18	3.14	3.07	2.96	2.90	2.71
10	4.96	4.10	3.71	3.48	3.33	3.22	3.14	3.07	3.02	2.98	2.91	2.80	2.74	2.54
11	4.84	3.98	3.59	3.36	3.20	3.09	3.01	2.95	2.90	2.85	2.79	2.67	2.61	2.40
12	4.75	3.89	3.49	3.26	3.11	3.00	2.91	2.85	2.80	2.75	2.69	2.57	2.51	2.30
13	4.67	3.81	3.41	3.18	3.03	2.92	2.83	2.77	2.71	2.67	2.60	2.48	2.42	2.21
14	4.60	3.74	3.34	3.11	2.96	2.85	2.76	2.70	2.65	2.60	2.53	2.41	2.35	2.13
15	4.54	3.68	3.29	3.06	2.90	2.79	2.71	2.64	2.59	2.54	2.48	2.35	2.29	2.07
16	4.49	3.63	3.24	3.01	2.85	2.74	2.66	2.59	2.54	2.49	2.42	2.30	2.24	2.01
17	4.45	3.59	3.20	2.96	2.81	2.70	2.61	2.55	2.49	2.45	2.38	2.26	2.19	1.96
18	4.41	3.55	3.16	2.93	2.77	2.66	2.58	2.51	2.46	2.41	2.34	2.22	2.15	1.92
19	4.38	3.52	3.13	2.90	2.74	2.63	2.54	2.48	2.43	2.38	2.31	2.18	2.11	1.88
20	4.35	3.49	3.10	2.87	2.71	2.60	2.51	2.45	2.39	2.35	2.28	2.15	2.08	1.84
22	4.30	3.44	3.05	2.82	2.66	2.55	2.46	2.40	2.34	2.30	2.23	2.10	2.03	1.78
24	4.26	3.40	3.01	2.78	2.62	2.51	2.42	2.36	2.30	2.25	2.18	2.05	1.98	1.73
26	4.23	3.37	2.98	2.74	2.59	2.47	2.39	2.32	2.27	2.22	2.15	2.02	1.95	1.69
28	4.20	3.34	2.95	2.71	2.56	2.45	2.36	2.29	2.24	2.19	2.12	1.97	1.91	1.65
30	4.17	3.32	2.92	2.69	2.53	2.42	2.33	2.27	2.21	2.16	2.09	1.96	1.89	1.62
40	4.08	3.23	2.84	2.61	2.45	2.34	2.25	2.18	2.12	2.08	2.00	1.87	1.79	1.51
60	4.00	3.15	2.76	2.53	2.37	2.25	2.17	2.10	2.04	1.99	1.92	1.78	1.70	1.39
120	3.92	3.07	2.68	2.45	2.29	2.18	2.09	2.02	1.96	1.91	1.83	1.69	1.61	1.25
∞	3.84	3.00	2.60	2.37	2.21	2.10	2.01	1.94	1.88	1.83	1.75	1.60	1.52	1.00

*The tabulated values are the values of F that cut off an upper tail with an area of 0.05 when v_1 and v_2 have the stated values.
Adapted from a reproduction from Fisher, R.A. and Yates, F. *Statistical Tables for Biological, Medical and Agricultural Research*, Oliver & Boyd, Edinburgh, in Udny Yule, G. and Kendall, M.G. (1950) *An Introduction to the Theory of Statistics*, Charles Griffin, London.

Table D6b

F-distribution: 1% points*

v_2 \ v_1	1	2	3	4	5	6	7	8	9	10	12	18	24	∞
1	4052	4999	5403	5625	5764	5859	5928	5981	6022	6055	6106	6192	6234	6366
2	98.49	99.00	99.17	99.25	99.30	99.33	99.36	99.38	99.39	99.40	99.42	99.44	99.46	99.50
3	34.12	30.81	29.46	28.71	28.24	27.91	27.67	27.49	27.34	27.23	27.05	26.75	26.60	26.12
4	21.20	18.00	16.69	15.98	15.52	15.21	14.98	14.80	14.66	14.55	14.37	14.08	13.93	13.46
5	16.26	13.27	12.06	11.39	10.97	10.67	10.46	10.27	10.16	10.05	9.89	9.61	9.47	9.02
6	13.74	10.92	9.78	9.15	8.75	8.47	8.26	8.10	7.98	7.87	7.72	7.45	7.31	6.88
7	12.25	9.55	8.45	7.85	7.46	7.19	6.99	6.84	6.72	6.62	6.47	6.21	6.07	5.65
8	11.26	8.65	7.59	7.01	6.63	6.37	6.18	6.03	5.91	5.81	5.67	5.41	5.28	4.86
9	10.56	8.02	6.99	6.42	6.06	5.80	5.61	5.47	5.35	5.26	5.11	4.86	4.73	4.31
10	10.04	7.56	6.55	5.99	5.64	5.39	5.20	5.06	4.94	4.85	4.71	4.46	4.33	3.91
11	9.65	7.20	6.22	5.67	5.32	5.07	4.89	4.74	4.63	4.54	4.40	4.15	4.02	3.60
12	9.33	6.93	5.95	5.41	5.06	4.82	4.64	4.50	4.39	4.30	4.16	3.91	3.78	3.36
13	9.07	6.70	5.74	5.20	4.86	4.62	4.44	4.30	4.19	4.10	3.96	3.72	3.59	3.16
14	8.86	6.51	5.56	5.03	4.69	4.46	4.28	4.14	4.03	3.94	3.80	3.55	3.43	3.00
15	8.68	6.36	5.42	4.89	4.56	4.32	4.14	4.00	3.89	3.80	3.67	3.42	3.29	2.87
16	8.53	6.23	5.29	4.77	4.44	4.20	4.03	3.89	3.78	3.69	3.55	3.31	3.18	2.75
17	8.40	6.11	5.18	4.67	4.34	4.10	3.93	3.79	3.68	3.59	3.45	3.21	3.08	2.65
18	8.28	6.01	5.09	4.58	4.25	4.01	3.84	3.71	3.60	3.51	3.37	3.13	3.00	2.57
19	8.18	5.93	5.01	4.50	4.17	3.94	3.77	3.63	3.52	3.43	3.30	3.05	2.92	2.49
20	8.10	5.85	4.94	4.43	4.10	3.87	3.70	3.56	3.46	3.37	3.23	2.99	2.86	2.42
22	7.94	5.72	4.82	4.31	3.99	3.76	3.59	3.45	3.35	3.26	3.12	2.88	2.75	2.31
24	7.82	5.61	4.72	4.22	3.90	3.67	3.50	3.36	3.26	3.17	3.03	2.79	2.66	2.21
26	7.72	5.53	4.64	4.14	3.82	3.59	3.42	3.29	3.18	3.09	2.96	2.71	2.58	2.13
28	7.64	5.45	4.57	4.07	3.75	3.53	3.36	3.23	3.12	3.03	2.90	2.65	2.52	2.06
30	7.56	5.39	4.51	4.02	3.70	3.47	3.30	3.17	3.07	2.98	2.84	2.60	2.47	2.01
40	7.31	5.18	4.31	3.83	3.51	3.29	3.12	2.99	2.89	2.80	2.66	2.42	2.29	1.80
60	7.08	4.98	4.13	3.65	3.34	3.12	2.95	2.82	2.72	2.63	2.50	2.25	2.12	1.60
120	6.85	4.79	3.95	3.48	3.17	2.96	2.79	2.66	2.56	2.47	2.34	2.09	1.95	1.38
∞	6.64	4.60	3.78	3.32	3.02	2.80	2.64	2.51	2.41	2.32	2.18	1.93	1.79	1.00

*The tabulated values are the values of F that cut off an upper tail with an area of 0.01 when v_1 and v_2 have the stated values.

Adapted from a reproduction from Fisher, R.A. and Yates, F. *Statistical Tables for Biological, Medical and Agricultural Research*, Oliver & Boyd, Edinburgh, in Udny Yule, G. and Kendall, M.G. (1950) *An Introduction to the Theory of Statistics*, Charles Griffin, London.

Table D7

Values of *T* for the Wilcoxon signed rank test

	Level of significance				
n	0.05	0.025	0.01	0.005	One-sided
	0.10	0.05	0.02	0.01	Two-sided
5	0	–	–	–	
6	2	0	–	–	
7	3	2	0	–	
8	5	3	1	0	
9	8	5	3	1	
10	10	8	5	3	
11	13	10	7	5	
12	17	13	9	7	
13	21	17	12	9	
14	25	21	15	12	
15	30	25	19	15	
16	35	29	23	19	
17	41	34	27	23	
18	47	40	32	27	
19	53	46	37	32	
20	60	52	43	37	
21	67	58	49	42	
22	75	65	55	48	
23	83	73	62	54	
24	91	81	69	61	
25	100	89	76	68	
26	110	98	84	75	
28	130	116	101	91	
30	151	137	120	109	
35	213	195	173	159	
40	286	264	238	220	
50	466	434	397	373	

Adapted from: Wilcoxon, F. (1949) *Some Rapid Approximate Statistical Procedures*, American Cyanamid Company, New York, p. 13.

Table D8
Acceptable values of *r* in the runs test*

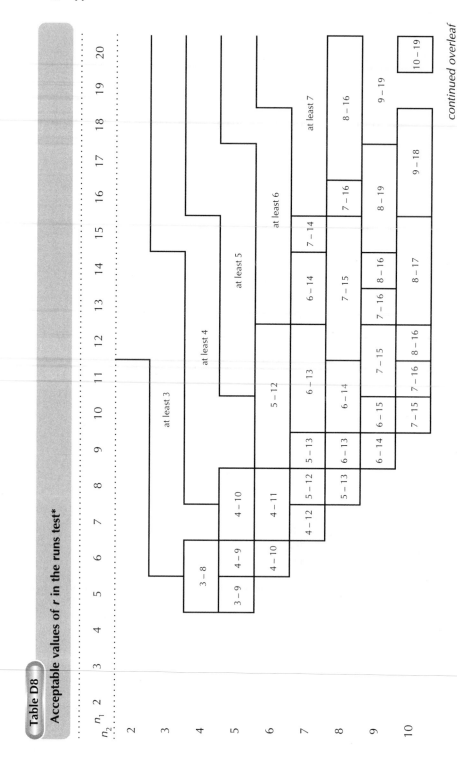

continued overleaf

Table D8 continued

n_2 \ n_1	2	3	4	5	6	7	8	9	10	11	12	13	14	15	16	17	18	19	20
11										8 – 16	8 – 17	8 – 18	9 – 18	9 – 19	9 – 19	10 – 19	10 – 19	10 – 20	10 – 20
12											8 – 18	8 – 19	9 – 19	9 – 19	10 – 20	10 – 20	10 – 21	10 – 21	11 – 22
13												9 – 19	9 – 19	10 – 20	10 – 20	11 – 21	11 – 21	11 – 22	12 – 23
14													10 – 19	10 – 20	11 – 21	11 – 22	11 – 22	12 – 22	12 – 23
15														11 – 21	11 – 22	12 – 22	12 – 22	12 – 23	12 – 23
16															12 – 22	12 – 23	12 – 24	13 – 24	13 – 24
17																12 – 24	13 – 24	13 – 25	14 – 25
18																	13 – 25	13 – 25	14 – 25
19																		14 – 26	14 – 26
20																			15 – 27

*Values of n_1 and n_2 are interchangeable. Entries in the boxes show the ranges within which r must lie to prevent rejection of the hypothesis at the 5% level.
Adapted from: Swed, F.S. and Eisenhart, C. (1943) Tables for testing randomness of grouping in a sequence of alternatives. Annals of Mathematical Statistics 14, 83–86.

Table D9a

Values of U for the Mann–Whitney U-test: critical values for 5% (2.5%) significance in a two-tailed (one-tailed) test

| n_1 | 2 | 3 | 4 | 5 | 6 | 7 | 8 | 9 | 10 | 11 | 12 | 13 | 14 | 15 | 16 | 17 | 18 | 19 | 20 |
n_2																			
2							0	0	0	0	1	1	1	1	1	2	2	2	2
3				0	1	1	2	2	3	3	4	4	5	5	6	6	7	7	8
4			0	1	2	3	4	4	5	6	7	8	9	10	11	11	12	13	14
5				2	3	5	6	7	8	9	11	12	13	14	15	17	18	19	20
6					5	6	8	10	11	13	14	16	17	19	21	22	24	25	27
7						8	10	12	14	16	18	20	22	24	26	28	30	32	34
8							13	15	17	19	22	24	26	29	31	34	36	38	41
9								17	20	23	26	28	31	34	37	39	42	45	48
10									23	26	29	33	36	39	42	45	48	52	55
11										30	33	37	40	44	47	51	55	58	62
12											37	41	45	49	53	57	61	65	69
13												45	50	54	59	63	67	72	76
14													55	59	64	69	74	78	83
15														64	70	75	80	85	90
16															75	81	86	92	98
17																87	93	99	105
18																	99	106	112
19																		113	119
20																			127

The table is symmetric about the diagonal; values of n_1 and n_2 are interchangeable

Adapted from: Owen, D.B. (1962) *Handbook of Statistical Tables*, Addison-Wesley, Reading MA; based on Auble, D. (1953). Extended tables for the Mann–Whitney statistic. *Bulletin of the Institute of Educational Research at Indiana University* **1**, 2.

Table D9b

Values of U for the Mann–Whitney U-test: critical values for 1% (0.5%) significance in a two-tailed (one-tailed) test

n_2 \ n_1	2	3	4	5	6	7	8	9	10	11	12	13	14	15	16	17	18	19	20
2																		0	0
3							0	0	0	0	1	1	1	2	2	2	2	3	3
4				0	0	0	1	1	2	2	3	3	4	5	5	6	6	7	8
5				0	1	1	2	3	4	5	6	7	7	8	9	10	11	12	13
6					2	3	4	5	6	7	9	10	11	12	13	15	16	17	18
7						4	6	7	9	10	12	13	15	16	18	19	21	22	24
8							7	9	11	13	15	17	18	20	22	24	26	28	30
9								11	13	16	18	20	22	24	27	29	31	33	36
10									16	18	21	24	26	29	31	34	37	39	42
11										21	24	27	30	33	36	39	42	45	48
12											27	31	34	37	41	44	47	51	54
13												34	38	42	45	49	53	57	60
14													42	46	50	54	58	63	67
15														51	55	60	64	69	73
16															60	65	70	74	79
17																70	75	81	86
18																	81	87	92
19																		93	99
20																			105

The table is symmetric about the diagonal; values of n_1 and n_2 are interchangeable

Adapted from: Owen, D.B. (1962) Handbook of Statistical Tables, Addison-Wesley, Reading MA; based on Auble, D. (1953). Extended tables for the Mann–Whitney statistic. Bulletin of the Institute of Educational Research at Indiana University 1, 2.

Table D10

Significance points of χ^2

p / v	.99	.98	.95	.50	.10	.05	.02	.01	.001
1	.00016	.00063	.0039	.46	2.71	3.84	5.41	6.64	10.83
2	.02	.04	.10	1.39	4.60	5.99	7.82	9.21	13.82
3	.12	.18	.35	2.37	6.25	7.82	9.84	11.34	16.27
4	.30	.43	.71	3.36	7.78	9.49	11.67	13.28	18.46
5	.55	.75	1.14	4.35	9.24	11.07	13.39	15.09	20.52
6	.87	1.13	1.64	5.35	10.64	12.59	15.03	16.81	22.46
7	1.24	1.56	2.17	6.35	12.02	14.07	16.62	18.48	24.32
8	1.65	2.03	2.73	7.34	13.36	15.51	18.17	20.09	26.12
9	2.09	2.53	3.32	8.34	14.68	16.92	19.68	21.67	27.88
10	2.56	3.06	3.94	9.34	15.99	18.31	21.16	23.21	29.59
11	3.05	3.61	4.58	10.34	17.28	19.68	22.62	24.72	31.26
12	3.57	4.18	5.23	11.34	18.55	21.03	24.05	26.22	32.91
13	4.11	4.76	5.89	12.34	19.81	22.36	25.47	27.69	34.53
14	4.66	5.37	6.57	13.34	21.06	23.68	26.87	29.14	36.12
15	5.23	5.98	7.26	14.34	22.31	25.00	28.26	30.58	37.70
16	5.81	6.61	7.96	15.34	23.54	26.30	29.63	32.00	39.29
17	6.41	7.26	8.67	16.34	24.77	27.59	31.00	33.41	40.75
18	7.02	7.91	9.39	17.34	25.99	28.87	32.35	34.80	42.31
19	7.63	8.57	10.12	18.34	27.20	30.14	33.69	36.19	43.82
20	8.26	9.24	10.85	19.34	28.41	31.41	35.02	37.57	45.32
21	8.90	9.92	11.59	20.34	29.62	32.67	36.34	38.93	46.80
22	9.54	10.60	12.34	21.24	30.81	33.92	37.66	40.29	48.27
23	10.20	11.29	13.09	22.34	32.01	35.17	38.97	41.64	49.73
24	10.86	11.99	13.85	23.34	33.20	36.42	40.27	42.98	51.18
25	11.52	12.70	14.61	24.34	34.38	37.65	41.57	44.31	52.62
26	12.20	13.41	15.38	25.34	35.56	38.88	42.86	45.64	54.05
27	12.88	14.12	16.15	26.34	36.74	40.11	44.14	46.96	55.48
28	13.56	14.85	16.93	27.34	37.92	41.34	45.42	48.28	56.89
29	14.26	15.57	17.71	28.34	39.09	42.56	46.69	49.59	58.30
30	14.95	16.31	18.49	29.34	40.26	43.77	47.96	50.89	59.70

Adapted from a reproduction from Fisher, R.A. and Yates, F. *Statistical Methods for Research Workers*, Oliver & Boyd, Edinburgh, in Yule, Udny G. and Kendall, M.G. (1950) *An Introduction to the Theory of Statistics*, Charles Griffin, London.

Table D11

Critical values of Kendall's τ and Spearman's r_S

	Significance level							
One-tail	0.05		0.025		0.01		0.005	
Two-tail	0.10		0.05		0.02		0.01	
v	τ	r_S	τ	r_S	τ	r_S	τ	r_S
2	1.000	1.000						
3	.800	.900	1.000	1.000	1.000	1.000		
4	.733	.829	.867	.886	.867	.943	1.000	1.000
5	.619	.714	.714	.786	.810	.893	.905	.929
6	.571	.643	.643	.738	.714	.833	.786	.881
7	.500	.600	.556	.683	.667	.783	.722	.833
8	.467	.564	.551	.648	.600	.746	.644	.794
10	.394	.506	.455	.591	.546	.712	.576	.777
12	.363	.456	.407	.544	.473	.645	.517	.715
14	.317	.425	.383	.506	.433	.601	.483	.665
16	.294	.399	.346	.475	.412	.564	.451	.625
18	.274	.377	.326	.450	.379	.534	.421	.591
20	.264	.359	.307	.428	.359	.508	.394	.562
22	.246	.343	.290	.409	.341	.485	.377	.537
24	.237	.329	.280	.392	.329	.465	.360	.515
26	.228	.317	.265	.377	.312	.448	.344	.496
28	.218	.306	.255	.364	.301	.432	.333	.478

Adapted from: Olds, E.G. (1938) Distributions of sums of squares of rank differences for small numbers of individuals. *Annals of Mathematical Statistics* **9**, 133–48; Olds, E.G. (1949) The 5% significance levels for sums of squares of rank differences and a correction. *Annals of Mathematical Statistics* **20**, 117–18; Neave, H.R. (1977) *Elementary Statistical Tables*, George Allen & Unwin, London.

Table D12

Coefficients for use in the Shapiro and Wilk test for normality

$n =$	2	3	4	5	6	7	8	9	10	11
a_1	.7071	.7071	.6872	.6646	.6431	.6233	.6052	.5888	.5739	.5601
a_2	–	–	.1677	.2413	.2806	.3031	.3164	.3244	.3291	.3315
a_3	–	–	–	–	.0875	.1401	.1743	.1976	.2141	.2260
a_4	–	–	–	–	–	–	.0561	.0947	.1224	.1429
a_5	–	–	–	–	–	–	–	–	.0399	.0695

$n =$	12	13	14	15	16	17	18	19	20
a_1	.5475	.5359	.5251	.5150	.5056	.4968	.4886	.4808	.4734
a_2	.3325	.3325	.3318	.3306	.3290	.3273	.3253	.3232	.3211
a_3	.2347	.2412	.2460	.2495	.2521	.2540	.2553	.2561	.2565
a_4	.1586	.1707	.1802	.1878	.1939	.1988	.2027	.2059	.2085
a_5	.0922	.1099	.1240	.1353	.1447	.1524	.1587	.1641	.1686
a_6	.0303	.0539	.0727	.0880	.1005	.1109	.1197	.1271	.1334
a_7	–	–	.0240	.0433	.0593	.0725	.0837	.0932	.1013
a_8	–	–	–	–	.0196	.0359	.0496	.0612	.0711
a_9	–	–	–	–	–	–	.0163	.0303	.0422
a_{10}	–	–	–	–	–	–	–	–	.0140

$n =$	21	22	23	24
a_1	.4643	.4590	.4542	.4493
a_2	.3185	.3156	.3126	.3098
a_3	.2578	.2571	.2563	.2554
a_4	.2119	.2131	.2139	.2145
a_5	.1736	.1764	.1787	.1807
a_6	.1399	.1443	.1480	.1512
a_7	.1092	.1150	.1201	.1245
a_8	.0804	.0878	.0941	.0997
a_9	.0530	.0618	.0696	.0764
a_{10}	.0263	.0368	.0459	.0539
a_{11}	–	.0122	.0228	.0321
a_{12}	–	–	.0000	.0107

Table D13

Percentage points of *W* for the Shapiro and Wilk test for normality

	\multicolumn{9}{c}{Level of significance}
	0.01	0.02	0.05	0.10	0.50	0.90	0.95	0.98	0.99
$n = 3$	.753	.756	.767	.789	.959	.998	.999	1.000	1.000
4	.687	.707	.748	.792	.935	.987	.992	.996	.997
5	.686	.715	.762	.806	.927	.979	.986	.991	.993
6	.713	.743	.788	.826	.927	.974	.981	.986	.989
7	.730	.760	.803	.838	.928	.972	.979	.985	.988
8	.749	.778	.818	.851	.932	.972	.978	.984	.987
9	.764	.791	.829	.859	.935	.972	.978	.984	.986
10	.781	.806	.842	.869	.938	.972	.978	.983	.986
11	.792	.817	.850	.876	.940	.973	.979	.984	.986
12	.805	.828	.859	.883	.943	.973	.979	.984	.986
13	.814	.837	.866	.889	.945	.974	.979	.984	.986
14	.825	.846	.874	.895	.947	.975	.980	.984	.986
15	.835	.855	.881	.901	.950	.975	.980	.984	.987
16	.844	.863	.887	.906	.952	.976	.981	.985	.987
17	.851	.869	.892	.910	.954	.977	.981	.985	.987
18	.858	.874	.897	.914	.956	.978	.982	.986	.988
19	.863	.879	.901	.917	.957	.978	.982	.986	.988
20	.868	.884	.905	.920	.959	.979	.983	.986	.988
21	.873	.888	.908	.923	.960	.980	.983	.987	.989
22	.878	.892	.911	.926	.961	.980	.984	.987	.989
23	.881	.895	.914	.928	.962	.981	.984	.987	.989
24	.884	.898	.916	.930	.963	.981	.984	.987	.989

Adapted from: Shapiro, S.S. and Wilk, M.B. (1965) An analysis of variance test for normality (complete samples) *Biometrika* **52**, 591–611.

General exercises

 (Chapter 3)

A1. Two cricketers each play 15 matches in a season, each match having two innings. The scores are shown below. Draw ordered bar charts allowing you (a) to compare the performances of the players and (b) to discuss any changes in performance over time and between first and second innings.

Mr Bash

Innings 1	6	0	22	0	24	7	58	4	9	88	107	35	135	86	95
Innings 2	104	117	0	59	93	146	6	36	163	123	86	9	97	93	148

Mr Bunt

Innings 1	103	57	51	48	0	91	66	83	14	83	72	42	26	63	31
Innings 2	55	47	63	49	36	26	101	57	91	111	35	92	80	42	61

A2. Drake Bicycles plc make bicycles at two factories. All wheels are tested each day over two months and the number of faulty wheels is recorded. The table shows the numbers of days with various numbers of faults. Draw frequency polygons and comment on the differences between factories.

Faults per day	0	1	2	3	4	5	6	7	8
Factory A	2	12	27	10	7	3	0	0	0
Factory B	1	7	19	18	10	3	0	2	1

A3. We show annual figures of government spending (a) in total and (b) on health (in £ billion). Draw a bar chart that shows the changes over time and compares (a) and (b).

Year	1985	1986	1987	1988	1989	1990	1991	1992	1993	1994	1995
Total spend	130.5	136.1	146.6	161.1	175.6	196.7	208.7	209.7	214.5	230.8	250.9
Health spend	17.9	19.2	20.9	22.9	25.2	27.8	31.2	35.0	36.8	38.7	40.8

A4. Statingham General Hospital has recorded the number of patients waiting to see a doctor at its A&E department at 3pm every day over many months. Draw a histogram.

Number of patients waiting	0–9	10–14	15–19	20–24	25–29	30–39	40–49	50+
Frequency	8	22	38	75	73	28	10	6

A5. Statingham Traction Ltd have timed the arrival of their buses at the town centre. Draw a histogram of the results.

Minutes late	0–	1–	2–	3–	4–	5–	10+
Frequency	25	20	19	11	7	12	6

A6. A local builder is renovating a canal and cannot work in the rain. The builder receives compensation according to the hours lost in a day: less than 1, nothing; 1–4, £200; more than 4, £400. Rain data for last year are shown below. Draw a cumulative histogram and find the number of days on which the builder received (a) no compensation and (b) £400.

Hours of rainfall	<1	1–	2–	3–	4–	5–	6+
Number of days	96	96	58	53	35	40	56

A7. The following data show a percentage regional breakdown of colliery output and employment in the United Kingdom in the late 1970s. Draw pie charts to compare the two distributions.

	North	Yorks and Humberside	East Midlands	West Midlands	Scotland	Rest of UK
Coal output (%)	12.4	27.4	30.9	9.2	8.4	7.6
Employment (%)	15.1	26.7	23.2	13.0	9.5	13.0

A8. Use pie charts and the data below to compare the employment structures of Chile and Canada. The figures are in thousands of workers.

	Agriculture and mining	Manufacture	Public utilities	Construction	Commerce	Social services	Other	Total
Canada	727	1 865	136	695	2 935	3 731	2 252	12 341
Chile	964	753	22	322	775	1 167	537	4 540

Group B (Chapters 5–12)

B1. One Saturday the average number of goals scored in the 43 matches played that day was 1.395. How many goals were scored?

B2. Shirley Middel drives to work five days a week. Over the last two months her recorded journey times in minutes were as shown below. (a) Calculate the arithmetic mean journey times (i) for each month and (ii) for both months taken together. (b) What is the variance of the February journey times?

February	27.1 26.2 33.6 28.7 37.8 25.4 31.8 27.5 26.4 41.9 30.6 28.3 36.8
	35.4 32.3 33.4 29.7 30.2 32.0
March	29.6 35.0 33.2 31.4 28.7 46.9 41.0 28.7 36.1 31.2 32.7 38.3 37.4
	35.6 29.8 28.2 36.3 38.1 40.2 32.6

B3. Nine doctors working at a health centre see 11 060 patients in 58 days. What is the average number of patients (a) seen by one doctor in one day and (b) visiting the centre in a day?

B4. The salaries (in £) paid to employees in the paint shop and the accounts department of a local firm are shown. Find the two arithmetic means and the two coefficients of variation, and comment on the result.

Paint shop	14 385 11 262 15 387 14 797 16 381 12 369 14 501 10 608 12 374
Accounts	7698 7698 9401 32 763 10 405 12 411

B5. Cashiers in a local store are paid an average of £267 a week. How would the average be affected if (a) wages were increased by 8% and (b) in one week all cashiers received a £50 bonus?

B6. A hospital has records of staff absences during 1996. From the data below calculate the arithmetic mean and standard deviation of the daily numbers of absences. What is the mode?

Number of staff absent	0–4	5–9	10–14	15–19	20–24	25–29	30+
Number of days	8	73	112	87	42	34	10

B7. From the table below calculate the average percentage unemployment rate of England and Wales (comprised of the five regions).

Region	N	M & EA	SE	SW	W
Economically active population (000)	4703.7	3413.7	5830.2	1215.6	790.5
Percent unemployed	6.6	5.4	4.2	6.4	7.4

B8. A trout lake is open to local anglers on 90 days in a year, at a charge of 75p for each fish caught. Using the data below, an assumed mean catch of 24.5 and units of 10 fishes, calculate the mean daily income of the owner.

Fish caught	0–9	10–19	20–29	30–39	40–49	50+
Number of days	2	12	29	23	17	7

B9. The landlord of the Skew Distribution informs the vice-chancellor of the number of students ejected on each night of the academic year. Find from the summary below the mean nightly number of ejections, and comment on this as an indicator of student rowdiness.

Students ejected	0	1	2	3	4	5	6–95	96	97+
Number of nights	114	56	12	15	6	6	0	1	0

B10. A variable X takes the values 1 to 6 with the frequencies shown below. Calculate as economically as possible the arithmetic means of X, $X - 3$, X^2 and $5X$.

X	1	2	3	4	5	6
f	9	14	17	18	2	0

B11. A garage conducts MOT tests at a fee of £25, and charges a repair cost to correct all faults. Receipts over the last year are summarised below. What is the average charge per owner? How would this have been affected if an additional test fee of £12.50 had been charged on repair costs of over £200? What is the quartile deviation of the charges shown in the table?

Repair cost (£)	0–49	50–74	75–99	100–149	150–199	200–299	300+
Number of cars	28	274	368	219	154	75	36

B12. The table below presents a hypothetical distribution of marks obtained by A-level candidates taking three subjects. Odd values and fractions are not possible. Making and stating any necessary assumptions, find the major and minor modal groups and the modal number of marks.

Mark	0–2	4–6	8–10	12–14	16–18	20–24	26–28	30
Percentage	10.0	13.0	17.5	13.5	27.5	8.5	6.0	4.0

B13. A farmer takes samples of 49 plums from each of many trees in an orchard and counts the damaged fruits. Find the modal groups from the table below and estimate the modal values. State all assumptions.

Number of damaged fruits	0–4	5–9	10–14	15–19	20–24	25–29	30–34	35–39	40–44	45–49
Number of trees		114	76	73	36	27	30	49	18	12	2

B14. Three cities have tabulated the ages of their shoplifters. Does the composite table suggest any differences between the modal ages?

Age	0–4	5–9	10–14	15–19	20–24	25–29	30–39	40–59	60+
Manchester	0	43	185	196	162	128	74	73	23
Melbourne	0	119	146	248	232	163	58	50	24
Detroit	76		417		318		134	129	57

B15. A table provided in the instructors manual, based on the 1991 census of population, shows the age distributions for males and females in Great Britain. Use graphical methods to find the model age for (a) males and (b) females. How would they be affected if the data were presented in (a) ten-year intervals starting with 0–9 and (b) intervals starting with 0–4 and then of width 10 years, 4–14, 15–24, etc.

B16. Fifty undergraduates get the statistics marks shown below. (a) Find the arithmetic mean, median and modal marks. (b) Combine the data into a frequency distribution with a five-mark interval starting at 35, then recalculate the mean, median and mode.

46	63	38	47	48	51	55	63	69	59
72	38	43	67	55	55	48	47	63	58
52	53	66	52	51	61	42	78	51	63
40	49	48	50	58	51	63	67	45	48
67	49	46	54	52	47	53	47	53	60

B17. Official statistics show the following distribution of unladen mass for road haulage vehicles. Find the median mass and the quartile deviation, carefully considering the number of decimal places.

Mass (tons)	<1	1–	2–	3–	5–	8+	Total
Vehicles (000)	600	572	132	152	127	102	1685

B18. Last year 468 houses were sold in Upper Sternum, with the percentage price distribution shown below. Use a graphical method to find the median selling price; careful about the number of decimal places. Determine the quartile deviation of the prices.

Price (£000)	<75	75–	100–	125–	150–	175–	200–	225–	250+
Percentage	5.9	13.4	17.9	10.6	14.2	8.4	11.1	8.9	9.6

B19. On 200 consecutive weeknights a restaurant served the following numbers of diners. Calculate the median and comment on its usefulness for estimating (i) takings over the year, (ii) the number of permanent table staff to be employed and (iii) the stock of cheese to be kept in the refrigerator.

Diners	0–9	10–19	20–29	30–39	40–49	50–59	60–69	70–79	80–90
Nights	36	14	26	28	12	13	10	20	41

B20. Calculate the harmonic and geometric means for (i) 5.5, 8.6, 11.5; (ii) 8, 11, 11, 12, 12, 12; (iii) 0.2, 0.6, 0.9, 1.1.

B21. A bank pays interest at the end of every quarter. In 1997 the quarterly interest rates were 5%, 6%, 6% and 8.5% per annum. A customer invested £200 on January 1. Find (i) the total interest received at the end of the year, (ii) the average interest rate per quarter and (iii) the average rate of interest received.

B22. A lorry travels from Birmingham to Carlisle with an intermediate stop. It averages 22 miles per gallon (mpg) over the first 88 miles, and 18 mpg over the remaining 126 miles. What is the average number of miles per gallon for the whole journey?

B23. Company A produces 10 railway coaches per week. Quality control tests in four consecutive weeks found 216, 196, 305 and 288 faults. Company B's test procedure is to count the number of coaches that yield 400 faults. Four tests yielded 14, 15, 15 and 19 coaches. Which company has the better record in terms of faults per coach?

B24. If the population of the West Midlands was 6.13 million in June 1986 and 6.87 million in June 1991, what were (i) the average annual rate of growth and (ii) the population halfway between these dates, assuming that it grew at a constant rate.

B25. A river authority tests water for impurities. At five different locations the volumes of water yielding 1 microgram of aluminium were 12.6, 11.8, 10.9, 10.9 and 13.7 litres. If the safe level is 0.085 micrograms per litre, is the water safe?

B26. The values of end-of-year stocks held by small builders' merchants were distributed as below. Find their standard deviation and quartile coefficient of skewness.

Value of stock (£000)	0–	15–	30–	45–	60–	75–	90–	125+
Number of firms	36	174	168	132	88	27	19	4

B27. The Scottish Institute of Rust Prevention Officers holds annual examinations. The top 20% of candidates are admitted to membership and the bottom 30% are not allowed to resit. Last year's results were as below. What was (i) the highest mark awarded to a candidate who will not be allowed to resit, (ii) the lowest mark awarded to a successful candidate and (iii) the mean deviation of all the marks?

Mark	0–49	50–99	100–149	150–199	200–249	250–299	300–349	350–400
Candidates	17	38	116	194	198	187	96	58

B28. Assembly line workers insert components into a circuit board. The numbers of components inserted before an error occurs have the distribution below. Use graphs to locate the first, second and third quartiles of the distribution, and calculate the semi-interquartile range.

Components inserted	0–49	50–74	75–99	100–124	125–174	175+
Frequency	15	35	78	127	136	74

B29. The times taken by 10 eight-year-old children to complete a numeracy test were recorded in two tests held four weeks apart. The results are below. Use the mean deviation to examine whether their variability has reduced.

Week 1	15.6	13.8	7.4	8.9	11.6	27.1	18.7	12.2	10.9	16.1
Week 2	13.9	13.7	8.7	7.6	19.1	16.2	12.7	11.6	12.3	14.9

B30. A fossil hunter divides a site into 900 ten-metre squares and counts the numbers of fossils (x) per square. If $\sum x = 7200$ and $\sum x^2 = 90\,000$ what are the mean, standard deviation and variance of the number of fossils per square?

B31. In 12 consecutive days the daily takings of a butcher (to the nearest pound) were £298, £316, £328, £246, £235, £512, £196, £279, £414, £297, £316 and £504. Calculate their mean and standard deviation.

B32. A firm supplies test tubes for research laboratories. Its quality control department has the following information about the numbers of batches with various numbers of defective tubes. A batch will be rejected if the number of defective tubes exceeds the mean by more than one standard deviation. Calculate (a) the mean number of defects per batch, (b) the standard deviation and (c) how many defective tubes a batch can have before it is rejected.

Defective tubes	0	1	2	3	4	5	6	7	8	9	10+
Number of batches	4	6	14	28	38	57	36	25	1	4	2

B33. Surveys in Manchester and Birmingham show that the mean office rents are respectively £265.00 and £287.50 per square metre. The standard deviations are £37.50 and £45.00. In which city do the rents have the greater coefficient of variation?

B34. A test of steel for impurities is conducted in each of two years, producing the results below. Calculate for each year the standardised values for 54, 162, 200 and 295 parts per million (ppm).

Impurities (ppm)	< 50	50–	100–	150–	200–	250–	300–	350–	400+
Steel samples (yr 1)	18	136	172	96	84	90	47	23	14
Steel samples (yr 2)	2	52	152	232	46	23	11	2	0

B35. The distributions of daily milk yields from two farms in 1996 were as below. (i) Calculate the variances for each farm. (ii) Would you use Sheppard's correction at all? What would it be? (iii) If the farmers were fined £50 for each day that the yield exceeded the mean by two standard deviations, how much did each farmer have to pay? (iv) If it was appropriate, how much difference did Sheppard's correction make?

Yields (litres)	<380	380–	400–	420–	440–	460–	480–	500–	520–	540–559
Farm A (days)	4	12	32	64	96	74	44	26	10	4
Farm B (days)	20	20	22	49	102	105	34	12	2	0

Group C (Chapters 13–15)

C1. Fifteen students gained the following marks out of 100 for examinations in economics and statistics. Plot a scattergram. One student scored 60 in economics but was absent from the statistics exam. Using a simple technique, what do you think is the lowest mark the student would have been likely to get in statistics? What is the student's highest likely mark?

Economics	43	46	53	58	54	48	55	64	59	56	63	69	68	75	73
Statistics	37	45	40	47	54	41	62	59	64	69	72	69	78	81	86

C2. The table compares the average rate of interest charged by a credit company for a 12-month period with the sales of kitchen appliances during the same period. Plot a scattergram. (i) If the sales figures for January had been unavailable, what value would you have guessed? Give your answer as a range. (ii) If the interest rate for December were unknown, what value might you have assumed from the sales figures?

Month	Jan	Feb	Mar	Apr	May	Jun	Jul	Aug	Sep	Oct	Nov	Dec
Interest (%)	16.28	15.45	14.36	16.72	14.20	15.62	15.80	17.05	16.20	17.81	17.05	17.62
Sales (£00m)	324	312	342	286	382	348	263	247	238	212	220	242

C3. Plot a scattergram for each of the data sets (a) and (b).

(a)	X	0.1	5.0	6.3	10.8	12.5	17.0	23.0	25.0	27.5	30.0	31.0
	Y	20	25	75	85	170	325	630	750	700	1010	860

(b)	X	4.8	4.0	17.2	15.0	34.1	48.0	65.0	75.0	114.8	133.0	140.0
	Y	1.00	1.75	2.30	3.30	3.20	4.12	3.80	4.31	4.40	4.85	5.12

Plot new scattergrams for Z and X where in (a) you define $Z = X^2$, and in case (b) you use your calculator to obtain $Z = \ln X$.

C4. A psychologist tested 60 children to see whether the number of hours spent watching television in a week (x) possibly explained the number of errors made in an English test (y). The data yielded $\sum x = 1420$, $\sum y = 2000$, $\sum x^2 = 40\,780$ $\sum y^2 = 86\,020$ and $\sum xy = 57\,180$. Determine the equation of an appropriate regression line.

C5. One hundred cows from each of six farms were injected with different doses of antibovine influenza. The numbers of cows subsequently contracting the disease are shown in the table. Do the data suggest a relationship between dosage and contracting the disease for these particular animals?

Farm	A	B	C	D	E	F
Dosage (cc)	2	4	8	9	12	16
Sick animals	15	7	11	3	5	1

C6. A car manufacturer records the daily numbers of absentees in the paint shop. They also record the numbers of vehicles that fail quality tests on those days. Over 10 consecutive working days the data were as below. Use a regression equation to see whether they suggest there are grounds for a more thorough enquiry into the proposition that absenteeism affects quality.

Absentees	3	4	7	7	8	11	10	15	15	3
Failures	8	16	19	26	13	21	34	31	40	20

C7. Using the data of Question C6, find the equation of the other regression line. Combining the two results, calculate the proportion of the variance in the number of failures in these days that is explained by absenteeism.

C8. The heights and masses of 10 fourteen-year-old boys are given in the table below. Calculate the correlation coefficient. If there are 2.5 cm to the inch, and 2.2 lb to the kilogram, what would be the correlation coefficient if height were measured in inches and mass in pounds?

Height (cm)	115	133	137	146	152	157	162	166	129	173
Mass (kg)	41	35	51	44	40	72	47	42	38	60

Group D (Chapters 17–20)

D1. A farmer has 9 bags of cattle feed, 4 from supplier A, 3 from Supplier B and 2 from Supplier C. In how many different ways can 3 bags be selected if (i) all the bags are different and (ii) all 9 bags from each supplier are identical.

D2. A driver passes through three traffic lights. The chance that the driver will have to stop at the first is 1/2, at the second 1/3 and at the third 1/4. Whether the driver is stopped at any light is independent of whether he or she is stopped at any other light. What is the chance that (i) the driver will make the whole journey without being stopped, (ii) the driver will be stopped only at the first and third lights, and (iii) the driver will be stopped at all three lights?

D3. A full set of 22 snooker balls, including 15 reds, is placed in a bag and three are drawn at random. What is the probability that all three are red if the first and second balls are (i) replaced and (ii) not replaced?

D4. A manufacturer of flashbulbs picks 20 bulbs at random to test. There is a 1% chance that any bulb will fail. What is the probability that (i) the first three bulbs fail, (ii) no bulbs fail and (iii) fewer than three bulbs fail?

D5. A hand of five cards is dealt to a poker player. What is the probability that (i) all five are of the same suit and (ii) all five are of the same suit and include the king?

D6. A builder is buying sawn timber by the batch. Five planks are selected at random, their lengths are measured and the numbers of knots are counted. A plank will fail if it is more than 5 cm short or has more than four knots. If two planks fail, the batch is rejected. The builder's experience suggests that 1 plank in 15 will be short and 1 in 12 will have more than four knots. (i) What is the probability that a batch will be rejected? (ii) How will this change if a plank must fail both tests before it is rejected?

D7. A rather old-fashioned football manager has 15 forwards, 8 halfbacks, 6 fullbacks and 3 goalkeepers. A team must consist of 5 forwards, 3 halfbacks, 2 fullbacks and 1 goalkeeper. In how many ways can the manager select a team from the available players?

D8. A bus has 32 seats. In how many ways can the occupied seats be distributed if there are the following numbers of passengers: (a) 1, (b) 7, (c) 22 and (d) 31?

D9. A doctor treating a patient for tonsillitis issues a prescription for antibiotics and provides for repeat prescriptions. The probabilities that the infection will be cleared by the first, second, third, etc., prescriptions are $P_1 = 0.523$, $P_2 = 0.216$, $P_3 = 0.198$, $P_4 = 0.063$. Draw a diagram of the probability distribution. What is the probability that (i) a fifth treatment will be required and (ii) the third treatment will not be required?

D10. Using the data from Question D9, determine the expected frequency distribution if during a tonsillitis epidemic the doctor treated 70 patients. How many prescriptions would the doctor expect his or her clinic to issue if there were 350 patients?

D11. A pack of cards having four suits, one of which is hearts, is cut and the suit of the face card is recorded. The process is repeated on four further occasions, the pack being shuffled between each trial. Write down the complete probability distribution for the number of hearts that will be obtained. Verify that the sum of the probabilities is equal to one.

D12. Statingham University buys eight new computers each with a Magic Super Chip. The probability that the chip will fail in any given year is 1/4. Find the probability for each of the following outcomes:

(i)   At the end of the year two computers will have needed new chips.
(ii) At the end of the year more than three computers will have needed new chips.
(iii) Fewer than five computers will have needed new chips by the end of the second year.

D13. Three laboratory assistants are producing molar sulphuric acid in 500 cc jars for a pharmaceutical laboratory. The probability that any assistant produces a substrength jar is 1/10. Each assistant produces one jar. What is the probability that two of the three jars will be substrength?

D14. A business executive takes clients out to lunch every day of the week except weekends. Experience suggests that the executive will get a table at his or her preferred restaurant 1 day in 3. What is the probability that in a three-week period the executive will obtain a table on six occasions? In one year the executive will work 200 days. Calculate the mean and variance of the number of days on which the executive will obtain a table.

D15. In an examination the mean mark was 73 and the standard deviation was 14. Three students had marks of 60, 84 and 73. What were their standardised scores?

D16. The numbers of letters delivered on a weekday to a large office is normally distributed with a mean of 832 and a variance of 324. On a particular day the number delivered was so low that the office manager would have attached a probability of 99.5% to there being more. What was the number of letters delivered?

D17. The mean mass of 500 male students is 70 kg, with a standard deviation of 6 kg. Assuming that the weights are normally distributed, how many students weigh (a) between 64 and 76 kg and (b) between 52 and 88 kg?

D18. The annual earnings of chartered superperiwinklers are normally distributed with a mean of £8034 and a standard deviation of £524. What level of income corresponds to the upper quartile?

D19. The time taken to travel from Statingham to Pwllheli by Lethargic Limousines is normally distributed about a mean of 7 hours 56 minutes with a standard deviation of 1 hour 23 minutes. What percentage of trips can be expected to take between 7 hours and 9 hours 19 minutes?

D20. Gestation times for horses have a mean of 337 days. If they are normally distributed with a standard deviation of 4.5 days, what gestation period do you expect to be surpassed by 2.5% of horses? Consider the precision of your answer.

Group E (Chapters 21–22)

E1. The quality control office for a company purchasing cattle cake is aware that recent batches have been contaminated with water. The officer has been told by the manufacturer that in a sample of one hundred 50 kg bags taken from a day's output of 10 000 bags the lowest quantity of water found was 1.5 kg, the mean was 2.8 kg and in the worst case there were 4.3 kg.

  (i) What is the probable error in the mean?
 (ii) Will the batch be rejected if the criterion is there should be a less than 5% chance that the mean mass exceeds 2.9 kg?

E2. A recent sample survey by the Southampton and Sunderland Building Society reveals that the mean price obtained by sellers of houses in Cornwall was £76 850. The standard error of this mean is quoted as £2640. What is the probability that the true mean (i) exceeds £82 130, (ii) is below £74 210 and (iii) lies between £68 930 and £84 770.

E3. In testing the times by mice to solve a maze, a psychologist discovers that the mice have eaten her results. The psychologist has managed to retrieve only the 95% confidence interval, which was 42.4 s. What was (i) the 99% confidence interval, (ii) the probable error and (iii) the probability that the difference between the true mean and the sample mean x was $-10.6 < \mu - x < 21.2$ s?

E4. A total of 105 passengers boarded a flight from London to New York. The expenditure per head (to the nearest pound) for 30 of the passengers on duty-free goods is shown below. Establish the 95% and 99% confidence intervals of the mean expenditure.

Expenditure (£)	15	8	75	48	0	43	63	114	127	96	15	23	52	55	80
	0	62	12	115	106	78	52	49	50	27	39	46	50	78	42

E5. Sugar is packed into bags by a machine guaranteed to deliver masses with a variance of 0.015 kg, but with a defective mean-setting control. If a random sample of 100 bags from a large consignment has a mean of 1.05 kg, find (i) a 99% two-tailed confidence interval for the mean and (ii) a minimum mean mass that the packers can guarantee with 99% confidence.

E6. The sugar packer in Question E5 scraps the defective machine and installs a new one with unknown operational statistics. The packer takes a random sample of 50 bags from a large consignment and finds a mean mass of 1.04 kg and a standard deviation of 0.013 kg. Find (i) a 99% two-tailed confidence interval for the mean and (ii) a minimum mean mass that the packers can guarantee with 99% confidence.

E7. What would be the answers to Question E6 if the sample was taken from a van load of 400 bags?

E8. An estate agent advertises in a local free newspaper. The estate agent questions 60 households and discovers that 27 throw the paper away without reading it. The newspaper claims a circulation of over 10 000 copies. (i) What are the 95% confidence limits for the proportion of households discarding the paper unread? (ii) The estate agent will cancel the advertisement if there is a 2.5% chance that the proportion of households reading the paper is below 40%. Will the estate agent cancel?

E9. A geneticist has been cross-breeding black and white mice. The geneticist's theories suggest that 20% of the offspring will be white. Of 88 births 22 are white. Will the geneticist's theory be accepted if the criterion is that the expected value lies within the 95% confidence interval of the sample proportion?

E10. In a survey 67 out of 120 policeman said they believed that the speed limit should be lowered to 20 mph. The Minister for Transport claimed there was clear evidence that the reduction was supported by more than half the police force. Consider what criterion might imply 'clear evidence'. Using this criterion would you agree with the minister?

E11. A stocking manufacturer claims that 'one size fits all'. The Stocking Standards Authority will agree, provided it is untrue for no more than 1 person in 1000. Tests on 100 customers reveal a standard deviation of 2.5 cm in foot size. (i) Calculate the standard error of the standard deviation. (ii) What are the 95% confidence limits for the population standard deviation? (iii) If the manufacturer wishes to be 95% sure of meeting the standard, what variation in length will it allow for?

E12. A random sample of 200 shops of all kinds showed that 56 had special offers. (i) Estimate the percentage of all shops that had special offers. (ii) Calculate a 95% confidence interval for this percentage. (iii) What sample size would have been needed if the estimate had to be correct to within 1% with 95% confidence?

E13. (a) If Z_1 is N(25, 6) and Z_2 is N(15, 9) what is $Z_1 - Z_2$? (b) If $\bar{x}_1$ is the mean of a sample of size n_1 and standard deviation s_1, and $\bar{x}_2$ is the mean of a sample of size n_2 and sample standard deviation s_2, write down an expression for the distribution of $\bar{x}_1 - \bar{x}_2$.

E14. On one day in Manchester, 26 motorists were fined a total of £3250 with a standard deviation of £20. The same day in Birmingham, 37 motorists were fined an average of £104 with a standard deviation of £25. Calculate the 95% and 99% confidence intervals for the difference between the two means.

E15. A sports goods manufacturer tests a sample of 145 tennis balls from each of its two factories: 36 fail in one factory and 25 fail in the other. Calculate the 95% confidence limits for the difference between the proportions failing.

E16. Statingham public library conducted a survey five years ago to compare the proportions of male and female borrowers who returned books late in August. The figure for men was 45% and for women 35%. They are about to repeat the experiment using samples of equal size. If the 95% confidence interval of the difference between the proportions is not to exceed 6%, what sample size is required?

Group F (Chapters 22–27)

F1. The Agricola Tyre Company produces heavy-duty tyres for agricultural vehicles. Laboratory tests over many years suggest a mean life of 11 000 miles with a standard deviation of 552 miles. The company is concerned that the quality of its raw material has been declining recently. A test on 36 tyres yielded a mean life of 10 800 miles. Does the company have reason for its concern?

F2. Viking Engineering has drawn a sample of 26 from the daily output of drive shafts. The mean diameter of the sample was 59 mm with a standard deviation of 2.1 mm. The process should deliver a mean diameter of 60 mm. Specify an appropriate significance level and determine whether, from the evidence, the process needs adjustment.

F3. The manufacturers of a pig feed additive claim tests show that on average pigs gained 7.5 kg in mass after two weeks of the treatment and the standard deviation of the results was 2.6 kg. A farmer complained that his 64 pigs gained only 7.05 on average. Was the farmer's complaint justified?

F4. A headteacher discovers a ten-year-old report by the previous head suggesting how the children at that time spent on average 11.5 hours a week doing homework. The standard deviation was 3.6 hours. The headteacher tests 100 of the current pupils and finds that the average time spent on homework is 10.1 hours. Does this suggest at the 1% significance level that the number of hours of homework has declined?

F5. Chariot Buses have increased their standard fare from 60p to 70p. Records for the last year before the change show that the average number of seats sold per day was 4515 with a standard deviation of 545. During the first 100 days after the price change, they sell an average 4472 seats per day. At the 1% level of significance does the evidence suggest a reduction in sales following the price change?

F6. An assayist is testing ore samples that are claimed to have come from the Neptune nickel mine. The 257 samples yield on average 3.6 g of nickel per kilogram of ore. The variance is 1.80 g. Neptune's geological surveys estimate the yield at 3.4 g per kilogram.

(i) Conduct a one-tailed test at the 1% level of significance to test whether the samples have a higher yield than those expected from Neptune.
(ii) Conduct a two-tailed test at the 1% level of significance to assess the likelihood that the samples came from the Neptune mine.

F7. In a test to determine whether rats have a sense of smell, a biologist places food in a maze. The mean time taken to locate the food from many trials is 87 s. A smell enhancer is added. Fifty more trials are conducted and the mean time to reach the food is 85.5 s with a variance of 39.5 s. Use a 5% level of significance to determine whether smell has an impact.

F8. Statingham Vegetable Gardeners grow cucumbers, which are packed into boxes of 100. The growers assert that the mean length of the cucumbers is 30 cm, measured along the outside curve from tip to tip, and the lengths have a standard deviation of 1.5 cm. An opened box of cucumbers is returned with the complaint that the average length is only 29.5 cm. Test the hypothesis that some of the cucumbers have been replaced with shorter ones.

F9. In recent months the mean breaking strength of Statingham Stitchers SuperFine Thread has been 19.54 g, with a standard deviation of 2.80 g. Yesterday a sample of 36 lengths showed a mean breaking strength of 17.92 g. Does this suggest the quality of the thread has deteriorated?

F10. An environmental lobby group has reported that 15% of all the households in the United Kingdom own two or more cars. Your own survey of 100 households in Cleveland produces only 12 such households. Using a 5% level of significance, does the evidence suggest (i) that the lobby group exaggerated, (ii) that Cleveland is not representative of the United Kingdom as a whole?

F11. A wholesaler buys biscuits packed in cartons of 200 from a manufacturer. He will reject the batch if the number of broken biscuits exceeds 10%. He selects one carton and counts 30 broken biscuits. Should he reject the batch if he uses a 1% significance level?

F12. What level of significance would be appropriate in the following cases?

(i) A political party investigates whether its proportion of the vote is likely to exceed 0.4.
(ii) A market researcher measures market penetration when 8% is critical to the success of the product, but the researcher is conscious of the research budget.
(iii) A professor of pharmacy is about to publish a paper claiming to have identified an anticancer drug that will cure one-third of all patients treated?

F13. The editor of the *Statingham Declaimer* quotes government statistics showing that one-third of all marriages end in divorce during the first 10 years. 'Not in my flock!' claims the Bishop. In fact, a survey in Statingham shows that of a sample of 50 couples married 10 years ago, only 12 had become divorced.

(i) Does the evidence support the Bishop at the 5% significance level?
(ii) If the divorce rate in Statingham is 24%, what sample size would have been required for there to be only a 2.5% chance of the Bishop being wrong?

F14. If in 1947 the median height of all adult women in Scotland was 160 cm, and in 1997 if 55 out of a sample of 100 adult women exceeded this height, does the evidence suggest that Scottish women are becoming taller?

F15. In Question F14 how many women in the sample would have needed to be taller than 160 cm for the result to be significant at the 10%, 5% and 1% levels?

F16. Doggomix Biscuits have recently been launched in all major supermarkets. To be successful the product must achieve a market share of 12%. One supermarket reports that Doggomix sales accounted for 5850 out of the first 50 000 sales of dog biscuits following the launch. Does the evidence suggest that the market target will be reached?

F17. As part of a traffic survey, 100 motorists record a mean journey time to work of 29.3 minutes with a variance of 19.6 minutes. Six months later the mean journey time is 30.7 minutes with a variance of 19.9 minutes. Does the survey suggest that journey times have increased? State your assumptions.

F18. A company makes ball-bearings in factories A and B. A random sample of 115 bearings made in A had a mean diameter of 7.6 mm. The population variance was 1.5 mm^2. A sample of 95 bearings from B had a mean diameter of 7.3 mm and the population variance was 0.8 mm^2. Is there a 5% significant difference between the mean diameters?

F19. Dexterity tests on two groups of patients award integer scores out of 100. A sample of 75 from group A had a mean score of 58 with a variance of 64. A sample of 60 from group B had a mean score of 52 with a variance of 72.25. Assuming that the two groups have a common population variance, do the results suggest a difference between the groups? State your assumptions.

F20. A fertility treatment is designed to improve the insemination success rate of pigs. A sample of 102 litters from pigs undergoing the treatment had an average size of 9.8 with a sample variance of 6.3. A sample of 86 litters from pigs not undergoing the treatment had a mean size of 8.2 and a variance of 6.8. Do the results support (at 5%) the hypothesis that a side effect of the treatment is an increase in the size of litter?

F21. A criminologist has been researching recidivism rates. The criminologist found that 63 out of 100 people serving first-offence prison terms of less than 3 years reoffended on release; 70 out of 140 with a first-term sentence in excess of 3 years reoffended.

(i)   Does the length of the first-term sentence affect recidivism rates? Use a 5% significance level.
(ii) Should you use a continuity correction? Would it affect the result?
(iii) Would the assumption of a common population variance have affected the result?

(**Group G**) (Chapter 28)

G1. A hovercraft makes the journey from Scotland to Northern Ireland 2000 times a year. The journey time T is N(37, 42.25). Passengers are compensated if the journey takes longer than 48 minutes. On how many journeys will passengers receive compensation? The company was fined for exceeding safe speed limits on 40 journeys. What journey time implies an unsafe speed?

G2. Statingham fire brigade has determined that the amount of water required to put out a fire is normally distributed with a mean of 3000 gallons and a standard deviation of 350 gallons. How much water must an engine carry if (i) the probability of having to call a second engine must be below 0.06, and (ii) the probability of putting out a fire without calling on other engines must exceed 0.85?

G3. Pneuromol is a steroid used for the treatment of asthma. The manufacturer's tests demonstrate that, following one dose, the time taken for the patient's breathing to return to normal is normally distributed with a mean of 8.6 minutes and a standard deviation of 36 seconds. A stockist tests 40 inhalers from the shelves and finds that the mean time to recovery is 8.8 minutes. What is the significance probability that the product is correctly marked as Pneuromol? What difference would it have made had the 36 seconds been the standard deviation of the sample rather than the population?

G4. The probability that a person will get a job within three weeks of being made unemployed is 0.42. If 150 people are interviewed three weeks after losing a job, what is the probability that (i) 60 will have obtained work, (ii) fewer than 56 have found work and (iii) fewer than 75 remain unemployed.

G5. The handicapper at Statingham racetrack is so good that the chance of the favourite winning a race is always 0.6. On a day on which there were eight races, what is the significance probability that the favourite won on only three occasions?

Group H (Chapter 29–30)

H1. Scaffold tubes are being cut to a length of 8 m but the equipment is worn and lengths vary. A sample of six tubes is shown to have a mean length of 8.07 m with a standard deviation of 0.06 m. Calculate the 95% and 99% confidence limits. How would they have changed if (i) the standard deviation had been 0.09 m and (ii) a sample of 16 tubes had yielded the same standard deviation?

H2. Using a 5% level of significance with each of the three cases in Question H1, should the hypothesis that the true mean is 8 m be rejected? Would your answer have been affected had a 1% level of significance been chosen?

H3. A sheep shearer in the outback knows that fleeces have masses that are normally distributed. The last two fleeces have masses of 2.5 and 3.3 kg. Calculate the 95% and 99% confidence intervals.

H4. A photograph of a set of supermarket shelves was projected onto a screen and 10 customers were asked to press a button on seeing a particular brand of baked beans. Their response times were approximately normally distributed with a mean of 6.5 s. The 95% confidence limits were 5.95 and 7.05 s. What was the variance of the response times?

H5. The test in Question H4 was repeated with a second group of 16 customers. The mean response time was 7.1 s with a variance of 0.6 s. Is there a difference between the reaction times of the two groups at the 5% significance level?

H6. Statingham chamber of commerce states that the average shop rent is £114.75 per square metre. Seven shops were recently let at £103.50, £119.25, £113.85, £117.45, £127.35, £107.75 and £122.85. Does this support the statement?

H7. The chamber of commerce also claimed that rents in Statingham were much lower than in Upper Sternum, where a sample of 11 recently negotiated rents showed a mean of £122.04 per square metre with a variance of 79.38. Were they correct?

H8. A company is choosing between two makes of safety valve. Wanting a valve that has a more or less predictable life, it will select the valve with the lower variance. If there is no difference between variances, it will go for the valve with the higher mean life. In all its tests it uses a 5% significance level. A sample of 61 A-valves has a mean life of 127 days with a variance of 49 days^2 whereas 31 B-valves have a mean life of 164 days with a variance of 73.5 days^2. Which valve will it choose?

H9. Every week samples of freshly produced carbon steel are tested for carbon content. In week 1, 30 specimens had a mean content of 6.1 g per kilogram, with a standard deviation of 2.57. In week 2, 25 specimens had a mean content of 6.0 g per kilogram with a standard deviation of 1.70. Has there been a significant reduction in the reliability of the production process?

H10. A person who shops every Saturday morning always arranges to be taken home by taxi. The shopper wants to give the taxi firm as accurate a collection time as possible, but this is difficult because of delays at the checkout. The shopper visits two supermarkets on each of six occasions and measures the standard deviations of the two sets of waiting times. If they are 4.8 minutes and 11.1 minutes, is there any evidence that queuing time in one supermarket is more predictable than the other? Use a 5% significance level.

Group I (Chapters 31–32)

I1. In an experiment to test the hypothesis that restaurant prices in the United Kingdom increase linearly with the distance between the centres of adjoining tables, a researcher determined that $r = 0.55$ from a sample of 30 restaurants. Examine the significance of r at the 1% level.

I2. An American university conducted a similar study to Question I1 in Philadelphia and obtained $r = 0.45$ from a sample of 25 restaurants. It claimed this implied that America was a more egalitarian society. Was the American correlation coefficient significantly lower than the UK coefficient? Does it support the American conclusions? State any assumptions that you make.

I3. Australian government statistics indicated a correlation of 0.62 between household income and expenditure on imported wines using 1991 data. In 1996 a survey of 100 households produced a value of 0.66. Does this suggest a change in the correlation over the five-year period?

I4. Using the data from Question C4 and assuming the children were a sample drawn randomly from the school, determine whether the values of b and r were significant at the 5% level and determine the 1% prediction limits for the estimated number of errors made by a child who watched television for 31.5 hours a week.

I5. Comment on the following results:

(a) In Utopia the correlation between the number of cots purchased and the number of marriages ending in divorce each year is 0.75.

(b) In a study of accidents on motorways, the correlation between the number of accidents ending in a fatality and the income of the driver was −0.62.

(c) Every January 4 a newspaper reports on the previous year's value of (i) UK exports and (ii) lorries purchased. An economist finds a correlation of 0.9% for the 21 years 1970–1990.

I6. A sociologist is investigating the relationship between the starting salaries of a sample of this year's graduates and the distance they are prepared to move away from the university in accepting a job. A similar study two years ago produced a significant correlation coefficient of 0.7. The sociologist does not believe that it will have changed by very much. What sample size is required to ensure this year's results will be significant at the 1% level? By how much would it change if significance was at the 5% level? What size would have been required for 1% significance if the correlation last time had been 0.45?

I7. Assuming that the farms in Question C5 were a sample of all the farms in Cheshire and Shropshire, was the regression coefficient significant at the 5% level?

I8. In a customer survey, a telecommunications company has measured the number of telephone calls made from each of 10 office blocks during a five-day period and compared the result with the number of people employed in each block. (i) What do the data reveal about the nature of the relationship between calls and employees? (ii) How many calls would you estimate would be made from an office with 87 employees?

Telephone calls	3350	1170	2250	2750	2825	2200	2525	2625	1650	3075
People employed	104	67	80	77	96	60	98	83	54	81

 Group J (Chapters 34–38)

J1. A university lecturer is aware that within the university approximately 40% of students would exceed a mark of 60% in any given paper. The lecturer reviews a set of results in the order in which they were marked. If *O* is a mark of 60% or more and *U* is a mark of under 60%, does the evidence suggest that the lecturer was influenced by his or her knowledge? State any assumptions.

O UUUU OO UUUUU OOOO UUUUU OOO U OO UU O

J2. Freeziveg Ltd pack frozen peas in 480 g bags. They weigh one pack every 10 minutes and record *O* or *U* according to whether or not it is overweight. At the end of an eight-hour shift, 16 bags were overweight and there were 10 runs. Do the data suggest that a systematic error is occurring?

J3. A local authority housing department was aware that one-third of all of its properties suffered from damp. A random sample of 120 houses revealed 50 that were damp. When the data were reviewed in the order in which inspections were carried out, it was found there were 41 runs. Stating all your assumptions, test the sample for randomness.

J4. Ten yachts circumnavigated the Isle of Wight in the following times in minutes:

165   187   188   171   199   182   173   192   178   188

If there were 200 yachts in the regatta, do the data support the hypothesis that the median time taken was 188 min?

J5. Nine press correspondents were asked to score Bloggs in a campaign for the presidency, giving 1 as lowest and 4 as highest. Arranged in alphabetical order, they gave scores of 1, 4, 3, 4, 4, 2, 3, 1 and 2. After being shown a campaign commercial, they gave new scores of 1, 3, 2, 2, 3, 1, 2, 2 and 1 (in the same order). Find at 5% whether the commercial had any effect.

J6. Following a forecast of a sugar shortage, a psychologist investigated whether women were more likely to stock up than men. The psychologist obtained for each of 50 stores the amounts of sugar purchased by one male and one female customer. On 33 occasions the female customer had purchased the greater quantity and on 2 occasions the purchases were the same. Do the data suggest greater prudence among women? Use a 1% level of significance.

J7. Ten households containing two children are selected at random and the pairs of children are randomly allocated to groups A and B. A drink containing a vitamin that is thought to aid concentration is given to group A. The children are asked to hit balls back over a net using a tennis racquet. The time taken to return 50 balls successfully is recorded. Use a Wilcoxon test with a 5% significance level to investigate the effect of the drink.

Pair	P	Q	R	S	T	U	V	W	X	Y
Group A	12	9	12	8	16	14	9	6	7	11 (mins)
Group B	18	17	11	13	13	8	>20	14	12	6 (mins)

J8. If in Question J7 there had been 50 households, 5 of which had equal ranks, and the sum of the positive differences on subtracting the group B ranks from the group A ranks was 70, would the result have been different?

J9. Fifteen clutch mechanisms were chosen at random from the stock of a parts distributor and tested to failure in a laboratory which recorded their equivalent driving miles. Nine of them, made by car manufacturers, had lives of 27, 52, 46, 66, 72, 55, 66, 64 and 63 thousand miles. Six made by specialist clutch manufacturers had lives of 55, 43, 11, 61, 41 and 47 thousand miles. Use a Mann–Whitney U-test at 5% to find whether durability depends on the kind of maker.

J10. Forty athletes were selected at random: 24 had been training at altitude for the past three months, the remaining 16 had trained at sea level. Following intensive physical tests in the laboratory, the pulse rates of the 40 athletes were measured and ranked. The sum of the ranks for the altitude-trained athletes was 411. Does this suggest that altitude affects the athletes' physiology?

J11. Following a ban on smoking in public places, a test was conducted of the association between smoking and the frequency of dining out. Seventeen smokers never dined out, 34 did so occasionally, 16 frequently and 8 very frequently. The corresponding data for non-smokers were 6, 34, 41 and 10. Is there evidence of an association at the 5% significance level?

J12. A shoemaker asks 273 randomly chosen men to choose one free pair of shoes. The colours are shown below, classified by age of chooser. Should the shoemaker take account of the age distribution of shoppers when deciding on output levels for each colour?

	Black	Brown	Grey	Other
Under 25	19	39	40	15
25–45	36	34	12	1
45–65	28	22	6	1
Over 65	8	11	1	0

J13. Ninety patients with sore throats were randomly divided into two equal groups: one to receive a new treatment and the other a placebo. Twenty-six of those getting the new treatment reported an improvement within one hour and another 10 in 1–2 hours. Of those getting the placebo, 16 reported improvement within an hour and another 18 in 1–2 hours. Examine the effectiveness of the treatment (at 5% significance). Assuming you had the opportunity to repeat the test with a different group of patients, would you? If so, why?

J14. Imperfect pumps are classified as (a) scrap, (b) repairable and (c) seconds. A manufacturer found that imperfect pumps made on a Monday could be noted as (a) 38, (b) 27 and (c) 35. Those made midweek were noted as (a) 50, (b) 65 and (c) 95. Friday's imperfections were (a) 62, (b) 58 and (c) 70. Are day of manufacture and classification associated? In a second test, group (b) had very few pumps, so groups (b) and (c) were combined. This gave $\chi^2 = 4.81$. How can this result be used? Stating your assumptions, what conclusions would you now draw?

J15. Twenty-six of 40 randomly chosen doctors, and 13 of 40 randomly chosen architects failed tests for hypochondria. Using 5% significance, decide whether doctors are more likely than architects to be hypochondriacs.

J16. Six men and one women in a post office queue said they were satisfied with the service. But two men and five women said they were not satisfied. Is there enough evidence to suggest that men are more easily satisfied than women? State your assumptions.

J17. Polygon Media Consultants test their new advertising campaign for Midnight Chocolate by surveying potential purchasers before and after the campaign. Before the campaign 79 people said they would buy, but after the campaign 48 of them had changed their minds. Also, before the campaign 79 people said they would not buy, but after it 46 changed their minds. Did the campaign affect intentions? (Use 5% significance.)

J18. Twenty-seven golfers in the club bar were asked if they approved the secretary's choice of refreshments. Eleven did, and the rest did not. Then it was explained that any change would lead to higher subscriptions. A new vote showed that four of the approvers had changed their minds, as had five of the disapprovers. Did the explanation influence opinion? (Use 5% significance.)

J19. Statingham's MP heard that the new tax on hamsters was proving unpopular in the Calculators' Arms. The MP visited Harry and talked to his customers. After the visit, 9 people continued to approve the new tax and 12 maintained their opposition; 19 were persuaded to change over to support the measure and 52 decided to change over to the opposition. Using a 10% significance level, assess whether the MP's speech made a difference.

J20. A potential TV advertiser wishes to assess the relationship between age and programme preference on the basis of the following listing of preferences. Under 25: Eastenders 37, Coronation Street 14, Brookside 22. Aged 25–39: Eastenders 26, Coronation Street 35, Brookside 22. Aged 40 and over: Eastenders 12, Coronation Street 22, Brookside 10. Use the contingency coefficient to help the potential advertiser.

J21. Crosstown Railways ask one passenger travelling on a peak time service and one travelling at off-peak to rank certain performance characteristics in order of importance. The results are below. Calculate Spearman's rank correlation coefficient and test its significance at the 5% level.

	Comfort	Safety	Security	Speed	Reliability	Frequency	Cleanliness
Peak time	5	4	7	3	1	2	6
Off-peak	3	7	5	1	4	2	6

J22. A company uses a standard typing test when interviewing secretaries. To assess the effectiveness of the test, they compare the ranks (1 = best) of the number of mistakes each of 10 current employees makes and the average mark they obtained in the secretarial examination. Using the Spearman rank correlation coefficient, do the data indicate any correlation at the 5% level?

Mark on test	83	72	45	86	57	52	58	80	80	50
Current rank	1	4	7	3	4	10	8	6	2	9

J23. Would the use of Kendall's rank correlation coefficient support the conclusion that you reached in Question J22?

J24. Eight buildings are being considered for the Great Architecture Award. Judge A ranks them 2, 1, 3, 4, 4, 6, 7, 8 and judge B, taking them in the same order, ranks them 3, 4, 6, 7, 2, 5, 1, 8. Is there agreement between the judges at the 5% level of significance?

 Group K (Chapters 39–43)

K1. A person travels to work on the same train every day. Over several months the worker has observed that the number of passengers with whom he or she shares a 12-seat carriage is uniformly distributed. What is the probability that (a) the worker travels alone, (b) the carriage is full and (c) the worker has between 3 and 6 travelling companions?

K2. A company employs 20 sales representatives. The probability of any given number exceeding their sales target on any day is uniformly distributed. What is the expected number who will exceed their target tomorrow? Calculate the variance of the number exceeding their target.

K3. A student drives home to get some laundry done every Saturday. The student must pass through 10 traffic lights. The probability of a light being on green when the student reaches it is 0.4. (i) What is the probability that the student will first be stopped at the fourth light? (ii) What is the probability that the student will make it home without being stopped? (iii) What is the mean number of times that the student can expect to be stopped?

K4. There is a 2% chance that the operator of an industrial press will not arrive for work on any day and a 5% chance that the machine will not start when switched on. Calculate the mean and variance of the number of consecutive days on which output can be expected.

K5. The number of customers visiting a chemist in one morning is uniformly distributed, taking values between 45 and 60. The number of drug company representatives calling is similarly distributed and takes values between 1 and 6. Calculate the mean and variance of the total number of people arriving at the shop.

K6. The casting process at a mine produces ingots of mass 25 kg. Errors are rectangularly distributed between 0.5 kg under and 1.6 kg over. Calculate the mean mass of an ingot and the variance of the errors.

K7. A stores supervisor receives orders at time intervals that are uniformly distributed between 3 and 30 minutes. What is the chance the supervisor can take a 10-minute tea break without interruption?

K8. A car mechanic performs the same task many times in a week. The time taken depends upon the condition of the vehicle. On a new car the work should take 20 minutes, and after 45 minutes the task is abandoned and the component is replaced. If the completion time probability falls uniformly, calculate the mean and variance of the distribution.

K9. The Payalot supermarket has 26 checkouts. In any 10-minute period the number of male customers arriving at all checkouts is randomly distributed with a mean of 57. The number of female customers is similarly distributed with a mean of 78.

 (i) Calculate the probability that more than 6 men will arrive at one checkout during the next 20 minutes.
 (ii) Calculate the probability that between 4 and 7 women will arrive at two specified checkouts during the next 10 minutes.
 (iii) Calculate the probability that fewer than 4 people will arrive at one checkout during the next 10 minutes.

K10. In an outbreak of measles among children under age 10, doctors report that the number of cases per square kilometre is randomly distributed with a variance of 3.5 cases. What is the probability that (i) there will be 3 cases in $1\,km^2$, (ii) there will be between 7 and 9 cases in $1\,km^2$ and (iii) there will be fewer than 2 cases in a $2\,km$ square?

K11. In Question K9 what is the modal value for (i) men arriving at one checkout in a 10 minute period and (ii) women arriving at one checkout in a period of half an hour?

K12. The number of letters delivered per address per day by Statingham General Post Office is randomly distributed with a mean of 4. The number of packages is randomly distributed with a mean of 1.6. (a) What is the probability that (i) one house will receive more than 3 items on one day and (ii) one house will receive 2 letters and 2 packages in one day? (b) What is the modal value of the total of letters and packages delivered to one address per day?

K13. A chemical company is testing an insecticide for the control of the greater corn beetle. They observe that the number of beetles per square kilometre in a cornfield is randomly distributed with a mean of 35. They mark four $1\,m$ squares in the field. Assuming that the treatment fails, calculate the probability that (i) one of the squares contains 22 beetles, (ii) one square contains 30–40 beetles and (iii) the four squares contain less than 130 beetles.

K14. A company manufacturing circuit boards for computers expects a 2.5% defect rate. In a random sample of 80 boards, what is the probability that 3 will be defective?

K15. In Question K9 what is the probability that in one 10-minute period (i) between 40 and 60 men arrive at the 26 checkouts and (ii) fewer than 60 women arrive at the 26 checkouts?

K16. If x is exponentially distributed with mean 0.1, calculate the probability that (i) $x < 3$ and (ii) $x > 5$.

K17. In Question K16 what is the value of x if $F(x) = 0.3$?

K18. A word processing operator notes that the package crashes five times an hour on average, the number of occasions being randomly distributed. What is the probability that (i) the next failure will occur within the next 6 minutes, (ii) there will be no failure within the next 15 minutes and (iii) the interval between the next two failures will be 10 minutes?

K19. On average an estate agent sells three houses and four flats in a six-day working week. (a) Calculate the probability that (i) the estate agent will sell a property within the next two days and (ii) the estate agent will not sell a house within the next three days. (b) What are the mean and variance of the time interval between sales?

K20. A biologist determines that the distance between plants infested by caterpillars is randomly distributed with a mean of 45 cm. Determine the probability that two infested plants are 60–70 cm apart.

K21. A haulage company has the contract to deliver all the products of a large brewery. Late deliveries are randomly distributed and occur on average five times during 40 working days. They pay compensation if two late deliveries occur within T days of each other. If the probability of a compensation payment is not to exceed 5%, what is the value of T?

 **Group L** (Chapter 44)

In all questions assume a 5% level of significance
L1. In cross-pollinating a certain type of plant, the varieties of offspring are hypothesised to occur in the ratio 8:4:2:1. In a test involving 300 offspring the observed frequencies were 148, 94, 32, 26. Do the data support the hypothesis?

L2. A maker of watering cans sees that last week the number of faults reported on different days were as follows: Monday 27, Tuesday 11, Wednesday 15, Thursday 13 and Friday 24. Do the figures support the maker's belief that Mondays and Fridays are bad days?

L3. An optician is testing whether short-sightedness is related to eye colour. The optician has evidence to suggest that eye colour is distributed according to the ratios 20 blue:18 brown:9 grey:3 green. The last 40 patients for whom the optician prescribed appropriate spectacles included 14 with blue eyes, 12 with brown, 9 with grey and 5 with green. Is there evidence of a relationship?

L4. The number of accidents per week on a 50-mile section of the M6 is assumed to be Poisson distributed. Do last year's records support the hypothesis?

Accidents per week	0	1	2	3	4	5	6	7
Frequency	6	8	16	13	4	2	2	1

L5. Groups of eight children are drawn at random from each of 220 schools and given a maths test. It is suggested that the numbers passing in each group will have a binomial distribution with a success proportion of 0.4. Do the following data confirm this?

Numbers passing	0	1	2	3	4	5	6	7	8
Frequency	21	35	58	42	36	22	4	2	0

L6. In a laboratory experiment, 150 groups of six mice were given a choice of two food sources, A and B. For each group the number choosing A was noted and used to prepare the following table. Does it support the hypothesis that the mice did not discriminate between the sources?

Mice choosing A	0	1	2	3	4	5	6
Frequency	6	9	19	25	47	36	8

L7. The time taken to complete the assembly of a carburettor by hand is assumed to have a mean of 45 s with a variance of $25\,s^2$. Do the data obtained from tests on 320 workers on an assembly line support the hypothesis that the time is normally distributed? State your assumptions.

Time taken (s)	< 30	30–33	33–36	36–39	39–42	42–45	45–48	48–51	51–54	> 54
Frequency	5	8	22	68	70	56	30	25	18	18

L8. The masses of 145 kittens taken at birth are as follows. Do the data support the hypothesis that the masses are normally distributed.

Mass (g)	< 80	80–85	85–90	90–95	95–100	100–105	105–110	> 110
Frequency	1	6	20	35	35	28	14	6

Answers

Group B

1. 60 **2.** (a) (i) 31.3, 34.5 (ii) 33.0 (b) 18.4 **3.** (a) 21.2 (b) 190.7
4. paint shop £13 562.67, 0.14; accounts £13 396, 0.66 **5.** (a) £288.36 (b) £317
6. $\bar{x} = 15.1, \sigma = 6.9$; mode $= 13$ **7.** 5.5% **8.** £23.63 **9.** 1.31
10. 2.83, -0.17, 8.93, 14.15 **11.** £143.37, £144.57, £35.51
12. major $= 16$–18, minor $= 8$–10; 16 **13.** 0–4 end mode; 30–34 minor mode with modal
value 32. **14.** without conversion: Man $= 16.2$, Mel $= 19.3$, Det $= 17.8$; with Detroit
ranges: Man $= 17.8$, Mel $= 20.0$ **15.** graphical solutions
16. (a) 54.0, 52, 63 (b) 53.9, 53.1, 49.5 **17.** 1.42, 1.00 **18.** graphical solution
19. 39 **20.** (i) 7.79, 8.16 (ii) 10.77, 10.9 (iii) 0.46, 0.59
21. (i) £13.05 (ii) 6.254% (iii) 6.525% **22.** 19.5 **23.** A 25.13, B 25.7
24. (i) 2.3% (ii) 6.49 million **25.** 0.084 micrograms per litre, water is safe
26. $\sigma = £23.2$, QC $= 0.09$ **27.** (i) 176 (ii) 293 (iii) 64.8 **28.** SIQ $= 31.7$
29. MD(1) $= 4.12$, MD(2) $= 2.49$ **30.** 8, 6, 36 **31.** $\bar{x} = 328.4, \sigma = 95.6$
32. (a) 4.85 (b) 1.94 (c) 7 **33.** Birmingham **34.** year 1: 1.31, -0.20, 0.20, 1.18;
year 2: -2.00, -0.01, 0.69, 2.45; **35.** (i) 1168.0, 1203.4 (ii) yes, no, 33.3 (iii) £550, £150
(iv) reduced σ to 33.68

Group C

1. 50–80 approx. **2.** (i) 230–330 approx. (ii) 15.9–17.6% approx. **3.** graphical
solution **4.** $y = 0.9 + 1.37x$ **5.** $y = 13.97 - 0.82x, r = -0.81$ **6.** $y = 6.62 + 1.74x$,
$r = 0.75$ **7.** $x = 2.0 + 0.32y, r = 0.75$ **8.** $r = 0.50$ in both cases

Group D

1. (i) 504 (ii) 9 **2.** (i) 0.25 (ii) 0.083 (iii) 0.042 **3.** (i) 0.317 (ii) 0.295
4. (i) 0.000 001 (ii) 0.818 (iii) 0.999 **5.** (i) 0.002 (ii) 0.000 77 **6.** (i) 0.156
(ii) 0.0046 **7.** 7 567 560 **8.** (a) 32 (b) 3 365 856 (c) 129 024 480 (d) 32
9. (i) zero (ii) 0.739 **10.** 36.61, 15.12, 13.86, 4.41; 690 **11.** 0.2373, 0.3955, 0.2637,
0.0878, 0.0146, 0.0001 **12.** (i) 0.3115 (ii) 0.1141 (iii) 0.9723 **13.** 0.027
14. 0.178; 66.7, 44.2 **15.** -0.93, 0.79, zero **16.** 785 **17.** (a) 341 (b) 498
18. £8387.7 **19.** 59% **20.** 346 days

Group E

1. (i) 0.047 (ii) no, crit lim $= 1.645$ s.e. **2.** (i) 2.28% (ii) 15.87% (iii) 99.74%
3. (i) 55.8 s (ii) 7.3 s (iii) 81.1% **4.** 95%: £43.1–64.9; 99%: £39.7–68.3

5. (i) 1.05 ± 0.32 kg (ii) 1.02 kg **6.** (i) 1.04 ± 0.0048 kg (ii) 1.0356 kg
7. (i) 1.04 ± 0.0045 kg (ii) 1.0359 kg **8.** (i) 0.45 ± 0.125 (ii) 2.5% one-tailed, 32.5%,
cancel **9.** 95% interval $= 0.25 \pm 0.09$, accept **10.** clear evidence is $p = 0.5$ lies outside
99% interval, critical limits include 0.5, reject statement **11.** (i) 0.177 cm
(ii) 2.5 ± 0.35 cm (iii) 7.08 cm below to 9.37 cm above mean **12.** (i) 28% (ii) $\pm 6.2\%$
(iii) 7746 **13.** (a) N(10, 15) (b) $N(\mu_1 - \mu_2, [s_1^2/(n_1 - 1)] + [s_2^2/(n_2 - 1)])$ where
$\mu_1 = E(\bar{x}_1), \mu_2 = E(\bar{x}_2)$ **14.** 95%: £21 $\pm$ 11.33; 99%:
£21 $\pm$ 14.91 **15.** 0.076 ± 0.1 **16.** 507

Group F

1. 95% upper limit $= 10\,980$ miles, so investigate quality **2.** upper 95% limit $= 59.87$ mm,
so adjust process **3.** 90% upper limit $= 7.59$ kg, so complaint not substantiated
4. one-tailed 99% upper limit $= 10.95$ h, suggests decline **5.** one-tailed 99% upper
limit $= 4600$ seats, so does not indicate reduction **6.** (i) lower limit $= 3.40$ g/kg, do not
reject H_0 (ii) CI $= 3.6 \pm 0.22$ g/kg, do not reject H_0 **7.** upper two-tailed limit $= 87.3$ s, so
do not reject H_0 **8.** one-tailed upper limit $= 29.75$ cm, so reject H_0 **9.** one-tailed 5%
upper limit $= 18.71$ g, so reject H_0 **10.** (i) one-tailed upper limit $= 17.4\%$, so do not
reject H_0 (ii) two-tailed upper limit $= 18.4\%$, so do not reject H_0 **11.** $z = 2.00$, so do not
reject H_0 **12.** (i) 10% (ii) 1% (iii) 0.1% **13.** (i) $z = 1.54$, so do not reject H_0
(ii) 81 **14.** 5% one-tailed $z = 1.00$, so do not reject H_0 **15.** 10%, 57; 5%, 59; 1%, 62
16. two-tailed 5% upper limit $= 11.98\%$, so reject H_0 **17.** 5% two-tailed $z = 2.22$, so
reject H_0 **18.** two-tailed 5% $z = 2.05$, so reject H_0 **19.** 5% two-tailed $z = 4.2$, so
reject H_0 **20.** one-tailed $z = 1.6$, so do not reject H_0 **21.** (i) two-tailed $z = 2.11$, so
reject H_0 (ii) yes, no (iii) $z = 2.05$, no effect on result

Group G

1. 91, 50.36 min **2.** (i) 3545 gal (ii) 3363 gal **3.** 0.0174, 0.0202 **4.** (i) 5.8%
(ii) 10.8% (iii) 98.1% **5.** 0.72

Group H

1. 95%: 8.00–8.14 m; 99%: 7.96–8.18 m (i) 7.97–8.17 m; 7.91–8.23 m (ii) 8.04 m–8.10 m;
8.03–8.11 m **2.** 5% $t = 2.61$, reject H_0; $t = 1.74$, do not reject H_0; $t = 4.52$, reject H_0; 1%
$t = 2.61$, do not reject H_0; $t = 1.74$, do not reject H_0; $t = 4.52$, reject H_0
3. 95%: ± 1.30 kg; 99%: ± 6.52 kg **4.** 0.53 s **5.** two-tailed $t = 1.93$, so do not reject H_0
6. two-tailed 5% $t = 0.34$, so do not reject H_0 **7.** one-tailed 5% $t = 2.58$, so reject H_0
8. $F = 1.5$, so do not reject H_0; $t = 2.87$, so reject H_0 **9.** use 1% level, $F = 2.285$, so do
not reject H_0 **10.** $F = 5.35$, so reject H_0

Group I

1. $t = 3.485$, so reject H_0 **2.** $t = 0.47$, so do not reject H_0 **3.** one-tailed 5% $t = 0.67$,
so do not reject H_0 **4.** $t = 6.35$, so reject H_0; $t = 11.9$, so reject H_0; 0–95 **6.** 11, 7, 32
7. one-tailed 5% $t = 1.61$, do not reject H_0 **8.** (i) $T = 391.6 \pm 26.38P$, $t = 0.81$ (ii) 2295

Group J

1. $N_{min} = 11$, so reject H_0 2. $z = -3.57$, so reject H_0 at 1%   3. $z = 3.36$, so reject H_0
4. $p = 0.2539$, so do not reject H_0 5. $p = 0.035$, so do not reject H_0 6. one-tailed
1% $z = 2.454$, so reject H_0 7. two-tailed 5% $T = 13.5$, so reject H_0 8. two-tailed 5%
$T = 5.05$, so reject H_0 9. 5% $U = 11.5$, so do not reject H_0 10. 5% $z = 2.237$, so
reject H_0 11. $\chi^2 = 14.95$ with $v = 3$, so reject H_0 12. $\chi^2 = 59.7$ with $v = 4$, so reject
H_0 13. $\chi^2 = 4.86$ with $v = 2$, so do not reject H_0 14. 5% level $\chi^2 = 8.16$ with $v = 4$,
so do not reject H_0; 5% level $\chi^2 = 12.87$ with $v = 6$, so reject H_0 15. $\chi^2 = 12.22$, so reject
H_0 16. $p = 0.1026$, so do not reject H_0 17. $\chi^2 = 0.043$, so do not reject H_0
18. $p = 0.5$, so do not reject H_0 19. $\chi^2 = 15.33$, so reject H_0 20. $C = 0.26$;
$\chi^2 = 15.15$ with $v = 4$, so reject H_0 at 5% level   21. $r_S = 0.46$, so do not reject H_0
22. $r_S = 0.78$, so reject H_0 23. $\tau = 0.59$, so reject H_0 24. either $\tau = 0.11$ or $r_S = 0.18$,
so do not reject H_0

Group K

1. (a) 0.0833   (b) 0.0833   (c) 0.3333   2. 10, 36.63   3. (i) 0.0864 (ii) 0.006 (iii) 4
4. 14.5, 195.6   5. 56, 24.17   6. 25.6 kg, 0.368 kg^2 7. 0.37   8. 28.3 min,
34.7 min^2 9. (i) 0.155 (ii) 0.593 (iii) 0.24   10. (i) 0.216 (ii) 0.062 (iii) 0.359
11. (i) 2 (ii) 8 and 9   12. (a) (i) 0.81 (ii) 0.038   (b) 5   13. (i) 0.006 (ii) 0.648
(iii) 0.199   14. 0.1784   15. (i) 0.669 (ii) 0.021   16. (i) 0.259 (ii) 0.607
17. −10.99   18. (i) 0.393 (ii) 0.287 (iii) 0.036   19. (a) (i) 0.903 (ii) 0.223
(b) 0.857, 0.734   20. 0.015   21. 6

Group L

1. $\chi^2 = 6.75$ with $v = 3$, so do not reject H_0 2. $\chi^2 = 11.1$ with $v = 4$, so reject H_0
3. $\chi^2 = 2.67$ with $v = 3$, so do not reject H_0 4. $\chi^2 = 1.30$ with $v = 4$, so do not reject H_0
5. χ^2 very large, so reject H_0 6. $\chi^2 = 22.2$ with $v = 5$, so reject H_0 7. $\chi^2 \gg 100$, so
reject H_0 8. $\chi^2 = 1.11$ with $v = 4$, so do not reject H_0

Answers to selected consolidation exercises

3.B (3) 30, 30, 36, 40, 18, say 3   **3.D** (1) 3, 5 cm; (2) 54°; (3) 3 cm; (4) 8000 h
5.A (1) 183.2 lb; (3) 43, 393, 1849, 48
6.A (3) (a) 1.45, (b) 56.4, (c) 24.6, (d) 324.6, (e) 19.7, (f) 7.39, (g) −0.77
7.A (3) mode is 2
8.A (5) (a) 4, (b) 29.74 yrs
9.A (7) 45 mph; (8) 24 dolls   **9.B** (1) 6; (2) (a) 1.074 or 7.4%, (b) 6
9.C (1) (a) 3.67, (b) 3.27, (c) 3.02   **9.D** (2) 4; (4) (a) HM 18/11, GM 1.82
(b) HM 36/17, GM 2.33
10.A (1) (a) 86 (b) 13
12.B (2) 2.5 (no units)
17.B (4) (a) 0.12, (b) 0.68   **17.C** (1) (a) $(1/54)^2$, (e) $(1/54)^6$; (2) 720; (3) 8!; (4) 35;
(5) 120, 5040, 60, 20160, 10080; (7) 1024
18.A (1) 36   **18.C** (2)(a) 32.4, (b) 207.36
19.B (5)(a) 42, (b) 57.84; (6) 45; (7) 10
20.A (5)(a) 18.87%, (b) 0.135%, (c) 0.5%, (d) 15.87%, (e) minute, (f) 2.5%, (g) 68.26%,
(h) 97%; (6) 34; (7) 2   **20.B** *see* 20.A (5)
21.A (4)(a) 127, (b) 127 ± 2.55   **21.B** (5)(a) (i) 28.43–31.57, (ii) −0.95 to +0.35;
(b)(i) 27.94–32.06, (ii) −1.15 to 0.55   **21.C** (2) 142.05–152.71   **21.E** (3) 21.05; (4) 20.09
21.F (4) 149; (5) 17.88, 61.61, 17.88 ± 0.97
22.A (4)(a) 1801, (b) 3109; (5)(a) 2401, (b) 4145; (6) 2085; (7) 37.86 ± 2.03%
28.C (2) 16, 9.6
39.A (2) (a) 5.5, (b) 35/12, (c) 2/3   **39.B** (3) (a) £7.00, £22.10, (b) £3.00, £22.10
43.B (1) (a) 1/8, (b) 8, (c) 0.472; (2) (a) 0.687 (b) 1 − 0.687

Index